The Mediterranean Region

The Mediterranean Region

Biological Diversity in Space and Time

Second Edition

Jacques Blondel, James Aronson, Jean-Yves Bodiou, and Gilles Boeuf

with the assistance of Christelle Fontaine

OXFORD
UNIVERSITY PRESS

Great Clarendon Street, Oxford OX2 6DP
United Kingdom

Oxford University Press is a department of the University of Oxford.
It furthers the University's objective of excellence in research, scholarship,
and education by publishing worldwide. Oxford is a registered trade mark of
Oxford University Press in the UK and in certain other countries

© Jacques Blondel, James Aronson, Jean-Yves Bodiou & Gilles Boeuf 2010

The moral rights of the author have been asserted

First published 1999
This edition 2010
Reprinted 2014

All rights reserved. No part of this publication may be reproduced, stored in
a retrieval system, or transmitted, in any form or by any means, without the
prior permission in writing of Oxford University Press, or as expressly permitted
by law, by licence or under terms agreed with the appropriate reprographics
rights organization. Enquiries concerning reproduction outside the scope of the
above should be sent to the Rights Department, Oxford University Press, at the
address above

You must not circulate this work in any other form
and you must impose this same condition on any acquirer

Published in the United States of America by Oxford University Press
198 Madison Avenue, New York, NY 10016, United States of America

British Library Cataloguing in Publication Data
Data available

Library of Congress Cataloging in Publication Data
Data available

ISBN 978-0-19-955799-8

We dedicate this book to all the children living today and those of tomorrow, all around the Mediterranean Sea. Whatever their country, religion, or language, may they live in peace, solidarity, and well-being.

With love from Jacques, James, Jean-Yves, and Gilles

Contents

Foreword by Peter H. Raven — x
Preface — xiii

1 Setting the Scene — 1
  1.1 The birth of the Mediterranean — 1
  1.2 The physical background — 5
  1.3 Climate — 12
  1.4 Mapping the limits of the region — 16
  1.5 Adjacent and transitional provinces — 19
  Summary — 21

2 Determinants of Present-Day Biodiversity — 23
  2.1 Drivers of biodiversity — 23
  2.2 Composition of the flora — 32
  2.3 The insect fauna — 38
  2.4 Vertebrates — 39
  2.5 Marine fauna and flora — 49
  Summary — 51

3 Present-Day Terrestrial Biodiversity — 52
  3.1 Flora — 52
  3.2 Invertebrates — 58
  3.3 Freshwater fish — 61
  3.4 Reptiles and amphibians — 63
  3.5 Birds — 68
  3.6 Mammals — 70
  3.7 Convergence and non-convergence among mediterranean-type ecosystems — 72
  Summary — 76

4 Present-Day Marine Biodiversity — 78
  4.1 Flora — 80
  4.2 Invertebrates — 87
  4.3 Fish — 91
  4.4 Marine birds — 94
  4.5 Whales — 95
  Summary — 98

5 Scales of Observation — 99
    5.1 A succession of life zones — 99
    5.2 Transects — 103
    5.3 Small-scale, within-landscape diversity — 113
    Summary — 117

6 A Patchwork of Habitats — 118
    6.1 Forests and woodlands — 118
    6.2 Matorrals — 122
    6.3 Steppes and grasslands — 125
    6.4 Old fields — 125
    6.5 Cliffs and caves — 125
    6.6 Riverine or riparian forests — 127
    6.7 Wetlands — 127
    6.8 Diversity of marine habitats — 133
    Summary — 136

7 Populations, Species, and Community Variations — 137
    7.1 East–west vicariance patterns — 137
    7.2 Life on islands — 140
    7.3 Community dynamics in heterogeneous landscapes — 146
    7.4 Adaptation, local differentiation, and polymorphism — 148
    7.5 Species turnover in time: migrating birds — 159
    Summary — 164

8 Life Histories and Terrestrial Ecosystem Functioning — 165
    8.1 Evergreenness and sclerophylly — 165
    8.2 Autumn-flowering geophytes: a strategy for surviving competition and drought — 169
    8.3 Annuals in highly seasonal environments — 171
    8.4 Herbivory and plant defences — 172
    8.5 Pollination — 175
    8.6 Fruit dispersal by birds — 179
    8.7 Decomposition and recomposition — 182
    Summary — 185

9 Life in the Sea — 186
    9.1 Marine life specificities — 186
    9.2 Pelagos — 187
    9.3 Benthos — 189
    Summary — 201

10 Humans as Sculptors of Mediterranean Landscapes — 202
    10.1 Human history and Mediterranean environment — 202
    10.2 Plant and animal domestication — 207
    10.3 Forest destruction, transformation, and multiple uses — 216

	10.4 In search of a long-lasting and convivial living space	224
	10.5 Traditional landscape designs	229
	Summary	233

11 Biodiversity Downs and Ups — 235
11.1 Losses — 236
11.2 Gains — 252
11.3 Fire: a threat and a driving force — 258
Summary — 261

12 Biodiversity and Global Change — 262
12.1 Human demography — 263
12.2 Habitat degradation and pollution — 264
12.3 Biological invasions — 265
12.4 Climate change — 281
Summary — 285

13 Challenges for the Future — 286
13.1 A microcosm of world problems — 286
13.2 Conservation sciences — 291
13.3 Steps towards sustainability — 299
13.4 Present threats and conservation efforts in the marine environment — 304
13.5 International cooperation — 309
13.6 Alternative futures — 311
Summary — 312

Glossary — 313
References — 318
Index — 357

Foreword

In this outstanding work, Jacques Blondel and James Aronson have improved greatly on their excellent first edition, which appeared a decade ago, adding Jean-Yves Bodiou and Gilles Boeuf as co-authors for the two new chapters on the sea itself. In doing so, and generally, the scope of the present volume has been broadened considerably and its interest and usefulness enlarged accordingly. I was most enthusiastic about what Blondel and Aronson accomplished their first effort; I am even more enthusiastic about the present volume, essentially a new book that retains the outstanding qualities of the earlier volume but adds a great deal to it. Altogether, three new chapters have been added, and the entire text has been significantly revised in the light of recent findings to provide a thoroughly up-to-date account of the environmental dynamics of the region—why it looks and 'works' the way it does at many different scales of space and biological integration.

With the addition of these excellent new chapters, Chapters 4 and 9, on the sea itself, this volume now covers the diversity of marine life in the Mediterranean and the ecosystems that it comprises. Over the tens of thousands of years of human interactions with the Mediterranean, the sea, like the lands around it, has been altered in countless ways, and the pressures on the vitality of its organisms, and the functioning of the sea itself have reached extreme levels that go beyond what could have been envisioned even a decade ago. The new treatment of marine systems is not limited to the chapters that have been added, but it is now thoroughly integrated throughout the general treatments of the area throughout the entire work. In addition, emphasis on the Iberian Peninsula, Italy, Greece, Turkey, and North Africa has been added, so that this new volume more comprehensively covers the entire region, one of intense interest to all who care about its individuality, its critical role in human history, and its place in the functioning of the global ecosystem as a whole.

The third new chapter mostly deals with biodiversity and global change. Indeed, the conservation challenges faced by the world as a whole and the Mediterranean area in particular have increased in severity over the past decade, particularly as our understanding of global climate change has improved. It now seems quite clear that we shall not be able to make our target of an increase of 2°C over pre-Industrial levels and that we shall be fortunate and have to take extraordinary measures to limit the increase to no more than 3°C, a point at which heat waves (such as the one in France when so many people died in 2003), extraordinary storm events, and widespread drought, the signs of which are already becoming evident, will become less manageable than they are now. The Mediterranean region itself has already warmed more than most of the rest of the Northern Hemisphere, and the consequences here can be expected to be more severe than elsewhere—although the ecosystems and sustainability of the whole world is now at risk. Combined with global warming, habitat destruction, invasive species (including pests and pathogens), as well as the selective overexploitation of individual species for particular purposes for food, medicine, wood, or other purposes will be major threats for a great number of the world's species in the forthcoming decades.

The effects of global warming on the survival of organisms in the Mediterranean, as in other areas of the world with summer-dry climates, are likely to be especially severe. All are subject to drought and all are rich in highly localized endemic species. Many of those species are restricted to mountaintops, cliffs, or other shaded and seasonally wet and cool habitats that are very likely to be eliminated as temperatures become higher. These locally moist habitats are areas where species of organisms persisted as the spreading polar ice caps created cooler currents offshore, over which the prevailing westerlies blew. Relatively cool, moisture-laden air then passes onto warm lands in the summer, the heat increasing its moisture-holding capacity and thus limiting precipitation and in some regions essentially eliminating summer rainfall. As these climates have developed and intensified over the past few million years, the distinctive plants and animals of the respective regions evolved into communities, and habitats, from which the original ecosystems were eliminated. Elements from those original ecosystems persisted in the locally, most-often, high-elevation habitats just mentioned, and it is precisely those relicts that are at the greatest risk now, thanks to several components of global change including global warming.

At the same time, the newly formed habitats into which evolutionary radiation has taken place are shifting kaleidoscopically, and many such habitats are expected to disappear over the course of the next few decades, as estimated independently for each of the areas of the world with a mediterranean climate. Thus about half of the endemic plant species of California including some relicts but also many of recent evolutionary origin, are likely to disappear over the next few decades as the particular conditions of the local habitats into which they evolved disappear. Around the Mediterranean Basin itself, the problem is likely to be extremely severe, and we must consider remedies if a reasonable percentage of these remarkable organisms are to be saved from extinction. The facts of the situation are carefully examined in this volume in a new Chapter 12, devoted to a systematic consideration of the factors bearing on conservation in the region. Although the resilience of these organisms has been proven over millennia of climate change, and in the face of intense human activities, the challenges they are facing now are unprecedented. Among other threats to environmental stability will be increased numbers and areas of wildfires in woodland communities, as experienced particularly in Greece in the summer of 2009 and elsewhere in recent years. Agriculture in the region, as throughout the world, will be hard-pressed to feed a rapidly-growing human population, and again, in the Mediterranean itself, the challenges will be enormous.

In order to deal effectively with these challenges, one must understand them and the characteristics of the environmental systems in which they exist. The complex assemblage of diverse habitats in a single region, with sharp gradations in temperature and precipitation, coupled with a complex geology, under a pattern of biological diversity that is as finely divided locally as any set of communities on earth. Add to that two million years of human occupation and intense exploitation, and one begins to see how a skilful treatment like that presented here is necessary to understand, interpret, and conserve what one sees in the area. It will be especially useful for people with some grounding in natural history, but it can inform anyone who wishes to know more about the living world of the region. Paradoxically, many organisms originally confined to the Mediterranean region are now making their way northward in Europe, under the new conditions associated with global warming, so that some of the biological patterns hitherto considered typically 'Mediterranean' may now be observed in some form at higher latitudes. This new book will also be very useful for understanding the comparable ecosystems found in and around California, in central and southern Chile, in south-western and southern Australia, and in the Cape Region of South Africa—all of them dominated in some areas by weedy plants originally from the Mediterranean itself (which has few such weeds from elsewhere!).

As conditions change rapidly in these areas, historically so important for human civilization and home to so many unique species of plants and animals, it will become increasingly important for us all to understand as well as we possibly can the reasons that the region has become what it is from a geological, climatic, and anthropogenic point of view. As it has shaped major portions of our civilizations, we have in turn shaped it—a kind of symbiotic relationship that has fashioned both the ecology of the Mediterranean region and the nature of its great civilizations, what they have been and what they are now. It is now clearly up to us to decide collectively what we would hope the region will be like in the future and to work to achieve it. In the final analysis, the authors are cautiously optimistic, in view of the many positive steps that are being taken in many Mediterranean nations, and internationally. The authors are to be congratulated on the gift that they have presented to us all in labouring to produce such an informative and well-written resource—a guide to the present and the past, as well as a key to the future we want for ourselves and those who will follow us and depend on us to maintain as bountiful a world as that which we enjoy now.

Peter H. Raven
Missouri Botanical Garden
St. Louis, Missouri, USA

Preface

In 1999, at the invitation of Oxford University Press, two of us (Jacques Blondel and James Aronson) published a book entitled *Biology and Wildlife of the Mediterranean Region*. While preparing this book, we were aware of the immensity of the task and of the huge number of omissions and possible misrepresentations we would inevitably make when undertaking such a gigantic subject. We tried and, perhaps, succeeded in shedding some new light on the biodiversity—in space and in time—of this endlessly fascinating region.

Ten years later, Oxford University Press asked us to prepare a new book with added text on the sea itself. Again, we hesitated but finally agreed to tackle this new challenge provided we could convince some experienced marine biologists to join us. Happily, Jean-Yves Bodiou and Gilles Boeuf, from the Marine Biological Station of Banyuls, near Perpignan, France, agreed to contribute their knowledge and insights concerning the diversity of marine life, as well as ecosystem functioning, biological invasions, and threats to marine biodiversity in the Mediterranean, including emblematic species, such as whales, marine turtles, and big fishes.

The four authors share responsibility for the whole book, but each of them, of course, contributed primarily in his own field of knowledge and research. James Aronson's research focuses on vegetation dynamics, plant biogeography, and the interactions between people and living systems, including the science and practice of ecological restoration of Mediterranean and other types of ecosystems. Jacques Blondel's main fields of interest are biogeography, community ecology, and the evolution and ecology of animal populations. Jean-Yves Bodiou is mostly interested in the characteristics of marine biota in relation to habitats and ecosystems, whereas Gilles Boeuf is a specialist of ecophysiology and several general aspects of marine life, marine models in scientific research, fisheries, and conservation.

Wading through the immense published material on biological diversity in the marine and terrestrial landscapes and seascapes of the Mediterranean was in and of itself an enormous undertaking that was a time-consuming and sometimes discouraging enterprise. It soon became apparent that we had to limit ourselves to selected aspects of biodiversity and make a number of somewhat arbitrary choices. Since the appearance of our 1999 book, an enormous number of fascinating studies have been published on a variety of ecological, evolutionary, and biogeographical issues concerning plant and animal species in the region. From the myriad of relevant studies that have been published in the first decade of this century, 446 new references have been considered in this book, thereby 'putting new and refreshing wine in old bottles'. This is to say that this book is really new, and much more comprehensive than the previous one. It includes 13 chapters instead of 10 and a variety of new data, ideas, and results covering a wide range of issues from the historical background of the establishment of floras and faunas to more recent problems related to the various components of global change, including biological invasions in both terrestrial and marine ecosystems.

Two points of great importance have been considered for understanding the patterns observed today, especially in terrestrial landscapes; namely, the historical components contributing to the changes and establishment of living systems and the preponderant role of humans in shaping and designing habitats and landscapes. Although these two points are more or less touched upon in all chapters, we devoted the first two chapters to the historical context, to set the scene, and a full chapter, towards the end of the book, to human influences on biological diversity, given how many peoples and societies have succeeded one another over the centuries and left their mark on different parts of the basin. It is true that such a region is inhabited for a very long time with dense human populations, obviously having, from the beginning, a strong impact on the ecosystems and also that it has been widely studied.

To avoid misunderstandings and prevent disappointments, we should point out that this book is neither an encyclopaedia nor a field guide. Nor is it in any way exhaustively representative of the vast literature on the subject. None of the four authors is an encyclopaedia by himself, so unfortunately entire fields are passed over more or less in silence. This is the case for most groups of insects and many groups of soil invertebrates, as well as entire groups, such as ferns, lichens, and fungi.

The book should be considered as an introductory text for ecologists, naturalists, students, scholars, and, more generally, for people with a natural sciences background. What makes the Mediterranean region a hotspot of biodiversity, as recognized by Norman Myers two decades ago, is not so much the abundance of living beings or the degree of menace hanging over them, as the simple fact of enormous species-level diversity. And this is true for both marine and terrestrial ecosystems. Except in particular habitats, such as freshwater and brackish wetlands or some lush forests, productivity of marine and terrestrial Mediterranean ecosystems is rarely high, but what fascinates the visitor is the diversity of life at any spatial scale. Our goal was to introduce the reader to the kaleidoscopic aspects of all living beings in this fascinating area where there is often more biodiversity in a single square kilometre than in any area 100 times larger in the northern parts of Europe. Topics include biological diversity at many spatial scales of space, from the entire basin to minute surface areas, and from entire floras and faunas to population variations. Rather than trying to report as much as possible on all that is known in the field of Mediterranean biodiversity, we selected some relevant examples to illustrate salient aspects of diversity at different scales of space and biological integration.

Many times, we have been frustrated because we had to restrain ourselves from developing certain aspects and giving more data. Our excuse for the many flaws, omissions, and mistakes that specialists will inevitably find in this book is that we *dared* to tackle an enterprise which has not been attempted previously from a modern biologist's point of view.

In the course of writing, we tried to adopt a user-friendly style, informative but informal, and to avoid technical jargon as far as possible. Terms that may be unfamiliar to some readers are given in bold type and are defined in a Glossary at the end of the book. Some sections will nevertheless appear too technical for some readers, while other parts will irritate specialists by their brevity and paucity of detailed references. Throughout the book, we will use the term Mediterranean to refer to the basin itself, and mediterranean when dealing with mediterranean-type systems or features that occur not only in the Mediterranean region, but also in four other regions of the world: central Chile, central California, the Cape region of South Africa, and two portions of southern Australia. Note that we include Macaronesia—the Canary Islands, Madeira, and the Azores—as part of the Mediterranean bioclimatic region.

We would like to acknowledge the skill and generous help of many friends and colleagues who provided us with material, encouragement, and comments on drafts of various parts of the book, including M.-C. Anstett, J.-C. Auffray, M. Barbéro, H. Bohbot, S. Bodin, C. F. Boudouresque, F. Catzeflis, A. Charrier, E. Crégut-Bonnoure, M. Debussche, E. Garnier, P. Geniez, B. Girerd, L. Hoffmann, L. Jones, F. Kjellberg, E. Le Floc'h, F. Lantoine, J. Lecomte, R. Lumaret, M.-J. Nègre-Bodiou, P. Quézel, J.-Y. Rasplus, F. Romane, P. Romans, J. Roy, S. Ruitton, T. Shulkina, and T. Tacket. Of particular value has been the help of M. Cheylan, A. Dafni, O. Filippi, F. Médail, P. Panoutsopoulos, D. Simberloff, J. Thompson, and F. Vuilleumier who read several chapters, and gave much valuable criticism. We also thank the various authors and publishers for permission to reproduce their photographs or previously published figures, as indicated in the text. We are also grateful to have had a patient editor in Helen Eaton who did not spare time in giving us suggestions and encouragement and in helping us to make the manuscript meet the standards of Oxford University Press.

Finally, a very special thank you to Christelle Fontaine, who has been instrumental in coordinating our work, conducting back-up research and final checks, and significantly improving the quality of every single chapter. She is more than an editorial assistant: we declare her an unofficial co-author of the book.

Jacques Blondel
James Aronson
Jean-Yves Bodiou
Gilles Boeuf
Montpellier, April 2009

CHAPTER 1

Setting the Scene

The Mediterranean Basin is one of the richest and most complex regions on Earth: geologically, biologically, and culturally. It is a living, moving mosaic that surprises, delights, and defies the imagination. There are over a dozen majestic mountain ranges, a kaleidoscope of forests, woodlands, and shrublands, a host of riparian, coastal, and other wetlands, and the sea itself, with its archipelago of many thousands of islands. With all this diversity of landscapes, seascapes, and organisms, the Mediterranean can be overwhelming. How can we best perceive and understand the biodiversity in this vast region?

If geophysical, climatic, historical, and ecological factors mostly contributed to the region's biological and ecological diversity, the weight of human factors is heavier in the Mediterranean region than in most other parts of the world and thus must be carefully considered. We thus begin this book with a historical, as well as a geophysical and climatic, overview of the Mediterranean lands, and an historical and hydrodynamic overview of the sea. Elsewhere, we will discuss the past and present human 'footprint' in the region. As a reflection of its ancient history of human occupation, over 460 million people, speaking over 50 languages and dialects, now inhabit the 24 different countries of the Mediterranean region. The Mediterranean is also the number-one destination worldwide for immigrants and the premier holiday destination as well. More than 265 million tourists came to the Mediterranean in 2005 (Arnold 2008), a figure which represents almost 30% of world tourism, and that number is expected to double by 2025 (De Stephano 2004). These 'summertime pilgrims' raise many problems and greatly increase the human footprint (see Box 13.1).

1.1 The birth of the Mediterranean

To analyse and understand Mediterranean biodiversity in space and in time requires a review of how and when the main physical features of the region have developed over geological times. To accompany this discussion, we have assembled in Fig. 1.1 some of the major events that have marked the history of the basin, with geological and climatic events for the Tertiary (Fig. 1.1a, left), and human events for the **Holocene** (Fig. 1.1a, right and b). We also refer the reader to the discussion of this history in Thompson (2005).

1.1.1 From the Tethys Ocean to the Mediterranean Sea

The Mediterranean region is one of the most geologically complex areas in the world and the only case of a large sea surrounded by three continents. We will provide only a brief overview of its history, which is still a subject of intense research and some controversy (see Biju-Duval *et al.* 1976; Rosenbaum *et al.* 2002a). Some 250 million years ago (hereafter, mya), at the end of the **Palaeozoic** era, all the world's land masses were joined together in a supercontinent, which was named Pangea by the German geologist and meteorologist Alfred Wegener. At the beginning of the Jurassic, some 200 mya, Pangea started to break up into two smaller supercontinents, Laurasia to the north and Gondwana to the south. These were separated during the whole Triassic (250–200 mya) by a single wedge-shaped ocean, called the Tethys or the Palaeotethys. This ancient ocean was transformed in the early **Mesozoic** (250–65 mya), because of the

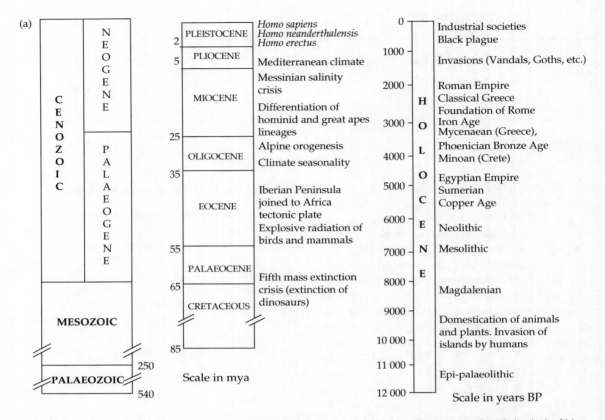

Figure 1.1 Synopsis of some landmark events in the Mediterranean: the left-hand side of (a) shows the Tertiary era; the right-hand side of (a) and (b) show the late Holocene: 35 centuries of major events, conflicts, and treaties. BP, before present; mya, million years ago.

northward movement of Gondwana and its collision with Eurasia.

As Pangea and the Tethys gradually split into several smaller units, the physical geography of the future Mediterranean area was transformed through continental convergence, collisions, and other shifts of tectonic plates. During the middle Jurassic (165 mya) and early Cretaceous (120 mya), the seafloor spreading created the Atlantic Ocean, between Africa and North America, and the Tethys—the ancestral Mediterranean Sea—between Africa and Eurasia. When Eurasia started to move away from North America in the late Cretaceous (80 mya) and early Tertiary, Africa moved eastwards, enlarging the Atlantic Ocean, and Africa and Europe moved closer together, triggering Alpine orogenesis, or uplifting. A second period of rafting and collision occurred during the Pliocene and Pleistocene, 5 mya–12 000 years ago, with additional vertical uplift and fracturing of the Alps. The interplay of Eurasian and African plates resulted not only in the rise of the Alps, but also in a progressive shrinking of the Tethys Ocean (Hsü 1971; Rosenbaum et al. 2002b). Finally, the Tethys closed definitively, during the **Cenozoic**, an era which began some 65 mya, when various fragments of Gondwana, including India and the Arabian Peninsula, finally collided with the rest of Eurasia. The last remnants of the ancient Tethys are the modern-day Mediterranean Sea and the smaller Black, Caspian, and Aral Seas, as we know them today (Rosenbaum et al. 2002a; see also Thompson 2005:16).

Since the beginning of the Tertiary, the history of the Mediterranean has been complicated by the isolation and individual movements of several

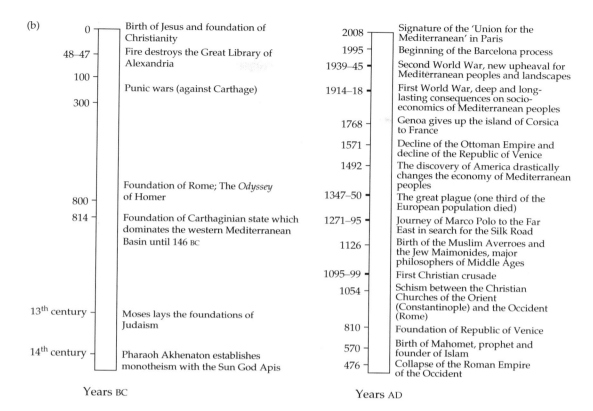

Figure 1.1 Continued

microplates, the most important of which are the Iberian Peninsula, Apulia (which includes Italy, the Balkans, and Greece), and the so-called Cyrno-Sardinian microplate (Biju-Duval *et al.* 1976; Rosenbaum *et al.* 2002a; Papazachos and Papazachou 2003). The Iberian microplate played a pivotal role in the evolution of the region because of its position between the African and the Eurasian plates. In the late Oligocene (28 mya), the south-eastwards motion of Africa relative to Europe caused the rotation of this microplate, which included all the large islands of the western Mediterranean and several crystalline blocks that were subsequently connected either to Africa or to Europe. Apulia was a continental crust connecting the continental masses of Africa and Eurasia, separating the eastern from the western basins of the Mediterranean Sea. Finally, the relevant dynamics of the Cyrno-Sardinian microplate date from the early Oligocene (35–30 mya), when it began to rotate south-eastwards, opening the Balearic basin. These histories have had important consequences on **endemism** and differentiation of plant and animals (see Chapter 3), and also helped provoke the frequent seismic and volcanic activities in various parts of the Mediterranean Sea, as described below.

The coming together of the African and Eurasian plates had two main consequences on the shaping of Mediterranean landscapes and seascapes. First, the Mediterranean Sea is now made up of a series of more or less individualized basins, as we will see later in this chapter. Second, as a result of the collision between the African and the Eurasian plates, there is a ring of mountains around the Mediterranean Basin, except in the south-eastern quadrant, between Tunisia and Egypt (see Fig. 1.2). The Romans named the sea *mediterraneus*, which means 'in the middle of the earth'. The Arabs and the Turks called the Mediterranean,

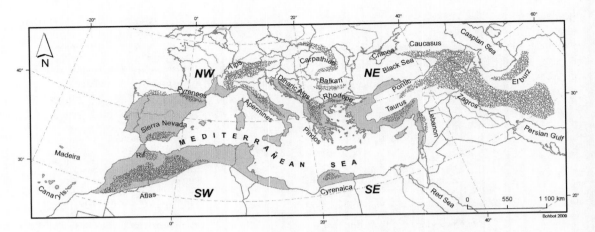

Figure 1.2 Approximate delimitation of the Mediterranean biogeographical area, including the coastal areas and some of the major mountain ranges, and the Macaronesian region as defined by Quézel and Médail (2003), consisting of three Atlantic volcanic archipelagos and a small enclave on continental Africa, in southern Morocco. NW, NE, SW, and SE relate to the north-western, north-eastern, south-western, and south-eastern quadrants, respectively.

the *Rumelian* (that is, the Romano-Byzantine) Sea (Matvejević 1999). However, a more appropriate name for the Mediterranean might be the 'Sea-among-the-Mountains'! From a biogeographic perspective, the Mediterranean region includes all the lands that stretch from the tops of the mountain ridges down to the shores, and to the bottom of the sea, where a surprising amount of life forms are found, as we shall see. It also includes the so-called Macaronesian region, off the Atlantic coast of Morocco (see Fig. 1.2).

1.1.2 The Messinian Salinity Crisis

The short but crucial period that followed the collision of Africa and Eurasia and the Mediterranean's enclosure occurred in the late Miocene and is called the Messinian Salinity Crisis (Duggen et al. 2003). It was one of the most spectacular geological events in the world in the entire Cenozoic, when the Mediterranean Sea dried up almost completely and became a desert with some scattered patches of hypersaline lagoons. This event took place starting c.5.96 mya and ended quite abruptly, some 630 000 years later, c.5.33 mya (Krijgsman et al. 1999; Rouchy and Caruso 2006). It is only recently that this landmark event came to light. Huge seabed salt deposits or 'evaporites'—in places over 1500 m deep—near Sicily, Calabria, and North Africa had long intrigued researchers, but it was not until the early 1970s that an international team using innovative deep-sea drilling methods were able to determine the contents of these thick deposits. The researchers found that these salt 'domes' contained not only sodium chloride but also many other evaporites like those found today in the Dead Sea. They also included fossilized remains of light-demanding algae (cyanobacteria), such as occur only in shallow water. This proved that the Mediterranean had dried up more or less completely during the Messinian Salinity Crisis, a brief but crucial period in the making of the Mediterranean world. A recent model implying two main stages of evaporite deposition that affected successively the whole basin, but with a slight diachronism, matches better the whole dataset (Rouchy and Caruso 2006). The distribution of the evaporites and their depositional timing were constrained by the high degree of palaeogeographical differentiation and by threshold effects that governed the water exchanges. The crisis included two different stages: the first one (lower evaporites) included the deposition of the thick homogenous halite unit with interbeds in the deepest basins, occurring in a glacial period between 6 and 5.6 mya; the second stage (upper evaporites) correlates with the interval of warming and global sea-level rise between 5.6 and 5.5 mya.

To help the reader comprehend the Messinian Salinity Crisis, it is worth pointing out that, today, annual water loss by evaporation from the Mediterranean is approximately 4500 km^3 per year, of which only 10% is replaced by rainfall and the influx of rivers. The remaining 90%, therefore, must come from the Atlantic Ocean through the narrow (14 km; 300 m deep) Straits of Gibraltar, where currents flowing in from the Atlantic Ocean are so strong that it would not be possible, even for a good swimmer, to get across. Without present water entry through Gibraltar, Mediterranean seawater level would decrease of more than 1 m per year. Thus, it is not surprising that, when the straits were temporarily closed, and prevailing climatic conditions in the lower Pliocene were much warmer than at present, less than 1000 years would have been required for the Mediterranean seafloor to dry up (Suc 1984).

The Messinian Salinity Crisis also had consequences for the Earth's crust at both the northern and southern shores of the Mediterranean Sea. Enormous fissures opened up, earthquakes shook the ground, and ancient volcanoes were reactivated, while several new ones were born. Emerging above the hypersaline flats, many coastal areas were transformed into isolated 'mesas' or islands, towering thousands of metres above the arid, increasingly saline flats below. Concurrently, great rivers, such as the Rhône and the Nile, continued to feed the nearly dry Mediterranean, shooting over high cliffs and gradually digging out deep gorges in the thick granitic crust and the limestone blocks at the sea's edges. One such gorge lies 900 m below sea level, at the mouth of the Rhône River, near Marseilles, while another is found in the ground more than 2000 m beneath Cairo, which itself lies more than 100 km inland and upstream from the Nile River delta near Alexandria.

Starting about 5.3 mya, a series of tectonic shudders shook the region anew, breaking open the land bridge between Morocco and Spain, and letting the waters of the Atlantic Ocean surge once again through the Straits of Gibraltar. Because of the drying up caused by the Messinian Salinity Crisis, this resurgence took the form of gigantic cascades 3000–4000 m high with a drop-off of more than 50 times the height of Niagara Falls. Each day, 65 km^3 of seawater poured in, which was enough to refill the basin in less than 100 years. Following the 'unplugging' of the Straits of Gibraltar and the refilling of the basin, the present-day size and shape of the Mediterranean, as well as its main physiographic and geomorphological features, were finally established, some 5 mya.

1.2 The physical background

At present, the Mediterranean Basin stretches over approximately 3800 km west–east and 1000 km north–south, between 30 and 45°N, bordering 24 different countries (see Fig. 1.2). A marked geographic boundary runs north–south through the Sicily–Tunisia 'sill', which is only 140 km wide and 600 m deep between the southern tip of Sicily and Cap Bon, Tunisia. This creates a very strong biogeographical contrast between the western and the eastern halves of the Mediterranean, the former being shifted somewhat north with respect to the latter. The boundary separating the two north–south ranges in each half of the basin runs approximately at the 36th parallel in the western half and at the 33rd in the eastern half. In the western half, west of the Sicily–Cap Bon line, biota are more **boreal** in character than in the eastern half, where they have more 'oriental' affinities; that is, with Asia minor and central Asia. The clear geographical and biological distinction between the two halves of the basin prompted De Lattin (1967) to recognize a western 'Atlanto-Mediterranean' subregion and an eastern 'Ponto-Mediterranean' one. An additional, purely physical feature that helps bring into focus the Mediterranean area as a whole is the striking contrast between its northern half, with its large Iberian, Italian, Aegean, and Anatolian Peninsulas, and the southern half of the basin, with its more or less rectilinear shorelines. As explained in the previous section, this contrast mostly results from the tectonic activity of the microplates that moved and evolved between the African and Eurasian main tectonic plates. In summary, for purely heuristic purposes, the basin can be divided further into four quadrants, the north-western, north-eastern, south-western, and south-eastern (Fig. 1.2). As we shall see, however, there are many nuances and regional peculiarities to take into account when studying Mediterranean biodiversity, both on land and at sea.

1.2.1 The sea

Many oceanographers have considered the Mediterranean Sea as a miniature ocean, based on its size and depth and deep-water circulation patterns. In fact, the Mediterranean Sea is a very special ecosystem with absolutely unique characteristics, including its hydrodynamic and **thermohaline** systems, as well as aspects related to water, temperature, and salinity (see Chapter 9). It is at first, a very interesting microtidal system.

1.2.1.1 SIZE AND SHAPE

The Mediterranean Sea is the largest inland sea in the world, stretching from Gibraltar to Lebanon (Box 1.1). In contrast to the European and Turkish shoreline, which is very irregular with massive peninsulas, the African shoreline is regular, relatively 'smooth', with a much warmer, drier climate than most of southern Europe, and includes only two perennial rivers, the Nile and the Moulouya at the border between Morocco and Algeria. Thus the inputs of fresh water to the sea are low and even lower than before, since the construction, in 1964, of the massive Aswan dam on the upper Nile.

Box 1.1. The Mediterranean Sea

Total length	3800 km
Maximal width	1800 km
Average width	700 km
Total area	2 500 000 km^2
Area of the western basin	850 000 km^2
Area of the eastern basin	1 650 000 km^2
Total coastline	46 500 km
Mainland coastline	22 000 km
Island coastline	24 600 km
Maximum depth	
Western basin	3731 m
Eastern basin	5121 m
Average depth	1430 m
Volume of the waters	3 700 000 km^3

In contrast, as mentioned above, the northern shorelines are jagged, partly due to the numerous chains of continental mountains and their complex geological history (see Fig. 1.2). These southern European shores of the Mediterranean are traversed and irrigated by important rivers like the Ebro, Rhône, Po, and several smaller rivers in the Balkans, which add further geophysical complexity. Connected with the Mediterranean Sea is the Black Sea, which, to the north and west, is the final destination of the Danube, Dniepr, and Don Rivers. The modern Mediterranean Sea includes several 'interior seas', or basins, which are closely related to the structural relationships between the major tectonic plates of Africa and Eurasia in the Oligocene and Miocene, and which are usually bordered by rather shallow sills. As shown in Fig. 1.3, the seas are: (1) the Alboran Sea, between Morocco and Spain; (2) the Balearic Sea, between the Balearic Islands and continental Spain, which opened southward to the main Mediterranean Sea around the upper Miocene, when the rotation of the Cyrno-Sardinian microplate reached its present position; (3) the Tyrrhenian Sea, between Corsica, Sardinia, and the Italian Peninsula, which is a young basin that began to open in the late Miocene; (4) the Ligurian Sea, between Corsica and the Gulf of Genoa; (5) the Adriatic Sea, between Italy and the Balkan Peninsula; (6) the Ionian Sea, between Italy, the Aegean Peninsula, and Libya; and (7) the Aegean Sea, between Greece and the Anatolian Peninsula to the east. The Aegean Sea is an arc-shaped basin located on the Aegean microplate. It overrides the old Mediterranean oceanic lithospheric plate, which is currently moving northwards. This results in an active subduction zone and much volcanic and seismic activity. The basins are: (8) the Algero-Provencal Basin, between the Gulf of Lion and the Algerian coast in the west; and (9) in the east, the Levantine Basin, which is considered to be a subsided portion of thin continental crust. Although now fully emerged, the island of Cyprus once lay at the bottom of the Tethys and rose up as a result of the collision of the African and Eurasian plates. Mt. Troödos and the Pentadactylos range rose above the surface of the sea a mere 20 mya.

A striking geomorphological feature of the basin as a whole is the surprising steepness of the

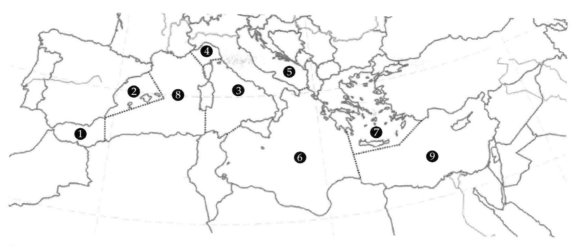

Figure 1.3 Geographical overview of the Mediterranean Sea. The numbers are explained in Section 1.2.1.1.

coast, which almost everywhere drops abruptly to a seafloor lying at 2500 m to as much as 4000 m deep. The average depth of the sea is only 1430 m, but much deeper troughs occur in some places, for example at the southern tip of the Peloponnese (5100 m) and south-east of Italy (4100 m). The coastal zone (0–200 m depth) represents only about 20% of the surface of the sea. The result is that tides in the Mediterranean are very small, except at the northern end of the Adriatic Sea and a few southern localities. In the shallow bays around the island of Djerba, Tunisia, for example, tides may reach 3 m in height, but this is very unusual for the Mediterranean. The continental shelf is also very narrow, except in the Gulf of Lion (southern France), between Tunisia and Sicily, and in the large Aegean area between Greece and Turkey.

A critical point to remember is that the Mediterranean Sea is divided into two deep basins: the western one, between Gibraltar and Sicily, 821 000 km^2, connected with the Atlantic Ocean by the Straits of Gibraltar; and the second basin, east of the Straits of Sicily (maximum 250 m depth). This eastern basin, extending to the coasts of Egypt, Israel, and Lebanon, is even larger, 1 363 000 km^2, and is naturally closed at its eastern end. It was therefore entirely dependent on the western basin for hydrological exchanges, until the digging of the Suez Canal in the mid-nineteenth century (see Chapter 2). The recent widening and deepening of the canal, in the 1970s, removed the former salinity barrier that existed between the Mediterranean Sea and the Red Sea. As a result, the Red Sea is now 10% saltier than the Mediterranean (4.2% as compared to 3.8% total dissolved salts) and 20% more than the Atlantic Ocean (3.4–3.5%).

Nevertheless, the global hydrological budget of the basin was little changed by the Suez Canal. Yet, the decrease of freshwater input from the Nile River, combined with the effects of global climate change, could lead to coastal ecosystems in the region being very heavily affected in the near future (see Chapter 2). The Suez Canal also represents an important entry point for alien invasive species (see Chapter 12).

The two sills that divide the western part from the eastern part of the Mediterranean Sea play a fundamental role as corridors, thanks to a shared two-layer system. Surface water, mainly of Atlantic origin, flows eastward, while deeper water coming from the eastern basin flows westward. The flows can vary with the seasons, but their volume is huge, invariably surpassing water flow 10^6 m^3 s^{-1} (1 **Sverdrup**). The two sills are also the sites of important exchanges characteristic of the unique hydrodynamic circulation system of the Mediterranean Sea (Box 1.2).

Box 1.2. Regulation of salinity in the Mediterranean Sea through Gibraltar

The combined water and salt budget of the Mediterranean Sea can be written as:

$$E - P - R = V_i - V_o$$

and

$$V_i S_i = V_o S_o$$

where E is evaporation, P is precipitation, R is river runoff, V is flow, and S is salinity across the Straits of Gibraltar (i for inflow, o for outflow) (Béthoux and Gentili 1999).
Combination of the two equations gives:

$$E - P - R = V_i - V_i S_i/S_o = V_i(1 - S_i/S_o)$$

and with the values of S_i and S_o (respectively 36.2 and 37.9 **psu**), in order to maintain the salinity and keep the sea level constant, the inflow V_i represents 21 times the water deficit and consequently V_o is 20 times the water deficit. This means that the flows over the sills are much more important than the water deficit. This function is a driver or forcing factor for the hydrodynamics of the whole sea.

1.2.1.2 HYDRODYNAMIC SYSTEM

The Mediterranean Sea is made up of three great water masses, starting (1) with a surface layer from 0 to 150 m depth that comes in from the Atlantic Ocean and flows east; (2) next, an intermediate layer occurs, 150–400 m deep, which arises from the eastern basin and moves westward through Sicily and the Straits of Gibraltar to the Atlantic, but only after being transformed or 'mediterranized' in the Levantine Basin, especially near the island of Rhodes, where cyclonic gyres mix water masses together in giant **eddies**; and (3) two deep-water masses, resulting from sea–air exchanges, characterized by strong and cold winds coming in from the North in winter (see Fig. 1.4).

Cooling and evaporation meanwhile significantly increase surface water density. Then, convective mixing of surface and intermediate waters create dense water, which sinks to the bottom and fills the two basins up to the level of the sills that is approximately 400 m below the surface.

This structure and the main water movements are a consequence of the negative water budget of the Mediterranean Sea. The intermediate position of the sea between subtropical climates to the south and east, and the much cooler temperate European climate on the northern European shores, creates a strong rate of evaporation, as mentioned already, corresponding to a water loss of 1.54 m each year. This loss is more than two and a half times the estimated gain derived from rain, river inputs, and the Black Sea outflows, which is estimated at a total of 60 cm per year (Béthoux *et al.* 1990; Klein and Roether 2001). However, a somewhat stable sea level is maintained, thanks to the upper current entering from the Atlantic through the Straits of Gibraltar. However, evaporation leads to a loss of fresh water and the Atlantic inflow brings huge quantities of salt. Thus, in order to keep the salinity roughly constant, a comparable quantity of salt must move westwards, towards the ocean. This does indeed occur, thanks to the deepest of the currents passing through the Straits of Gibraltar, carrying an outflow of water that is highly saline, for reasons explained in Box 1.2.

The Atlantic water inflow, with a salinity of about 36.2 psu and a temperature of 15.4° C, moves eastward. Due to the **Coriolis force**, it passes on the right along the coast of North Africa, sending some branches towards the north in the western basin, but the majority passes through the Straits of Sicily in the Ionian Sea and the Levantine Basin, with a salinity of 37.15 psu, always being pushed to the east (Fig. 1.4). The inflowing waters become more saline as a result of strong evaporation rates along the African coast, but they always remain

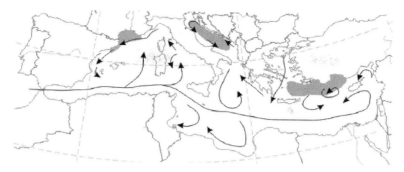

Figure 1.4 Surface currentology and deep-water production areas (grey shading).

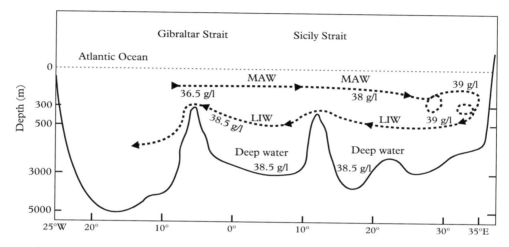

Figure 1.5 Surface, intermediate, and deep-sea water circulation (after Lacombe and Tchernia 1960 with modifications). Values in g/l indicate salinity.

less saline than the surrounding Mediterranean waters.

Figure 1.5 shows the movements of the three water masses in the Mediterranean Sea with the Modified Atlantic Water (MAW) at the surface or subsurface and the **Levantine Intermediate Water** (LIW) in intermediate position, progressing above the deep waters of the two basins, crossing the Straits of Sicily at the lower part of the water column.

The hydrodynamics of Mediterranean Sea waters are of particular interest because they are initially conditioned by evaporation. The 'sucking up' of Atlantic water to the atmosphere has repercussions on the Mediterranean Sea as a whole and creates a particularly active thermohaline circulation within the two basins. These repercussions are not limited to the Mediterranean, because LIW outflow at Gibraltar provides very saline waters to the Atlantic. This in turn may play a role in the global circulation of the North Atlantic Ocean.

In conclusion, it is not appropriate to consider the Mediterranean Sea as a miniature ocean, despite its thermohaline circulation, deep convections, and dynamics of water-mass formations. It is unique, because of the lack of important tides, the characteristics of the deep waters, and the global hydrodynamic system deriving from its topography and its position between largely tropical Africa and Europe, with its cooler and more irregular climate. It is also very unusual in terms of the deep water, which is never really cold in the Mediterranean; that is, an average of 13° C compared with 2.5° C in all the deep oceans. With warmer global climate in the future, deep-water species will be less likely to adapt effectively to even a small increase in water temperature. As noted by Zaballos *et al.* (2006), there are very pronounced ecophysiological differences between bacteria and other prokaryotic organisms living at great depths in the Atlantic Ocean and those found in the Mediterranean. Indeed, the highly contrasted temperatures prevailing at great depths in these two bodies of water have lead to the differentiation of very different assemblages of prokaryotes. However, whatever the temperature in the deepest zones of a given sea or ocean, it is always extremely stable and only a small increase resulting from the global warming could be catastrophic for species not able to endure temperature fluctuations, the so-called stenotherms.

1.2.2 Islands and archipelagos

With 11 879 islands and islets, 243 of which harbour permanent human populations (Arnold 2008), the Mediterranean possesses one of the largest archipelagos in the world, after the Caribbean, Indonesia, and the South Pacific. The geodynamics of the region are such that the majority of islands—more than 9800 in all—are located in the Aegean Sea. Greece is by far the most important island country in the Mediterranean. It ranks first by number of islands (9835, 123 of which harbour a permanent human population), coastline length (12 000 km), and area of islands (250 000 km^2). The total coastal length of Mediterranean islands is 24 622 km, only 15% less than the mainland coastline. No other region of the world has so many land–sea interactions so that a kind of symbiotic relation results, as a combination of physical, bioclimatic, and geobotanical features. Most of the larger islands have long been entirely disconnected from any continent, at least since the Messinian Salinity Crisis, *c*.5.5 mya. Also, as happens in many regions of the world, islands are associated with high seismic and volcanic activity. Several of them are ancient submarine volcanoes. For example, Mt. Troödos, in eastern Cyprus, emerged as an oceanic island at the end of the Cretaceous, 60 mya (Gass 1968). As noted already, the off-shore Macaronesian region is also part of the Mediterranean region, from a biogeographical perspective (see Chapter 3).

To the list of true islands and archipelagos must be added a series of 'biological islands', such as the Peloponnesian Peninsula, artificially isolated at the end of the nineteenth century by the opening of the Corinthian Canal. In addition, many so-called biological islands occur in most or all of the mountainous ranges that encircle the Mediterranean realm. For example, an exceptionally isolated biological island is the Djebel Akhdar, 700 m above sea level, in Cyrenaica, north-eastern Libya. Cyrenaica receives an annual average of 400 mm rainfall, as compared to less than 25 mm in the rest of the country. Other striking examples are Mt. Athos and Mt. Olympus in Greece, and the five distinct mountain ranges of northern Morocco.

1.2.3 Topography

The topographic diversity of the Mediterranean results in large part from the numerous mountain chains (see Table 1.1) that define the basin's various continental contours, except in the south-east (Egypt and Libya), where the lowland desert meets the sea. These mountain ranges, whose geography and history were vividly described by Braudel (1949), Houston (1964), and McNeil (1992), include the Alps, Pyrenees, Apennines, Caucasus, Pontic, Pindos, and Taurus Mountains of Anatolia, the

Table 1.1 Some of the highest mountain peaks of the Mediterranean Basin presented in a clockwise fashion from southern Spain

Mountain	Country	Altitude (m)
Sierra Nevada	Spain	3478
Canigou	France	2785
Bégo	France	2873
Etna	Italy	3260
Olympus	Greece	2920
Taurus	Turkey	3920
Lebanon	Lebanon	3090
Troödos	Cyprus	1950
Chambi	Tunisia	1540
Hodna-Aures	Algeria	2330
High Atlas	Morocco	4172
Teide	Tenerife, Canary Is.	3718

mountains of Lebanon, and the Rif, Kabylie, Atlas, and Anti-Atlas ranges of North Africa, not to mention the many cordilleras of the Iberian Peninsula. Mountain ranges of high elevation are found in Spain's Sierra Nevada (the Baetic Cordillera) and on some of the larger islands, including Majorca, Corsica, Crete, and Cyprus. Some of these peaks reach 4000 m or more, and completely isolate upland basins or plateaux, as well as creating 'hidden' valleys opening only towards the sea.

The various Mediterranean mountain systems provide the source of most of the rivers that traverse the Mediterranean lands, then spread and meander through vast alluvial plains to extensive deltas. Examples are the Guadalquivir, Ebro, Rhône, Po, Evros, and Nestos Rivers and watersheds. Among the main rivers traversing Mediterranean lands, two of the largest—the Nile and the Rhône—are 'non-native' since their headwaters arise in tropical East Africa and the Alps, respectively.

1.2.4 Volcanic activity and earthquakes

Tectonic remodelling, stimulated by microplate dynamics and the Alpine uplifts, as well as by the seismic cataclysms provoked by colliding African and Eurasian plates, which both continue to move towards each other at a rate of about 2 cm per year, all contribute to a complex geomorphological situation. In many parts of the basin, tectonic events have been accompanied by renewed bouts of volcanic activity, such as the eruption of Vesuvius in AD 62, which resulted in the vast, ash-covered mausoleum of Pompeii and Herculaneum. Other well-known examples of still active or near-active volcanoes are Stromboli and Mt. Etna, which is the highest volcano in Europe. In fact, the Mediterranean is a huge seismic 'hearth', especially in its eastern part. The eruption of Santorini, in the Greek Cyclades, some 3000 years ago, was one of the most significant recent natural 'catastrophes' in the Earth's history, comparable in its impact to the eruption of Krakatoa in 1883. The Santorini eruption probably helped precipitate the collapse of the Minoan civilization that flourished in Crete during the Bronze Age, 5000–3000 years ago (see Fig. 1.1a).

In addition to volcanic eruptions, very few places in the Mediterranean region are entirely free of seismic risk since the effects of large earthquakes may occur several hundreds of kilometres from the epicentre. In Turkey alone, no fewer than 60 000 people have been killed and 380 000 houses destroyed by earthquakes since 1930 (Kolars 1982). Historical records also reveal the occurrence in the region of at least 100 devastating earthquakes during the past thousand years. At the site of the ancient temple of Olympia, in south-western Greece, there are rows of nearly identical discs carved from granite and marble in the first millennium BC to form the columns of a temple, which were all knocked over by an earthquake in AD 426. Still visible today, lying just where they fell, these columns give a vivid impression of the destructive strength of earthquakes. More recently, Lisbon was hit in 1877, Al Asnam (formerly Orléansville), Algeria, in 1954, Agadir, Morocco, in 1960, Skopje, Republic of Macedonia, in 1963, and Ardebil, Iran, in 1997. The most recent volcanic eruption in the Mediterranean region occurred in August 1997, partly destroying the historic frescos of the cathedral of Assisi, in northern Italy, and the devastating earthquake destroyed the city of l'Aquila and several villages on 6 April 2009 in the Abruzzo, central Italy, killing more than 300 people.

1.2.5 Soils

The last physical components we will briefly consider are soils. Soils are closely related to geological substrates: the 'memory' of the Earth, which plays a pivotal role in the structure and dynamic of ecological communities. Many Mediterranean soils and substrates are limestone of marine origin, and one can find fossil seashells well above the timberline in most Mediterranean mountains. However, unusual soil types and discontinuous geological substrates including volcanic soils also contribute to the local and regional diversity of habitats. Metamorphic granitic and siliceous (acidic) parent rocks occur locally, as do also occasional ultrabasic rock outcrops in Cyprus, continental Greece, Serbia, Croatia, and Montenegro. As lime content and degree of alkalinity have a great influence on plant growth, it is not surprising that different vegetation types occur on calcareous compared with non-calcareous substrates. High rates of endemism are often related to unusual rock types, particularly when they are embedded in a geological matrix of a different nature (see Chapter 2). For example, recent studies in the Iberian Peninsula show that many endemic species of plants there are restricted to acid soil 'islands' in a surrounding geobotanical 'matrix' of limestone substrates (Medrano and Herrera 2008).

Many soil types, especially in the northern part of the basin, are ferrugineous brown soils, known as *terra rossa*, but dolomite (from degraded calcites), clayey marls, rendzines, loess, regisols, lithosols, and alkaline and gypsum outcrops also occur more or less sporadically in many regions. The latter are very poor in nutrients and often harbour endemic plant species, just as serpentine substrates do. In some parts of the basin, especially in Spain, along the Adriatic coast of Croatia, Montenegro, and Albania, and in Anatolia, large karstic outcroppings occur, where rainfall infiltrates rapidly and then reappears far away as Vauclusian springs at the foot of mountain ranges. These springs are the outcome of complicated networks of underground water resulting from the dissolution of thick calcareous deposits.

This brief tour of the physical setting reveals how the Mediterranean region, with its islands, coastal lands, rivers, and high mountains, provides a veritable cornucopia of habitats, all finely distinguished by local topographies and soil types, an intricate filigree of microclimates related to altitude, rainfall, and exposition of slopes.

1.3 Climate

The Mediterranean climate is transitional between cold temperate and dry tropical. But the bimodal Mediterranean climate regime we know today only began to appear during the late Pliocene, about 3.2 mya, as part of a global cooling trend (Suc 1984; see also Thompson 2005:18–26). But it is only about 2.8 mya that today's prevailing climate became established throughout the region. Since then, the contrast between the alternating hot, dry, and cold wet seasons has intensified steadily up to the present day. However, this most recent period has itself been punctuated by alternating glacial and interglacial periods, with the former far outlasting the latter.

Given this unique combination of hot and dry summers, and cool (or cold) and humid winters, little surface water is available during the months of maximal solar irradiation. Accordingly, the short spring and autumn seasons are critical periods for plant growth. Apart from the mountains, snow falls rarely in the Mediterranean, but periods of hard frost are not infrequent and when they occur immediately after a period of balmy weather they can cause much damage and mortality.

1.3.1 Regional and local variation

Despite its strongly bimodal pattern, Mediterranean climate shows much wider variation in temperature and rainfall than adjacent areas north, east, or south. Mean annual rainfall ranges from less than 100 mm at the edge of the Sahara and Syrian deserts to more than 4 m on certain coastal massifs of southern Europe. The climate on the northern shores is much harsher than on the southern shores, with cold winters and strong winds blowing in from adjacent continental areas. Timing of rainfall also varies much more from one subregion of the Mediterranean to another than in any other mediterranean-climate area (Cowling *et al.* 1996; Dallman 1998:169–72; Médail 2008a; Fig. 1.6).

For at least 2 months each year in the western Mediterranean and 5–6 months in the eastern half, when there is no precipitation at all, most plants and animals experience a water deficit to which they must respond with ecophysiological or behavioural adaptations (see Chapter 8). For living organisms in the region, heat and drought are generally more stressful than low temperatures. Thus, the long dry summer periods are the unfavourable period of the year for plants and animals, contrary to regions further north where the unfavourable period is winter.

Although mean annual temperature by itself is not of great biological significance, its geographical variation gives a good idea of the range of climatic conditions. In the Mediterranean Basin, mean annual temperatures range from 2–3° C in certain mountain ranges, such as the Atlas and

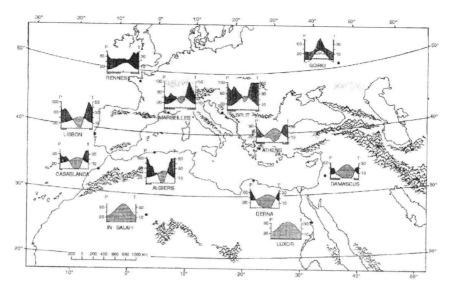

Figure 1.6 The range of local climate patterns occurring around the Mediterranean Basin and in a few adjacent regions, as illustrated by a series of ombrothermic diagrams drawn according to Gaussen (1954) (data from Walter and Lieth 1960). P, precipitation, mm; T, temperature, °C.

the Taurus, to well over 20°C at certain localities along the North African coast. At a local scale, the Mediterranean is well known for pronounced climatic differences over very short distances. Such variability is under the influence of factors that include slope, exposition, distance from the sea, steepness, and parent rock type. As the innovative colony of English and French gardeners on the Côte d'Azur noticed at the end of the nineteenth century, this area has 'one thousand and one microclimates' crowded cheek by jowl.

Perhaps the most useful system for characterizing Mediterranean bioclimates was proposed by Gaussen (1954), who first had the idea to compare annual changes in temperature (T, in °C) and precipitation (P, in mm) on the same graph. The resulting ombrothermic (precipitation and temperature) graph is drawn with $P \times 2$ on one axis and T on the other. The rationale for multiplying P by a factor of 2 is that plants are assumed to suffer from drought when the rainfall curve 'falls' beneath the temperature curve during the driest times of year. This is admittedly a very empirical approach, but the classification retained by UNESCO (1963) proved to be quite useful. It is based on this system which yields a sum of monthly indices of dry months, wherein a 'dry' month's precipitation is less than or equal to twice the average temperature for the same month. Thus, by definition, in a dry month $P \leq 2T$. Such indices make it possible to construct the highly useful ombrothermic diagrams, a few examples of which are given in Fig. 1.6. This system of classification has its critics, who prefer to define the limits of the dry period in a mediterranean-climate region with the ratio P/ETP (P is precipitation, ETP is potential evapotranspiration). A ratio of 0.35 is defined as the threshold behind which dryness does not allow the growth of plants. However, this ratio gives very similar results to Gaussen's scheme, as shown in Fig. 1.6.

1.3.2 Wind

The Mediterranean is a windy area. Wind regimes in the northern side of the basin are mainly northerly, as determined by seasonal temperature differences between land masses and the sea. In winter, the anticyclone of Siberia moves down towards central Europe and sends cold-air masses towards the Mediterranean Sea. The most important winter winds are the *mistral* (northern wind) and the *tramontane* (north-western wind), blowing in the Gulf of Lion, France, the *bora* (north-north-eastern), which flows in the Adriatic Sea, and the *gregal* (north-eastern wind), which blows in the Gulf of Genoa, and the Tyrrhenian and Ionian Seas. The *gregal* is the most feared wind in Malta. The *meltem* is a seasonal wind of the north sector that blows every summer in the Aegean Sea, which is sandwiched between the anticyclone of the Azores and the depression of Pakistan, what makes the cold air of the north of Eurasia sweep down towards the Mediterranean Sea. This air warms itself on the lands which it crosses and arrives warm and dry to the sea. It ensures stable good weather and regular winds all summer on the Aegean Sea and the Cyclades islands.

The evaporation induced by these northern winds plays a considerable role in the hydrology of the Mediterranean Sea because they are at the origin of deep waters: *mistral* and *tramontane* for the western basin, *bora* and *gregal* for the eastern. Even in winter, this important evaporation due to the strength of the winds has the double consequence of cooling the surface water and increasing its salinity. These two factors together increase the density of the water masses, which sink on the continental slope towards the depths, keeping their surface temperature at around $12°C$. They constitute the deep waters, which fill the two basins to the depth of the straits. The sites of formation of the deep waters are strictly localized: the Gulf of Lion for the western basin, and the Adriatic Sea, and to a lesser degree the Aegean Sea, for the eastern basin (Fig. 1.5). Both these so-called catabatic continental winds can provoke sudden springtime cold spells, including diurnal temperature swings of $10°C$ or more.

When the climate gets warmer, the prevailing wind is the *sirocco*, which blows from Africa over the Mediterranean Sea. It is a warm and dry wind of desert origin, following the continental winds blowing from the south on North Africa, the *khamsin* in Egypt, the *ghibli* in Libya, and the *chili* in Algeria. When it extends over the sea during the occasional spells of the hot dry Sahara winds (*harmattan*, *foehn*), it soaks up humidity while passing over the sea and brings rains loaded with ochre-coloured dusts of the desert on the north coasts as far inland as the highest peaks of the Alps. Particularly present during the equinoxes, it is active throughout the summer in the western basin, where it is accompanied by other southerly winds (*marin*, *gharbi*). It loses most of its influence in the eastern basin, which is swept by the *meltem*.

Somewhat less common but locally important throughout the north-western quadrant of the Mediterranean are humid winds from the east (*levant*) or the west (*ponant*). The *levant* blows in summer in the occidental basin and it often combines with the thermal breezes. It can also rise in winter, accompanied with a strong swell. It is then very violent and affects the Spanish coasts of the western basin. In winter, the *ponant* is a wind of monsoon strength blowing from Spain when the temperature of the sea is higher than that of the land. If a depression settles down on the north of Spain, it can become very strong on Gibraltar, and the Alboran and Algerian Seas.

Wind strongly increases evaporation, hence aggravating the effects of drought and high temperatures in summer. In areas with strong dominant winds, direct effects on vegetation include morphological effects, such as wind-oriented flag-like structure of trees, and many effects on the physiology and life-history traits of plants and animals.

Winds, sea evaporation, and coastal topography all contribute significantly to local precipitation events and storms, explaining the intensity and the violence of some coastal rains.

1.3.3 Unpredictability

A wide range of climatic variations, from long-term cycles of several centuries to extreme sudden unpredictable events, undoubtedly play an important role in shaping life cycles and distributions in many Mediterranean organisms. From one year to the next, or between seasons of a given year, and even within the course of a single day, temperature extremes, precipitation, winds, and other climatic factors can vary dramatically. The wide range of diurnal temperature fluctuations at certain seasons, the violence of certain winds, and calamitous short-lived rainfall events make the Mediterranean climate notoriously capricious and unpredictable. During such violent storms what are normally insignificant streams can be transformed in a matter of hours into devastating torrents, capable of carrying away houses and their terrified inhabitants. In 1876, 128 people were drowned at Saint Chinian, in southern France, by the flooding of a stream so small that even many locals were unaware of its name (the Verzanobre). As a consequence of an exceptional rainfall in October 1969, the Rio de Las Yeguas, Spain, overflowed its bed and flooded the small village of La Roda de Andalusia with mud 1 m deep. In the eastern Pyrenees, no fewer than six or seven exceptional rainfalls, totalling 200–600 mm of water within one or a few days, have occurred in the last 25 years. On 22 September 1992, torrential rain turned the small Ouvèze River into a devastating torrent, flooding the small city of Vaison-la-Romaine, southern France. Sixty people drowned and several bridges were destroyed. The only one which resisted the flood was a very old Roman bridge. In the summer of 2002, major floods again took place in the Gard region of southern France, with much damage resulting.

Storms may be harmful for animals as well. In June 1997, a violent 1-day storm with approximately 300 mm of rainfall occurred in a small valley on the island of Corsica, which, incidentally, is an experimental site for a long-term study of the population biology of birds. The rain washed away the caterpillars, which were feeding on the leaves of the trees. This resulted in a food shortage for breeding blue tits (*Cyanistes caeruleus*). As many as 21 of 57 nest boxes, where blue tits had their broods, were deserted by the adults, and the young died in the nest. Not a single fledgling succeeded in leaving the nest so that the population crashed and stayed low for several years (Blondel *et al.* 1999). In coastal lagoons, where sterns, waders, and flamingos breed in large colonies, hundreds of nests may be flooded by a sudden rise of water levels.

Sudden rises in river levels may be all the more devastating when enormous quantities of land material are carried downstream, aggravating the effects of flooding. Erosion is a major problem in many Mediterranean territories because of steep slopes and deep and narrow valleys. Moreover, many mountain ranges are not solid constructions based on resistant rock, but rather accumulations of soft, crumbly materials that are sensitive to erosion. The violence of the rivers and streams during rare cataclysmic events is all too often exacerbated by the greatly increased surface runoff, resulting from uncontrolled deforestation of entire watersheds, even on the steepest hillsides. More recently, the precipitous increase in asphalt, macadam, and concrete, as well as soil compaction resulting from the use of heavy machinery and the excessive use of pesticides, tend to destroy or reduce the biological activity of soils and thereby to increase the risks of flooding after exceptionally heavy rains.

Vaudour (1979) estimated that soils are being eroded at a rate of 1 m per millennium in the Madrid region. The rate of soil ablation due to erosion can be as high as 1.4 mm per year (1.4 m per millennium) in many catchments of the Mediterranean Basin (Dufaure 1984), especially in North Africa, where Sari (1977) cited several examples in the Ouarsenis region of Algeria, where catchments lose 1000–2000 tons/km^2 per year! In the eastern and southern parts of the basin, most of the region's landscapes are so highly dissected, complex, and unstable, with steep slopes and shallow rocky soils, that the former protective covering of vegetation has been destroyed. Once exposed, the shallow underlying mantle of soil quickly falls prey to erosion. In the more degraded regions of the eastern half of the basin and most of North Africa as well, there is a mosaic of heavily degraded, eroded sites, with only a very few well-maintained sites remaining. Gully erosion and landslides of mountain slopes result in 'badlands' of no use for humans and also very poor in biological diversity. It was to fight against the devastation caused by flooding that, in the second half of the nineteenth century, in France, there began the immense public works of replanting mountains (RTM, *Restauration des Terrains en Montagne*), of which some of the most famous are those of Mt. Aigoual and Mont-Ventoux (see Chapter 13). Similar efforts have been undertaken more recently in many other parts of the Mediterranean.

Finally, unusually prolonged droughts also frequently occur. The summers of 1976, 1989, 2003, 2005, and 2007 were exceptionally dry throughout the region, and a large number of catastrophic fires took place in these years. For example, in 2003 more than 400 000 ha of forest burned in Portugal alone, and a single devastating wildfire destroyed

180 000 ha of forest and killed 10 firemen there on 24 August 2005. At long intervals, catastrophic cold spells may strike as well, such as the winter of 1709, when the entire harbour of Marseille and the Rhône River itself froze over. Nevertheless, the Mediterranean is above all a land of abundant sunshine with nowhere less than 2300 h of sunshine per year and more than 3000 h in most of the eastern and southern parts of the basin. Combined with nearly 250 rain-free days per year, these basic climatic features allow the cultivation of spring wheat, olives, and grapes, the immemorial trio of Mediterranean agriculture (see Chapter 10).

In the next section, we discuss various ways to delimit the Mediterranean geographically, from biological, climatic, and finally 'bioclimatic' perspectives.

1.4 Mapping the limits of the region

Defining and mapping the Mediterranean realm from a biological point of view has been a subject of hot debate among biogeographers for more than a century. There are no sharp borders with neighbouring regions and many factors must be considered, including vegetation, climate, latitude, and altitude. Any approach adopted will thus always be somewhat arbitrary. In the European portion of the Mediterranean, the main areas of discussion, in so far as limits are concerned, are the higher zones of the mountain ranges. To the east and south, similarly controversial alpine zones occur in Turkey and North Africa. Moreover, in those regions there are vast areas covered by steppe vegetation that some authors include in the Mediterranean biogeographical region while others adamantly exclude them.

Ecologists, historians, and geographers all agree that what provides unity to the region and also its particularity is its climatic pattern of hot, dry summers and humid, cool, or cold winters, a climate type that is also found in parts of California and Chile, the Cape Province of South Africa, and two disjunct regions of southern and south-western Australia (Dallman 1998; Médail 2008a; see Chapter 3). In the Mediterranean Basin, this bimodal weather pattern represents a sharp discontinuity, or anomaly, in the sequence of climate types occurring from the Equator to the North Pole.

Within the Mediterranean Basin itself, climate also changes with rising altitude in mountain ranges and when travelling from west to east. On the whole, a sharp gradient exists between the colder, wetter north-western and north-eastern quadrants of the basin and the hotter, more arid, south-eastern and south-western parts in North Africa and the Near East, as described below. Depending on the definition used, 'the Mediterranean area' comprises between approximately 2 million km^2 (Médail and Myers 2004) to as much as 9.5 million km^2 (Daget 1977). Following Quézel (1976a), Quézel and Médail (2003), and Médail (2008a), we use 2.3 million km^2 in this book. At its outer limits, the area shows great biogeographical and climatic complexity where it meets the boreal forests in central and northern Eurasia, the vast steppe regions of central Asia and north-western Africa, or the hot subtropical deserts of north-eastern Africa and the Middle East. As noted, the Mediterranean realm also includes the so-called Macaronesia, which encompasses Madeira and the Canary Islands, and a few smaller volcanic archipelagos located off the west Saharan coast of north-west Africa (see Fig. 1.2).

1.4.1 Which bioindicators?

One historical approach to mapping and defining the region has been to rely on so-called bioindicator plant species thought to provide a reliable index to mediterranean-type ecosystems; that is, they can survive, even thrive, on long hot and dry summers and cool wet winters, but are unable to survive prolonged periods of frost. The main candidate for bioindicator—and, more popularly, as a regional emblem—is the olive tree (*Olea europaea*). Pliny the Elder (AD 23–79) was probably the first to use the area of distribution and cultivation of the olive tree to define the limits of the Mediterranean. This idea has merit and it has persisted. For example, the international Blue Plan programme (www.planbleu.org/; see Chapter 13) also emphasizes the close similarity of the olive tree's distribution area when defining the Mediterranean region.

Furthermore, not only does the olive tree's distribution cover most or all of the region, the tree's leaf type and biological type, or **growth form**, are also characteristic as it seems particularly well adapted to the long dry summers of the Mediterranean. To wit, the thick, waxy leaves that remain on the trees for 2–3 years or more represent an outstanding water-saving system that is shared by many evergreen trees and shrubs in the basin and elsewhere. This so-called **sclerophyllous** leaf structure occurs in many species in the basin and all other mediterranean-climate regions (see Chapters 3 and 8), such as the bay tree (*Laurus nobilis*), strawberry tree (*Arbutus unedo*), lentisk (*Pistacia lentiscus*), and many others. Deep frost, however, acts like a sword of Damocles for olive groves in the northern parts of the basin, sometimes decimating plantations over huge areas. The olive tree's remarkable regenerative abilities after severe stress (fire, cutting, disease, etc.) are also typical of many sclerophyllous Mediterranean trees and shrubs, including many 'evergreen' or perennial-leaved oaks.

Indeed, the primary alternative bioindicator for the Mediterranean realm, first proposed by the geographer Drude in 1884, is the holm oak (*Quercus ilex*), whose stiff, long-lived leaves are somewhat similar to those of the olive tree. The sclerophyllous holm oak and several congeners, including *Quercus coccifera* and *Quercus calliprinos*, dominate huge expanses of vegetation around the Mediterranean Basin (see Chapter 7). However, here again, problems exist. The holm oak is absent from large portions of the eastern half of the basin, where it is restricted to warm plains and foothills near the coast. It also extends well outside the Mediterranean Basin in some areas, as for example along the Atlantic coast of France and along the River Rhône almost to Lyon. Furthermore, recent studies indicate that the holm oak, like many or most oaks in fact, is a rather complex botanical entity rather than a clear-cut single species in the popular sense of the term. It is best seen as part of a hybrid swarm involving at least three other species (Michaud *et al.* 1995), including the cork oak (*Quercus suber*), which is not even considered to be closely related to it, **phylogenetically** (Toumi and Lumaret 1998; Lumaret *et al.* 2002; Lopez-de-Heredia *et al.* 2005; see also Box 7.1 and Chapter 10).

To a modern biologist, however, there are problems with this approach. For one thing, it is problematic to use *any* cultivated plant, such as the olive tree, to delimit a biogeographical area. As for many long-cultivated plants, it is very difficult to determine the real origins of the olive tree, and it may not even be native to the Mediterranean region (Besnard *et al.* 2007; see also Chapter 10). Further, it is almost impossible to distinguish wild from cultivated forms of this tree genetically. The holm oak is better, in that it is not cultivated. But, the genetics of the holm oak are complicated, as noted above, and we have to add the related eastern species, *Q. calliprinos*, to achieve a full coverage of the basin (see Chapter 7). Indeed, any single bioindicator seems inadequate to define an entire region from a scientific point of view. By contrast, in Chapter 5 we will find that bioindicators are very useful in recognizing and distinguishing between **life zones** within a given region.

1.4.2 Plant associations as indicators

The next approach, historically, to mark the boundaries of the Mediterranean was to take certain plant 'associations' that include the holm oak, or other similar evergreen oaks, as markers. Historically, the French botanists Emberger (1930a), Flahaut (1937), and followers considered the presence of the evergreen formations dominated by holm oak, or closely related species as being 'diagnostic' of a mediterranean-type bioclimate. Like the olive tree, these oaks are long-lived, resprout readily after fire or cutting, and show many ecophysiological adaptations that recur among many dominant woody plant species in mediterranean-type climate areas around the world (see Chapter 8). They are slightly more tolerant to extreme cold than the olive trees, as witnessed by the fact that they do not freeze during very cold winters such as that of 1956. Together these long-lived, sclerophyllous, and highly resilient plants constitute a characteristic dense, evergreen woodland, often human-transformed into shrubland, the likes of which are not found elsewhere except in other mediterranean-climate regions (see Chapter 8).

However, a broad range of **life forms** other than evergreen sclerophyllous also occurs in the Mediterranean Basin, not only at higher altitudes, but also within the evergreen formations themselves (see Chapters 5 and 6). Thus, the basin's vegetation cannot be defined solely on the basis of evergreen oak woodlands and shrublands, even if these do indeed represent one of the most remarkable and most characteristic vegetation structures in the circum-Mediterranean region, as well as in all other mediterranean-climate regions (e.g. Dallman 1998; Médail 2008a). Many types of plants and associations in fact share the terrain, as is readily apparent both in autumn and winter, when leaves of the many deciduous species present change colours and fall, just as in temperate forests. Thus, this approach to mapping the Mediterranean region is also unsatisfactory.

1.4.3 A climatic approach

A climate is generally considered to be 'mediterranean' when summer is the driest season, during which there is a prolonged period of drought (see above). Several authors have attempted to use climatic data to delimit the Mediterranean realm. At one extreme, some scientists have emphasized temperature and the range of mean annual rainfall. But this leads to very narrow lines being drawn around a littoral band, characterized by mild winters, and excludes all the high and mid-altitude mountain areas, despite the long-term presence there of typical Mediterranean flora and vegetation. At the other extreme, in the so-called 'isoclimatic' Mediterranean definition given by Emberger (1930b) and mapped by Daget (1977), the only criterion is that summer should be the driest season and that there should be a period of effective physiological drought during that season. This approach encompasses not only the Mediterranean mountaintops but also the vast adjacent steppe regions where there is no appreciable rainfall in summer. Using this approach, about 8 million or even 9.5 million km^2 are brought together under the term 'Mediterranean area', including the central Asian steppes, as far east as the Aral Sea and the Hindus Valley, all the northern half of the Sahara desert, and most of the Arabian Peninsula. This becomes entirely meaningless when considering the ecology or biology of individual organisms or communities. Something intermediate is clearly needed.

1.4.4 A 'bioclimatic' approach

Strictly biological or climatic approaches prove to be inadequate to our purposes when trying to delimit the Mediterranean region. A more realistic approach combines both climatic (temperature and precipitation) *and* biological—specifically vegetation—factors, as first advocated by Gaussen (1954) (see Fig. 1.6). In addition to climatic analysis, typical plant assemblages are identified in this 'bioclimatic' approach, by indicating two or more dominant tree or shrub species, whose combined presence invariably characterizes one of a series of altitudinal vegetation zones, which replace each other, according to altitude, latitude, and slope exposition; that is, wetter, north-facing compared with drier, south-facing slopes. This approach makes it possible to identify enclaves of typical Mediterranean vegetation hundreds or even thousands of kilometres from the basin itself. Examples of such enclaves may be found in the mid-Saharan mountains, the isolated Hoggar (3003 m) and Tibesti (3415 m) ranges, the foothills of the Alburz Mountains along the southern Caspian shores of northern Iran, and the highland regions on either side of the southern Red Sea. A series of bioclimatic types or life zones are defined in this bioclimatic approach, as we will describe in Chapter 5. Although it is tempting to seek direct correspondences between a given bioclimatic zone and a given altitudinal life zone, variation occurs as a result of latitude, exposure, and soil types. Furthermore, there is no satisfactory answer to the question of what is a 'Mediterranean mountain', as compared to a mountain range simply marking a regional boundary. What about ranges, or parts of ranges, where the bioclimate on one side is clearly Mediterranean, but not on the other side?

A certain amount of ambiguity and controversy also exists in areas where vegetation has been drastically altered by people, either very early in the Holocene, or much more recently. In such areas, where very little remains today of the original Mediterranean habitats, sufficient evidence exists to suggest that four to five millennia ago a typical Mediterranean vegetation did exist over large areas. However, following many authors, we would rather recognize habitats on the basis of their present-day composition, and not on an inferred hypothetical and controversial 'historical climax'. Thus we exclude the Cape Verde islands from Macaronesia, even though, bioclimatically speaking, it was long considered to belong with the Canaries, the Azores, and Madeira.

Several field guides to the Mediterranean flora and fauna draw a limit at about 1000 m altitude, but this is not justified from a broad historical/ecological perspective. In our view, delimitation of the Mediterranean phytogeographical territory should include not only the 'basal' zone, which is usually dominated by evergreen shrubland formations, but also the altitudinal zones above it. According to this approach, the extreme eastern Pyrenees and most of the Apennine ranges, for example, are included in the Mediterranean area, even if they are not uniformly endowed with a Mediterranean climate. The western Pyrenees and the northernmost part of the Apennines, however, are excluded but the Canary Islands, including the Salvage Islands, and Madeira—the Macaronesian region (see Fig. 1.2 and Table 1.1)—benefit from a largely mediterranean-type bimodal climate. In addition, the important insights their extant vegetation provides into ancient plant lineages and even forest types, which were formerly widespread around the Mediterranean Basin in Pliocene times, make these archipelagos of utmost importance for understanding the history of Mediterranean biota.

Using bioclimatic criteria, the Mediterranean territory covered in this book is about 2 300 000 km^2 in all and spread over 24 countries (Table 1.2). Cer-

Table 1.2 Area (km^2) of or within each of the 24 countries and territories lying within the Mediterranean bioclimate region, and the contribution of each (%) to the total. Note that areas given do not necessarily correspond to entire national territories

Country	km^2 × 1000	%	Country	km^2 × 1000	%
Spain	400	17.3	Turkey	480	20.7
Portugal	70	3.0	Cyprus	9	0.4
France	50	2.2	Syria	50	2.2
Monaco	0.002	8.10^{-7}	Lebanon	10	0.4
Italy	200	8.6	Israel	22	0.9
Malta	0.3	1.10^{-4}	Jordan	10	0.4
Slovenia[1]	0.4	2.10^{-4}	Palestine[2]	6	0.3
Croatia[1]	17.2	0.7	Egypt	50	2.2
Bosnia-Herzegovina[1]	0.1	4.10^{-5}	Libya	100	4.3
			Tunisia	100	4.3
Montenegro[1]	1	4.10^{-4}	Algeria	300	12.9
Albania	28	1.2	Morocco	300	12.9
Greece	100	4.3			

[1] Toni Nikolić, personal communication.
[2] Palestinian territories (Gaza and West Bank).
Source: after Quézel 1976b and Toni Nikolić, personal communication.

tain areas, for example parts of Libya, and the total absence of certain countries, such as Iraq, are questionable. In both these countries, small but important areas of mediterranean-type woodland exist. If the broad isoclimatic region first recognized by Emberger were adopted, then *all* of Libya and *all* of Iraq would be included in the Mediterranean. By contrast, the narrow approach cited above would lead us to reduce to about 500 km^2 the portion of Libya included, but also to include some of the highlands of northern Iraq. In sum, the bioclimatic approach we adopt constitutes a compromise, which excludes the steppe areas outside the first ring of Mediterranean mountains. The approximate results are shown in Table 1.2 and in Fig. 1.2. Practically, this delimitation roughly coincides with the 100 mm isohyet. It also corresponds with the approach taken by Quézel and Médail (2003) and Médail (2008a).

1.5 Adjacent and transitional provinces

The next step in 'setting the scene' is to identify important 'neighbours' of the Mediterranean region (Fig. 1.7), with which so many important

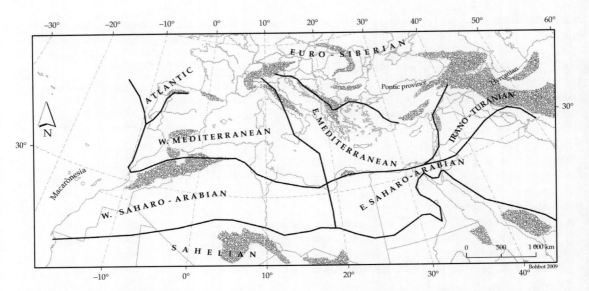

Figure 1.7 Subdivisions of the Mediterranean area and delineation of the major adjacent biogeographical regions and provinces (after Quézel 1985, and Zohary 1973, with modifications).

interactions—both biological and cultural—take place in space and time.

1.5.1 The Pontic province

The so-called Pontic biogeographical province, named after the Pontic Mountains of northern Turkey, comes into contact with the Mediterranean and the vegetation types of the two intermingle in a variety of ways. Most remarkably, along the southern edge of the Black Sea, there occurs an assemblage of forest types that resemble formations that apparently covered cooler parts of southern Europe in the late Tertiary. Within this mosaic, there is a disjunct series of 'Mediterranean enclaves' in a very narrow belt from sea level to 200–300 m, particularly on thin soils and southern exposures (Davis 1965). Elsewhere around the Black Sea (sometimes called the Euxine province), several typically Mediterranean shrubs and small trees occur as colonizers wherever the indigenous temperate forest has been destroyed. Conversely, there are numerous enclaves of non-Mediterranean Pontic vegetation further south in the Anti-Taurus, the Syrian Amanus range, and even parts of northern Lebanon.

About 500 km to the south-east of the Pontic province, the so-called Hyrcanian province occupies the flanks and summits of the coastal mountains at the southern extremity of the Caspian Sea. This province apparently occupied a considerably larger area during the late Tertiary and Pleistocene (Zohary 1973), and descended much further south towards Iraqi Kurdistan and northern Syria, as shown both by fossil records and extant relics subsisting outside this territory today. The deciduous species of maple (*Acer*), birch (*Betula*), hazelnut (*Corylus*), beech (*Fagus*), ash (*Fraxinus*), lime or linden (*Tilia*), and elm (*Ulmus*) that occur in the northern Mediterranean regions, as well as parts of the fauna there, all show clear Arcto-Tertiary origins. A certain number of the evergreen trees commonly found here also occur in wetter parts of southern Europe, despite their strong Euro-Siberian affinities. Examples include boxwood (*Buxus sempervirens*), holly (*Ilex aquifolium*), and yew (*Taxus baccata*). Conversely, just as in the Pontic province, in the Hyrcanian, there are many enclaves of Mediterranean elements, which colonize disturbed forest areas and may become locally dominant. The Pontic and Hyrcanian provinces are sufficiently similar and overlapping that some authors treat them as a single province of the much larger Euro-Siberian region.

1.5.2 The Irano-Turanian province

To the east of the Mediterranean Basin lies the large Irano-Turanian region, which includes the semi-arid steppes, or wooded steppes, dominated by silvery, fragrant wormwoods (*Artemisia*) stretching eastward and north from Iraq, central Turkey, and central Iran to central Asia. Winters here are extremely cold and summers are extremely hot, with very low annual rainfall (<200 mm). It is noteworthy that many faunistic and floristic elements of predominantly Irano-Turanian distribution are also prominent in low rainfall areas of the Mediterranean Basin, not only in the eastern half of the basin, but also in southern Iberia and much of North Africa (see Chapters 2 and 3). A long history of inter-penetration and east–west migrations has occurred with revealing **refugia** occurring notably in the Balkans and regions further north (e.g. Bennett *et al.* 1991; Djamali *et al.* 2008a, 2008b; Magyari *et al.* 2008).

1.5.3 The Saharo-Arabian province

To the south of the Mediterranean Basin lies the immense zone called the Saharo-Arabian biogeographical region. It extends about 4800 km from Mauritania eastward to the Red Sea and is about 2000 km wide. With 9.6 million km^2, the Sahara is the largest desert in the world and more than three times the size of the entire Mediterranean Basin. Desert conditions in the Sahara appeared as early as the late Miocene, *c.*7 mya, which explains why there has always been relatively little faunal and floristic interchange between these desert regions and the Mediterranean. Today, the northern fringes of the Sahara show a mediterranean-type climate, with rainfall occurring between autumn and spring, while its southern fringe has a tropical climate, with rainfall occurring in summer when temperatures are the highest. Beyond the Red Sea, this gigantic region extends much further eastward, across the Arabian Peninsula and on to parts of southern India, with gradual replacements of some but not all dominant taxa (e.g. *Acacia, Zizyphus, Zygophyllum, Atriplex*).

As with the Irano-Turanian region to the east, between the Mediterranean and the Saharo-Arabian regions there have been ebbs and flows in climate and biota and in adjacent East Africa, and the Horn of Africa as well. As mentioned above, there are revealing enclaves and even endemics of typically Mediterranean elements in the high mountains of the Sahara, as there are also in the Horn of Africa and Yemen. But, in general, and unlike the oriental steppes, the vast region has primarily acted as a barrier between the so-called Afro-tropical and Mediterranean (and northern European temperate) biogeographical regions, except during certain periods of the Pleistocene and Quaternary (see Chapter 2).

Summary

The birth of the Mediterranean region has been a long process, characterized by a suite of violent tectonic events, which occurred on a wide range of scales of space and time in the course of the past 250 mya. Some catastrophic events, such as, for example, the almost complete desiccation of the sea, the Messinian Salinity Crisis, had many consequences on living systems, both terrestrial and marine. Other catastrophic events, which moulded many parts of the region, include earthquakes and volcanic activity. A brief overview of the physical setting reveals how the Mediterranean region, with its myriad islands, coastal lands, rivers, and high mountains, is a legacy of this turbulent history, providing a mosaic of habitats, with an astonishing heterogeneity of local topographies, soil types, and microclimates related to altitude, rainfall, and exposition of slopes. A wide range of winds, originating from the north, south, west, and east, also have major impact on living systems.

This, combined with the geographical location of the Mediterranean at the intersection of three major land masses—Europe, Africa, and Asia—resulted in the emergence of a very distinctive fauna and flora. The Mediterranean Sea is the largest inland sea in the world and is divided in two major basins, including several minor ones, which mostly result from the tectonic activity of several microplates. A high heterogeneity of key factors, such as salinity, currents, and temperatures, characterize this sea, which has almost no tides and a very narrow continental belt. It is also the warmest deep sea on the planet.

The climate of the region, with its long, hot, and dry summer and wet and cool winters, is characterized by a high unpredictability in space and time. Climatic extremes may have harmful effects on living systems and accelerate gully erosion.

A complex and important issue is to define and delimit the Mediterranean region from a biological perspective. Certain plant species may be used as bioindicators but the best tool for this purpose is certainly a bioclimatic approach. This approach allows us to recognize several adjacent biogeographic provinces with transitional floras and faunas.

CHAPTER 2

Determinants of Present-Day Biodiversity

Although Mediterranean ecosystems can be considered 'young' because of the relatively recent appearance of a mediterranean-type climate, they are composed of species originating in almost all known biogeographic realms of the world and these 'colonizers'—many of which are ancient or 'primitive' taxa—constitute the vast majority of present-day species. Extant faunas and floras of the Mediterranean are thus a complex mixture of elements, some evolving *in situ* and others arriving from other regions in the distant or recent past. From historical cues, including fossil remains of the late Tertiary, we gain insight about the origin and **turnover** of Mediterranean biota over the past few million years. As we saw in Chapter 1, tectonic plate movements, volcanism, climate, and the resulting complexity of topography and geology played important roles in the birth and physical shaping of the region. Here we discuss four additional factors, or drivers, that make the region so exceptionally rich in biological diversity and provide brief introductions to the composition of the flora and some of the major faunistic groups of vertebrates and invertebrates, both on land and at sea.

2.1 Drivers of biodiversity

Whether we consider populations, species, habitats, or landscapes the four main drivers that affect these components of biodiversity in the Mediterranean belong to the realm of biogeography, geological history, landscape ecology, and human history.

2.1.1 Biogeography

Located at the crossroads of Europe, Asia, and Africa, the Mediterranean Basin has long served as a meeting ground for species of dramatically varying origins. As we saw in Chapter 1, the region's physical environment and climate have changed radically since the Mesozoic, with the result that biological composition of the different portions of the basin and migration routes of colonizing species have changed repeatedly. As a result, there are species originating from Siberia, South Africa, and many other places; there are even some relictual taxa among the soil-dwelling fauna whose closest relatives all occur in Antarctica (di Castri 1981). The Mediterranean can also be considered as a huge 'tension zone' (Raven 1964), lying between the temperate, arid, and tropical biogeographical regions that surround it (see Chapter 1), a zone where intricate interpenetration and interaction of taxa, hybridization, and **speciation** have been particularly favoured and fostered, as compared to more homogeneous regions further north and south. Certain subregions of the Mediterranean, such as Morocco and Turkey, as well as ancient tectonic microplates, such as Iberia, illustrate this situation particularly well.

One striking feature of Mediterranean biodiversity is the relatively low number of elements of Afro-tropical origin, except some North African mammals and plants, and several fish and insect species, including widespread species and some endemics (see below). Several obstacles prevented most Afro-tropical species from colonizing the basin with the same rate of success as Euro-Siberian and Irano-Turanian taxa. Clearly, the massive Afro-Arabian deserts, whose aridity dates back at least to the Miocene/Pliocene boundary (c.6 mya), and also the east–west orientation of the region's massive barriers to dispersal such as seas and mountain chains impeded south–north biological migrations. Moreover, the huge Saharo-Arabian province contributed far fewer plant species to the Mediterranean than either the Euro-Siberian or the much smaller Irano-Turanian region (see below), whereas the Mediterranean 'donated' a large number of taxa to the Sahara (Quézel 1978, Quézel 1985)!

One way to gain insight into the situation described above is to compare the trends of biodiversity in temperate biomes of Europe with those of the two other warm temperate regions of the northern hemisphere, namely eastern North America and eastern Asia. The size of these three massive forested blocks is roughly similar (approximately 10 million km^2), and they share the same history since they all belonged to the ancient northern supercontinent, Laurasia, prior to its break-up in the Jurassic. In sharp contrast to Europe, the biota of eastern North America and eastern Asia were never entirely cut off from potential source and refuge areas further south, at least since the early Tertiary. Accordingly, both regions' biota have many more tropical-derived families than do the ecologically equivalent European ones. For example, Latham and Ricklefs (1993) recorded three times more tree species in the mesic forests of eastern Asia (729) than in those of eastern North America (253), and almost six times more than in European forests (124). According to Huntley (1993), the comparatively low taxonomic diversity of tree species in European forests is linked to the limited area available during glacial episodes. In other words, many forest taxa that occur today in the Nearctic and in eastern Eurasia disappeared in Europe during Pleistocene glacial periods for lack of refugia within the continent. However, remnants of these plant and animal communities of the western Palaearctic survived in the Mediterranean region, as we will see below.

Now, if differential extinction rates of plant species occurred on a larger scale in Europe than in the two other temperate blocks of the Northern Hemisphere, one would also expect more differential extinction rates in animals as well, which would explain for example why European forest bird faunas are poorer in species than their Nearctic counterparts (Mönkkönen 1994). Finally, the tropical floras and faunas that disappeared from Europe at the end of the Pliocene, in the face of climatic deterioration, were not able to recover when climate improved, because the Sahara desert and the Mediterranean and Black Seas acted as barriers to dispersal. As a result, all the major groups of extant biota in the Mediterranean Basin are much more heavily derived from Holarctic 'stocks' than from the vast tropical lands to the south.

The combined role of history and habitat heterogeneity in the Mediterranean Basin helps explain current trends of biodiversity. Using a procedure for downscaling European species atlas distributions across Europe from a 10 km × 10 km grid resolution, Araújo et al. (2005) demonstrated that hotspots of richness for plants and what they call herptiles (i.e. reptiles and amphibians) occurred in the Mediterranean part of Europe, but that of birds occurred in temperate Europe, not in the Mediterranean. This is a consequence of the historical events just discussed previously, which is another means to illustrate the poor rate of speciation in birds within the limits of the Mediterranean. However, in the line of expected trends of spatial turnover of species, which provide an estimation of the beta diversity (see Fig. 5.10), hotspots of rarity are much higher for the three groups of organisms—plants, birds, and herptiles—in the Mediterranean than further north (Fig. 2.1).

2.1.2 Geological history: from plate tectonics to Pleistocene glaciations

The first and prominent determinant of biodiversity in the Mediterranean is the turbulent geological history of this region, squeezed between the huge land masses of Africa and Eurasia.

2.1.2.1 PLATE TECTONICS

The repeated splitting and joining of land masses has had a decisive influence on the evolution of plants and animals in the Mediterranean Basin throughout the Tertiary (see Chapter 1). The many cases of differentiation, **vicariance** (see Chapter 7),

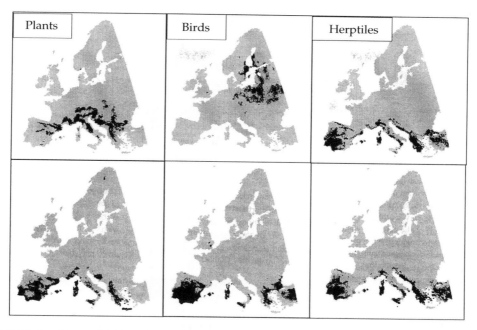

Figure 2.1 Distribution of downscaled richness hotspots (above) and rarity hotspots (bottom) for plants, birds and herptiles (reptiles and amphibians) across Europe. Richness and rarity hotspots coincide for plants and herptiles but not for birds (see text) (after Araújo et al. 2005).

and endemism in various groups of plants and animals are a legacy of the long and violent history of the region since the Mesozoic. Many speciation events in the basin are apparently much more ancient than was formerly thought and may have been associated with complex splitting and rejoining of tectonic microplates during the **Palaeogene** and early **Neogene**, 65–25 mya. Repeatedly, during these periods, isolation of land masses within the Tethys trapped species that evolved in isolation, giving rise to the many palaeoendemics we find today in all major groups of plants and animals. Most regions of especially high endemism within the basin are those that have been isolated in the past, such as the many tectonic microplates that were scattered in the Tethys during the early Tertiary (e.g. Apulia and the Ibero-Mauritanian and Cyrno-Sardinian microplates), or the many islands and upland areas that served as refugia during the various glacial periods of the Quaternary. A large body of evidence from palaeontological and molecular studies shows that large climatic changes occurring as far back ago as the Miocene/Pliocene produced biological refugia favourable for species differentiation. Oosterbroek and Arntzen (1992) observed a recurrent pattern of differentiation associated with isolation on tectonic microplates during the Tertiary in seven unrelated groups of invertebrates and vertebrates, including butterflies, flies, scorpions, frogs, and newts. However, because of the difficulty in establishing the species phylogeny and because the so-called **molecular clock**, upon which depends the rate of evolutionary change, varies so widely among phylogenetic groups, there is still much uncertainty about when and where extant taxa differentiated. We can only state that sometime in the far past, and at various places in the basin, geographical isolation on 'palaeo-islands' led to differentiation of many plant and animal species that still survive today. Many others, of course, have disappeared long ago (see Chapter 11).

2.1.2.2 PLEISTOCENE CLIMATIC UPHEAVALS

Although the modern Mediterranean climate was established some 2.8 mya (see Chapter 1), a large series of fluctuations occurred in the Pleistocene, which induced large-scale distributional shifts of species and communities back and forth across Europe. Over 30 years ago it was recognized that changes in the Earth's orbit, the so-called **Milankovitch cycles**, are the fundamental causes of Quaternary climatic oscillations (Hays et al. 1976). The Earth's orbital parameters—that is, orbital eccentricity, obliquity, and precession—have periods of 100 000, 41 000, and c.20 000 years, respectively (Taberlet and Cheddadi 2002). The first massive ice sheets associated with these phenomena in the Northern Hemisphere started to grow about 2.5 mya and these major astronomic changes resulted in climatic oscillations, with cold periods being much longer than warm periods. They occurred during most of the Pleistocene with a dominant 100 000-year cycle (Webb and Bartlein 1992). For these reasons, the Quaternary should probably be viewed as 'a cold epoch interrupted periodically by catastrophic warm events—the brief interglacials with climate similar to that of today' (Davis 1976). In particular, the climate changes that Europe experienced during the late Pleistocene, 50 000–30 000 years before present (years BP), were abrupt and severe. Therefore, the relative climatic stability recorded during the last 8000 years seems to be the exception rather than the rule. Both the long-term and the short-term climatic fluctuations had dramatically affected the distribution and abundance of species. Present-day biogeographical patterns are those of an interglacial period of the kind that has been rather rare since the beginning of the Pleistocene. During the entire Pleistocene, continuous remodelling took place in the structure, functioning, and composition of both local and regional floras and faunas, as a result of these shifts and upheavals. Any attempt to interpret biogeographical patterns requires investigating past dynamics of the biota at appropriate scales of space and time. At timescales of 10^3–10^5 years, the distribution and composition of living systems have changed continuously with each glacial/interglacial alternation throughout the Pleistocene. Much insight has been provided in recent years by palaeobotanical studies that reveal the transitory nature of environments and the profound changes wrought over geological time in ecological communities (Pons 1984).

Pollen grains are particularly useful in establishing a site or region's particular history because each plant species (or genus) is characterized by a unique pollen design that is clearly visible under the microscope. Fossil pollen layers accumulate through time wherever soil conditions are favourable, especially in moist and soft soils, such as bogs, riverbeds, and marshes. Using data provided by fossil pollen combined with radiocarbon dating, palaeobotanists can draw 'isopoll maps' (maps with lines of equal relative pollen abundance for a taxon), showing the dynamic distribution of various species over time. These maps indicate the long-term turnover of plant communities in a given region over time.

For example, oaks (*Quercus*) and hazelnut have progressively migrated from their southern refugia to the north in the course of the last 12 000 years, as illustrated in Fig. 2.2.

A similar but opposite shift in distribution occurred repeatedly during the Pleistocene, each time that a glacial period forced biota to 'escape' or 'migrate' to the south. At timescales of tens of millennia, by the way, it is perfectly acceptable to use the term migration in the same way as we refer to the seasonal migration of animals. Just as some birds and mammals 'track' annual weather and climate, so over the long term plant and animal taxa 'track' and adjust to long-term climate change. As Huntley and Birks (1983) put it, 'Just as the knot [*Calidris canutus*] must be capable of migrating 5000–8000 km or more in two or three stages and at average speeds of up to 70 km per hour in order to track annual climate changes, so trees must migrate similar distances, again at rates which allow their populations to track long-term climate changes (50–2000 m per year)'. During the periods when almost no arboreal vegetation survived north of the mountain chains that delimit the Mediterranean region, much of southern Europe was occupied by landscapes where steppe formations were dominant or at least present, with the same shrub species, for example wormwood (*Artemisia herba-alba* complex), as occur today in semi-arid steppe areas of the Near East and North Africa.

Generally speaking, forest floras and associated faunas of the western Palaearctic migrating towards the equator to escape the cold were blocked by the Sahara desert and, for some groups, the Mediterranean Sea itself. Thus, these migrant taxa had to concentrate around the Mediterranean Sea,

Figure 2.2 Isopoll maps illustrating the northward expansion of oaks (*Quercus*) and hazelnut (*Corylus*) over the last 12 000 and 10 000 years, respectively. Three classes of pollen density are shown: stipple 2–5%, horizontal stripe 5–10%, black fill >10% (after Huntley 1988).

especially in the larger peninsulas. During the coldest phases, average temperatures of sea water must have been about 8°C lower than prevailing temperatures today. The snow limit was at 1200 m in Galicia, north-western Spain, and in Liguria, western Italy, and at 2000 m in the Atlas Mountains of Morocco.

A schematic representation of the distribution of major vegetation belts at the present time and during the most severe phase of the last glaciation (Würm, 30 000 years BP) is given in Fig. 2.3.

Analyses of fossil pollen deposits have shown that even during the most extreme glacial periods many refugia existed for plants and animals on certain mountain slopes, in the large peninsulas (Iberian, Italian, Aegean, Anatolian), on islands, as well as in the valleys of large rivers. More recently, Leroy and Arpe (2007) used sea-surface temperature for evaluating the areas where trees might have survived the harsh climatic conditions of the Last Glacial Maximum (LGM). By down-scaling the simulated data, using the difference between the LGM and present-day simulations on a high-resolution climatology, these authors found different scenarios for refuge areas, depending on the species of trees. For typical warm deciduous or **summergreen** trees, such as several species of *Quercus*, *Castanea*, and *Ostrya*, suitable areas during the LGM included several areas of the Iberian Peninsula, Italy, the Balkans, the Pindos Mountains of central Greece, east of the Rhodopes, at some places along the southern and eastern coasts of the Black Sea, and along the southern coast of the Caspian Sea. For **cool summergreen** trees, such as *Carpinus*, *Corylus*, *Fagus*, *Juglans*, *Tilia*, *Ulmus*, and *Frangula*, the area was larger, including much of southern France, the Balkans, Hungary, and large areas north of the Black Sea. For these historical reasons these are areas of high biodiversity. Incidentally, the inclusion of the Black and the Caspian Sea regions, which are usually excluded from European maps, is especially important because these regions have certainly had an influence on the European genotypes of the temperate summergreen trees (Leroy and Arpe 2007). In addition, the Mediterranean area was much larger and even more ecologically complex during the LGM than it is today, as a result of the worldwide sea level being some 100–150 m lower than today. The sea-level **eustatic** fluctuations driven by glacial/interglacial

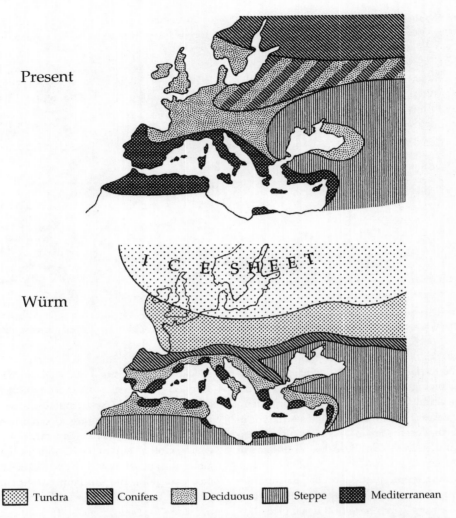

Figure 2.3 Schematic representation of the main vegetation belts in Europe and North Africa at the present time and during the most severe phase of the Würm glaciation (30 000 years BP) (after Brown and Gibson 1983, in Blondel 1995).

cycles of the Pleistocene have repeatedly increased and decreased the areas of the islands and coastal areas, shortening the distances among islands and between islands and mainland areas, favouring colonization events. Many opportunities thus arose for species to differentiate as repeated fragmentation of initial areas of distribution took place over time. Recent works demonstrated, however, that extant elements of the post-glacial biota of north-western Europe persisted through glacial periods not only in the 'classic' Mediterranean southern refugia, but also farther north in small pockets of favourable microclimates. Recently, **phylogeographic** studies have provided evidence on the existence of hitherto unknown glacial refugia outside the Mediterranean Basin. Called 'cryptic refugia' by Provan and Bennett (2008), they were first suggested by the presence of genetic lineages of mammals, reptiles, and amphibians, some of them widespread across northern Europe, that did not correspond to the lineages present in any of the Mediterranean peninsular refugia. These cryptic refugia were located in the area around the Carpathians and were characterized by mixed deciduous and coniferous woodland, often on south-facing slopes. Bhagwat and Willis (2008) found that species which were the most likely to survive during full glacial episodes in these refugia were wind-dispersed, habitat-generalist trees with the ability to reproduce vegetatively, and habitat-generalist mammals, which are northerly distributed today. Incidentally, these recent findings must be carefully considered for using the so-called 'bioclimatic envelope models' to forecast future species responses to global climate change (Araújo and Guisan 2006).

For most species, which had to survive in southern refugia during glacial episodes, they presumably came together to form faunas of many different communities, without any clear geographic delimitation between Mediterranean and non-Mediterranean communities. For example, fossil faunas in deposits of southern France dating back to the late glacial maximum, *c.*18 000 BP (Würm glaciation), include disparate species assemblages with both 'northern' species, such as reindeer (*Rangifer tarandus*) and snowy owl (*Nyctea scandiaca*), and Mediterranean thermophilous species, such as Hermann's tortoise (*Testudo hermanni*). This indicates that climatic conditions allowed for the persistence of both boreal and Mediterranean species. Taxa of primarily Euro-Siberian distribution, such as whortleberries (*Vaccinium*) and several species of willow (*Salix*), the tundra-dwelling snowy owl, and the reindeer, all occurred widely in the Mediterranean region during glacial periods. Pollen in cave sediments, as well as charcoal analysis from Gorham's cave, Gibraltar, which was inhabited by Neanderthals during the LGM, reveal a highly diversified landscape with oaks, pines (*Pinus*), juniper (*Juniperus*) savannas, forest, wetlands, grasslands, and Mediterranean-type coastal scrubs. Both 'cold' and thermophilous floras and faunas co-occurred in a very diversified environment around the Rock of Gibraltar (Finlayson and Carrión 2007). What makes the response of species and communities to climatic cycles difficult to interpret is that each species reacted independently and differently to climate changes in terms of persistence in a refugium, migration rates, and colonization routes (Taberlet *et al.* 2008).

Similarly, many of the huge colonies of seabirds that breed in northern Europe as far north as Spitzberg had to find refuge in the south during glacial times. Many cliffs and islands of the Mediterranean Sea were populated by the large colonies of seabirds that are so characteristic of steep cliffs in the northern part of the Atlantic Ocean today. A number of fossil sites have provided evidence that gulls (*Larus*), auks (*Uria*), and gannets (*Sula bassana*) used to breed along the coasts of Iberia, France, and Italy during glacial episodes of the Pleistocene. The flightless and now extinct great auk (*Pinguinus impennis*) was widespread as a breeding species in the Mediterranean during most of the Quaternary. Fossil records of this species exist from as far east as Calabria, Italy (see Box 2.1). The last pair of great auks was hunted out in Iceland in 1844.

> **Box 2.1. The wonderful bestiary of ornate caves**
>
> Two caves with extraordinary paintings have been discovered in southern France, the Chauvet cave in 1994 in Ardèche and the Cosquer cave along the coast in 1996. The Chauvet cave contains some of the oldest (about 30 000 years old) and most beautiful cave paintings ever found in the Mediterranean Basin (Chauvet *et al.* 1995). A vast bestiary is portrayed, including 300 or more different animals, such as bison (*Bison*), horse (*Equus*), bear, deer, mammoth (*Mammuthus*), hyena (*Crocuta*), panther (*Panthera*), lion (*Panthera leo*), rhinoceros (*Dicerorhinus*), reindeer, giant deer (*Megaloceros giganteus*), auroch (*Bos primigenius*), and ibex (*Capra ibex*), as well as the only known representation in Palaeolithic art of large birds, such as the eagle owl (*Bubo bubo*). In the underwater Cosquer cave near Marseilles, paintings depict many mammals, such as horses, bisons, and ibexes, and extraordinarily detailed rendition of the great auk, dating back 20 000 years BP. To get an idea of how much has changed the shores of the Mediterranean since Palaeolithic times, note that the entrance of the Cosquer cave is now 36 m below sea level!

During interglacial periods, the plants and animals spread north again without leaving the Mediterranean region. Thus, the survival of species of boreal origin in these refugia during the glacial periods added to the contemporary Mediterranean biota many species of northern origin. Species with very large modern-day distributional ranges, such as the chaffinch (*Fringilla coelebs*) or the chiffchaff (*Phylloscopus collybita*), expanded their ranges to the north during interglacial periods without leaving the region, just as deciduous oaks and their seed-dispersing companion bird, the jay (*Garrulus glandarius*), that occurred in southern Europe during both glacial and interglacial periods, expanded into most of Europe during interglacials. They tended to alternate between small isolated populations, during unfavourable glacial periods, and large, widespread populations, during more favourable interglacial periods. Many species differentiated to some extent when they were dispersed among several Mediterranean refugia, as we will show below and as discussed in detail for vascular plants by Thompson (2005).

During interglacial periods, such as the present one, large parts of the Mediterranean were covered with forests with the exception of high mountains and some high plateaux in Iberia and Anatolia. From the end of the last glacial period, however, some 12 000 years ago, deciduous forests have spread progressively from the Mediterranean area to the north at rates of about 1 km per year (Huntley and Birks 1983). Thus, the Quaternary history of the most important European forest belts and their associated faunas has been one of a series of massive migrations back and forth across the western Palaearctic (Sánchez Goñi *et al.* 2005). As we saw above, migrations from the east (especially the Irano-Turanian region) and the south concurrently affected the Near Eastern and African portions of the Mediterranean area. However, although the basin was much more extensively forested in the past than in most of modern times, it appears that shrublands and heathlands did occur spontaneously in many places, creating a mosaic landscape due to climate and geology. Fossil pollen analyses demonstrate that low-growing shrub formations occurred throughout the last 2 million years, most probably in quite patchy spatial distribution (e.g. González-Sampériz *et al.* 2005). Likely regions for the persistence of shrubby formations were coastal and inland sandy areas and mountainous portions of the larger islands (Reille 1992), where microclimate and/or limited soil reserves prevented the development of forest. These formations appear to have been dominated by the same sclerophyllous shrubs that occur today, such as lentisk, false olive (*Phillyrea angustifolia*), junipers, and the remarkable rope-like shrub *Thymelaea hirsuta* (Reille *et al.* 1980), as well as drought-tolerant shrubs of arid and semi-arid steppes, including wormwood, *Ephedra*, various grasses and Asteraceae, such as *Centaurea*, and saltbushes (*Atriplex*). In some of these areas, pollen has also been found of canopy trees, such as oaks, beech, and firs (*Abies*), which are currently found in the Mediterranean region only in high mountains, and fragments of primary-type forests. This kind of assemblage was probably concentrated in humid habitats near the banks of rivers (Kaniewski *et al.* 2005).

At a macroecological scale, Pleistocene climatic fluctuations and habitat changes induced large-scale distributional shifts of species and communities back and forth across Europe. This resulted in gene flow among populations, preventing **allopatric speciation** in many groups of animals, for example birds, which largely explains the nature and origin of extant species assemblages in the Mediterranean. Extinction events have been particularly severe in the western Palaearctic because of massive east–west barriers that prevented species to find refuge in tropical areas during glacial episodes (see Chapter 11). This is particularly so in the case of the Sahara and Arabian-Syrian deserts that originated during the late Miocene/Pliocene (6 mya), and have acted since then as barriers to dispersal between North Africa and tropical Africa to the south and west, and western Asia to the east. Crete and Cyprus are examples of islands that partially escaped these extinction events (Thompson 2005; Médail and Diadema 2006), along with certain sheltered microregions in the Balkans Peninsula and the Eastern Pyrenees (Siljak-Yakovlev *et al.* 2008).

These extinction events have been partly compensated by differentiation processes that occurred at the population level in many parts of the Mediterranean Basin, especially in the major Mediterranean peninsulas (see below).

In this context, much work has been done on the genetic diversity of organisms in relation to their migration back and forth between northern Europe and the Mediterranean, according to the alternation of glacial and interglacial episodes. Using genetic markers for investigating the genetic consequences of the splitting of populations in Mediterranean refugia, it has been shown that many tree species show higher levels of polymorphism in the Mediterranean refugia than in the re-colonized areas and that only a subset of the genetic variation present in refugia occurs at higher latitudes (Petit et al. 1997; Thompson 1999). This is a nice example of how colonization processes modify genetic diversity. In addition, the different refugia often show marked differentiation in gene frequencies. The mechanisms responsible for this differentiation remain unclear; they could result from random events or from drift due to isolation or else to adaptive differentiation in the different refugia. However, genetic markers sometimes allow distinctions to be made between variation in contemporary patterns of gene flow and historical events linked to glaciations. For example, comparing nuclear allozyme and chloroplast DNA in the Mediterranean annual ragwort (*Senecio gallicus*), Comes and Abbott (1998) showed that population differentiation of this species, which is common in the Iberian Peninsula and southern France, occurred when the populations were restricted in Pleistocene coastal refugia. The spatial structure of chloroplast DNA markers of this species is more a result of populations sharing chloroplast DNA profiles as a result of historical plant associations and re-colonization from particular glacial refugia than a result from contemporary genetic processes. In any case, these patterns of genetic diversity indicate that the Mediterranean region is not only a biodiversity hotspot for species, but also a reservoir of genetic diversity. For this reason, the basin is rich in the so-called evolutionarily significant units (Moritz 1994); that is, populations that are likely to be able to respond to global changes thanks to their high levels of genetic diversity. Such populations which are good candidates for conservation issues deserve special care and conservation (Petit et al. 1998). We will come back to these phylogeographic issues later in this chapter.

2.1.3 Landscape ecology

Changing scale now, the astonishing biodiversity in the Mediterranean is readily perceived in the 'mosaic effect' so typical of the basin's subregions and, within them, the individual landscapes. This is not a recent feature, but rather a long-standing essential feature in the Mediterranean. It plays a critical role in generating and maintaining diversity at the scales of populations and species. Thanks to its kaleidoscopic topographical, climatic, and geo-pedological complexity (see Chapter 1), the Mediterranean is exceptionally rich in regional or local endemics of plant and animal genera, species, and subspecies (see Chapter 3). Along with the long narrow peninsulas and isolated mountains, the huge Mediterranean archipelago is an outstanding framework for speciation to occur in populations isolated by geographical and ecological barriers, as recently described for plant species by Thompson (2005) and Médail (2008a). Almost every island in the Mediterranean has its own subset of unique native species (Médail 2008b). Ecologically similar to islands, Mediterranean mountains typically harbour many endemics, for example with up to 42% endemism among higher plants (Médail and Verlaque 1997; Hampe and Arroyo 2002; see also Chapter 3). In Chapter 5 we will provide more details and discussion of this important issue with the aid of altitudinal and latitudinal gradients.

Additionally, within regions, landscape heterogeneity is remarkably high, forming patchworks of habitats, as we will describe in Chapter 6. The juxtaposition of many different habitats and of 'landscape units', with an even larger number of possible pathways and stages of degradation or regeneration, occurring together and changing over time, thanks to local disturbance regimes, yield the 'moving mosaic' pattern so characteristic of Mediterranean landscapes.

For plant and animal populations and for communities and ecosystems, this pattern of landscape-scale 'patchiness' has profound consequences on species and populations, which are patchily distributed, as we will show in Chapter 7.

2.1.4 Human history

The fourth important determinant of Mediterranean biodiversity is anthropological. Variations in human land-use patterns and site-specific histories of resource management, which often resulted in overexploitation and resource depletion, have had profound impact on living systems throughout the basin (see Chapter 10). Evolutionary consequences of this factor can be seen in the structure and composition of the vegetation and the life-history traits of many species (see Chapter 8). Both vegetation structure and individual species show a wide array of adaptations to human perturbations including fire-setting, clear-cutting, heavy browsing and grazing by herds of domestic livestock, and ploughing. The exceptional richness of annual plant species in the Mediterranean flora is also to some extent the result of long-standing but constantly changing human activities.

We will now discuss the various processes and biogeographical divisions that lead to the setting in place of present-day flora and fauna. In other words, when and how did extant species first arrive, become established, and, subsequently, evolve in the basin? We will take examples primarily from the two most extensively studied groups of organisms—vascular plants and vertebrates—even though special sections are devoted to insects and marine flora and fauna as well.

2.2 Composition of the flora

The Mediterranean flora today is a complex mixture of taxa whose biogeographic origins, respective age, and evolutionary histories vary enormously. Climate changes that occurred from the Pliocene onwards have been responsible for the demise of most subtropical plant species (e.g. Lauraceae, Myrtaceae, Palmae, etc.), which covered most of the western Palaearctic during the Palaeogene and early Neogene, the so-called Madrean-Tethyan flora (Axelrod 1975). Among the species of this thermohygrophylous flora, species of the genus *Laurus* (Lauraceae) are emblematic relics of the Tethyan subtropical flora and have been used to infer the response of this flora to the Plio-Pleistocene climatic deterioration. Modelling the relationships between climate and *Laurus* distribution over time, using both present and fossil species from the mid-Pliocene (3 mya), when the climate was still much warmer and moister than today (Haywood and Valdes 2004), Rodrigues-Sanchez and Arroyo (2008) found that *Laurus* species preferentially occupied warm and moist areas with low seasonality. Models fitted to Pliocene conditions predicted the current species distribution, confirming that the large suitable areas for these species were considerably reduced during the Pleistocene. Only some humid refugia enabled their long-term persistence until present times within the Mediterranean Basin, Transcaucasia, and Macaronesian islands. It is possible that future climate conditions will re-open formerly suitable areas for these species. Interestingly, these authors demonstrated strong niche conservatism over the last 3 mya, which suggests largely deterministic range dynamics allowing predictions of future range dynamics as a response to global warming, a topic that we will address in Chapter 12.

Today, each region has had its own unique turnover sequences and interplay of plant species among biogeographical elements, but analyses and comparisons are simplified by the widely accepted designation of five main groups dominating the basin's flora. These five groups differ in their biogeographical origins, which the student can often guess by comparing suites of life-history traits that often recur within a group. The five groups are (1) Afro-tropical, which includes several different subgroups; (2) Holarctic, which corresponds to the Euro-Siberian region in Fig. 1.7; (3) Irano-Turanian; (4) Saharo-Arabian; and (5) indigenous, corresponding to species which have apparently differentiated *in situ* in the basin, including both ancient, or palaeoendemics, and recent, or neoendemics.

2.2.1 Afro-tropical components

This first historical group comprises plants that differentiated in the dry tropics of continental Africa and adjacent regions in the era of the Tethys Sea, before continental drift had separated the New

World from Eurasia. Related taxa, especially among **hard-leaved evergreen** species, can be seen in central and southern California and other dry parts of North and South America. Axelrod (1975) has termed such links Madrean-Tethyan and examples include the **evergreen** oaks, and cypress, but also a large number of annuals as well (see Fritsch 2001; Médail 2008a).

Among the Afro-tropical elements of ancient lineage, the so-called palaeotropical relicts are the evergreen genera *Asparagus, Capparis, Ceratonia, Chamaerops, Jasminum, Nerium, Olea*, and *Phillyrea* (Quézel 1985). Other 'younger' African elements also occur, especially in small disjunct populations in montane or pre-montane areas of North Africa and the Near East. Plant families, which are today distributed mostly in the tropics, but which have one or a few representatives in Mediterranean forests and shrublands, include Aquifoliaceae, Arecaceae (the palm family), Aristolochiaceae, Fabaceae (legumes), Moraceae, Myrtaceae, Salvadoreaceae, and Vitaceae. In warmer periods of the Miocene/Pliocene and the Pleistocene, all of these, as well as many Afro-tropical trees, shrubs, and vines, were common throughout the Mediterranean. Today, with a few exceptions, they are limited to wet habitats in frost-free regions of the basin. The Mediterranean taxa of this group were either present in the area long before the onset of the mediterranean-type climate in the Pliocene, about 3.2 mya, or else arrived recently, since the last glaciation. If they are 'survivors', they persisted despite the climatic upheavals of the late Tertiary and the Quaternary, including successive waves of prolonged Pleistocene glaciations. This scenario is confirmed by fossil records of such palaeotropical species found in southern European floras of the upper Eocene (e.g. *Chamaerops, Smilax*), the Oligocene (e.g. *Olea*), and early Miocene (e.g. *Phillyrea*) (Palamarev 1989).

One interesting and biologically important regional exception in the large-scale extinction crisis that resulted from the Pleistocene glaciations is the flora of the Macaronesian region consisting of several isolated archipelagos, off the coast of Morocco (Fig. 1.2), of which the Canary Islands and Madeira are the most important. The Cape Verde islands are sometimes included in Macaronesia (e.g. Kim *et al.* 2008). However, if including the Cape Verde Islands in Macaronesia makes sense bioclimatically, on faunistic and floristic and broad biogeographical grounds it is tenuous and unhelpful today (A. Machado, personal communication; Carine *et al.* 2004; Vanderpoorten *et al.* 2008).

Although the flora of the Canary Islands has affinities with many other biogeographical areas (Juan *et al.* 2000), it undoubtedly has a Mediterranean 'signature' (Sunding 1979; Shmida and Werger 1992). For that reason, these islands present a fascinating window on the past of Mediterranean flora and landscapes. In other words, the Canary Islands acted as a repository of the palaeofloras and vegetation formations that had prevailed in much of the Mediterranean Basin during the Miocene, but have since mostly disappeared (Quézel and Médail 2003). This special flora includes taxa and notably evergreen trees and shrubs, in many families of palaeotropical origin (see Chapter 3). There are still sizeable patches of mixed forests on several of the easternmost and wettest Canary Islands (La Palma, Tenerife, and Gran Canaria) that give an idea of what many lowland Mediterranean forests probably looked like in the Miocene and Pliocene eras (Suc 1984). These 'archaic plants' and the forest fragments they occupy are survivors of a more tropical, arboreal Mediterranean flora of Tertiary times (see Chapter 3) and include members of tropical plant families, such as the Arecaceae, Sapotaceae, Rubiaceae, and Theaceae, as well as a large number of taxa in the laurel (Lauraceae) and olive families (Oleaceae). Many of these trees played important parts in the Tertiary floras of southern Europe, in the Miocene/Pliocene period, but are now entirely or mostly absent in the Mediterranean region outside of Macaronesia, and a few steep canyons of the Anti-Atlas Mountains, southern Morocco (Barbéro *et al.* 1981; Backlund and Thulin 2007).

Further, there are several plants species, for example tree heath (*Erica arborea*) and two grasses (*Andropogon distachyos* and *Hyparrhenia hirta*), that show a geographical distribution pattern similar to the olive tree (Besnard *et al.* 2007). They are present in the Mediterranean Basin, Macaronesia, the Saharan mountains, tropical Africa, and Arabia, which clearly indicates a shared history of plant distribution between Africa and the Mediterranean (Quézel 1978). Recent studies on the genus *Erica* (McGuire and Kron 2005) and the aroid family (Araceae) (Mansion *et al.* 2008) provide further examples of these fascinating and complex tales of the region's 'deep' history.

Apart from these obvious tropical links, there is also evidence of relationships between the Mediterranean flora, and that of the semi-arid and arid formations that occur intermittently across Africa, from Mediterranean North Africa all the way to the Cape Province of South Africa (Raven 1973). This ancient Rand-flora component, as it is often called, includes the olive tree complex and the endemic argan tree (*Argania spinosa*) in southwestern Morocco. Migration routes followed by this flora were at times restricted to the mountain ranges of Africa, which acted either as stepping stones or as refugia for such groups as *Erica*, *Olea*, *Salvia*, and *Helichrysum*. A third subgroup among the Afrotropical components is sometimes called Sudanian.

It is represented in a few southern parts of the basin by 'tongues of penetration' in some parts of North Africa and along the Dead Sea valley, shared by Israel and Jordan (Zohary 1973; see Fig. 2.4). The area of distribution of the so-called Sudanian belt in Africa stretches from the Atlantic coast in southern Mauritania right across sub-Saharan Africa to the Red Sea coasts and reappears in certain parts of the Indian subcontinent. Several dozen species of this biogeographical group are found both in the sub-Saharan belt of savannas and also narrow parts of the southern Mediterranean quadrants.

The fact that there is little endemism among the Sudanian elements in the Mediterranean area and a relatively high frequency of adaptations for long-distance seed dispersal, and/or usefulness for people, led Shmida and Aronson (1986) to argue for a recent arrival for many of them; that is, since the last glacial period. In many cases, they occupy quite different habitats, for example in the Dead Sea area or the dry riverbeds of North Africa, from

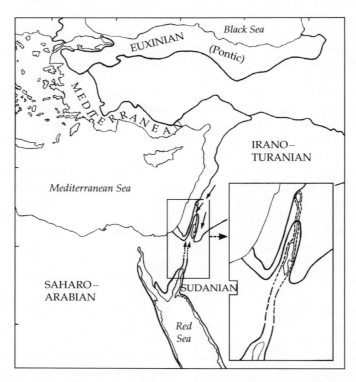

Figure 2.4 Phytogeographical subdivisions of the eastern Mediterranean and, in box, of Lebanon, Israel, and north-western Jordan, showing Irano-Turanian and Sudanian 'tongues of penetration' into the Mediterranean area (after Zohary 1973; Shmida and Aronson 1986).

those they occupy in the tropical savannas south of the Sahara. Yet, to the best of our knowledge, they are the same species, showing a disjunct distribution. Problematic cases include several species of the emblematic *Acacia* trees of the African savannas, which also occur in the southern fringes of the Mediterranean, north of the Sahara, and in the Dead Sea area, along with trees like *Ziziphus*, *Balanites*, and *Salvadora persica*, and a number of vines from strictly or primarily tropical families, like Menispermaceae and Asclepiadaceae. The date palm (*Phoenix dactylifera*) may also be considered part of this group.

2.2.2 Holarctic components

This second category includes many species and families of clearly 'northern' extra-tropical origin, which are mainly Holarctic. Some of these were already established in the Mediterranean by the late Pliocene, before the first glaciations, and have persisted mostly in the colder, wetter life zones of the north-western and especially north-eastern quadrants of the basin. This group includes the Oriental plane tree (*Platanus orientalis*), walnut (*Juglans regia*), hazelnut (*Corylus avellana*), and beech. Many of the plants of boreal or Holarctic origin also participate in the flora of the temperate zone that constitutes most of the vegetation found today in western Eurasia. This group includes deciduous broad-leaved tree genera, such as *Acer*, *Alnus*, *Betula*, *Fagus*, *Quercus*, and *Ulmus*, but also many herbaceous taxa, such as *Aquilegia*, *Doronicum*, and *Gentiana*. The mountain ranges of the northern Mediterranean played a prominent role in the survival of these taxa during glaciations, as they served as refugia and allowed many species to persist.

Many of these Holarctic elements (e.g. various species of *Epimedium*, *Rhododendron*, *Pterocarya*, and *Zelkova*) are most common or even endemic in the north-eastern quadrant of the basin, where prevailing climatic conditions are most suitable for them. Indeed, at the northern frontiers of the north-eastern quadrant of the basin, both the Pontic and Hyrcanian provinces of the Euro-Siberian region (see Fig. 1.7) harbour a large number of taxa and formations that intermingle with and contribute to contemporary Mediterranean vegetation, especially in areas disturbed by human activities. The Pontic province is very rich in evergreen genera, such as fir, pine, spruce (*Picea*), and *Rhododendron*. The southern Caspian, Hyrcanian province, by contrast, lacks these elements, but is rich in deciduous trees, among which the monotypic ironwood (*Parrotia persica*) of the witchhazel family (Hamamelidaceae) is most notable. This Arcto-Tertiary relict formerly occurred widely in various life zones throughout the eastern Mediterranean. Today, it can be seen only at mid-altitude slopes of the Alborz Mountains south of the Caspian Sea, along with hop hornbeam (*Ostrya*), *Zelkova*, various oaks, and *Rhamnus*. Together these trees and shrubs colonize and occupy large areas of badly degraded beech forests. They are examples of taxa from a more xeric flora (i.e. the Mediterranean basal zone) replacing a more mesic one, after the montane habitat of the beech woods has been destroyed by humans (Djamali et al. 2008a). The alpine flora of the Pontic and Hyrcanian provinces also shows many links with the Oro-Mediterranean life-zone vegetation in the various eastern Mediterranean mountains (Djamali et al. 2008b).

2.2.3 Irano-Turanian components

The third group is part of the Arcto-Tertiary **Mesogean** flora and provided many 'old colonists' to our area. It includes hundreds of Irano-Turanian elements, such as *Artemisia*, *Ephedra*, *Haloxylon*, *Pistacia*, *Salsola*, and *Suaeda*, whose centres of diversity and, no doubt, of origin are located in the semi-arid steppes of central Asia, where summers are exceptionally hot and winters exceedingly cold and dry. The so-called forest-steppes occur in this vast region since early Tertiary times or even before, but they are badly degraded through long centuries of mismanaged land and overexploitation of natural resources. As a result, the area is mostly characterized today by vast steppes, whose trees and large shrubs have mostly been removed for firewood, and whose former top soil has blown away.

Some arboreal elements—mostly deciduous—of this Irano-Turanian flora do survive in patches and appropriate habitats throughout the Middle East and the eastern Mediterranean, however, and provide another insight into the deciduous components of the Mediterranean flora, especially in the eastern parts. Examples include the Judas tree (*Cercis siliquastrum*) and the storax tree (*Styrax officinalis*), both of which do also occur in the western Mediterranean quadrants, but only sparingly (Fritsch 2001). By contrast, in the north-eastern quadrant, both trees are abundant enough to be used as bioindicators of the thermo-Mediterranean life zone (see Chapter 5).

Zohary (1973) considered that, since the Late Tertiary, there must have been waves of interpluvial penetrations of Irano-Turanian elements into the Mediterranean region. Indeed, recent palaeobotanical evidence from the Thracian plain of the present day Balkans appears to confirm this idea. Magyari *et al.* (2008) show that predominantly open vegetation occurred during the Weichselian late glacial (Würm; *c*.10 000–15 000 years BP), although macrofossil remains of woody taxa, including many Rosaceae (see below), as well as a *Celtis*, *Fraxinus*, and *Alnus*, in particular, demonstrate the persistent presence of patches of wooded steppe and gallery forest. These authors argue convincingly that 'the "oriental" element of the Balkan flora reached south-east Europe from Turkey prior to the Holocene, probably via the Thracian Plain during a late Quaternary glacial stage but no later than the late Weichselian'.

However, in the Holocene, when human impact grew much greater (see Chapter 10), this pattern would have intensified in line with the ecological trend whereby human transformation and simplification of ecosystems generally aid plants from more xeric habitats to colonize more mesic ones rather than the opposite. We shall call this 'Zohary's law', though Zohary himself (1962) wrote in terms of the 'expansion drive of desert plants'. The example given above of the ironwood and other woody Mediterranean migrants or 'colonizers' in the disturbed forests of the Pontic and Hyrcanian provinces suggests that this 'law' applies to Mediterranean, as well as to steppic Irano-Turanian, elements and perhaps far better than to strictly 'desert' flora, as discussed below with reference to the Saharo-Arabian component.

Zohary's law may not only explain why the Irano-Turanian region has contributed many more taxa to the Mediterranean region than the opposite, it also probably explains why an abruptly inter-digitating pattern is observed at their borderlands (see Fig. 2.4). It would be interesting indeed to be able to chart how that frontier pattern has evolved in the last 30 or 40 years, since the impacts of global warming have begun to intensify. However, a more direct human factor must also be considered in the intrusion of Irano-Turanian elements to the Mediterranean flora: almost all of the region's cultivated fruit and nut trees are of Irano-Turanian origin, mostly in the apple family (Rosaceae). These include hawthorn (*Crataegus*), apple and almond (*Prunus*), pear (*Pyrus*), mountain ash (*Sorbus*), and at least eight other genera. These taxa have spread—and been carried by people—throughout the basin. A very high level of intra- and interspecific hybridization may account for the several dozen taxa in these groups now found in Europe and North Africa (e.g. Aldasoro *et al.* 1998). But the role of humans cannot be excluded from any discussion of differentiation and rapid evolution in this group and several other groups, as we will see in Chapter 10.

It is also worth noting that there are a far greater number of deciduous oak species in the eastern Mediterranean than in the western part (see Chapter 7). They all show strong affinities with the cluster of oaks found further east, in the Irano-Turanian region and in the foothills of the Himalayas, where some 20 species of deciduous oaks occur. In this context, it is striking that pines are totally absent from the Irano-Turanian region, while some 16 species and subspecies of *Pinus* occur in the Mediterranean region and play prominent roles in many vegetation types, usually in association with oaks (Barbéro *et al.* 1998). The same is true of oaks, for which at least 25 species occur in the region. Thus, pines and oaks may be considered as valuable bioindicators of the Mediterranean realm in this rough grain, inter-regional context (see Chapter 1). It is noteworthy that human determinants in the distribution of those two groups of trees in the Mediterranean cannot be discounted (see Richardson *et al.* 2007). However, recall that both pines and oaks are very widespread in the northern hemisphere as well, so oversimplification must be avoided.

2.2.4 Saharo-Arabian components

The fourth category encompasses taxa from the vast Saharo-Arabian deserts and semi-deserts that significantly contribute to local floras and landscape diversity in the frost-free arid regions of the southern and south-eastern margins of the Mediterranean Basin. This xerophytic desert component of the flora, along with the steppic one, appears to be quite ancient, dating back at least to the Tertiary. However, Zohary's law notwithstanding, very few Saharo-Arabian elements have succeeded in extending their distribution northwards and establishing themselves as part of contemporary Mediterranean ecosystems. Among those which have succeeded, there is a mixture of species that apparently penetrated as a direct result of long-term climatic fluctuations during the Pleistocene and those arriving in more recent times. As for the Sudanian elements discussed above, it is difficult to discern between the two groups. These elements include members of the saltbush family (Chenopodiaceae), the lignum vitae family, Zygophyllaceae (*Balanites*, *Peganum*, *Tribulus*, *Zygophyllum*), a few perennial grasses, and others (Zohary 1973; Quézel 1985). Their distribution roughly coincides with the 150 mm isohyet that marks the limit of the Mediterranean-desert frontier zone. With global warming, it may be that their ranges will extend further north, a topic we reserve for Chapter 12.

2.2.5 Indigenous components

This category includes several thousand autochtonous elements of the so-called Mesogean flora. They differentiated after the beginning of the Oligocene within the coastal regions around the Tethys Sea, especially on the many tectonic microplates, which were scattered between the African land mass to the south and the Eurasian supercontinent to the north (Quézel 1985; see Chapter 1). Within this Mesogean flora, the strawberry tree and the various evergreen sclerophyllous oaks are considered to be indigenous Mediterranean taxa, or palaeoendemics, along with the native species of *Helianthemum*, *Lavatera*, *Salvia*, *Cupressus*, *Pinus*, and *Juniperus* (Quézel 1985). As Thompson (2005:31) points out, both the Iberian and Balkan Peninsulas are particularly rich in genera that have diversified 'rampantly' in recent times (e.g. *Genista*, *Narcissus*, *Linaria*, *Thymus*, and *Teucrium* in Iberia, and *Silene* and *Stachys* in the Balkans). Palaeoendemic genera also occur in these same areas, further highlighting the complexity of plant evolution in the Mediterranean. They provide evidence for ancient isolation events in the setting up of current distribution patterns.

Another important portion of this small but crucial component is the anthropogenic elements: that distinct sub-flora of about 1500 **segetal** and **ruderal** annuals that evolved locally (Zohary 1973) and occur in varying associations in fields, pastures, and along the roadside. Surprisingly, several dozen taxa of this autochtonous group actually show very limited distribution, despite their long history as weeds. As many as 200 of them are endemic to the Mediterranean and the Middle East. Among these, a large number of endemic genera are represented by one or just a few species, such as *Bunias* and *Calepina* (Brassicaceae), *Cardopatium* and *Ridolfia* (Asteraceae), and *Bifora*, *Exoacantha*, and *Smyrniopsis* (Apiaceae), and many others.

As a result of the basin's complicated history, palaeo- and neoendemics commonly occur side by side. Neoendemics belong to lineages that appeared in the region by immigration and subsequently differentiated (examples are serviceberry (*Amelanchier*) and the many species of rockrose (*Cistus*), as well as a wide variety of Asteraceae, e.g. *Centaurea*).

From the foregoing, we can summarize the history of the Mediterranean flora as follows. In the first part of the Tertiary, until the Oligocene, the flora was typically tropical, with a mixture of forest and savanna landscapes that were very different in structure and composition from those prevailing today. After the beginning of the Oligocene and especially from the end of the Miocene and the Pliocene, with the establishment of a mediterranean-type climate (c.2.8–3.2 mya) bounded by more arid and semi-arid areas to the south and east and more mesic ones further north, the typical Mediterranean flora progressively developed and differentiated in the form and fashion we know today.

From the Messinian Salinity Crisis to the end of the Pliocene and the beginning of the Pleistocene, a large turnover of floras resulted in the disappearance of most tropical species. This tropical flora was progressively replaced by modern floras, which include such disparate historical elements, as Saharo-Arabian taxa (e.g. *Retama, Lygos*), steppe species (*Artemisia*, various Chenopodiaceae), the many autochtonous Mediterranean (*Arbutus, Ceratonia, Quercus, Pistacia, Myrtus,* and *Cistus*), and boreal elements (*Alnus, Fraxinus, Tilia,* and *Ulmus*). About 2.3 mya, a drying trend led to the development and expansion of steppe associations (e.g. *Artemisia, Ephedra*), as well as those of typical Mediterranean xerophytes, such as *Phillyrea, Olea, Cistus,* holm oak, and lentisk (Suc 1980). For more discussion, see Bocquet *et al.* (1978) and Gamisans and Marzocchi (1996).

2.3 The insect fauna

Invertebrate faunas of the Mediterranean show much closer affinities with Palaearctic or Holarctic assemblages than with those of any other biogeographical region (Casevitz-Weulersse 1992), a feature which is shared with many other groups of animals. Many boreal species and groups colonized the Mediterranean during glacial periods, partially replacing pre-existing palaeotropical fauna. Good examples are provided by butterflies in the genera *Parnassus, Colias,* and *Pieris* that live in montane and lowland habitats. Many forest-dwelling species, which are usually **phytophagous**, are also boreal species, which followed their host trees as their distribution ranges repeatedly shifted north and south in response to glacial cycles. Examples are the beetle *Rosalia alpina* (see Plate 4a) and the moth *Aglia tau*, both of which are tightly linked to beech. Many other groups of boreal origin are to be found among beetles, dragonflies, diptera, and other invertebrate groups.

As in all other groups of terrestrial animals except mammals, invertebrate species of tropical origin are scarce in the Mediterranean and have only marginal biogeographical importance. For example, not a single species of Mediterranean ant is of tropical origin, and most chironomids (81%; Laville and Reiss 1992) and butterflies (90%; Larsen 1986) have a Palaearctic distribution. Among the 321 butterfly species in the Mediterranean, only 24 species (7.5%) have tropical affinities. They can be grouped into one of four categories: Afro-tropical (by far the most important category with 11 of 24 species), Oriental (two species), Palaeotropical (nine species with a wide distribution in the Old World tropics), and Neotropical (with only one species, the monarch, *Danaus plexippus*). The monarch entered the basin recently and has resident populations in the Canary Islands and probably in the Jordan valley. It is a regular visitor to the eastern Mediterranean and may sometimes be seen in the north-western quadrant of the basin, in Italy, France, and the Iberian Peninsula (Larsen 1986). The few non-Palaearctic elements either represent the last remnants of original tropical faunas or else have entered the basin in various epochs.

Some examples of the former group may be found in Lepidoptera; for example, the beautiful two-tailed pasha (*Charaxes jasius*) (Fig. 5.6). Most of the non-Palaearctic species of tropical origin are highly mobile and migratory with only five of these species being sedentary. In addition, there are very little morphological differences between the Mediterranean populations of tropical butterflies and their closest tropical relatives. Only the uncommon nettle-tree butterfly (*Libythea celtis*) and the two-tailed pasha are specifically distinct and probably archaic, suggesting a relict status from a Pliocene or early Pleistocene (Larsen 1986). Other good examples are found among carabid beetles and the Diptera (Psychodidae, Chironomidae).

Among the 213 species of butterflies occurring in the Iberian Peninsula, only five species of Anthocarinae and some species of Polyommatinae are considered of African origin (Martin and Gurrea 1990). Many Mediterranean butterfly species of tropical origin share certain common characters. Most of them are very mobile migrants and occur only sporadically in the basin, fluctuating widely in the numbers present from one year to the next. They are often limited to very special ecological conditions at the southern limits of the basin and do not differ even at the subspecific level from other populations in the tropics proper (Larsen 1986). Other pre-Quaternary elements in the families Carabidae, Staphylinidae, Buprestidae, Chrysomelidae, and Scarabaeidae are forest-dwellers initially tied to subtropical forests, mainly laurisilva in the Canary Islands (see Chapter 3). They are found today in soils and forest litter, as well as in caves. Distribution patterns of many species (e.g. chironomids and butterflies) strongly suggest that the tiny fraction of species of tropical origin reached the Mediterranean in various epochs, either by an eastern route following the Nile River valley, or rather by a western route, along the shores of the Atlantic Ocean. Examples of the latter group are some very rare species of chironomids of Afro-tropical origin (*Dicrotendipes collarti*, *Paratendipes striatus*) that occur only in the *khettaras* of the region of Marrakech, Morocco, which are part of a traditional irrigation system in arid areas where groundwater is brought to the surface and kept constant at a temperature of approximately 19–23°C (Laville and Reiss 1992).

2.4 Vertebrates

We will make a special focus on vertebrates because we are more familiar with them than with invertebrates and also because there is much more insight on their history than on that of other groups. As we saw for the region's flora, contemporary vertebrate faunas of the Mediterranean Basin are a legacy of repeated waves of immigration, extinction, and *in situ* differentiation that go as far back as the late Oligocene/early Miocene. The history of mammals is associated with land connections between Asia, Europe, and Africa across the Tethys and the Paratethys seas. For example, the 'Levantine corridor' allowed many faunal exchanges during the Neogene and the Quaternary (Tchernov 1992; see Chapter 1). On account of its geographical location, the Levantine region has been used many times as a land bridge between Eurasia and Africa, especially in relation to the Pliocene/Pleistocene climatic fluctuations. The first important faunal interchange in this region occurred after the closing of the Tethys at the eastern part of the Mediterranean in the lower Miocene, 20 mya. This explains why many rodents in arid and semi-arid habitats of North Africa are modern representatives of ancient Asian lineages. Although the geography of the region has stabilized since the Miocene, there was an extensive turnover in the faunas of birds and large mammals associated with the Pliocene/Pleistocene climatic changes.

2.4.1 Speciation from the Miocene to the Pliocene

Although it has long been thought that most modern species of vertebrates are recent and have evolved during the Pleistocene, recent palaeontological findings as well as molecular systematics strongly support the idea that many species are in fact much more ancient. For example, the closely related marbled (*Triturus marmoratus*) and crested (*Triturus cristatus*) newts were long considered to have diverged recently through geographic isolation. They are indeed largely separated geographically, the former occurring in the Iberian Peninsula and western France, and the latter in the rest of Europe (Fig. 2.5). But even where they do co-exist, as in western France, reproductive incompatibility prevents their hybridization. This suggests an ancient evolutionary separation. Thus, based on fossils and a variety of biochemical and molecular data, it now appears that the two species must have differentiated much earlier, i.e. between 7 and 12 mya (MacGregor *et al.* 1990).

An interesting palaeobiogeographic configuration is the disjunction and rotation of the Cyrno-Sardinian microplate from the Pyrenean region in the Miocene. Several taxonomically diverse groups of organisms with closely related species (sister taxa) currently occur on each of the three land masses. One example is that of the vicariant distribution of newts of the genus *Euproctus*, which inhabit mountain streams in the Pyrenees (*Euproctus asper*), Corsica (*Euproctus montanus*), and Sardinia (*Euproctus platycephalus*). Active or passive dispersal of these species is virtually nil so that the time of the **cladogenetic** events within this group can be assumed to coincide with the tectonic events. This situation allowed calibration of molecular clocks for a time period of 15–30 mya (Caccone *et al.* 1994). Mitochondrial DNA sequence variation of these species has been examined using another species of *Euproctus* (*Euproctus waltl*) as an **outgroup**. Results of this approach are consistent with those from other cues (palaeontological, morphological, immunological, and allozyme), which

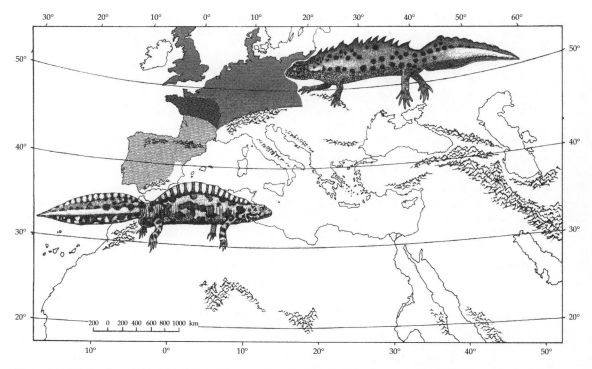

Figure 2.5 Isolation of populations in the Miocene/Pliocene gave rise to differentiation at the species levels in the crested and marbled newts (after MacGregor *et al.* 1990).

date back to the splitting of these species as the three land masses separated from each other. Thus speciation events within this group are 18–21 mya old.

A third example of ancient differentiation among closely related species is that of the Mesogean nuthatches, *Sitta*. Three species of narrowly endemic nuthatches occur in the basin: the Corsican (*Sitta whiteheadi*), the Kabyle (*Sitta ledanti*), which inhabits a small region of Algeria, and the Kruper's nuthatch (*Sitta kruperi*), only found in Turkey. Since the three species are very similar, one might expect that they differentiated recently, during the Pleistocene. However, molecular studies have revealed that in fact they belong to quite separate lineages that diverged at the beginning of the Pliocene, *c*.5 mya (Pasquet 1998).

A fourth example is found in partridges (*Alectoris*). Four species occur in the basin and are largely **allopatric**, with some cases of hybridization. Randi (1996) provided evidence from molecular data (mitochondrial DNA, cytochrome *b*) that speciation events among these partridges occurred between 6 and 2 mya, probably as a consequence of lineage dispersal and isolation of allopatric populations.

2.4.2 Pleistocene differentiation

There are in fact very few examples of Mediterranean vertebrate species that differentiated at the species level during the Pleistocene. One of them is the warblers (genus *Sylvia*; see Plate 6a). These small birds are abundant and characteristic in Mediterranean shrublands; indeed 14 of 19 species in this genus are Mediterranean endemics. Although molecular phylogenetic techniques have shown that the differentiation of some of these warblers started as far back as the end of the Miocene, 6.3–6.8 mya, with the splitting of the two large warblers, the blackcap (*Sylvia atricapilla*) and the garden warbler (*Sylvia borin*), the burst

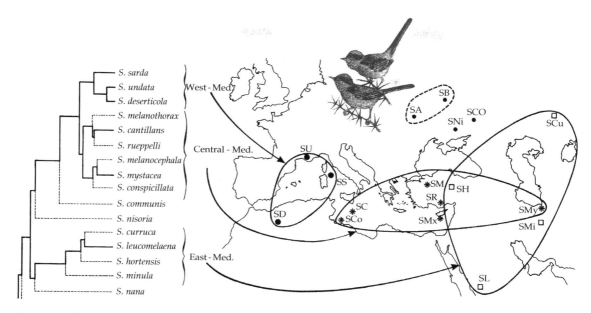

Figure 2.6 Relationships between the three main groups of Mediterranean warblers (*Sylvia*) and their geographical range. Symbols refer to the 'centre of gravity' of their distribution range. Differentiation within these three groups is supposed to have occurred as a result of repeated episodes of isolation and re-expansion of matorral-like habitats in lowland areas of these three regions during the Pleistocene. SU, *S. undata*; SS, *S. sarda*; SD, *S. deserticola*; SC, *S. cantillans*; SCo, *S. conspicillata*; SM, *S. melanocephala*; SR, *S. rueppelli*; SMx, *S. melanothorax*; SH, *S. hortensis*; SL, *S. leucomelaena*; SMy, *S. mystacea*; SMi, *S. minula*; SCu, *S. curruca*. The 'centres of gravity' of the four mid-European species (SA, *S. atricapilla*; SB, *S. borin*; SNi, *S. nisoria*; SCO, *S. communis*) are also shown. The dashed lines in the phylogenetic tree refer to relationships that entail some uncertainty (after Blondel *et al*. 1996).

of radiation of the most closely related species—that is, those that are tightly linked to Mediterranean shrublands—occurred in the Pleistocene, between 2.5 and 0.4 mya (Blondel *et al*. 1996; Shirihai *et al*. 2001). Three principal centres of speciation have been proposed for this genus on the basis of an analysis combining a biogeographical approach and the molecular phylogeny. The first is located in the western Mediterranean (three species), the second in the Aegean Peninsula (six species), and the third in the Near East (five species) (Fig. 2.6). The hypothesis that a series of separate speciation events may have occurred in shrubland habitats is supported by palaeobotanical analyses, which show that the spatial extent of these shrublands has varied with fluctuating climatic conditions (Pons 1981).

For the majority of Mediterranean vertebrates, however, Pleistocene differentiation has mostly been restricted to the subspecific level. In birds, a burst of recent studies using molecular systematics (mitochondrial DNA) provides evidence that many species are much more ancient, dating back to the Pliocene (Klicka and Zink 1997). This does not mean, however, that Pleistocene environmental changes had little impact on extant avian diversity. Repeated changes in selection pressures on small populations that were isolated in refugia during glacial periods resulted in considerable microevolutionary genetic diversification (Avise and Walker 1998). Reconstructing patterns of genetic differentiation and past colonization routes across continents is possible from molecular techniques, a scientific discipline called phylogeography. Many species display significant geographically oriented phylogeographic units that are genetically distinct. Pleistocene scenarios of differentiation in glacial refugia and subsequent range expansions of two or several phylogeographic units have been proposed for explaining these patterns in several groups of plants and animals (Petit *et al*. 1997; Taberlet *et al*. 1998; see Fig. 2.7). These findings

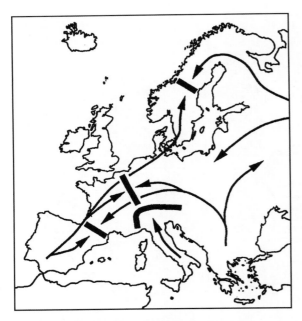

Figure 2.7 Post-glacial colonization routes of some European animals and plants. Arrows indicate direction of northward expansion routes. The thick black lines indicate contact zones between previously separated populations emerging from their respective refugia (after Taberlet *et al.* 1998).

suggest that speciation events are gradual processes over long-time spans, rather than a point event in history. This sheds new light on the importance of Pleistocene events in accelerating speciation processes that begun much earlier, in the Pliocene.

One example of such patterns among mammals is that of the brown bear (*Ursus arctos*), in which much genetic differentiation occurred during glacial periods when the species was split into several isolated populations in each of the main Mediterranean peninsulas (Fig. 2.8). Genetic analyses (Tablerlet and Bouvet 1994) help reconstruct the course of re-colonization of Europe by these animals as climate improved during the Holocene. For example, the present-day Swedish population of brown bears is genetically closer to that of the Cantabrian Mountains of northern Spain than either is to the Italian population in the Apennines (Saarma *et al.* 2007). This strongly suggests that the extant Scandinavian populations are derived from an Iberian stock. Based on similar evidence, the brown bear populations of central Europe are probably derived from the Italian Pleistocene refuge. Nevertheless, all European brown bear populations clearly belong to the same biological species. However, recent studies on ancient DNA in temperate species have cast some doubt on the idea that populations were isolated from each other in refugia during glacial maxima (Provan and Bennett 2008). Brown bears are now extinct in central Europe, but analyses of ancient DNA from fossils indicate high genetic diversity and gene flow in areas north of the traditional peninsular refugia. These findings provide insight into population persistence and dynamics beyond the southern refugia and challenge the assumption that extant gene pools are representative of those found in refugia during glacial episodes (Valdiosera *et al.* 2007).

The brown bear is just one of the many species that have been studied using this kind of molecular approaches. Other examples involved the water vole (*Arvicola sapidus*), several species of newts (*Triturus*), and the white-toothed shrew (*Crocidura leucodon*) (Taberlet *et al.* 1998). More recently, a phylogeographic study of the pool frog (*Rana lessonae*) in Italy, using mitochondrial cytochrome *b* gene fragments, showed that three phylogroups differentiated in northern Italy, the whole Italian Peninsula, and Sicily (Canestrelli and Nascetti 2008). This study suggests that the extant genetic structure of the pool frog has been shaped through a number of evolutionary processes in three separate refugia with recent population expansions and secondary contacts. This case study is one example among many others of the kind of evolutionary processes which make the Mediterranean region, and especially the larger peninsulas, a matrix for genetic diversity. The rationale is that populations that have survived throughout glacial maxima in refugia had a longer demographic history than populations that have evolved during the post-glacial, hence a higher level of genetic diversity (Taberlet *et al.* 1998). This rationale gave rise to a scenario of 'southern richness and northern purity' (Hewitt 1999) because re-colonizing populations are usually composed of subsets of the genetic diversity present in the source populations that persisted in refugia. The founding process of re-colonizing populations resulted in a series of sequential founder effects and

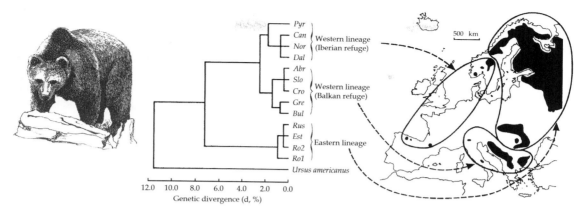

Figure 2.8 Isolation of populations of the brown bear during glacial periods gave rise to the differentiation of three main lineages that evolved in isolation in the main Mediterranean peninsulas and in eastern Europe. Each lineage includes well-defined **haplotypes** (*Pyr*, Pyrenees; *Can*, Cantabric cordillera; *Nor*, Norway; *Dal*, Dalmatia; *Abr*, Abruzzo; *Slo*, Slovenia; *Cro*, Croatia; *Gre*, Greece; *Bul*, Bulgaria; *Rus*, Russia; *Est*, Estonia; *Ro1*, *Ro2*, Romania. The American bear (*Ursus americanus*) is used as an outgroup (after Taberlet and Bouvet 1994).

bottlenecks. Higher genetic diversity in southern populations resulted from long-term isolation of populations within geographically separate refugia, leading to genetic differentiation due to drift. However, the recent discovery of glacial refugia outside the Mediterranena Basin in the so-called cryptic refugia mitigates this scenario (Provan and Bennett 2008).

2.4.3 Pleistocene mammal assemblages

The ancient large-mammal fauna of the late Pliocene, 2.8 mya, which persisted into the early Pleistocene on the northern shores of the basin, included both tropical and boreal faunal elements. Tropical species included a cheetah (*Acinonyx*), several large felids (*Homotherium*, *Megantereon*), a panda (*Parailurus*), a raccoon-dog (*Nyctereutes*), and a tapir (*Tapirus*), as well as several gazelles (*Gazella*) and antelopes (*Alcephalus*). But the end of the Pliocene permitted the appearance of 'temperate' faunas, including a very large bovid (*Leptobos*), the first true horse (*Equus*), the first mammoth (*Mammuthus*), and several large carnivores, including the contemporary wolf (*Canis lupus*). Climatic degradation of the beginning of the Pleistocene resulted in a progressive decline of the tropical species, so that at the end of the early Pleistocene (*c*.1.5 mya) many of them went extinct.

In the middle Pleistocene (1 mya), a major faunal turnover, associated with large climatic cycles, led to the disappearance of all tropical species, several faunal-turnover episodes, and, finally, the settling-in of modern faunas. The first true bison (*Bison schoetensacki*) replaced *Eobison* and several new cervids appeared: the reindeer, the red deer (*Cervus elaphus*), and the roe deer (*Capreolus capreolus*). Besides the reindeer, mammals of 'cold origin' included the woolly rhino (*Coelodonta antiquitatis*) and the musk ox (*Ovibos pallantis*). The primitive boar (*Sus strozzi*) was replaced by the modern wild boar (*Sus scrofa*), the mammoth (*Mammuthus meridionalis*) by a more evolved species (*Mammuthus trogontherii*), which co-occurred with the elephant (*Palaeoloxodon antiquus*). Large carnivores included the Etruscan bear (*Ursus etruscus*), which subsequently evolved as the well-known cave bear (*Ursus spelaeus*), and the modern wolf.

A testimony of this magnificent late-Pleistocene mammal fauna is provided by many fossil sites in southern Europe (Table 2.1), as well as by the superb wall paintings in many ornate caves of southern France and northern Spain. Beautiful examples are those of the Cosquer cave and the recently discovered Chauvet cave (see Box 2.1). Some survivors of this ancient fauna, which included a surprisingly large number of large predators, disappeared only recently. The lion (*Panthera (leo) spelaea*) and

Table 2.1 List of fossil remains of large mammals found in various middle- and late-Pleistocene deposits of southern France (+ means present; 0 means absent)

Families and species	Mid	Late
Canidae		
Canis lupus	+	+
Vulpes vulpes	+	+
Alopex lagopus	0	+
Cuon alpinus	0	+
Ursidae		
Ursus thibetanus	+	0
Ursus arctos	+	+
Ursus spelaeus	+	+
Hyaenidae		
Crocuta spelaea	0	+
Felidae		
Panthera (Leo) spelaea	+	+
Panthera pardus	+	+
Lynx spelaea	+	+
Felis sylvestris	+	+
Mustelidae		
Gulo spelaeus	0	+
Meles meles	+	0
Proboscidae		
Palaeoloxodon antiquus	+	0
Mammuthus primigenius	+	0
Rhinocerotidae		
Coelodonta antiquitatis	+	+
Dicerorhinus hemitoechus	+	+
Equidae		
Equus	+	+
Equus germanicus	+	0
Suidae		
Sus scrofa	+	+
Cervidae		
Rangifer tarandus	+	+
Megaceros giganteus	0	+
Cervus elaphus	+	+
Capreolus capreolus	+	+
Bovidae		
Bos primigenius	+	+
Bison priscus	+	0
Hemitragus	+	0
Hemitragus cedrensis	+	0
Capra	+	+
Capra ibex		
Rupicapra	0	
Rupicapra rupicapra	0	+
Sciuridae		
Marmotta marmotta	+	+
Castor fiber	+	+

These deposits correspond to both glacial and interglacial periods.
Source: After Defleur *et al.* (1994).

elephant survived until historic times in Greece and Syria, respectively, while the porcupine (Hystricidae family) and the Barbary macaque (*Macaca sylvanus*) are still present in southern Italy, and southern Spain and North Africa, respectively.

In the eastern Mediterranean, many species of Eurasian origin took advantage of the gradual improvement in climate during the Riss–Würm interglacial period (*c*.110 000–70 000 years BP) to expand their areas of distribution (Tchernov 1984) and colonize North Africa through the Levantine corridor. Examples are close relatives of the fallow deer (*Dama dama*), red deer, roe deer, wild boar, auroch, goat (*Capra hircus*), and ibex. Then, the last cold episode of the Pleistocene, the Würm glaciation (*c*.30 000 years BP), forced many mammals into southern Mediterranean refugia. Some, such as the weasel (*Mustela*), roe deer, and fallow deer, survived in the Levant at least until the first millennium BC; that is, several millennia after the major warming at the end of the last glacial (Dayan 1996). The current presence of the weasel in Egypt could be a relict of its widespread distribution in the eastern Mediterranean during the Holocene, perhaps because it became **commensal** with the dense human population in the Nile delta. The roe deer and the fallow deer both disappeared early in the twentieth century, probably as a result of hunting (Yom-Tov and Mendelssohn 1988). The progressive desiccation of the Eastern Mediterranean during the Holocene was the main causal factor in the extinction of many Afro-tropical and Palaearctic faunal elements, further isolating the faunas of tropical Africa and Eurasia (see Chapter 11).

In North Africa too, a large turnover of the mammal fauna occurred during the Pleistocene. After a short period of direct contact with Europe during the Messinian Salinity Crisis, which did not result in massive faunal interchange between the African and the Eurasian land masses, the mammal fauna of the Maghreb evolved in relative isolation. This fauna was clearly African in character during the early and middle Pleistocene with several species of antelope, an elephant, and many species of rodents (e.g. *Ellobius, Meriones, Arvicanthis, Gerbillus*). Savanna-like mammal assemblages of African character, including goats, antelopes, elephants, white rhinos (*Ceratotherium simum*), hares

(*Lepus*), jerboas, and jackals were enriched during glacial periods of the late Pleistocene by Eurasian species that colonized North Africa, using the 'eastern route', the narrow belt of Mediterranean habitats that stretched along the seashore prior to the northward extension of the Sahara desert in what is now Libya and Egypt. These Palaearctic elements included the brown bear, aurochs, and deers (*Megaloceros algericus, Cervus elaphus*). The species richness of large-hoofed mammals and carnivores increased from 17–20 species to 29, between the late Riss (110 000 years BP) and the late Würm (14 000 years BP) ice ages. As in the northern shores of the Mediterranean, this period was characterized by an impressive number of large carnivores: dogs (two *Canis* species), lycaon (*Lycaon*), fox (*Vulpes*), brown bear, genet (*Genetta genetta*), hyenas (two species), cats (two *Felis* species), lynx (two *Lynx* species), lion, and panther (two species). This rich fauna has been drastically reduced at the end of the Pleistocene, as elsewhere in the Mediterranean and indeed almost everywhere in the world, with the extinction of species of both Palaearctic and tropical origin.

2.4.4 Human-induced persecution and introductions

As in many other parts of the world, the former, extraordinarily rich 'megafauna' of the Mediterranean Basin was drastically reduced in the late Pleistocene/Holocene times through the combined effects of a changing climate and of the various types of pressure exerted by prehistoric humans. The overkill hypothesis suggested by Martin (1984), whereby prehistoric people were largely responsible for the mass extinction of large mammals in the late Pleistocene, probably applies quite well in the Mediterranean region, especially the islands.

Throughout the Mediterranean Basin, uninterrupted forest clearing, burning, hunting, persecution, and finally both deliberate and accidental introduction of exotic species have all combined to alter the pre-existing faunas. In the Near East, as in southern Europe and North Africa, uninterrupted human pressure from the beginning of the Holocene (*c*.10 000 years BP) has led to the extinction of the majority of large species and especially ungulates (Tchernov 1984). We will come back to these issues in Chapters 11 and 12.

Concurrently with this wave of extinction, the Mediterranean Basin has experienced many post-Pleistocene colonization events by mammals. Some species colonized Mediterranean Europe from the Middle East, such as the marbled polecat (*Vormela peregusna*), the jackal *Canis aureus*, and Guenther's vole (*Microtus guentheri*), whereas others came from North Africa; for example, the mongoose (*Herpestes ichneumon*) and the genet. It is possible, however, that the last two species were introduced in Europe by the Arabs after the collapse of the Roman Empire. In the Near East, the post-glacial immigration from south-western Asia of several species of rodents and of the desert hedgehog (*Hemiechus auritus*) accompanied a deterioration of Mediterranean habitats (Tchernov 1984).

Finally, humans have intentionally introduced a number of species, for example the rabbit (*Oryctolagus cuniculus*) and the red deer from Europe to North Africa and the Middle East, while several species benefited from human migrations to expand in the Mediterranean Basin as 'camp-followers'. This is the case for three commensal murid species, the house mouse (*Mus musculus*), the black rat (*Rattus rattus*), and, much later, in the Middle Ages, the Norway rat (*Rattus norvegicus*). The two former species colonized the Near East from Asia, making use of human settlements as stepping stones. Finally, several species, for example the coypu (*Myocastor coypus*), the muskrat (*Ondatra zibethicus*), and the cottontail rabbit (*Sylvilagus floridanus*), were intentionally introduced from the Americas for their fur or for hunting. In most cases they do not seem to have become pests, except perhaps the muskrat, which in some places causes damage to the dikes of canals and to fishing or hunting ponds.

2.4.5 History of the bird fauna, or why are there so few indigenous species?

Compared to the more homogenous bird fauna of northern and central Europe, the Mediterranean avifauna is a collection of cold boreal, semi-arid steppe, and indigenous Mediterranean elements, resulting in a mixture of disparate faunas. Of the 370 or so breeding species of birds found in the Mediterranean, only 64 can be considered as being indigenous; that is, having evolved within the limits of the region (Blondel and Farré 1988). Most of these species occur in shrublands rather than in forests.

2.4.5.1 BIRD COMMUNITIES IN MATURE FORESTS: A BIOGEOGRAPHICAL PARADOX

Unexpectedly, the taller, more natural (i.e. native), and more architecturally diverse or complex the trees in a given Mediterranean habitat, the fewer bird species of Mediterranean origin are found there (Blondel and Farré 1988). In a long-established forest dominated by the typically Mediterranean holm oak, there is not a single bird species that is not also found in the forests of central and northern Europe. In fact, most European forest-dwelling bird species occur more or less uniformly across the continent, including the Mediterranean islands and the forested parts of North Africa as well. Such a high degree of uniformity in the avian species of mature forests of the western Palaearctic has led to the prediction that bird communities throughout the region would resemble each other more closely, as they move up vegetation gradients towards maximum vegetation biomass and complexity. Evidence supporting this hypothesis has been obtained (Blondel 1995) by comparing avian species along five different vegetation gradients in the western Palaearctic: three in the Mediterranean region (Provence, Corsica, and Algeria) and two further north, in temperate climate areas of Europe (Burgundy and Poland). In each of the five regions compared, the species present tended to be native when the vegetation was relatively undeveloped, with many species of definitely Mediterranean origin in Provence, Corsica, and Algeria. As the vegetation develops in complexity, however, a remarkably high degree of convergence was detected in the composition of the bird communities of the five regions (Fig. 2.9). All include very similar bird assemblages composed of forest-dwelling species of boreal origin. A similar pattern occurs in other groups of animals, for example snails (Magnin and Tatoni 1995) and plants, but few detailed studies have been carried out to verify this in these groups.

2.4.5.2 BIRD COMMUNITIES AND PLEISTOCENE CLIMATIC UPHEAVALS

Given the remarkably high diversity of habitats and the many barriers to immigration and gene exchange among populations that should have given rise to new species, one is led to wonder why there are not more endemic bird species in the mature forests of the basin. Endemic forest-dwelling species in the Mediterranean include the sombre tit (*Parus lugubris*), the Syrian woodpecker (*Dendrocopos syriacus*), three nuthatches (Corsican, Kabyle, and Kruper's; see above), and two pigeons (*Columba bollii* and *Columba junionae*). The very low number of forest birds of obvious Mediterranean origin can be explained by the Pliocene/Pleistocene history of vegetation belts and their associated biota, as described above. Three main points stand out, as follows.

First, Mediterranean biota did not survive during recent glacial periods in what is today the Sahara-Arabian desert. Instead, most forest biota of Europe survived in refugia within the Mediterranean area. Forest communities of birds in the Mediterranean Basin have therefore never been geographically isolated from those at higher latitudes, which would have been a prerequisite for local differentiation, according to the model of allopatric speciation.

Second, during the relatively brief interglacial periods of the Pleistocene, the lowlands of the Mediterranean Basin were covered with deciduous oak forests (e.g. downy oak (*Quercus humilis*) in the west, and Tabor oak (*Quercus ithaburensis*) in Turkey, Syria, and Lebanon), while evergreen species, such as the holm oak, were restricted to a limited number of relatively dry habitat types (Pons 1981). The bird faunas of these regions no doubt had a central European character quite different from what one expects to encounter in Mediterranean shrublands. Therefore, it is not surprising that the many floristic and faunistic species of boreal origin that found refuge in the Mediterranean region during glacial periods did not leave the area when the climate improved and thus are today part of the Mediterranean forest biota.

Third, the study of fossil pollen has shown that shrubland formations occurred more or less

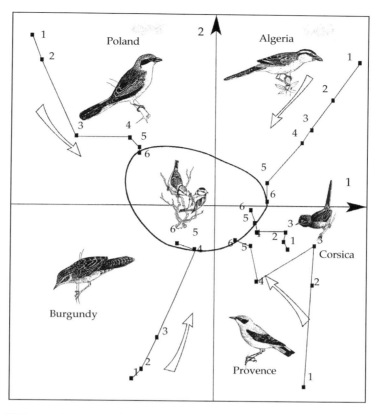

Figure 2.9 Similarities of bird communities in mature forests as compared with those in habitats with less-developed vegetation in five regions of the western Palaearctic. Numbers on the black lines (1–6) correspond to gradients of increasing biomass and complexity of vegetation, with 6 being a full-grown forest. Two of these regions are non-Mediterranean (Poland and Burgundy) and three are located in the Mediterranean region (Algeria, Corsica, and Provence). Location of the different values in the figure has been determined by multivariate statistics (correspondence analysis) (after Blondel 1995).

isolated throughout the whole Pleistocene in many parts of the basin, where climatic and/or soil conditions did not allow forest to develop. Such long-standing shrubland formations have been localized for instance in the Baetic Cordillera, Spain, in several islands such as Corsica, as well as in many parts of the Near East. Because these shrublands were much reduced in size, as compared to the present day situation, and scattered in several disjunct areas throughout the basin, active differentiation processes were likely to occur among avian taxa adapted to this type of habitat. The best example of speciation among shrubland birds is probably that of the warblers, which was described earlier in this chapter.

2.4.6 Freshwater fish

With more than 230 species, among which 148 local endemic species (Reyjol *et al.* 2006), the Mediterranean Basin is surprisingly rich in freshwater fish species. Freshwater fish are particularly interesting because, unlike terrestrial animals, in the absence of human intervention, their dispersal relies entirely on the geomorphological evolution of hydrographical networks. This is particularly the case of the so-called primary freshwater fishes (Darlington 1957), which, unlike peripheral fishes such as most salmonids, are unable to naturally disperse over lands and seas, which explains that they are totally absent from all the Mediterranean islands. Distribution patterns of primary freshwater fishes may give an interesting insight in the history and palaeogeography of hydrographic systems because individual river basins are in some ways biological islands. The freshwater fish fauna of the Mediterranean Basin includes two groups of species: a series of species that are widespread in the whole of Europe, the 'Danubian' species, which constitute a rich homogeneous ichthyofauna in a socalled Danubian district (Bianco 1990), and several assemblages, which are specific to more restricted districts around the basin and share few elements in common, either with one another or with the Danubian district. On the basis of the modern distribution of the species, 12 districts have been defined around the Mediterranean Basin by Bianco (1990). Several scenarios have been proposed for explaining their origin and history. Almaça (1976) and Banarescu (1973) postulated that many species originated from central Europe though river connections in the Miocene or earlier and that most of those that occurred north of the Mediterranean went extinct with glacial episodes, whereas species in the Mediterranean survived in refugia. However, no fossil data support this scenario. Another scenario, advocated by Bianco (1990) suggests a 'sea' dispersal event of primary freshwater fishes rather than a process of river captures involved in the first scenario. From the geography and history of the Mediterranean Sea during the Middle Miocene, a vast Parathethys basin, the so-called freshwater Lago-Mare, occurred north of the Mediterranean, which was much more saline and eventually almost completely desiccated during the Messinian Salinity Crisis. A series of seaways between this Parathethys Lago-Mare and a Lago-Mare freshwater phase of the Mediterranean before the Messinian resettlement of marine conditions with the coming back of Atlantic saline waters allowed primary freshwater species from the Parathethys to invade the Mediterranean region and spread in the various districts. Thus the lacustrine conditions of the Lago-Mare phase of the Mediterranean played an essential role in the penetration of Parathethyan fish elements in the various hydrographical basins that encircle the Mediterranean. The fact that among the 12 districts, four of them (Tunisian, Algerian, southern Iberian, western Greece) do not share any cyprinoid species with the Danubian district is an indication that these species are Mediterranean survivors of ancient species which went extinct in their former range of distribution in the Danubian district. This scenario is supported by the fact that fossil data show that while, in the Middle Miocene, most of the primary fish genera were already established in Europe, in several southern peninsular countries and in north-western Africa they were apparently absent and occurred after the Messinian Salinity Crisis (Bianco 1990). The Lago-Mare stage presumably played an essential role for the early penetration by Parathethyan primary freshwater fishes and for their dispersion in peri-Mediterranean river systems. The distribution of endemic taxa and the presence in southern peninsular Europe and north-western Africa of several elements which are unrelated with Danubian primary freshwater fishes indicates a penetration of ancestors via the 'Mediterranean sea way' rather than by 'river captures' from northern Europe. This scenario is supported by the age of fossils which are not older than the Messinian and by the particular kind of endemic fauna which is often unrelated to typical extant European freshwater fish elements (Bianco 1990). Examples are the monotypic genus *Anaecypris* in southern Iberia and the monotypic genus *Aulopyge*, which lives in a few endoreic rivers of the karst area of the Dalmatian district (Bianco 1986). In any case, the uniqueness of nearly all endemic cyprinid fishes in several peri-Mediterranean districts suggests a long isolation and independent evolution. They are considered as Parathethyan ancestors isolated since the Lago-Mare event, about 5 mya, which is the only known phase of brackish or freshwater environment in the Mediterranean history (Hsü *et al.* 1977). In view of the large central European area drained by the modern Black Sea, the salinity dilution during glacial/interglacial ice-melting phases of the Aegean Sea were possibly much more pronounced than that of the present-day Baltic Sea, where the river discharges produce a low saline condition allowing primary freshwater fishes to live and disperse (Ekman 1957). From a thorough study of the fish fauna of 406 hydrographical networks in Europe, Reyjol *et al.* (2006) came to the same conclusion that the Lago-Mare hypothesis explains the specificity of the peri-Mediterranean fish fauna as well as the history of re-colonization of Europe from the Ponto-Caspian Europe following the LGM. Thus peri-Mediterranean Europe and Ponto-Caspian Europe must be considered as biodiversity hotspots for European riverine fish with an extant freshwater fish fauna which is mostly determined by contemporary climate regimes and historical events dating back to the Miocene (Oberdorff et al. 1999).

2.4.7 Reptiles and amphibians

From a biogeographical perspective, the Mediterranean reptile and amphibian faunas became established as early as the late Eocene to mid-Miocene (38–15 mya) from several biogeographical regions, including the Euro-Siberian, Saharo-Arabian, and Turano-Caucasian. Biochemical studies have shown that divergence among major groups of lizards (e.g. Lacertidae) apparently took place during the Oligocene/early Miocene. Thereafter, Pliocene/Pleistocene climatic fluctuations remodelled this fauna through extinctions and new waves of speciation. Although certain reptiles, like the ringed snake (*Natrix natrix*) or the Schokar sand snake (*Psammophis schokari*), are of central European and Saharo-Arabian origin, respectively, most Mediterranean reptiles originated in western Asia, notably the Caucasian region, which is a 'hotspot' for this group (Meliadou and Troumbis 1997), and in North Africa. Indeed, the desert belt that limits the Mediterranean to the south favoured the differentiation of many groups of reptiles, for example *Acanthodactylus*, which penetrated to some extent in the Mediterranean area.

The ecophysiology of amphibians, which need moist habitats and bodies of water for their reproduction, helps explain why regional species richness is so uneven in the basin (see Chapter 3). This explains why Mediterranean North Africa has only 12 species of amphibians (Lescure 1992), several of which colonized the area from the north after the closing of the Straits of Gibraltar at the end of the Miocene, about 6 mya. Examples of 'recent' immigrants include the fire salamander (*Salamandra salamandra*), the common toad (*Bufo bufo*), the midwife toad (*Alytes obstetricans*), and probably also the stripeless tree frog (*Hyla meridionalis*) (Oberdorff et al. 1999).

2.5 Marine fauna and flora

The concept of 'hotspot', most often applied to terrestrial regions, is appropriate for Mediterranean marine biota as well, especially when the degree of endemism and current threats are considered, as we will see in Chapters 4, 11, and 12.

The disproportionately high biological diversity of Mediterranean marine fauna, compared to that of most oceans, can be explained in part because it is probably the most studied and the best known, having been the first to interest occidental naturalists and scientists since ancient times. Fredj et al. (1992) estimated that the Mediterranean fauna is composed of 67% taxa of Atlantic origin, 5% of eastern origin, having invaded the sea in modern times through the 150-year-old Suez Canal (the so-called Lessepsian species; see below and Chapters 4 and 12), and 28% of endemics. Consequently, there is a sharp drop in species richness from west to east: 92% of the combined species occur in the western basin and only 54% in the eastern basin (Fredj and Laubier 1985). During the Triassic epoch, the Tethyan fauna was composed of warm-water species affinitive to the actual indo-pacific fauna. In the Oligocene (30 mya), as we have seen in Chapter 1, the Tethys shrunk, and the Isthmus of Suez was formed during the Miocene (10 mya), definitively separating the Mediterranean Sea from the Indo-Pacific Ocean. Most of the tropical species remaining in the Mediterranean Sea were then driven to extinction during the Messinian Salinity Crisis (see also Chapter 1), apart from a few survivors in highly protected areas, due to the fact that the communication with Atlantic Ocean was never completely closed (Jolivet et al. 2008). The quantity of water entering the sea from the Atlantic was greatly inferior to the evaporation, but that was enough to preserve here and there, and at different levels, marine environments where fauna and microfauna could survive during the crisis. This surviving tropical contingent, called the Palaeomediterranean by Pérès (1985) derives from *in situ* evolution and can also be called palaeoendemic, the same term we used when discussing the terrestrial flora earlier.

When oceanic water suddenly returned, at the beginning of the Pliocene (5.33 mya), the

Mediterranean Sea was repopulated by so many Atlantic species that it now constitutes an Atlantic province from a faunistic point of view. The importance of the palaeoendemics is low compared to the entire extant marine biota.

Further, the position of Gibraltar led to an Atlantic repopulation taking place with two different source regions in the Atlantic Ocean, namely the Lusitanian region, from Gibraltar north to the British Isles, and the Mauritanian region, from Gibraltar south to Cape Blanc. The influence of the former was stronger during the cold periods of the Quaternary and that of the latter was stronger during the warm periods. During the whole Quaternary, the alternation of ice ages and warm interglacials resulted in alternative immigration waves of boreal and subtropical Atlantic species, from Lusitanian and Mauritanian regions respectively. Pérès (1985) quotes the arrival of 'Senegalian' species during the Tyrrhenian, some of them being still present in the warmest parts of the eastern basin. On the other hand, the Würm glaciation left a rich fauna of boreal species, all of them being now extinct in the western basin. One may suppose that the cooling of the water was less severe in the eastern basin, because only one of these boreal species, the mollusc *Chlamys septemradiata*, has been found there issuing from this period.

If the post-Messinian fauna of Atlantic origin predominates in the Mediterranean Sea, it is complemented with a strong proportion of endemic elements. At first, tropical forms issued from the Tethys: most taxa disappeared at the species level, but genera are still present and suggest that an *in situ* evolution resulted in the differentiation of endemic species related to the palaeomediterranean fauna. Pérès (1985) noted a distinct group of genera that are only represented in the Atlanto-Mediterranean region and in the western Indo-Pacific: Cnidaria, Crustacea, Echinodermata, the whole family of Sepiolidae, and several genera of fishes. He noted also the case of 'twin' or vicariant species, like the shrimps, *Lysmata seticaudata* (Mediterrranean) and *Lysmata ternatensis* (Indo-Pacific), the crabs *Dromia vulgaris* (Mediterranean) and *Dromia dromia* (Indo-Pacific), and *Octopus macropus* (Mediterranean) and *Octopus variabilis* (Japan), and, among marine plants, *Posidonia oceanica* (Mediterranean) and *Posidonia australis* (South Australia).

The second origin of endemics traces to environmental conditions of the organisms remaining after the Messinian Salinity Crisis. The marine domain does not present the kaleidoscopic topographical and climatic complexity of the terrestrial environments, but the topography of the coasts and bottoms still establishes geographical and ecological barriers, particularly in the shallow bottoms, where isolated populations can evolve independently. After the Pliocene, beginning of the speciation of the neoendemic species, the alternation of ice ages and warming up of the Quaternary produced important changes in temperature and salinity of the sea. Each climatic period resulted in the disappearance of most of the species brought by the former and in the arrival of new forms. These must adapt themselves to the Mediterranean conditions such as the little tide and the rarity of continental shelves.

This set of evolutionary pressures first affected the coastal biotopes and it is effectively the place where the majority of endemic species occur. More than half of them live between 0 and 50 m depth (Fredj and Laubier 1985). Neoendemism is clearly shown by the high proportion of specific endemism (28%), as compared to a very low generic endemism (2%).

The benthic algal vegetation shows the same distribution with a palaeoendemic group of few species, reflecting an earlier connection with the Indo-Pacific Ocean, a majority of temperate elements in common with the Atlantic Ocean, neoendemic elements, also of Atlantic origin, and Lessepsian species. Atlantic forms stay preferentially in the western basin and Lessepsian in the eastern. Many of them with a geographical restricted distribution are considered as biogeographic indicators for the Adriatic Sea, eastern Mediterranean, or Balearic Islands (Cinelli 1985).

The last group of Mediterranean marine species dates only from the digging of the Suez Canal (1859–67) and derives from the Red Sea and more easterly biota. We will come back to these colonizing species, often called Lessepsian species, in Chapter 12.

Summary

This chapter examines the determinants and drivers of biodiversity in the Mediterranean region, including the many processes of immigration, extinction, **sorting processes**, and regional differentiation. The main determinants of terrestrial and marine biodiversity are related to the complex tectonic dynamics of the basin and orographic heterogeneity of a region which has been squeezed between the major continental land masses of Africa and Eurasia. In addition, climatic features, biogeographical issues, and anthropogenic drivers and factors contribute as well. The major climate deterioration that occurred in the Pliocene and that culminated with the periodic oscillations of glacial and interglacial episodes during the whole Pleistocene played a key role in moulding and shaping present-day communities, species, and populations. Indeed, over the entire western Palaearctic, much of contemporary genetic diversity—both intraspecific and interspecific—is primarily a legacy of the differentiation processes that occurred wherever and whenever biota were restricted and separated in isolated refugia, notably the large peninsulas of the northern shores of the basin, but also in many isolated localities where local microclimates allowed for the persistence of temperate biotas. The Ibero-Mauritanian plate, the Cyrno-Sardinian and Apulian microplates, and the Anatolian plate in particular have been fertile matrices for differentiation in all groups of plants and animals. This has resulted in the exceptionally high levels of endemism that characterize many groups of terrestrial plants and animals. At the same time, historical events also explain why some groups, for example birds, include relatively few local endemics. The long-lasting processes that determined the diversity of plant and animals produced a biological legacy that has been deeply modified by humans through transformation of ecosystems, degradation of habitats, and persecution of animals. With respect to the Mediterranean Sea, modern-day marine biota is composed of a few palaeoendemics and many more taxa derived from species originating from the Atlantic Ocean, some of which evolved *in situ* to produce neoendemic species. Marine environments also include some species which are ice-age remnants, as well as interglacial remnants of subtropical species and an increasingly important group of Red Sea immigrants arriving via the 150-year-old Suez Canal. But contrary to the terrestrial fauna and flora, endemism among the marine biota is not exceptionally high.

CHAPTER 3

Present-Day Terrestrial Biodiversity

Our purpose in this chapter is to describe the major biogeographic and ecological trends of species diversity in some selected groups, pointing out features that are typical of Mediterranean biota. We cannot give a detailed account of all the species of plants and animals which occur in the Mediterranean Basin for three reasons: (1) such an endeavour would be out of the scope of this book, (2) we are unfamiliar with many groups, especially invertebrates, to make a thorough review of their status in the Mediterranean, and (3) more importantly, our aim is to give a dynamic account of biological diversity, considering processes and evolutionary trends rather than just describing patterns and giving lists of species. In addition, while knowledge of biological diversity is fairly good for some groups, surprising discoveries are made every year. For example, even in such well-studied groups as vertebrates several new species were recently discovered, including a shrew (*Crocidura*) in Gozo, an islet quite near Malta, two frogs (*Alytes muletensis*, in Majorca, and *Discoglossus montalentii*, in Corsica), a nuthatch (*Sitta ledanti*) in Algeria, and a land tortoise (*Testudo weissingeri*) in Peloponnesus, Greece. Even the 'common' house mouse appears to be much more complex in the region that was thought. Recent genetic studies using new biochemical techniques such as molecular phylogenies (Rajabi-Maham *et al.* 2007) reveal that three or even four distinct species occur in the region. Similar analyses of other small mammals, such as voles, and of course invertebrates, will also certainly lead to the discovery of new species. In a range of plant groups that were thought to be well understood, heretofor unsuspected complexity is coming to light thanks to careful cytogenetic studies and experimental crosses.

One reason for this 'hidden' diversity is the high level of endemism at small scales of space in most groups of terrestrial plants and animals in the Mediterranean (Table 3.1). Thanks to the region's complex climate, history, geology, and topography, thousands of biological isolates occur in islands, peninsulas, and mountain ranges, where local differentiation gives rise to endemic species and subspecies. Indeed, levels of endemism here far exceed those found in any other part of Europe, the Near East, or Africa, except in the Cape Province (Cowling *et al.* 1996; Médail and Quézel 1999; Mittermeier *et al.* 2004). They are particularly high in islands. For example, the Maltese islands have some 20 endemic species or subspecies of Tracheophyta, two of Bryophyta, seven of Mollusca, 11 of Arachnida, five of Isopoda, 37 of Coleoptera, 20 of Lepidoptera, six of Diptera, four of Hymenoptera, five of other insects, one reptile, and one mammal.

Thanks to its high biological diversity, and degree of endemism, the Mediterranean Basin has been recognized as one of the world's 34 major biodiversity 'hotspots' (Myers *et al.* 2000; Mittlemeier *et al.* 2004; see also Chapter 13). Many factors have contributed to make the basin something like a 'matrix' of species diversity. Perhaps the most important is the Pliocene/Pleistocene history of the western Palaearctic, as explained in the previous chapter, but other factors play an important role, as discussed in this chapter and throughout the entire book.

3.1 Flora

The flora of the Mediterranean area is one of the richest in the world with respect to its size. It includes more than 25 000 species of flowering

Table 3.1 Numbers of species, levels of endemism, and percentage of the species richness known worldwide for vascular plants and several groups of animals in the Mediterranean Basin

Group	Number of species	Endemism (%)	Known species (%)	Sources
Vascular plants	≈25 000	50	10	Quézel (1985)
Freshwater fishes[1]	250	63.5	–	Smith and Darwall (2006)
Reptiles[2]	355	48	2.5	Cox et al. (2006).
Amphibians[2]	106	64	1.5	Cox et al. (2006).
Mammals	197	25	4.2	Cheylan (1991)
Birds	366	17	3.8	Covas and Blondel (1998)
Insects[1]	≈150 000	–	0.6	Baletto and Casale (1991)
Butterflies	321	46	–	Higgins and Riley (1988)

[1] For the northern banks of the Mediterranean Basin only.
[2] Tentative figure, probably largely underestimated.

plants (Vogiatzakis *et al.* 2006; Médail 2008a), or about 30 000 species and subspecies (Quézel 1985; Greuter 1991; Médail 2008a), as well as 160 or more species of ferns. That is about 10% of all known plant species on Earth, a figure estimated to be between 238 000 and 260 000 (Greuter 1994), although the area is only about 1.5% of the emerged lands. Compare this regional richness to the mere 6000 species of higher plants found in Europe north of the Mediterranean Basin, an area that is about three to four times greater in size! Approximately 290 tree species contribute to the various forests of the Mediterranean Basin, as compared to only 135 species in all of central and northern Europe (Médail 2008a). These striking differences, which also occur in several groups of animals, including birds, has been attributed by Latham and Ricklefs (1993), among others, to differential extinction of tree species of the northern latitudes during glacial periods (see Chapter 2). The difference in arboreal flora between the Mediterranean and the northern/central Europe is even greater for endemic species: 201 in the Mediterranean region as compared to 46 in temperate Europe (Médail and Diadema 2006).

3.1.1 Endemism

Indeed, the main reason for high plant species richness of the Mediterranean is the remarkable number of endemics, many of which are restricted to a single or a few localities, especially in sandy areas, islands, geological 'islands' of unusual soil or rock type, or geographical islands isolated in mountain ranges. As shown in Table 3.1, approximately half of the plant species of the Mediterranean are endemic, and no fewer than four-fifths of all European plant endemics are Mediterranean (Gomez-Campo 1985; Vogiatzakis *et al.* 2006). Thus, the Mediterranean area is an important reservoir of plant diversity, most comparable in fact to California and three parts of the southern Hemisphere, which also have mediterranean-type conditions, the Cape Province of South Africa, central Chile, and two parts of south-west Australia (see below). Surprisingly, the tropical third of the African continent harbours just about the same number of vascular plant species and endemics as the Mediterranean region, even though it is about four times larger, and enjoys a year-round growing season as compared to the highly bimodal seasons in the Mediterranean (Médail and Quézel 1997).

Ten regions in the basin with particularly high numbers of species and large percentages of endemics are shown in Fig. 3.1. These regional hotspots cover 22% of the basin's total area and harbour no fewer than 5500 endemic plant species; that is 44% of the endemics for the entire region (Médail and Verlaque 1997; Médail and Quézel 1999).

As for overall biodiversity, levels of endemism generally increase with increasing altitude and on islands. In Mediterranean mountain ranges, whether continental (Atlas, Taurus, Lebanon, Anti-Lebanon) or insular (Corsica, Sardinia, Crete), the contribution of endemic species can exceed 25%. For example, among the 400 endemic plant taxa in

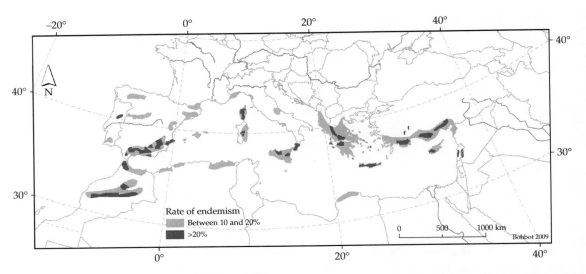

Figure 3.1 Hotspot areas for plant species diversity in the Mediterranean Basin, including the Canary Islands and Madeira. Modified from Médail and Quézel (1997) and Médail (2008b).

Andalusia, 125 (31%) are restricted to the mountains, and levels of endemism can reach 50% in certain Spanish mountain ranges of the Baetic Cordillera, the Sierra Nevada, and the Serrania de Ronda (see Fig. 3.1). The largely upland Iberian Peninsula as a whole harbours more than 1200 endemic species and subspecies of vascular plants (Gomez-Campo and Herranz-Sanz 1993) that, given a total of 4839 species listed for Portugal and Spain in the Med-Check-List, amounts to an extraordinary 24.8% rate of endemism for the entire peninsula. The Anatolian Peninsula, in Turkey, with its succession of life zones from sea level up to more than 5000 m altitude, is the eighth richest region in the world for plant endemism, with 9000 endemic plant species (Küçük 2008).

It is interesting to decipher the respective roles of ecology and history as drivers of endemism in hotspots. Analysing the ecological and historical factors explaining the patterns of 115 endemic plant species in the hotspot of Maritime and Ligurian Alps near the border of southern France and Italy, Casazza et al. (2008) showed that local concentrations of endemics result from a combination of thermoclimatic belts and different types of substrates, with a good congruence between areas of endemism and the corresponding specific bedrocks. This means that at the small scale of regional hotspots, ecological factors are more important than historical factors for explaining endemism rates. In this particular case at least, glaciations seem to have had a lower influence on plant distribution and their effects, if any, were alleviated by post-glacial migrations. On the other hand, glaciations had a strong influence on richness at larger scales. Thus, interaction between ecological local features and historical events confirms that biogeographical studies should be multi-scaled and cover both ecological and historical components as recently recommended by several authors (e.g. Morrone 2001).

Mediterranean island floras typically also show high percentages of endemism. Examples include Corsica with 13% (316 endemics from a total of 2325 species), as compared to only 7.2% endemism for the nearby continental area of south-eastern France. Similarly, Crete has about 12% plant endemics (209/1735), Sicily 11% (321/2793), and the three larger Balearic Islands, approximately 10% (173/1729) (Médail 2008b). Unusual geological substrates that are often found on islands are particularly conducive to endemics—for example gypsum or dolomite substrates—as well as ultrabasic (serpentine) formations in Cyprus, which include four species of *Alyssum* (Brassicaceae), which are low, spiny shrubs. Elsewhere this large genus of about

175 species, mainly found in Mediterranean Europe and Turkey, is represented almost exclusively by strictly annual species (Mengoni *et al.* 2003). An additional oddity of this genus is that some 73–75 species, mostly of the eastern Mediterranean, can soak up unusually large quantities of nickel and other heavy metals from the soil (Frérot *et al.* 2006). This is probably an adaptation to unusual edaphic conditions, a fact that is being exploited for purposes of **bioremediation**. The genus is just one of many large genera of the Old World that have produced exceptionally large numbers of species in the Mediterranean area, especially on islands (Thompson 2005).

3.1.1.1 MACARONESIA

As an oceanic volcanic archipelago, it is not surprising that the Canary Islands and Madeira far surpass the islands of the Mediterranean Sea itself in terms of endemism. Endemic plant species in these islands are found in at least 15 genera, in 12 different families, with 10 or more endemic species per genus (Quézel 1995). Conclusions on the numbers differ however, which reflects the need for more field and laboratory studies. Juan *et al.* (2000) recognized approximately 1000 native vascular plant species for the seven Canary Islands and estimated 27% endemism among them, as well as 50% endemism for the terrestrial invertebrate fauna of approximately 6500 species. In contrast, Francisco-Ortega *et al.* (2000) and Santos-Guerra (2001) estimate the number of native species of Canarian plants to be approximately 1425, of which 570 species (approximately 40% of native flora) are endemic. They also note that about 20% of these endemics are endangered and listed in the endangered (E) category established by the IUCN.

The flora of the Canary Islands is also particularly interesting because it includes a series of 'archaic plants' that were formerly widespread around the Mediterranean Basin in Miocene/Pliocene times, some 7–25 mya (see Chapter 2). For example, some patches of mid-altitude (400–1500 m) forest of the Gran Canaria, La Palma, and Teneriffe islands are called laurisilva forest because they are dominated by four endemic, broad-leaved evergreen trees, which are members of four different genera of the tropical laurel family (Lauraceae). These are *Persea indica*, a relative of the avocado, which is native to the Amazon; *Apollonius barujana*, whose closest relatives occur in tropical India, *Ocotea foetens*, and *Laurus azorica*. Other endemic canopy trees of the laurisilva also belong to tropical families, including *Sideroxylon* of the Sapotaceae, *Myrica faya* (Myricaceae), *Visnea mocanera* (Ternstroemiaceae), and *Picconia excelsa* (Oleaceae). For the most part, these families—except for three genera of Oleaceae and the bay tree—have disappeared from the Mediterranean Basin proper since the climatic deterioration of the Plio-Pleistocene wiped out this flora from most parts of the basin.

One conspicuous, indeed legendary, endemic species of the Canaries is the so-called dragon tree (*Dracaena draco*) (Marrero *et al.* 1998), which was the source of the 'dragon's blood' much prized by early European seafarers as a source of varnish, medicine, incense, and dye (Fig. 3.2). The presence of this tree in the Canary Islands (and one or two mountain canyons in southern Morocco) is very striking, as the other 40-odd species in the genus *Dracaena* (Agavaceae) all occur in tropical woodlands of Africa, Madagascar, and Asia.

Another fascinating feature of the vegetation, especially the cliff-dwelling cohorts, in the Canary Islands is the woody or arborescent life forms found in such genera as *Limonium*, *Echium* (Marrero-Gómez *et al.* 2000), *Aeonium* (Crassulaceae), *Kleinia*, and *Sonchus* (Asteraceae), and *Jasminum odoratissimum*, which elsewhere in the Mediterranean and adjacent regions are entirely herbaceous (Jorgensen and Olesen 2001). When other shrubby species exist, for example in the genus *Kleinia*, and Crassulaceae, they are found in the frost-free zones of Africa and Madagascar.

Shmida and Werger (1992) studied the range of life forms among endemic plant species in the Canary Islands. They found more than 70% endemism among tree and shrub species, as compared to only 5.6% among the annual plants (see also Kim *et al.* 2008). Indeed, in a broad survey of oceanic island floras, Carlquist (1974) noted that the high number of woody species in taxa that are otherwise herbaceous on mainland areas occur in the Canary Island flora, which is a typical feature for island floras. But in the context of the

Figure 3.2 A remarkable stand of *Dracaena draco*, the famous dragon tree of the Canary Islands. Photo kindly supplied by A. Machado.

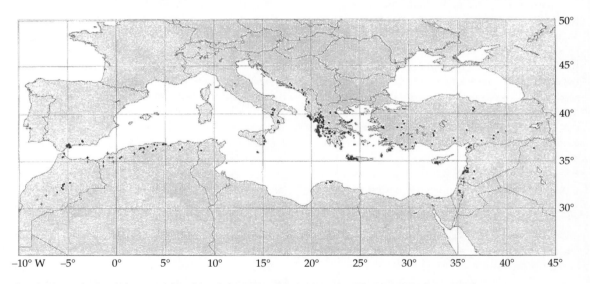

Figure 3.3 Distribution of *Plocama calabrica* (•) and *Plocama brevifolia* (+) (reproduced from TAXON with permission).

Mediterranean flora they represent another group of 'archaic lineages' of particular interest.

A last example to consider is the small tree *Plocama pendula* in the largely tropical family Rubiaceae, which is endemic to the Canary Islands. It was long thought the species was phylogenetically isolated, constituting a monotypic genus whose presence and endemic status in the Canary Islands was a mystery. However, based on very careful research, the contours of the genus *Plocama* have been dramatically revised (Backlund *et al.* 2007) and the genus now includes 34 species, distributed in Macaronesia, several parts of Africa, Socotra, the Arabian Peninsula, and south-western

Asia east to Pakistan and Punjab. Furthermore, two additional, primarily Mediterranean species have been described (Backlund and Thulin 2007), one quite widespread and one limited to North Africa (Fig. 3.3).

3.1.1.2 EXAMPLES OF ENDEMIC PLANTS AND THEIR DISTRIBUTION

Patterns of limited species distributions and varying degrees of species/genus ratios are common within the Mediterranean flora, as can be illustrated with three examples. The first is that of the Northern Hemisphere fir tree (Pinaceae), with 39 species in all and nine endemic species in isolated mountain ranges throughout the basin (Table 3.2).

The second example is that of the arar or Barbary thuja (*Tetraclinis articulata*), a monotypic conifer genus found only in the thermo-mediterranean life zone of North Africa and a few southern Spanish mountains. This long-lived, fire-resistant tree is a palaeorelict, whose closest living relatives are 14 species of *Callitris*, found in subtropical forests of south-eastern Australia and nearby islands, and New Caledonia. This peculiar situation deserves serious study and intensified conservation efforts, as the few remaining populations of the Barbary thuja are highly endangered by overexploitation of the wood and the burl. Once used for top-quality furniture during the Roman Empire, the vestiges are now used to manufacture small craft objects, like boxes and plates, for sale to tourists in a few towns in Morocco. An important replanting and reintroduction program is underway in Morocco, with almost 4300 ha already planted since 1995 (M. Abourouh, personal communication).

The third example concerns the genus *Arbutus*, which is represented in the Mediterranean by two relatively widespread species, which may well be a vicariant pair, and two geographically isolated ones. Strawberry trees, as these species are called in reference to their edible red-orange and mottled fruit, generally occur on non-calcareous soils, like all the numerous heaths and heathers in this same family (Ericaceae). In addition to the most widespread species, *Arbutus unedo*, absent only in Libya and Egypt, there is a closely related but largely allopatric eastern species, *Arbutus andrachne*, and two endemic species, one in the Canary Islands (*Arbutus canariensis*) and one (*Arbutus parvarii*) in the isolated coastal mountain of Libya, called Djebel Akhdar (Fig. 3.4). The other 10 members of this genus occur in woodlands of southern California, Arizona, and northern Mexico.

In part, the situation of the Mediterranean *Arbutus* seems similar to that of the Mediterranean firs and the Barbary thuja, in that early and successful adaptation to rather special and difficult habitats, especially in mountains or islands with strong geographical boundaries, apparently led to a low level of adaptive radiation. However, two of the four strawberry tree species are widespread (and interfertile, since hybrids of the two have been repeatedly produced by horticulturists). In this context, it is pertinent to note that people have selected, modified, and introduced the strawberry tree in various parts of the Mediterranean area. In the Balkans, large-fruited, 'elite' cultivars of *A. unedo* have long been widely propagated for commercial plantations. In Portugal, the fruits are preserved and used in liqueurs. The tree occurs in warmer coastal areas of the western British Isles, where it is considered part of the 'Lusitanian' flora. It could well be that the geographical range of this strawberry tree, and perhaps others, has been intentionally expanded by people and that its genetic makeup was modified, just as was so for many other fruit trees. In the case of the Barbary thuja, by contrast, human exploitation has been focused on wood and the attractive burl used for furniture-making, as described above. Even with this tree and many other trees, the question of human-mediated

Table 3.2 Endemic firs in the Mediterranean Basin (Aussenac 2002)

Species	Region of endemism
Abies maroccana	Rif (Morocco)
Abies pinsapo	Southern Spain
Abies numidica	Algeria (Babor mountain range)
Abies nebrodensis	Sicily
Abies cephalonica	Greece (mostly in Peloponnesus)
Abies borisii-regis	Northern Greece and Bulgaria
Abies equi-trojani	Turkey
Abies bornemulleriana	Pontic region, northern Turkey
Abies cilicica	Southern Turkey, Lebanon

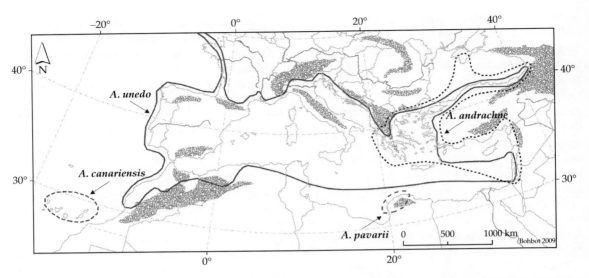

Figure 3.4 Distribution of the four Mediterranean species of strawberry trees (genus *Arbutus*), showing that two species are widespread and two are highly restricted (after Sealy 1949).

dispersal is worth addressing. For example, in the Hoggar and Tibesti Mountains, in the middle of the Sahara, there is an endemic cypress, *Cupressus dupreziana* (Abdoun *et al.* 2005). People may have played a role in getting it there. Another species that has undoubtedly been selected, moved about, and genetically 'improved' by humans is the cork oak (see Chapters 10 and 11). Historical data show that the cork oak benefited from a sharp increase in density in the twentieth century because forest management has favoured this species (mostly for cork production) at the expense of the Algerian oak (*Quercus canariensis*) (Urbieta *et al.* 2008).

3.1.2 Mediterranean specificities

Some features that are typical of (but not exclusive to) the Mediterranean flora are worth mentioning. Annual species in general, and ruderals and segetals in particular, are well represented in the basin's flora, as this life-history strategy is highly adapted to various kinds of perturbations, including ploughing, grazing, and the considerable stress of 2 or as many as 5–6 months of summer drought.

In addition, an impressive number of taxa have shown particular success in adaptive radiation and speciation, such as *Allysum* or *Centaurea* (Asteraceae). The latter genus has an estimated 450 species in the Mediterranean area and the Near East. That figure includes approximately 170 species in Turkey, including 105 endemics, as well as 44 endemic species in Greece and 78 in Iberia (Quézel and Médail 1995), all showing a remarkably wide range of growth forms, life-history strategies, and other ecological adaptations. Additional examples of this kind of adaptive and geographical radiation are the Mediterranean members of the vetches (*Astragalus*) and the spurges (*Euphorbia*). An excellent example is the succulent spurges of Spain, Morocco, and Macaronesia (Molero and Rovira 1998), which show a respectable subset of the truly remarkable range of life forms found in this genus, from tiny creeping annuals to majestic trees.

3.2 Invertebrates

It would be beyond the scope of this book to give a detailed account of the diversity of invertebrate faunas in the Mediterranean, especially the soil-borne organisms, which appear to reach the highest peaks of diversity in the world here, even higher than in the tropics (di Castri and di Castri 1981). Instead, we will provide a few highlights and an overview.

3.2.1 Species richness and endemism

The Mediterranean area is by far the richest region in Europe in terms of invertebrate species diversity. Three-quarters of the total European insect fauna are found in the basin (Baletto and Casale 1991). At the scale of Europe, very high species diversity in the Mediterranean for most groups of insects fits the latitudinal trends of increasing diversity from boreal to tropical regions (Pianka 1989). However, a reverse trend occurs in some Mediterranean peninsulas, for example in the case of butterflies in the Iberian Peninsula (Martin and Gurrea 1990), which is presumably due to the so-called peninsular effect, a biogeographic pattern also found in many other parts of the world.

Baletto and Casale (1991) estimated as approximately 150 000 the number of insect species in the Mediterranean Basin, but this is a rough estimate because probably no more than 70% of insect species had been described and named at that time. Figures for most groups of invertebrates are increasing rapidly as scores of new species are being described each year. Dafni and O'Toole (1994) estimated as 3000–4000 the number of bee species in the Mediterranean, which makes of the basin a prominent centre of diversity for this group. As many as 1500–2000 species of bee species occur in Israel alone (O'Toole and Raw 1991). In the Chironomidae, a large family of dipterans, 703 species have been reported in the basin, 97 of which (14%) are exclusive to this area (Laville and Reiss 1992).

Levels of endemism are also high for most groups of insects. In some isolated mountains and larger islands, endemics may account for 15–20% of the insect fauna, or even 90% in some caves. As for most other groups (see Chapter 2), classic refuge areas are mountain chains, such as the Kabyle, Atlas, and Rif chains in North Africa, the central cordilleras of Iberia, the Pyrenees, the Alps, the Balkans, and the Taurus Mountains of southern Turkey. In some remote mountain ranges, extremely limited populations of 'odd' or archaic species are sometimes found. For example, several very unusual carabid beetles have recently been found in moist relictual habitats of the High Atlas in Morocco (e.g. *Relictocarabus meurguesae*), as well as in the high mountains of Kurdistan.

Percentages and regions of endemism vary greatly among groups of insects, according to their particular histories and evolutionary potential. Endemism is generally fairly high in beetles, stoneflies, flies, and several groups of spiders that evolved locally from an ancient Tertiary fauna, especially on Mediterranean islands and high mountains (see Chapter 2). Especially high concentrations of species are found in the Atlas Mountains, the Pyrenees, the southern Alps, including the Dinaric chain, and the Apennines. The Mediterranean countries with highest percentages of endemism in butterflies are Spain—with 16 species, or 7.5% endemism—and Greece and Italy—with 13 species each (9%). This pattern supports the contention that the presence of large peninsulas has had an important role in the development of endemic insect taxa in the Mediterranean Basin, just as for plants and many other groups of animals (Dennis *et al.* 1995). Although richness in butterflies on Sardinia, Corsica, and Sicily is large compared to that in other Tyrrhenian islands, the butterfly faunas of these islands are clearly impoverished when compared to nearby mainland regions of the same size, especially those of Corsica and Sardinia, which are more distant from the mainland than Sicily (Dapporto and Dennis 2008).

Several groups of insects, including butterflies, show a surprisingly low level of endemism, even on islands. For example, three of 46 species in Sardinia (*Papilio hospiton, Hipparchia neomiris, Maniola nurag*) and two of 52 species (*Papilio hospiton, Hipparchia neomiris*) in Corsica are island endemics (Dapporto and Dennis 2008). It may be that, as for smaller Italian and Aegean islands, most of the butterflies of Sardinia and Corsica are part of highly dynamic system of populations where migration and gene flow proceed continuously from and to neighbouring mainland areas (Dapporto and Dennis 2008). A similar situation probably also occurs in large raptors, which seem to be overrepresented on islands, given the unusually high ratio of raptor species richness over area in several islands, especially Crete (see Chapter 7). It is also possible that some insects were introduced to islands in recent times by humans, as was the case for some mammals and plants. For example, in Corsica, only three of the 83 native species of ants (3.6%) are considered endemic to the island and only 23 species (27.7%) are considered to be Mediterranean in their distribution (Casevitz-Weulersse 1992). Thus, well over half of the species present come from very distant centres of origin, but note that not a single species is of Afro-tropical origin (see Chapter 2). Some of the highest diversity values of Mediterranean butterflies occur in the southern Alps and, somewhat less so in the Italian and Iberian Peninsulas. High numbers of species of Pieridae and Hesperiidae also occur in the Atlas Mountains of North Africa, whereas Satyridae and Nymphalidae are particularly well represented in the southern Alps.

An example of a butterfly genus that has had an explosive adaptive radiation in the mountain ranges of southern Europe is the genus *Erebia* (Satyridae). Most taxa in this genus occupy narrow subalpine habitats at altitudes between 1200 and 1500 m. In groups such as *Erebia*, evolutionary divergence is more likely to occur in the three-dimensional relief of mountains with finely dissected landscapes than in groups that are distributed in more uniform areas, such as the Lycaenidae in Iberian upland regions. In the Papilionidae, Lycaenidae, and Hesperiidae, peak values of species richness are found in the Balkans and Iberia. On the other hand, particularly low species numbers are found on Mediterranean islands.

3.2.2 Mediterranean ground beetles

In the genera *Calosoma*, *Carabus*, *Pamborus*, *Ceroglossus*, and *Cychrus*, more than 1000 species of ground beetles have been described, with 850 species in the genus *Carabus sensu lato* alone (Deuve 2004). Several of these are found only in the Mediterranean Basin, where they are terrestrial carnivores, living on a diet of worms, molluscs, snails, maggots, insect larvae, and caterpillars, among other prey. Most of them are active at night and can generally be found hiding beneath rocks and other objects during the day.

A few lineages are particularly interesting from biogeographical and ecological perspectives, of which we will give just a few examples. With five species, the genus *Macrothorax* is endemic to the western Mediterranean, each with a very distinct distribution: *Macrothorax planatus* (Sicily), *Macrothorax rugosus* (southern Spain, northern Morocco), *Macrothorax celtibericus* (central Spain and Portugal), *Macrothorax morbillosus* (southern Italy, Sicily, Corsica, Sardinia, Balearic Islands, and northern Algeria), and *Macrothorax aumonti* (northern Morocco). At low altitudes, they are all active throughout the year, and reproduce after the onset of autumn/winter rains, usually in October. At high altitudes, by contrast, they are only active during the spring (Darnaud *et al.* 1981). A newly described subspecies of *M. rugosus* was recently discovered in southern France in the coastal Pyrenees close the Spanish border, in a very dry environment (Forel and Leplat 1995).

The striking genus *Chrysocarabus* is famous among collectors of ground beetles for its unusual ornamentation. It has eight recognized species (Deuve 2004), probably originating from an area today immerged below the Gulf of Lion. Two species are confined to the Mediterranean coast: *Chrysocarabus solieri* in the subalpine zone of southeastern France and on the border with Italy (Darnaud *et al.* 1978a), and *Chrysocarabus rutilans* between Barcelona, in north-eastern Spain, and the eastern Pyrenees and Ariège, in south-western France (Darnaud *et al.* 1978b). The other six species migrated to the west (*punctatoauratus*, *splendens*, *lineatus*, and *lateralis*) or to the north (*hispanus* and *auronitens*).

The genus *Procerus* is an eastern group, with four large dark species which are among the largest in the Carabidae: *Procerus gigas* from Italy, the Balkans, Bulgaria, and northern Greece, *Procerus duponcheli* from southern Greece, *Procerus scabrosus* from Turkey, and *Procerus syriacus* from Syria and Lebanon (Darnaud *et al.* 1984a; see Fig. 3.5). *Megodontus caelatus* and *Megodontus croaticus* are endemic from the coastal Balkans (Darnaud *et al.* 1984b). Finally, *Hadrocarabus genei* is endemic to Corsica and Sardinia and *Archicarabus alysidotus* lives both on the coast and in low mountains between Marseille and Roma (Forel and Leplat 1995).

Many species of ground beetle which are highly emblematic of the western Mediterranean are threatened by fires, including *Carabus rutilans* and *Carabus solieri*, which today is protected in France and Italy and off-limits now to capture and trade.

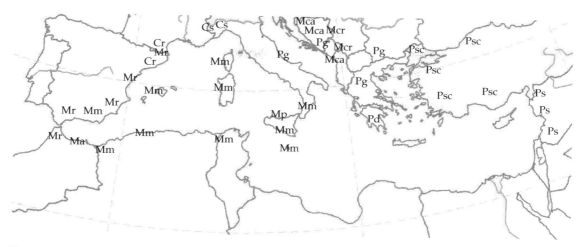

Figure 3.5 Carabid distribution in the Mediterranean. Cr, *Chrysocarabus rutilans*; Cs, *Chrysocarabus solieri*; Ma, *Macrothorax aumonti*; Mm, *Macrothorax morbillosus*; Mp, *Macrothorax planatus*; Mr, *Macrothorax rugosus*; Mca, *Megodontus caelatus*; Mcr, *Megodontus croaticus*; Pd, *Procerus duponcheli*; Pg, *Procerus gigas*; Ps, *Procerus syriacus*; Psc, *Procerus scabrosus* (adapted from Darnaud *et al.* 1978a, 1978b, 1981, 1984a, 1984b).

Carabus olympiae is only known to occur naturally in the southern Alps, in the Val Sessera in Italy, but it was recently reintroduced in the Mercantour Park in France, and it is strictly protected now in both countries. Another narrowly endemic Mediterranean taxon, *Carabus clathratus* subsp. *arelatensis*, is in fact a subspecies of a much more widely distributed species of northern Europe and Asia. Only found in a few littoral lagoons of southern France, it is the only winged and flying species in our region. All these beetles are excellent indicators of habitat changes that are occurring today. They are also very sensitive to pesticides and insecticides.

3.3 Freshwater fish

In terms of freshwater systems, we can define the Mediterranean Basin as all river basins flowing through the region into the sea, except the upstream portions of the Nile and Rhône, which originate far outside the Mediterranean biogeographical realm (see Chapter 1). We also include the whole Iberian Peninsula, including Portugal, which is justified on palaeogeographical grounds because, as a result of the barrier of the Pyrenees, the entire fish fauna of the peninsula is endemic to varying degrees and is of Mediterranean origin.

3.3.1 Diversity and endemism

The number, diversity, and geographical isolation of watersheds in the Mediterranean Basin has fostered a remarkably high species richness among freshwater fishes, among which as many as 148 species are endemic (Smith and Darwall 2006). A remarkable feature is that 98 of these endemic species occur in the Balkans region and 41 in a single site within that region (Crivelli and Maitland 1995). Most species occur in mountain rivers, natural lakes, and lowland rivers, but just a few in marshes and coastal lagoons. These species belong to 15 families and 49 genera. Centres of species richness include the Po River in northern Italy, the lowest Orontes in south-western Turkey, Lake Kineret (Tiberiad) in Israel, and the lower Guadiana in southern Spain. Not strictly part of the Mediterranean Basin, a similar level of species richness and endemism is also found in the Tagus River in Portugal, and the coastal basins of the Gulf of Cádiz and the Guadiana River in Spain. Most of these endemics (63%) belong to the Cyprinidae, but endemic species also occur in other families, such as the Cobitidae (11%), Gobiidae (8%), Salmonidae (6%), and Cyprinodontidae (5%). To these must be added two endemic freshwater species of lampreys (Petromyzontidae) and one of sturgeon (Acipenseridae). As for many other groups, levels of endemism are especially high in the large Iberian, Italian, and Aegean Peninsulas, as well as in Turkey (which is peninsula-like), presumably because these regions acted as refugia during glacial periods (see Chapter 2).

With 110 species, Greece harbours the largest number of fish species of any region in the basin (Economidis 1991). Of these, 21 **euryhaline** species occur in freshwater and brackish water lagoons. Excluding 11 introduced species, 78 species (71% of the total) are indigenous and 37 of these are endemic. However, as in many other groups of plants and animals, regional endemism levels for Mediterranean fishes are highest in the Iberian Peninsula, as 25 of the 29 species and subspecies of indigenous Cyprinidae, Cyprinodontidae, and Cobitidae that occur there are endemics (Corbacho and Sánchez 2001). Most of these species live in springs, mountain torrents, lakes, and lowland rivers, while a few are restricted to marshes and coastal lagoons.

Fish communities of the more arid regions of the basin are much less rich in species. For example, the native freshwater fish fauna of Israel includes only 32 species and subspecies and 14–16 introduced species (Goren and Ortal 1999). This fauna belongs to eight families originating from Africa (28% of the species), Eurasia (24%), the Levant (19%), the Mediterranean (15%), and the Red Sea (14%). Many of these species are endemic to single catchments. The richest fish fauna occurs in the Jordan River with 26 species, 19 of which are found in Lake Kineret (Tiberiad), but three of the native species became extinct after the drainage of Lake Hula. In Tunisia, there are no more than 12 species, four of which have been introduced recently (Kraiem 1983). Some endemic species are restricted to very small areas. Examples are several species of the genera *Noemacheilus* (Cobitidae) and *Aphanius* (Cyprinodontidae) in very small springs, streams, and ponds near the Dead Sea, and three species of Cichlidae in the oases around the Chott el Djerid, Tunisia. Although most species of the Mediterranean fish fauna belong to Palaearctic lineages, some are of Afro-tropical origin (Cichlidae, Clariidae), or are relicts of the Tethys Sea (several species of Cyprinodontidae). The Cyprinidae species that succeeded in colonizing North Africa presumably did so from Iberia in the Miocene when the two land masses were connected by a narrow isthmus. As noted in Chapter 2, southern European rivers are hotspots of fish diversity in Europe (Reyjol *et al.* 2006). As we will see in Chapter 12, many of those species are threatened by land-use changes and global changes, including climate warming, desiccation, and introduction of invasive species (see Table 3.3).

Islands are also severely impoverished in freshwater fish because marine environments are strong barriers to dispersal for such species. For example, only 12 freshwater fish species are native in Corsica (Roché 1988), but many species have been introduced in most of the larger islands (see Chapter 12). Some interesting cases of 'capture' of previously marine species by freshwater streams may sometimes occur. This is the case of the freshwater blenny (*Salaria fluviatilis*), which inhabits Mediterranean rivers northward to Lake Annecy in the northern French Alps. This small fish, 8–12 cm in

Table 3.3 Number of families, genera, and endemic species of freshwater fishes in the Mediterranean Basin, and percentage of threatened species from the IUCN categories ('critically endangered', 'endangered', and 'vulnerable')

Family	Genera	Species	Percentage threatened
Acipenseridae	1	1	100
Balitoridae	2	15	67
Blenniidae	1	2	50
Cichlidae	3	6	33
Clupeidae	1	2	100
Cobitidae	1	21	86
Cottidae	1	1	100
Cyprinidae	25	164	54
Cyprinodontidae	1	8	50
Gasterosteidae	1	1	100
Gobiidae	3	12	58
Percidae	1	2	50
Petromyzontidae	2	2	50
Salmonidae	3	13	38
Siluridae	1	1	n.a.
Valenciidae	1	2	100

n.a., not available.
Source: From Crivelli and Maitland (1995) and Smith and Darwall (2006).

size, belongs to the Bleniidae family, which is of marine origin. The body is much elongated with a long dorsal fin. This carnivorous fish with powerful jaws and teeth lives in lakes and small streams and springs with pebble and stones. It colonized coastal rivers of the Mediterranean some 4 mya. In Corsica, each river has its own population of blenny with river-specific patterns of phenotypic and, presumably, genetic variation. Magnan *et al.* (unpublished work) showed that the morphology of the populations is mostly explained by a spatial component and habitat characteristics. The spatial component may reflect genetic differences associated with patterns of colonization of each river. The speed of the current and the substrate are the habitat components associated with local environments. The variation of these populations results from their geographical isolation because of the marine barrier to dispersal.

3.3.2 Phylogenetic uncertainties

Although the phylogenetic status and time of differentiation of most fish species in the Mediterranean Basin are poorly known, some interesting cases are worth mentioning. For example, the complex systematics of barbels (*Barbus*; Cyprinidae) have recently been clarified through molecular studies (Tsigenopoulos and Berrebi 2000). This genus includes one group of large species (*Barbus barbus* type), which live in the large rivers and lakes of central and southern Europe, and a second group (*Barbus meridionalis* type), which includes smaller species that live in small mountain rivers of the Mediterranean. Occasionally, species of the two groups co-occur in the same river and hybridize, e.g., *B. barbus* and *B. meridionalis* in France, *B. barbus* and *B. meridionalis petenyi* in Slovakia, and *B. barbus* and *B. meridionalis* subsp. *peloponnesius* in Greece. Although these taxa seem closely related, in fact they derive from highly distinct lineages.

Some very interesting endemic Mediterranean trout species are also worth noting, such as the endangered *Salmo trutta macrostigma*, restricted to upper streams in Corsica, and the Prespa trout (*Salmo trutta peristericus*) in the Prespa lakes of western Greece and Albania. Another is the marble trout (*Salmo trutta marmoratus*), which is an endangered freshwater fish of the Adriatic Basin, including the catchments of the Po River, its alpine tributaries, and some rivers of the Dinaric Alps that empty into the Adriatic. Although the current status and biology of the various forms of *Salmo trutta* were studied by Povz *et al.* (1996), there is still much uncertainty concerning its taxonomy (see Presa *et al.* 2002; Apostolidis *et al.* 2008). This beautiful fish is the second largest trout in Europe, with large individuals weighing up to 30 kg and a total length of 140 cm, second in size to the huchen (*Hucho hucho*), called the Danube salmon, which lives in the Danube River and some of its tributaries. This species can reach 50 kg in weight. The marble trout differs in colour and shape from the brown trout (*Salmo trutta fario*). Diagnostic characters are the absence of red spots or blotches that characterize the brown trout, the whitish or yellowish belly, and the marble patterns of its olive-brown to olive-green skin with a copper-red tint. Populations of the marble trout show little genetic variation, with heterozygosity ranging from 0 to 1%, as compared to 5–7% in other Mediterranean trout species. This fish is restricted to upland streams, where summer water temperatures do not exceed 15° C and winter water temperatures are of the order of 2–3° C. All these trout species are threatened because of hybridization with the brown trout, which has been repeatedly introduced into their range for sport fishing. Only in some upper reaches of rivers in Slovenia and Corsica can pure marble trout and Corsican trout be found today.

3.4 Reptiles and amphibians

In the region as defined in this book, 355 species of reptiles in 22 families are found, as are 106 species of amphibians in ten families (Delaugerre and Cheylan 1992; Cheylan and Poitevin 1994; Cox *et al.* 2006; Sindaco and Jeremcenko 2008). We shall now discuss their distribution and diversity in space and time.

3.4.1 Contrasted patterns of species richness and distribution

Reptiles are 'at home' in the dry and warm Mediterranean area and are much more abundant and diverse than amphibians, which usually

Table 3.4 Regional numbers of species and numbers of endemic species (percentage in parentheses) of reptiles and amphibians in various regions of the Mediterranean area. Data are given for 179 and 62 species of reptiles and amphibians, respectively

Region	Reptiles		Amphibians	
	Total	Endemics	Total	Endemics
Mediterranean area	179	113 (68)	62	37 (59)
Mainland regions				
Iberia	33	8 (24)	22	7 (32)
Italy	20	0	17	6 (35)
Balkans	45	11 (24)	17	4 (24)
Near East	84	26 (31)	15	2 (13)
Cyrenaica (Libya)	25	0	2	0
Maghreb (north-west Africa)	59	26 (44)	12	2 (17)
Insular regions				
Balearic Islands	10	2 (20)	4	1 (25)
Corsica	11	3 (27)	7	2 (29)
Sardinia	16	3 (19)	8	5 (63)
Sicily	18	1 (6)	7	0
Crete	12	0	3	0
Cyprus	21	1 (5)	3	0

Source: From Cheylan and Poitevin (1994).

live in moister habitats (Table 3.4). Reflecting the contrasted ecology and physiology of the two groups, reptile species diversity increases from north to south and from west to east, along clear gradients of aridity. For example, there are only 20 species of reptiles on a surface area of 141 500 km^2 in Italy, as compared to 25 species in the Cyrenaica region of northern Libya, a mere 27 000 km^2. Better still, in Cyprus (9250 km^2), there are as many as 21 species of reptiles.

In contrast to reptiles, species richness of Mediterranean amphibians increases from south to north and from east to west. Since the north–south and west–east aridity gradients favour reptiles in the southern and eastern parts of the basin, whereas more humid climates favour amphibians in the northern and western parts, regions that are rich in reptiles tend to be poor in amphibians, and vice versa. Thus, in North Africa, the number of amphibian species declines from west to east, with 11 species in Morocco, seven in Tunisia, and only two in Tripolitania, western Libya, namely the green toad (*Bufo viridis*) and the marsh frog (*Rana ridibunda*), both species that have very wide distributional ranges. Differential distribution patterns between reptiles and amphibians suggest that historical effects differed greatly between the two groups. From this arose various regional specificities and several cases of vicariance with east–west species replacements (see Chapter 7). Examples of such species pairs include the marbled and crested newts (Fig. 2.5), the palmate (*Triturus helveticus*) and common newt (*Triturus vulgaris*), the Iberian (*Pelobates cultripes*) and common spadefoot (*Pelobates fuscus*), and the stripeless and European tree-frog (*Hyla arborea*).

The many factors which have moulded the Mediterranean Basin and its biota are reflected in levels of endemism, which are very high in the two groups (Table 3.4)—48% for reptiles and as much as 64% for amphibians (Cox et al. 2006)— but regional levels of endemism sharply contrast between them. In reptiles, levels of endemism amount to 44% in North Africa, 31% in the Near East, and 24% in the Iberian Peninsula, whereas in amphibians, endemism is highest in Italy (35%), then Iberia (32%), the Balkans (17%), North Africa (17%), and the Near East (13%) (Cheylan and Poitevin 1994).

The reptile fauna of the Mediterranean Basin as a whole includes snakes, lizards, tortoises (see Box 3.1), and even tropical relicts, such as two species of chameleon (*Chamaeleo chamaeleon*, in North Africa and southern Iberia where it has been introduced, and *Chamaeleo africanus* in Greece, Crete, Cyprus, and the Near East; see Plate 4b). The most speciose reptile families are the Lacertidae (30% of the world species), then the Trogonophidae (16.6%), the Testudinidae (8%), the Viperidae (7.4%), and finally the Anguidae (5.3%).

Other tropical relicts include a tortoise (*Tropnyx triunguis*) and the giant lizard (*Varanus griseus*), in Anatolia and the Near East, as well as several species of snakes. The crocodile *Crocodylus niloticus* occurred in southern Morocco until the 1950s (see Chapter 11). Many neoendemic species of reptiles in the genera *Podarcis*, *Lacerta* (see Plate 4c), *Chalcides*, and *Vipera* evolved in the basin as a result of intensive adaptive radiation in localized areas. In the Lacertidae, the genera *Algyroides* and *Psammodromus* (four species in each of them) are typical relict Mediterranean endemics. Thus, to get an idea of the situation of endemism in reptiles and amphibians,

> **Box 3.1. The European pond terrapin (cistude)**
>
> Among the six tortoises that occur in the Mediterranean Basin, three are terrestrial: the spur-thighed tortoise (*Testudo graeca*), Hermann's tortoise, and the marginated tortoise (*Testudo marginata*). The other three are mostly aquatic; that is, the Spanish terrapin (*Mauremys leprosa*), European pond terrapin (*Emys orbicularis*), and the stripe-necked terrapin (*Mauremys caspica*).
>
> The European pond terrapin, also called the cistude, occurs in a large part of Europe, but is much more common in Mediterranean countries than further north. This secretive animal spends most of its life at the bottom of shallow ponds, marshes, and slow-running rivers and canals. Adults are 20–30 cm long, with a black shell decorated with yellow spots and stripes. Males are much smaller than females. The cistude becomes active at water temperatures above about 28° C, walking slowly along the bottom of its pond in search of food, but coming up from time to time to breath at the surface of the water. It can also swim quickly using its flat palmate legs as oars. On warm and sunny days, when the ambient temperature rises to several degrees above that of the water, the animal climbs on any thing hard that floats to heat itself, for example dead trees or stumps, piles of dead reeds, or ancient muskrat huts. Up to 30 animals may sit together, basking in the sun, but quickly go back in the water when disturbed.
>
> A carnivore par excellence, the cistude feeds mostly on aquatic insects and molluscs, but also on small fishes, newts, and carrion. The shell of this slow-growing animal becomes hard only after 4–5 years, but life expectancy is exceptionally high and some individuals may live for at least 40–50 years. When ponds and small rivers dry up in summer, the cistudes bury themselves in the ground until water returns in autumn. They spend the winter, from October to March, sunk 20–30 cm deep in the mud at the bottom of the pond. A bearing female typically produces six large eggs that she will bury in spring, 10–12 cm deep in fields or pastures, sometimes as much as 600 m away from the nearest body of water. After 3–4 months of incubation, the young tortoises, which are very small at birth (less than 2 cm long), spend their first autumn and winter hidden in their burrow. Then, early the following spring, they take their first, perilous trip to the nearest body of water. This is risky indeed: around 50% of them will be taken by predators before they reach the pond.

it is necessary to consider subregions within the basin, as shown in Table 3.4.

Among amphibians, most endemic species belong to archaic lineages that have remained relatively unchanged morphologically since their first appearance in the Eocene (55 mya) (see Chapter 2). Examples include two genera of toads—*Pelobates*, with three species in the basin and *Discoglossus*, with five species—and salamanders of the genus *Euproctus* (three species). The nearly eyeless *Proteus anguinus* of the Dinaric **karsts** and caves is another example of an ancient lineage of tropical origin. This species is the only representative in the Old World of the remarkable Proteidae family that also includes five species found only in North America (Box 3.2).

3.4.2 Impoverished but highly endemic insular faunas

Compared to mainland areas of similar size, species impoverishment—that is, decline in species numbers—in Mediterranean islands is 43% in reptiles and 60% among amphibians (see also Chapter 7). However, their patterns of diversity and distribution have been significantly disrupted by human-mediated introductions, and also extinctions of endemics caused by humans since the beginning of the Holocene (see Chapter 11). This may explain why there are, on average, few endemic reptiles and amphibians in Mediterranean islands, especially in those of the eastern half of the basin. As in mainland areas, there are more

> **Box 3.2. The blind cave salamander, *Proteus anguinus***
>
> One of the most fascinating Mediterranean amphibians is *Proteus anguinus*, an almost eyeless and highly specialized salamander-like animal found only in caves and underground rivers of the Dinaric karsts, from the border between Italy and Slovenia south to Bosnia-Herzegovina (see Plate 5a). This species is patchily distributed in more than 250 localities along the Dalmatian coast and on offshore islands, where genetic isolation has led to a close matching between ecotypic adaptations and habitats (Sket 1997; Goricki and Trontelj 2006). It probably derives from an ancient lineage that arose in the region when tropical climates prevailed during the Oligocene-Miocene and then colonized the area as limestone karstic formations progressively developed.
>
> *P. anguinus* is large for an amphibian, reaching 25–50 cm in length, and it bears large feathery gills whose pink colour contrasts dramatically with the animal's pigment-free, bleach white skin. This **neotenic** animal has only two toes on each hind leg and reproduces without ever reaching an 'adult' stage of body development; instead it retains a larval characteristics throughout its life cycle, a feature that is common among amphibians. The cave salamander lives in large bodies of slow-running, underground rivers with temperatures nearly constant at 8° C, or else in deep caves. Its metabolic rate is very low, in keeping with the fact that organic matter and especially animal prey are scarce in these unlighted habitats. By contrast, its hearing is extraordinarily acute and it also finds prey through chemical sensory means. A rare beast indeed this nearly blind amphibian!

amphibian species in islands of the western half of the basin than in those of the eastern half: seven species in Corsica, eight in Sardinia, and seven in Sicily, as compared to only three each in Crete and Cyprus.

Table 3.5 shows the lists of amphibians and reptiles that are endemic to the larger Mediterranean islands. Examples of highly endemic species in the western islands are two species of salamander (*Euproctus montanus* in Corsica and *Euproctus platycephalus* in Sardinia). These are archaic species, which—contrary to the closely related newts—do not have lungs. They breathe instead through the mouth and skin. Moreover, their breeding biology is unique for their family. After having laid her eggs under a stone in a small river, the female takes much care of them, actively watching them until they hatch. In the Balearic Islands (Mallorca), the recently discovered small toad (*Alytes muletensis*) of the Discoglossidae family is the last survivor of an endemic fauna, which has been enriched by nine species of reptiles and three species of amphibians, all introduced by humans (Delaugerre and Cheylan 1992). Not surprisingly, the Canary Islands have very high endemism rates in reptiles, which include 14 endemic species (87% endemism). Active differentiation occurred in these islands in the genera *Tarentola* (four species), *Gallotia* (seven species), and *Chalcides* (three species). The subfamily Gallotiinae, which differentiated 16–12 mya from the North African genus *Psammodromus*, includes several giant lizards of more than 1 m long. They have been for long considered as extinct but were recently rediscovered on Hierro, Tenerife, and La Gomera islands. Notably, some of these lineages are found only in the Mediterranean region, for example the subfamily Gallotiinae in the Canary Islands, the genus *Archaeolacerta* in Corsica and Sardinia, and the genus *Euleptes* (e.g. *Phyllodactylus*) on several Tyrrhenian Islands.

Cyprus is an exception to all the rules. If one considers comparable habitats elsewhere, one would expect to find several endemic reptiles and amphibians in Cyprus, but there is only one endemic species of snake (*Hierophis cypriensis*) and 10 endemic subspecies of reptile (Böhme and Wiedl 1994). This level of endemism is very low for an island of volcanic origin that has been geographically separated from the mainland since the Pliocene. One possible explanation is that repeated colonization events on natural rafts have caused a swamping effect and prevented local evolution

Table 3.5 Endemic species of reptiles and amphibians in the larger Mediterranean islands

Taxa	Family	Distribution
Amphibia (Urodela)		
Salamandra corsica	Salamandridae	Corsica
Euproctus montanus	Salamandridae	Corsica
Euproctus platycephalus	Salamandridae	Sardinia
Speleomantes genei	Plethodontidae	Sardinia
Speleomantes flavus	Plethodontidae	Sardinia
Speleomantes supramontis	Plethodontidae	Sardinia
Speleomantes imperialis	Plethodontidae	Sardinia
Amphibia (Anura)		
Discoglossus sardus	Discoglossidae	Tyrrhenian Islands
Discoglossus montalentii	Discoglossidae	Corsica
Alytes muletensis	Discoglossidae	Balearic Islands
Rana cerigensis	Ranidae	Aegean Islands
Rana cretensis	Ranidae	Crete
Sauria		
Tarentola angustimentalis	Gekkonidae	Canary Islands
Tarentola delalandi	Gekkonidae	Canary Islands
Tarentola gomerensis	Gekkonidae	Canary Islands
Tarentola boettgeri	Gekkonidae	Canary Islands
Galliotia atlantica	Lacertidae	Fuerteventura, Lanzarote[1]
Galliotia caesaris	Lacertidae	Hierro, La Gomera[1]
Galliotia galloti	Lacertidae	Tenerife, La Palma[1]
Galliotia simonyi	Lacertidae	Hierro[1]
Galliotia bravoana	Lacertidae	La Gomera[1]
Galliotia intermedia	Lacertidae	Tenerife[1]
Galliotia stehlini	Lacertidae	Gran Canaria[1]
Anatololacerta troodica	Lacertidae	Cyprus
Archaeolacerta bedriagae	Lacertidae	Corsica and Sardinia
Podarcis wagleriana	Lacertidae	Sicily
Podarcis raffonei	Lacertidae	Eolie Islands[2]
Podarcis filfolensis	Lacertidae	Malta and Pelagian archipelagos
Podarcis gaigeae	Lacertidae	Skyros archipelago (Sporades, Greece)
Podarcis milensis	Lacertidae	Milos archipelago (Greece)
Podarcis tiliguerta	Lacertidae	Corsica and Sardinia
Podarcis pityusensis	Lacertidae	Ibiza and Formentera (Balearic Islands)
Podarcis lilfordi	Lacertidae	Minorca and Mallorca (Balearic Islands)
Algyroides fitzingeri	Lacertidae	Corsica and Sardinia
Chalcides simonyi	Scincidae	Fuertebentura, Lanzarote[1]
Chalcides viridanus	Scincidae	Tenerife, La Gomera, Hierro[1]
Chalcides sexlineatus	Scincidae	Gran Canaria[1]
Ophidia		
Hierophis cypriensis	Colubridae	Cyprus
Macrovipera schweizeri	Viperidae	Cyclades Islands, Greece

[1] Of the Canary Islands.
[2] In the Tyrrhenian Sea off the north coast of Sicily, Italy.
Source: Blondel and Cheylan (2008).

of the indigenous taxa. But a more likely explanation is that several species, intentionally or inadvertently introduced by humans, have directly or indirectly pushed the endemic species to extinction. According to this more persuasive argument, differences in species richness and endemism on islands in the western, as compared to the eastern, half of the basin could result from the islands of the western Mediterranean having been colonized by humans much later than those of the eastern half (see Chapter 10).

3.5 Birds

Judging from the geographical, topographical, and ecological heterogeneity of the Mediterranean Basin, the diversity of birds in the Mediterranean should be very high. Indeed, as many as 370 or so species may be found in the basin, compared to only 500 for all of Europe (Covas and Blondel 1998). However, the contribution of each biogeographical region is quite uneven and three groups are clearly dominant. The largest one includes 144 species of northern, boreal origin, which are characteristic of forests, freshwater marshes, and rivers all over western Eurasia.

The second group consists of 94 steppe species, most of which presumably evolved in the margins of the current Mediterranean area, notably in the **'eremic'** Sahoro-Arabian region, which extends from Mauritania—where the Sahara meets the Atlantic Ocean—eastwards across Africa, the Red Sea, the Arabian Peninsula, and on to the semi-deserts of southern Asia. This belt has almost always—at least since modern species evolved—isolated the Palaearctic from the Afro-tropical and Oriental realms (see Chapter 2). The importance of this faunal element in the Mediterranean Basin has been favoured, both in geographical distribution and in population sizes, by the generalized, human-induced retreat of forest cover since the Neolithic (see Chapter 10). As a direct result of this deforestation, many species that were formerly rare are now widespread and common throughout the basin.

The third group encompasses all the species that are more or less linked to shrubland habitats, the so-called **matorrals** (see Chapter 6). Good examples are the partridges (*Alectoris*) and the many species of warbler (*Sylvia*, *Hippolais*). Given the extent and the high diversity of shrubland formations in the Mediterranean Basin, the number of species in this group is surprisingly small (42 species or 11.5% of the region's overall avifauna).

To these three principal elements should be added two smaller groups of birds of great biogeographical and ecological interest. The first is the rock-dwelling species found on steep cliffs, screes, and rock outcroppings of hills or mountainous Mediterranean regions and, indeed, throughout the southern Palaearctic as a whole. Examples are the lammergeier (*Gypaetus barbatus*), the blue rock thrush (*Monticola solitarius*; see Plate 6b), and the chough (*Pyrrhocorax pyrrhocorax*), which bird watchers come from all over the world to see. The second small group, known as **sarmatic**, includes bird species inhabiting lagoons and coastal swamps. This group includes a flamingo (*Phoenicopterus roseus*; see Box 11.8), the white-headed duck (*Oxyura leucocephala*), marbled duck (*Anas angustirostris*), slender-billed gull (*Larus genei*), and Mediterranean gull (*Larus melanocephalus*).

As in most other groups of animals, only very few Mediterranean bird species are of Afro-tropical origin. Examples of species that presumably reached the Atlantic shores of Morocco from tropical latitudes are a goshawk (*Melierax metabates*), the Cape owl (*Asio capensis*), and the guinea fowl (*Numida meleagris*).

3.5.1 Homogeneity and few endemics

The bird fauna of the Mediterranean differs from the other groups of animals, and from the flora, in two main points that were raised in Chapter 2. First, despite their disparate biogeographical origins, many species and groups of birds are rather homogeneously distributed across the basin, so that regional variation in the composition of local bird assemblages is not very marked. Of course, there are several endemic species and regional variations, but on the whole the vast majority of species are widespread throughout the basin. This is especially true for forest-dwelling birds. From statistical tests devised to detect regional differences in species composition, Covas and Blondel (1998) were able

Table 3.6 Numbers and examples of bird species that presumably evolved in forest, steppe, and shrublands of the Mediterranean area

Habitat type	Forest	Steppe	Shrubland	Other (sea, fresh water, rocks)
Number of species	16	19	20	9
Examples	Laurel pigeon	Barbary partridge	Moussier's redstart	Marbled teal
	Corsican nuthatch	Red-necked nightjar	Cetti's warbler	Slender-billed gull
	Sombre tit	Dupont's lark	Sardinian warbler	Audouin's gull
	Spotless starling	Berthelot's pipit	Marmora's warbler	Pallid swift
	Syrian woodpecker	Black-eared wheatear	Black-headed bunting	Rock nuthatch

to discriminate only some of the most conspicuous differences, for example between the south-eastern and the north-western quadrants of the basin. This homogeneity of regional bird faunas results partly from the long distances that birds are able to cover.

Secondly, and more surprisingly, there is very little endemism in the Mediterranean bird fauna. Only 64 species (17% of the total) appear to have evolved within the geographical limits of the Mediterranean Basin (Table 3.6). On exception is the Spanish imperial eagle (*Aquila adalberti*), which is endemic in the Iberian Peninsula (Box 3.3). Levels of endemism are low on islands too, except on the Canary Islands where eight species of land birds are endemic: two pigeons (*Columba bollii* and *Columba junoniae*), a swift (*Apus unicolor*), a pipit (*Anthus berthelotii*), a stonechat (*Saxicola dacotiae*), a blue tit (*Cyanistes teneriffae*), the blue chaffinch (*Fringilla teydea*; see Plate 6c), and a canary (*Serinus canaria*). There are two endemic bird species in Cyprus: the Cyprus wheatear (*Oenanthe cypriaca*) and the Cyprus warbler (*Sylvia melanothorax*); and only one in Corsica, the Corsican nuthatch (see Chapter 2). Another warbler, the Marmora's warbler (*Sylvia sarda*) occurs in several islands of the western part of the basin (see Chapter 2).

One wonders why more speciation events did not occur among birds in the highly fragmented, heterogeneous landscapes and regions of the basin. One explanation is that the present extension

---

Box 3.3. The Spanish imperial eagle

This beautiful and impressive eagle used to be considered as a subspecies of the imperial eagle from central and eastern Europe, but it has been recently recognized as a full species, originating in the Iberian Peninsula about 1 mya, when it split from the eastern imperial eagle (*Aquila heliaca*). This species is typical of one of the most common habitats in Spain, the **dehesa** (see Chapter 10). This is a very large bird of prey, measuring 70–83 cm in length and weighing 2.5–3.5 kg with females being larger than males, which is the rule in diurnal raptors. In contrast to many other species of eagle, the Spanish imperial eagle is fairly productive, laying a clutch of three or four eggs, usually producing two or three fledglings as compared to one or exceptionally two in the golden eagle (*Aquila chrysaetos*). The Spanish imperial eagle is resident with an extensive home range covering as much as 30 000 ha. The nest is built in large oaks in *dehesas*, sometimes in subalpine pine forests. The highest densities are found in plains and **sierras**, where its preferred prey, the rabbit, is common. The species is considered as Endangered on the IUCN Red List because its population sharply decreased until the 1970s as a consequence of several causes including persecution, habitat destruction, poisoning, electrocution by power lines, and the collapse of its preferred prey due to various diseases. However, from no more than 50 pairs in the 1970s, the population steadily recovered to nearly 200 pairs thanks to important efforts of conservation, including LIFE projects from the European Union.

of scrub and shrublands is secondary, resulting directly from human activities and perturbations, and therefore species have not yet had time to differentiate. However, palaeobotanical data have shown that more or less isolated patches of matorral have existed in the area at least since the Pliocene. Accordingly, successive episodes of expansion and contraction of these shrublands have presumably favoured differentiation in at least a few groups (see Chapter 2).

3.5.2 Subspecies variation

Contrary to the rate of endemism at the species level, at the subspecific level, morphological changes in island-dwelling birds have led taxonomists to recognize a great many subspecies. For instance, on Corsica more than half the species of birds are considered to be endemic subspecies. Such high levels of intraspecific variation result from the high geographical diversity of the basin with its many islands, peninsulas, and other geographic and ecological barriers to dispersal. Bird species that occur as breeders in the area are represented by an average of 5.4 subspecies per species at the scale of their worldwide distribution, 2.3 subspecies per species at the scale of the Palaearctic, and 2 subspecies per species in the Mediterranean Basin itself. Weighting these figures by the sizes of surface areas involved in order to obtain an index of regional subspecific variation, the proportion rises from 0.56 for the entire Palaearctic realm to 6.7 in the Mediterranean Basin (Blondel 1985). This represents a truly remarkable range, as compared to other parts of the world. Examples of resident birds that exhibit a particularly large amount of subspecific variation are the jay and the blue tit.

3.6 Mammals

Approximately 197 species of mammals occur in the Mediterranean Basin, 25% of which are endemic to the area (Cheylan 1991) (see Table 3.7). Three main factors influence the composition of the non-volant mammal fauna (i.e. excluding bats): (1) multiple biogeographic origins, due to the proximity of and contribution from three continental land masses, (2) repeated faunal turnover provoked by climatic variation during the Pliocene/Pleistocene period, including numerous intercontinental exchanges, and (3) more so than for any other group of animals, species richness and distribution have been influenced by local human history, especially persecution and hunting, since the early or mid-Palaeolithic (see Chapter 2). One characteristic of the Mediterranean mammal fauna is the large number of browsers, such as asses, goats, gazelles, hamsters, gerbils, and jerboas.

3.6.1 Regional specificities

In contrast to birds, mammal faunas of the four quadrants of the basin differ sharply. The eastern part of the Mediterranean is richest in mammals (106 species), with as many as 23 species of Asian origin that do not occur elsewhere in the basin. The second richest region is North Africa (south-western quadrant), with 84 species, followed by the north-eastern quadrant (80), and finally western Mediterranean Europe (north-western quadrant), with 72 and 77 species in Italy and Iberia, respectively (see Table 3.7).

Non-flying mammals are more sensitive to physical barriers than bats or birds, which means that the mammal faunas of southern Europe, the Middle East, and especially North Africa, are quite distinct. Although only 14 km wide, the Straits of Gibraltar have effectively isolated Europe from Africa for all non-volant mammals since its opening after the Messinian Salinity Crisis, which hindered exchanges between the two continents. As a consequence, mammal faunas on the northern side of the sea are basically Euro-Siberian in origin, with wild boar, deer (*Cervus*, *Capreolus*), and the brown bear, as typical elements. Apart from a few exceptions, such as the porcupine (*Hystrix cristata*) in southern Italy, the Barbary macaque in southernmost Spain, and some rodents and shrews (*Mastomys*, *Xerus*, and *Crocidura*), mammal species of tropical origin were eradicated from the Euro-Mediterranean regions by the beginning of the Pleistocene (see Chapter 11).

Although many mammals of North Africa are of Palaearctic origin (41%), in contrast to most other groups of animals and plants, a large number of species are Afro-tropical or Saharo-Sahelian. Colonization of North Africa by typical Afro-tropical

Table 3.7 Numbers of mammal species in different regions of the Mediterranean Basin (excluding islands)

Group	North Africa	Iberia	Italy	Balkans	Anatolia	Near East
Hedgehogs (1, 4)	1	2	1	1	1	1
Shrews (1, 14)	6	7	8	7	4	1
Moles (1, 4)	0	4	3	2	1	1
Bats (7, 39)	25	26	27	25	34	9
Primates (1, 1)	1	1	0	0	0	0
Canids (1, 3)	2	2	2	3	3	3
Ursids (1, 1)	0	1	1	1	1	0
Mustelids (1, 9)	5	5	6	5	6	7
Genet/mongooses (2, 3)	2	2	0	0	2	1
Cats/felids (1, 7)	4	2	1	2	5	5
Hyraxes (1, 1)	0	0	0	0	1	0
Pigs (1, 1)	1	1	1	1	1	1
Deers (1, 3)	1	2	2	2	2	3
Goats/sheep/gazelles (1, 9)	3	2	2	2	5	3
Squirrels (1, 5)	2	1	1	2	1	1
Glirids (1, 5)	1	2	4	5	5	1
Porcupines (1, 2)	1	0	1	0	1	1
Mole-rats (1, 3)	0	0	0	1	2	0
Mice/rats (1, 16)	8	6	6	10	11	8
Gerbils/hamsters (1, 26)	15	0	0	2	9	9
Voles (1, 19)	0	8	5	8	7	4
Jerboas (1, 4)	2	0	0	0	3	2
Gundis (1, 1)	1	0	0	0	0	0
Hares/rabbits (1, 3)	2	3	1	1	1	1
Elephant shrews (1, 1)	1	0	0	0	0	0
Total	84	77	72	80	106	62

Figures in parentheses indicate families and species for the entire basin, respectively.
Source: After Cheylan (1991).

elements, such as large felids, elephants, or rhinos presumably occurred long ago when the Sahara region was more humid than it is today. More recently, at both the western and eastern extremities of the desert, along the shores of the Atlantic Ocean and through the Nile River valley, various mammals of tropical origin also entered the Mediterranean. At the western end of the desert, small shrews and rodents (*Crocidura*, *Mellivora*, *Xerus*, *Mastomys*, and *Acomys*) colonized western North Africa, mainly Morocco, from the south. Examples of species of tropical origin that colonized the Near East through the Nile valley are the genet, a mongoose (*Herpestes*), the hyrax (*Procavia*), the bubal antelope (*Alcephalus busephalus*), a spiny mouse (*Acomys*), as well as a large fruit-eating bat (*Rousettus aegyptiacus*) and the ghost bat (*Taphozous nudiventris*). The macaque is the only native primate of the basin, living in disjunct populations in Algeria and Morocco. Occasional vagrants in North Africa of species living in the Saharo-Arabian region include the hunting dog (*Lycaon pictus*), the leopard (*Acinonyx jubatus*), and several gazelles (*Addax nasomaculatus*, *Gazella dama*, and *Oryx damah*).

During periods when desert conditions did not reach the shores of the Mediterranean Sea, a belt of Mediterranean-like habitats, the so-called eastern route, extended along the shores of the sea, allowing the colonization of North Africa as far west as Morocco by species of boreal and southeastern Asian origin. Later, when the Sahara desert extended northward to the sea, this belt disappeared and was replaced by arid habitats which made a link between North Africa and the Near East. Examples of species adapted to such habitats co-occurring in both regions, thanks to this ancient link, are asses, antelopes, gerbils, and jerboas.

3.6.2 Few endemic mammals

Today, the mammals of the Mediterranean Basin include few local or regional endemics, apart from the amazing Etruscan shrew (*Suncus etruscus*) (Box 3.4). The Spanish hare (*Lepus granatensis*), the spiny mouse (*Acomys minous* in Crete and *Acomys cilicicus* in Turkey), and one porcupine in Cyprus are among the rare endemic species among mammals. In Mediterranean islands, after the mass extinction of the endemic mammal fauna that started in the late Pleistocene (see Chapter 11), only two endemic species of shrews are left, one in Sicily (*Crocidura sicula*) and one in Crete (*Crocidura zimmermani*). Endemic rodents include several gerbils and no fewer than eight species of voles (*Pitymys*). This last group, of boreal origin, provides us with the best documented example of mammal speciation in Mediterranean refugia (Chaline 1974; see Chapter 2).

3.7 Convergence and non-convergence among mediterranean-type ecosystems

A long-standing hypothesis upheld by many scientists is that similarity of bioclimates in different parts of the world should result in similar adaptations arising among phylogenetically unrelated organisms. This theme received much attention in the 1960s and 1970s, in the framework of the International Biological Programme (see e.g. Cody and Mooney 1978). However, in the case of the Mediterranean biota, the specific and unique history which

---

Box 3.4. The smallest mammal in the world

Among many other oddities, the Mediterranean region harbours the smallest mammal in the world, the Etruscan shrew, which is no more than 3.6–5.2 cm long and weighs from 1.6 to 2.4 g (Jürgens *et al.* 1996; Jürgens 2002). This typically Mediterranean species, with very long hair for a shrew, especially on the tail, does not extend beyond the limits of the thermo- and meso-mediterranean life zones and is widespread wherever it finds suitable habitats that are warm enough for it to survive over winter. Suitable habitats include fallow terraces, stone walls, stone piles, ruins, and traditional human settlements, all of which provide this species with shelter, food, and protection against predators (Fons 1975). A buffered microclimate between or under stones makes this kind of habitat particularly suitable for this mammal, as well as for two other shrews (*Crocidura suaveolens* and *Crocidura russula*), which occur in the same type of environment. Although the species is active all year round, it is in summer months, from early June to late September that population sizes are highest as a result of high reproductive rates in this season. In spite of its small size, the Etruscan shrew is a fearsome predator for any living prey of sufficiently small size. The shrew eats insects, worms, myriapods, and snails in large volumes and at very high rates, because the survival of this mammal requires incessant feeding.

As could be expected for such a small endothermic organism, the Etruscan shrew also exhibits some of the world's records for physiological performance in mammals. Electrocardiogram recordings have shown that the mean heart rate of resting individuals at ambient temperature (22° C, which is the mean temperature experienced by this species while resting in its nest within stone walls), is approximately 835 min^{-1}, a value that can reach 1093 min^{-1} in active animals. The mean resting respiratory rate is 661 min^{-1} and the highest value recorded was 894 min^{-1}. The muscles contract at up to 900 min^{-1} for breathing, 780 min^{-1} for running, and 3500 min^{-1} for shivering. This species also has the highest mass-specific energy consumption of all mammals. At ambient temperatures, the Etruscan shrew consumes 267 ml O_2 kg^{-1} min^{-1}; that is, 67 times as much as resting humans (Jürgens *et al.* 1996; Jürgens 2002)!

all groups of organisms have undergone has indelibly marked their evolutionary trajectories and ecological attributes. Therefore, the relative merits of the arguments for and against convergence are worth considering briefly (see di Castri and Mooney 1973; di Castri *et al.* 1981; Blondel *et al.* 1984; Dallman 1998; Thompson 2005; Cowling *et al.* 2005; Médail 2008a).

Apart from the Mediterranean Basin, four other areas in the world are characterized by 'mediterranean-type' climate and ecosystems. These are central Chile, southern and central California, the Cape Province of South Africa, and disjunct parts of southern and south-western Australia (Fig. 3.6). As a result of general atmospheric circulation patterns and cold off-shore ocean currents, these areas are almost all located on the western shores of continents, between 30–35° and 40–43° latitude, and have an adjacent arid region to the south, north, or east of them.

If we compare the floristic diversity of the Mediterranean Basin with that of other areas with mediterranean-type climate, taking into account the much smaller surface areas involved there, then the Mediterranean appears as the poorest in number of plant species per unit area (Table 3.8). But it is second only to the Cape Province of South Africa in the percentage of endemic species. Despite their considerable difference in size, the floras of the two mediterranean-climate areas of the Northern Hemisphere (the Mediterranean region and California) have lower proportions (50 and 61%, respectively) of endemic plant species than those of both the Australian and South African regions, which are considerably smaller in size (Table 3.8).

Furthermore, the human histories of the five regions differ dramatically, as well as their impact on local floras and associated vegetation. Table 3.9 compares the relative duration of human occupation in the various areas. One of the prominent features of the Mediterranean Basin, as compared to the other mediterranean-type regions, is the considerably longer period of human occupation and intensive exploitation of resources (see Chapter 10). Consequently, human as opposed to non-human determinants of biodiversity must have been of paramount importance in the Mediterranean Basin. This idea can be presented in the form of an 'impact factor' (right-hand column of Table 3.9).

3.7.1 Convergence or serendipity

Given the particular features and constraints of the Mediterranean bioclimate, major patterns of ecosystem structure and function have been thought to involve different kinds of species and communities that independently acquire similar sets of adaptations in these different areas. The number of publications on **evolutionary convergence** between and among phylogenetically unrelated taxa and species assemblages is enormous (e.g. Cody and Mooney 1978; di Castri *et al.* 1981; Schluter and Ricklefs

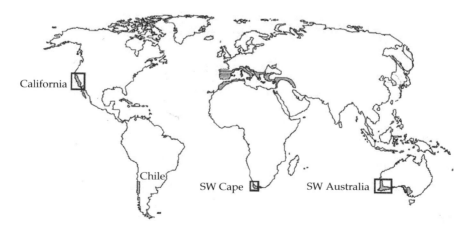

Figure 3.6 The four areas of the world with a Mediterranean bioclimate. Reproduced (slightly modified) with permission from Cowling *et al.* (2005).

Table 3.8 Plant species diversity (S), species/area ratios (S/A), number of endemic species (E), percentage of endemism (E/S), and endemism as a function of surface area (E/A) in the five mediterranean-climate areas of the world

Area	Area (A) (10^3 km^2)	Number of species (S)	S/A	Number of endemics (E)	Percentage of endemics (E/S)	Endemism rate (E/A)
Mediterranean	2300	25 000	10.9	12 500	50.0	5.43
California	320	3488	10.9	2128	61.0	6.65
Cape Province	90	9086	101.0	6226	68.5	69.18
Central Chile	140	3539	25.3	1769	50.0	12.64
South and south-west Australia	310	8000	25.8	6000	75.0	19.35

Source: After Médail (2008a).

Table 3.9 Human occupation histories of the major mediterranean-climate regions

Region	Arrival of indigenous people	Indigenous agriculture, livestock husbandry	First European settlement	Introduction of cultivated cereals and non-native livestock	Impact factor (time+habitat change)
Mediterranean	>500 000	10 000	n.a.	10 000	++++++
California	10 000	Scarce	239	150	+++
Cape Province	>200 000	Scarce	310	260	+++
Central Chile	10 000	Scarce	420	470	+++
South and south-west Australia	40 000	Nil	132	110	++

All figures are in years BP.
n.a., not available.

Source: Fox and Fox (1986).

1993; Hobbs *et al.* 1995; Verdú *et al.* 2003 among many others). Convergence in form and function has been unequivocally demonstrated in different mediterranean-type phyla, especially in plants, reptiles, birds, and mammals. Convergence at the community-wide level is more difficult to demonstrate and more controversial.

Perhaps the two most striking resemblances among the different mediterranean-type ecosystems are their remarkable floristic richness in relation to their geographic size and also the preponderance of evergreen sclerophyllous trees and shrubs belonging to unrelated plant families. This leaf type and the generally low shrubby vegetation structure usually associated with it are arguably a logical outcome of the bimodal Mediterranean climate, with the dry period falling at the hottest season of the year. Yet, as we discuss in Chapter 8, there are several ways to consider the evolutionary significance of sclerophylly. Many studies support the view that the morphology of plants and the overall physiognomy of plant communities are quite similar in the five regions with mediterranean bioclimates (e.g. Verdú *et al.* 2003).

However, other factors must also be considered, such as different histories of the biota and the effects of disturbances, such as fire and grazing, which vary from one region to another. For example, natural fire is a critical ecological factor in some but not all of the regions. It is much less common in California than in South Africa or southwestern Australia and uncommon in central Chile (Cowling *et al.* 1996, 2005; Médail 2008a). Similarly, grazing pressure over the millennia has not had the same importance in the five regions. Transformation of native ecosystems and biota has proceeded exceedingly fast in Chile, California, South Africa, and Australia, since European-style agriculture and livestock husbandry were introduced from 100 to 500 years ago. Yet the ecological and evolutionary impact of livestock grazing and browsing has been incomparably more profound in the Mediterranean Basin than in the other mediterranean-type ecosystems (Seligman and Perevolotsky 1994). Both the fires and grazing imposed over 10 millennia by pastoralists and agriculturalists have certainly contributed in important ways to the floristic diversity and to the shrubby vegetation structures of the Mediterranean Basin. The same cannot be said for the other mediterranean-climate areas.

If a certain degree of convergence can be seen for the dominant evergreen trees and shrubs, no such trend is found for the understorey plants for which soil types appear to play a more important role than climate. In South Africa and south-western Australia, small coastal strips of sedimentary limestone occur in a larger matrix of metamorphic parent rocks. In central Chile and California, there is hardly any limestone at all, but in the Mediterranean Basin limestone is by far the major type of parent rock. These **pedological** and **lithological** differences among the different mediterranean-climate regions are reflected in differences in the vegetation. For example, the soils of the Australian and South African regions are extremely poor in phosphorus and other essential elements. As a result, plants there have evolved diverse mechanisms in their root systems, and diverse mutualistic relationships with fungi and bacteria to facilitate the absorption of nutrients (Lamont 1982).

3.7.2 The end of a myth?

In spite of the enthusiastic impetus given to studies of convergence, many authors have challenged the generalization of convergent evolution in regions with mediterranean-type climates, pointing out the many divergences among the regions (e.g. Shmida and Whittaker 1984; Blondel et al. 1984; Babour and Minnich 1990; Hohmann et al. 2006). As shown above, these differences may be related to factors, such as soil fertility, fire, topographic and climatic heterogeneity, human occupation histories, and, perhaps more importantly, the evolutionary history of the taxa involved. Indeed, two non-human historical determinants must also be considered; that is, sorting processes and phylogenetic constraints, or the conservation of ancient traits that may have lost their adaptive significance. Two examples will be given to illustrate the importance of these historical effects in explaining composition patterns and community structure in mediterranean-type ecosystems.

3.7.2.1 SORTING PROCESSES IN WOODY PLANT TAXA

Basically, the idea of convergence of living systems among the different mediterranean-type regions of the world leads one to predict that organisms would evolve characters that are adapted to a mediterranean climate, whatever their origin and history. The convergence hypothesis predicts that the large variety of growth forms and life-history traits that characterize extant mediterranean-type floras results from evolutionary responses to a mediterranean-type bioclimate. Alternatively, if phenotypes and life-history traits of plant taxa that evolved *before* the establishment of a mediterranean climate still persist *after* the large climatic and ecological changes associated with the appearance of this climate, then historical factors must be considered of prime importance. Extant Mediterranean assemblages of plants include species that originated at different geological times and, for many of them, under tropical conditions. The question thus arises: do these co-occurring taxa sets share morphologies and life-history traits specifically adapted to present-day Mediterranean conditions? If so, extant plant assemblages would be the result of sorting processes, keeping only Mediterranean-adapted species among plants of different origins.

Analysing variations in life-history traits (e.g. summergreenness, **spinescence**, sclerophylly, sexual reproductive systems, and seed dispersal) among the woody plant flora of Andalusia, southern Spain, Herrera (1992) found no evidence of differential extinction events among pre-Mediterranean genera in the regional flora he analysed. This means that certain traits that are observed today already existed among the set of woody plants present in the area when the climate was tropical in the late Miocene and early Pliocene. Life-history traits, which evolved under tropical conditions, have largely survived as 'ecological phantoms' despite a dramatic shift in overall climatic regime. In contrast, all 'new' taxa that differentiated after the establishment of a mediterranean climate clearly evolved adaptations to Mediterranean climatic conditions. This example is an illustration that observed traits are not always adaptive and do not necessarily fit the 'adaptationist programme', as pointed out by Gould and Lewontin (1979). The conservation of ancient traits in modern plant species casts a doubt on the hypothesis of convergence as an evolutionary response to similar bioclimatic conditions.

3.7.2.2 CONVERGENCE AMONG BIRD COMMUNITIES

Evolutionary convergence among bird communities of different areas with a mediterranean bioclimate was studied by Cody (1975) for California and Chile, and by Blondel et al. (1984) for Mediterranean France, as compared to California and Chile. Blondel and colleagues compared **ecomorphological** configurations of the bird species, in particular their size and shape in relation to foraging habits, among different mediterranean-type regions. The basic assumption in this study was that ecomorphology reflects the different means by which a given species or guild of species utilize food resources. Comparisons were made among mediterranean bird communities along matched habitat gradients of increasing complex vegetation structure, from shrubland to forest. These data, involving 31 species in France, 31 in California, and 38 in Chile, were then compared to a non-Mediterranean gradient in France (Burgundy, 42 species), which was chosen as a control group. Convergence would be demonstrated if there were more overlapping or similarities among the ecomorphological 'space' of mediterranean communities than between any of them and those of non-Mediterranean Burgundy. Statistical analyses revealed considerable overlap among all four regional 'spaces'. This result was interpreted to mean that the Mediterranean communities do not resemble each other any more than the non-Mediterranean control. Only the hummingbirds of Chile and California scored differently on the morphological 'space', but this is not surprising as this group has no equivalent in western Europe. The lack of convergence was attributed to differences in the phylogenetic origin and the biogeographic history of the different sets of species. Just as for woody plants in Spain, morphological, physiological, and behavioural constraints on lineages of different origins and history presumably had more influence on species assemblages and species-specific habitat requirements than their sharing a similar type of environment.

However, conclusions from tests of convergence depend on the level of similarity and the choice of variables used. Clearly, convergence may exist for some community attributes, such as total numbers of species (Schluter and Ricklefs 1993), but not for others, such as those that involve large evolutionary changes in species and genera. Although convergence in morphology, structure, and, presumably, ecological function is most likely to take place among groups of organisms that depend strongly on climatic variation and seasonal patterns of nutrient cycling (e.g. plants, invertebrates, and lizards), it is less likely for homothermous vertebrates that rely more on the structural attributes of the ecosystems in which they live. Thus, convergence in such attributes as species richness and community structure may be an epiphenomenon of similar patterns of resource-sharing in relation to the structure of habitats. As for plants, historical and phylogenetic constraints may limit adaptation such that some animals may not always evolve life-history traits tightly adapted to the particular environment in which they now occur.

Having here reviewed the vast panoply of terrestrial biodiversity, in the next chapter we will discuss present-day marine biodiversity and the environment in which it occurs.

Summary

Trying to summarize the diversity of plant and animal species which are known to currently occur in the Mediterranean Basin was not the aim of this book, which instead focuses on processes rather than on patterns. Therefore we tried to point out the factors that best explain the Mediterranean specificities of some important groups. For example, why are there so many endemic species in several groups, such as phanerogams, fish, reptiles, amphibians, and many groups of invertebrates, and so few in other groups such as birds? Why has only a tiny fraction of the bird fauna differentiated within the limits of the Mediterranean Basin and thus can be considered as endemic to the region? The answer lies in the interaction of many factors among which the history of evolutionary differentiation in relation to dispersal and geographical aspects of the basin played a crucial role. For example, the fact that Mediterranean forests have not been isolated from temperate forests in the western Palaearctic during the alternation of glacial and interglacial episodes explains that no opportunities occurred for allopatric speciation in birds.

Dispersal capacities of organisms obviously explain much of the differences that are found among groups, in terms of distribution and biogeographic origin. For example, differential dispersal capacities explain such typical features as an exceptionally high species impoverishment of freshwater fish in most Mediterranean islands, or the fact that the only groups of vertebrates which include a large number of species of Afro-tropical origin is the mammal fauna in North Africa.

Since other regions on Earth enjoy a mediterranean-type bioclimate, an interesting question is to test the popular hypothesis of evolutionary convergence of biotas at the scales of species and communities among these different mediterranean-type ecosystems of the world. Although clear cases of evolutionary convergence are indisputable at the level of species, studies on convergence at the scale of communities have provided much more mitigated results.

CHAPTER 4

Present-Day Marine Biodiversity

Just as is the case in many terrestrial groups of Mediterranean organisms, biological diversity of marine plants and animals in the Mediterranean Sea is extremely high (Bianchi and Morri 2000), with between 20 and 25% endemism in plants (Cabioc'h et al. 2006), 11% among fish, and between 18 and 50% among invertebrates depending on the groups. The Mediterranean Sea is one of the richest seas in the world in terms of biodiversity, harbouring 5.6% of the world's marine animal taxa and 16.9% of the marine flora, in a relatively small area equal to only 0.82% of the surface of the 'world ocean' (Bianchi and Morri 2000). The number of animal species in the Mediterranean Sea ranges from 2.2% of world species for echinoderms to 22% for sea mammals, according to current estimates. One bias which could overemphasize these figures, however, is that the Mediterranean is much better explored and studied than most seas or oceans, with well-established marine research stations in Italy, France, Spain, Greece, Crete, Israel, Algeria, and Tunisia, among others. Summarizing data from various sources, Bianchi and Morri (2000) estimate there are roughly 8500 macroscopic marine organisms in the Mediterranean Sea today (Table 4.1), but Briand (2002) proposes the much higher figure of 12 000 species.

As in terrestrial biota, Mediterranean marine biota are composed of species with many different biogeographic origins: Atlanto-Mediterranean, pan-oceanic, palaeoendemic of Tethyan origin, neoendemic, and subtropical. An additional group invaded the eastern part of the sea when the newly dug Suez Canal connected the Mediterranean Sea to the Red Sea at the end of the nineteenth century (see Chapters 2 and 12). Thus, the sea is also a crossroads for myriad marine life forms, just as the lands surrounding it are for land biota.

Several historical factors have contributed to the high biodiversity of the Mediterranean Sea. These include the variety of climatic and hydrological conditions in the western half of the sea relative to the eastern half. In the western half, primarily temperate-zone biota are found, while in the latter there are many subtropical species. However, although the Mediterranean Sea is a remnant of the warm equatorial Tethys Sea, the Messinian Salinity Crisis, which led to a nearly complete desiccation of the sea and resulted in mass extinction of the former marine biota, caused the disappearance of many or most palaeotropical elements (see Chapter 1). It was not until the re-opening of a connection with the Atlantic Ocean, about 5.3 mya, that the Mediterranean was repopulated by biota of Atlantic origin. Thus, from a biogeographic point of view, the Mediterranean Sea is fundamentally part of the Atlantic province described by Briggs (1974). But a travelling **pelagic** (open-sea) larva of an Atlantic **benthic** (near-bottom) species passing into the Mediterranean Sea will find quite different life conditions with regards to salinity, temperature, and currents. Further, it must adapt to this new environment or die (see Chapter 9). The location and the dimensions of the passageway, or corridor, to this new 'inland' sea environment will greatly affect the larvae's chances for survival.

A mere 14 km wide and 300 m deep, the Straits of Gibraltar are nonetheless a true barrier, or rather a physical threshold, which separate the oceanic domain from the Mediterranean one, and account for many of its particularities. This results in a large number of features found nowhere else in the

Table 4.1 Species richness of some groups of aquatic plants and animals in the Mediterranean Sea

Group	Number of species	Percentage of the world species richness
Plants		
Red algae	867	16.5
Brown algae	265	17.7
Green algae	214	17.8
Seagrasses	5	10.0
Vertebrates		
Cartilaginous fishes	81	9.5
Bony fishes[1]	532	4.1
Reptiles	5	8.6
Mammals	21	18.4
Invertebrates		
Sponges	600	10.9
Cnidarians	450	4.1
Bryozoans	500	10.0
Annelids	777	9.7
Molluscs	1376	4.3
Arthropods	1935	5.8
Echinoderms	143	2.2
Tunicates	244	18.1
Other invertebrates	≈550	4.1

[1]Golani *et al.* (2002) estimated that there are 650 fish species.
Source: Bianchi and Morri (2000).

world ocean, such as a water budget operating at constant deficit due to the high rates of evaporation, which is just the opposite of the situation prevailing during the ice ages. Other features of high importance are the dual climatic influence (temperate-continental to the north and subtropical in the south), highly irregular coastlines in the north and much more homogeneous ones in the south, important average depth of the water, and general absence of tides (see Chapter 1). In this context, it should be emphasized how Aristotle was fascinated by the Mediterranean and its inhabitants (Box 4.1).

As they pass into the Mediterranean across the Gibraltar threshold, pelagic larva discover a coastal environment with limited continental shelves, no tides or tidal currents, and often steep slopes just a few metres from the shore. Coming from the deep Atlantic, these larvae will not find the low temperatures of their natal habitat, but, if they succeed in surviving, their potential for vertical extension of

Box 4.1. Aristotle (384–322 BC), father of zoology

Son of a rural doctor, Aristotle was a philosopher, logician, political thinker, and biologist, with special interest in medicine, physiology, and zoology. From ages 17 to 37, he enrolled in Plato's school of philosophy, the Lyceum in Athens, where he studied the mathematical sciences, medicine, and biology. After Plato's death in 347 BC, he left the Academy to travel for a dozen years, during which time he spent 2 years on the Aegean island of Lesbos, studying natural history, especially marine biology, with Theophrastus. This interest, shared with fishermen and other people 'of the sea', stayed with him until the end of his life.

After the accession to power of his former pupil, Alexander the Great, in 335 BC, Aristotle returned to Athens to become director of the well-named Peripatetic School, a position he held for the last 12 years of his life. In his published treatises, the part devoted to biology constitutes almost a third. His great innovation was that in each field of inquiry he approached, he combined observation with experimentation, and always followed what today we call a scientific procedure. In *De Partibus Animalium*, he was the first author to define the attributes common to whole groups of animals and attempted to provide a logical explanation to the recurrent 'forms', and other common attributes. Aristotle may be considered as 'father' of zoology just as Theophrastus was the father of botany (see Box 8.1). Aristotle's system of classification of living creatures persisted almost unchanged until Count de Buffon proposed a better one, in the mid-eighteenth century. Aristotle's greatest regret was to be unable to explain the mechanisms of marine tides, something that fascinated him. As the ecologist R. Margalef (1985) put it, Aristotle had exceptional knowledge about the creatures of the Mediterranean, a knowledge that no doubt incorporated the wisdom of his forerunners, his teachers, and his friends among fishermen.

their habitat will be very large indeed, thanks to the homogenous temperature of deep Mediterranean waters.

Thus the pelagic and benthic animal populations of the Mediterranean Sea have to be flexible enough to cope with or to become adapted to the new biotic and abiotic conditions they find there. If the connection with the Atlantic Ocean has allowed a steady input of marine organisms, strengthened by geological history (e.g. the Messinian Salinity Crisis; see Chapter 1), the ecological peculiarities of the Mediterranean Sea have favoured the development of specific populations, especially by benthic organisms, as illustrated by the abundance of neoendemic species, both in the fauna and the flora. The diversity of coastal environments generates numerous localized habitats (see Chapter 6) and contributes to a high level of endemism, which is favoured by the isolation of populations.

Despite the very high species richness of most groups of marine organisms (Table 4.1), including sea turtles (Box 4.2), population sizes are usually low in Mediterranean Sea biota (cf. Box 11.6 on the monk seal). For example, one never sees here the kind of spectacular seabird colonies that are so characteristic of steep cliffs in the North Atlantic; the same is also true for dolphins. This is due to the relatively low productivity of the sea and relative narrowness of the continental shelves, which also explain why most of the basin's fisheries are relatively minor as compared to those of the Atlantic.

4.1 Flora

Plants need light for their photosynthesis and therefore they only occur in zones which are penetrated by light. In the marine environment, the **euphotic** or well-lit zone is vertically restricted and rarely extends more than 100 m below the surface (see Chapter 6). Thus all primary production takes place in this relatively narrow zone of the sea. The marine flora of the euphotic zone can be divided into the pelagic and the benthic cohorts or guilds, which present very different life-history strategies (see Chapter 9). However, we have to keep in mind the fact that a part of the biomass produced here is 'exported' to the deeper, no-light (aphotic) zones near the sea bottom. These zones are thus trophically dependant on the euphotic area for organic matter. Considering that 90% of primary production—that is, plant biomass—is recycled in the euphotic layer, it is only a small portion of the primary and secondary pelagic biota which, after a long process of sedimentation, reaches the deep, no-light zones, which represents 95% of the volume of the sea! As this detritus constitutes the unique organic supply for the deep-sea zones, actual biomass production there is poorer than the Saharan desert, in terms of dry weight of living matter per

---

Box 4.2. Sea turtles

Five species of sea turtles occur more or less commonly throughout the Mediterranean Sea, among which the caouan (*Caretta caretta*) and the green sea turtle (*Chelonia mydas*) are the most abundant. They feed in deep waters upon a variety of animals, mostly deep sea invertebrates (salps) and jellyfish, but they also venture into more shallow waters where they prey upon crabs, urchins, and molluscs. The large lute turtle (*Dermochelys coriacea*), which can weigh up to 500 kg, formerly bred in the Mediterranean, but is only a rare visitor there today. Two other rare visitors are the hawksbill sea turtle (*Eretmochelys imbricata*) and the Kemp turtle (*Lepidochelys kempii*). The caouan still regularly breeds in the eastern Mediterranean, the Ionian Islands, Peloponnesus, southern Turkey, Cyprus, and Israel, as well as in Libya and possibly Tunisia. Since the beginning of the twentieth century, however, many breeding sea turtle sites have been deserted or destroyed, as a result of tourist and industrial encroachment combined with fishing accidents. For both these species, the destruction of breeding sites, for example in Malta, Sicily, Sardinia, and Corsica, has led to a sharp decline in populations (Delaugerre 1988). Only the green sea turtle still breeds in fairly large numbers on Turkish beaches, and to a lesser extent in Israel and Cyprus.

square metre or hectare (Cabioc'h et al. 2006). We now examine the two components of the marine flora in more detail.

4.1.1 The pelagic flora

Pelagic primary production is carried out by huge quantities of microscopic unicellular organisms floating just under the sea surface. The largest organisms—20–200 µm in diameter—are called phytoplankton, the intermediate-sized organisms—between 2 and 20 µm—are the nanoplankton, and the smallest organisms include the picoplankton, which range between 0.2 and 2 µm in size. Only discovered in the late 1970s, the picoplankton represents 10–70% of the biomass of the marine vegetal plankton (the annual mean is 43%) (Lantoine 1995). In this group occurs the smallest (0.8 µm) known free eukaryote, *Ostreococcus tauri*, discovered recently in the Thau lagoon, west of Montpellier, France (Courties *et al.* 1994; Chrétiennot-Dinet *et al.* 1995). The lower the nutrient quantities, the more picoplankton is abundant, generally speaking. Thus, as the Mediterranean Sea has fewer nutrients per volume of water—is more **oligotrophic**—in the eastern than in the western half, the abundance of picoplankton is greater in the eastern half.

Given their very small size and the very large numbers of cells involved, the density of phytoplankton is generally estimated using a proxy, namely the content of **chlorophyll *a*** — which of course is the most common pigment of the plant kingdom—present in a fixed volume of water. Measured in micrograms per litre or milligrams per cubic metre, chlorophyll *a*—often denoted Chl *a*—concentrations indicate the relative quantity of phytoplankton in seawater, either through direct chemical dosages or, quite frequently nowadays, by remote sensing. The areas richest in phytoplankton, with a Chl *a* content higher than $0.5\,\text{mg m}^{-3}$, are all coastal waters, such as the north-western part of the Alboran Sea, between the south-eastern coast of Spain and the Mediterranean coast of Morocco, as well as the Gulf of Lion, the Gulf of Gabès, the north-western portion of the Adriatic Sea, the northern Aegean, and the Nile delta. Values for the western half of the basin range from 0.15 Chl *a* mg m^{-3} in the centre, to $0.5\,\text{mg m}^{-3}$ near the coasts. In contrast, apart from the specific situations mentioned above, all the waters of the eastern part of the basin have less than $0.15\,\text{mg m}^{-3}$ and the values for the south of the Ionian Sea and the Levantine Basin are still lower, between 0.04 and $0.06\,\text{mg m}^{-3}$ (Jacques 2006).

To understand the global significance of phytoplankton in the seas, it is important to realize that 'phytoplankton accounts for only 1–2% of the total global biomass, but may fix between 35 and 45 Gt (Gigatons, i.e. billion tons) of carbon each year, i.e. not less than 30–60% of the global fixation of carbon on Earth' (Sakshaug *et al.* 1997). These impressive figures concerning the disproportionate role of phytoplankton in global carbon budgets give an indication of the huge potential productivity of these organisms and their importance in marine food chains, and, by extension, for the numerous planetary processes. With this in mind, we can also better understand that the oligotrophy of the Mediterranean waters, especially in the eastern half, has a strong impact on the biological processes of the sea itself.

Biodiversity is also affected by this problem because less primary production leads to a decline in small grazers, which are preyed upon by young predators, and so on. The seaweeds (see below), which take their nutrients directly from the water, are also affected. Another matter of concern is that this process can also open up ecological niches that invading Lessepsian species (see Chapter 12) can occupy. The strong diversity of calanoid copepods in the Levantine Basin is an example. But it is much more dangerous for the **benthos**, where there can be a competition for space.

4.1.2 The benthic flora

There are four types of benthic marine flora in the Mediterranean Sea, which is much more diverse than the pelagic flora. The first two are lichens and unicellular diatoms, which we treat rather briefly. The third and fourth groups are the Mediterranean algae, known by the common name of seaweeds, and various vascular phanerogams, all of which are monocotyledons and known collectively as seagrasses. We will discuss these latter two types in more detail.

Lichens grow in abundance in the sea-spray zone, where no other kind of plant survives. Being independent of the tides, lichens are found on all the European coasts with the same or vicariant species to those found along the Mediterranean coasts, where the most well known is *Verrucaria amphibian*, which makes black patches on the rocks. They are very present and widely represented in the Mediterranean region with continental and coastal species (Mies and Feige 2003).

The second benthic plant type consists of unicellular diatoms, called microphytobenthos. Contrary to seaweeds, which are limited to hard substrates, these microscopic organisms are present on all the sea bottoms of the **phytal** zone, as well as on the sandy, muddy, and rocky substrates. Sadly, the only context in which these organisms are visible to the human eye is on the dirty windows of neglected aquariums where they form a brownish or greenish veil. The biodiversity of these benthic diatoms is great (Haubois *et al.* 2005), and their productivity in terms of photosynthesis and carbon fixation is indeed enormous. In the shallow waters, benthic diatoms combine with pelagic diatoms after re-suspension by turbulence and together they can constitute half of the microalgae present in the water column (Guarini *et al.* 2004). As a result, these micro-organisms play a major role in global carbon cycles and also have great value as ecological indicators of global climate change. Colonizing a small fraction of the total Earth's surface (smaller than the accuracy of the ecosphere surface estimate), they could however generate a flux corresponding to the 'missing carbon sink' (Guarini *et al.* 2008).

4.1.2.1 THE SEAWEEDS
In all marine environments of the planet, the coastal hard bottoms are colonized by algal 'seaweeds', limited in their growth in a downward direction by decreasing quantities of sunlight and in an upward direction by the need for constant moistening. These particularly common and familiar seascapes vary in terms of taxonomic composition, biomass, and specific distribution in response to the specificities of their settling site. Being **sessile**, seaweeds need a support, which is generally a rocky substratum, but can also be other seaweeds, Neptune grass (*Posidonia oceanica*) leaves, or even the shells of living molluscs, such as limpets, or spider crabs. An important detail to note is that they have no roots, only clamps, and they take their nutrients directly from the water.

In the Mediterranean, the algal flora evolved through the successive geologic eras from the Tethys Sea epoch until the present, including upheavals such as the Messinian Salinity Crisis (see Chapter 1), when the cascade of incoming Atlantic waters brought in huge quantities of spores and stem-segments of Atlantic algae species that were joined later by boreal or subtropical species during glacial and interglacial episodes. The major part of the benthic flora is of Atlantic origin. The Indo-Pacific forms of seaweed are scarce, apart from one common species of the intertidal zone, *Rissoella verruculosa* (Cinelli 1985). All present-day species have by definition become adapted to life in the Mediterranean, including the quasi-absence of tides, unusually low nutrient content in the water, and relative stability of temperatures along the vertical **thermocline**, contrasting with significant differences in surface temperature from one season to the next.

A striking characteristic of this benthic flora is the strong percentage of endemic species, amounting to 20–25% of the species as mentioned above. Considered to be neoendemics, for the most part they are the result of adaptive radiation that has taken place in the unique and varied Mediterranean biotopes. A second special feature of marine Mediterranean environments is the quasi-absence of very large species of seaweeds such as they commonly occur on the shores of the world's oceans. This is a consequence of limited intertidal areas and above all the low quantities of nutrients in the water. In fact, a large proportion of the Mediterranean seaweeds are very small species, with the exception of *Laminaria ochroleuca*, which may reach 6 m in length, and *Saccorhiza polyschides*, which may attain 4 m. These 'giants' are both quite localized in distribution: in the Alboran Sea which is under direct Atlantic influences and in the Straits of Messina, between Sicily and the Italian mainland, where unusually strong currents produce hydrological conditions similar to those found in the Atlantic. However, species diversity is very high indeed, with 540 species near the French–Spanish border (Boudouresque *et al.* 1984), 505 along the coasts of

Corsica (Boudouresque and Perret-Boudouresque 1987), and 468 along the Algerian coast (Perret-Boudouresque and Seridi 1989). Boudouresque and Perret-Boudouresque (1979) report similar diversity for Italy (542 species), but much lower numbers in Greece (370 species) and Turkey (173 species), presumably as a result of the decreasing west–east nutrient gradient of the sea.

The most spectacular seaweed diversity occurs near the shores in places where continental supplies are abundant and the waves permanently renew both nutrients and oxygen in the seawater. Below the level where the waves beat against the shore and in the sheltered areas, seaweeds are always present with lower densities and biomasses. The species are rather flimsy for the most part, one of which, *Acetabularia acetabulum* (ex *Acetabularia mediterranea*), is particularly interesting. The thallus and stalk of this little parasol-shaped species consists of a single cell including the nucleus, making of it a very useful and convenient subject for cell biologists and other scientists. Notably, in the 1930s, the German biologist J. Hämerling used this unicellular alga to elucidate the function of the cell nucleus. This alga is sub-endemic to the Mediterranean Sea, being found also along coastlines in the Canary Islands.

Seaweeds are highly diverse in their morphology, with morphotypes being clearly correlated with the characteristics of their habitats. They can be filamentous (*Chaetomorpha*), hollow and tube-shaped (*Enteromorpha*), with a soft and thin thallus (*Dictyota, Ulva, Porphyra*), more or less stiff and more or less branched (*Cystoseira, Codium*), striped (*Laminaria*), in the form of little vesicles (*Valonia*), or pad-shaped (*Colpomenia*). They can be incrusted on the bottom, but not calcareous like the seaweeds of the intertidal belts (*Ralfsia, Nemoderma*). Another important group in the Mediterranean is the so-called calcareous algae, with some species showing moderate calcification—for example, *Halimeda* and *Corallina*—and others with a remarkably stone-like thallus. Some of these particularly hard species produce vegetal bio-accretions that constitute two biotopes that are unique to the Mediterranean. The first is found in the intertidal zone and the other in the low-light zone; both consist of limestone remains of dead thalli superimposed one upon the others.

The habitat of the intertidal zone (see Chapter 6) occurs on the vertical rocky shores along exposed coasts, at the lower part of the zone. It comes from the development of a calcareous alga, *Lithophyllum byssoides* (ex *Lithophyllum lichenoides*), that forms small, very adhesive, hemispherical cushions, a bioconstruction called by oceanographers trottoir (the French word for pavement or sidewalk), due to its pavement-like appearance. In the course of many years, dead thalli accumulate to form a ledge up to a metre thick.

The trottoir is composed of a superficial part of living *Lithophyllum*, which covers a thin layer of dead algae not yet compacted. At its core, the bulk of the ledge is made up of a compact structure of cemented dead thalli (Fig. 4.1). Growing only along exposed coasts, it is constantly moistened by the waves at its lower part. The water then rises by capillarity in this porous environment from the bottom of the trottoir to the top, where it evaporates off the surface. This movement maintains a constant moisture level and reduces temperature variations. The lower part is often attacked by wave erosion or by organisms which excavate the limestone. Protected from direct light, it is colonized by low-light-zone seaweeds and one can sometimes find small coralligenous concretions (Plate 7a).

This particular habitat is quite widespread in the western basin, but seems to be absent or very rare in the eastern basin, presumably because sea temperatures are too high. The quasi-absence of tide allows *L. byssoides* to elaborate this unique habitat, which is a meeting place for terrestrial and marine faunas. Fortunately, its ecological requirements keep it in locations where human activities are not too threatening, but with the development of tourism, they are a point of easy landing for the small boats and walkers can alter its surface. The danger also comes from big cities and from their chemical effluents because its porous structure makes it very sensitive to surface pollution (oil, cleaning agents, heavy metals). The trottoir declines sharply in the vicinity of big cities.

In the low-light zone, another bio-accretion called coralligenous is produced mainly by the activity of three species of red seaweeds from the Corallinaceae family—*Pseudolithophyllum expansum, Neogoniolithon mamillosum,* and *Mesophyllum*

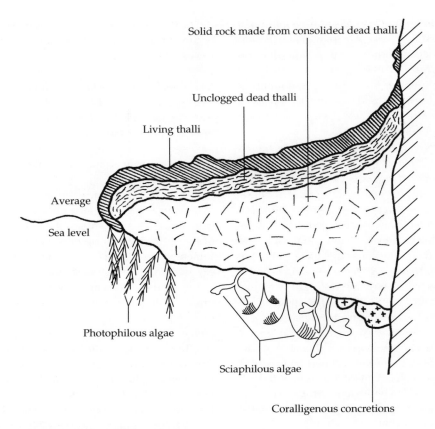

Figure 4.1 Section of a trottoir of *Lythophyllum byssoides*.

lichenoides—each of which is made up of approximately 85% pure carbonates. Another prominent family, the Peyssoneliaceae, is also present with algae whose carbonate content of approximately 75%. The hard thalli of all these species accumulate on top of each other successively when they die. Together with the remains of the fauna in place, they constitute calcareous bio-constructions on the hard substrates of the low-light zone. There are often small, isolated coralligenous nodules occurring on isolated rocks, but the really important coralligenous formations are found in the form of large terraces several meters thick, dissected only by narrow spaces. These formations can cover large surfaces of several hectares in size if ecological conditions are favourable; that is, when currents are low and constant. Laubier (1966) studied two large terraces, 0.34 and 0.58 km^2 in size, in the south of the Gulf of Lion, very near the French–Spanish border, where they are sheltered from the direct influence of the Liguro-Provencal current by the so-called Bear Cape. The terraces are covered at their top by the red encrusting algae mentioned above. They continue to elaborate them, taking advantage of the low ambient light. As for other bio-accretions, the central part of the coralligenous concretion becomes compact, but a layer about 50 cm thick remains porous, thanks to a large network of waterways, crevices, and cavities where life is possible (see Plate 7a).

A third group of hard calcareous algae occurs in the high-light zone, the most common being *Lithophyllum incrustans*. It does not build any structure but is present wherever it finds a place to settle. It is naturally more abundant in the non-exposed places where the density of seaweeds is low. It occupies all the available space on the substratum and gives pink colours to the sublittoral rocks, pebbles, and the bottom of pools. This species does not tolerate emersion, and many white calcareous spots seen on the rocks correspond to little pools of water having evaporated during summer, entailing the death of these algae.

Finally, an unusual type of calcareous algae is found on the soft bottoms of the low-light zone, where very young individuals of two different calcareous species, *Lithothamnion calcareum* and *Lithothamnion coralloides*, settle above a little hard particle and surround it completely when growing. These algae are free, not sessile, and 'sit' on the sea bottom, forming a mass somewhat similar to the biogenic substrate called *maërl* on the Brittany coast of France. They are not rare and can be found on detritic bottoms under strong currents (Pérès 1967), near the French and Italian coasts, in the northern Adriatic and the Aegean seas, and in the Tunisian Gulf of Gabès (Augier 1973).

The algae are an important food resource in coastal marine ecosystems, although no more than 10% of the biomass of seaweeds is directly consumed by **herbivores**. For the other 90%, they rejoin the detritic food web, thereby contributing indirectly to the animal biomass (Graham and Wilcox 2000). The secondary production is essentially carried out by small organisms; that is, members of the little macrofauna—crustaceans and annelids—and the meiofauna.

Recent immigrants invaded these habitats (see Chapter 12). The Lessepsian species invading the Mediterranean Sea from the Red Sea are the best known, but other taxa have been introduced recently as unwanted companions to cultivated oysters: 90% of the vegetal biomass of the Thau lagoon mentioned above is in fact of Japanese origin. Three Japanese species, *Sargassum muticum*, *Laminaria japonica*, and *Undaria pinnatifida*, are all currently escaping from the lagoon into the open sea. Finally two tropical species must be mentioned, *Caulerpa taxifolia*, which escaped from the public aquarium in Monaco, in recent years, and *Caulerpa racemosa*, which is of Lessepsian origin. These species are both great threats for the Neptune grass meadows which can be suffocated by their abundance. The small seaweeds (*Acetabularia* and many others) are threatened by these aggressive invasive algae, of which we will have more to say in Chapter 12.

4.1.2.2 SEAGRASSES

Marine phanerogams derive from marshland ancestors which returned to the marine domain more than 100 mya. After having evolved adaptations to thrive in marine environments, they remained very stable and retained their structure of vascular plants with a stem bearing typical roots and leaves and a strategy of sexual reproduction, including flowers, seeds, and fruits. They live and bloom under the surface of water. Their most striking feature is that their pollen is emitted in sticky filaments carried away from the plant's male flower parts by the current. They have also some adaptations for aquatic life such as the absence of stomata in the leaves. They possess a network of air-filled micro-vacuoles, called the aerarium, which provide buoyancy.

Belonging to one of four small families (Posidoniaceae, Zosteraceae, Hydrocharitaceae, or Cymodoceaceae), close to the more familiar Potamogetonaceae and Juncaceae families, they are organized with a stem called a rhizome, which bears roots and leaves arranged in sheaves (Fig. 4.2). They are commonly called seagrasses because of their long, narrow, and usually green leaves, and because they often form large stands or meadows that resemble terrestrial grasslands. Being photosynthesizers, they have to grow in the **photic** zone and tend to grow anchored in sand or mud bottoms in shallow coastal waters, where their foliage provides a large number of inviting shelters and ecological micro-habitats of particular importance for the smaller members of the benthic fauna.

Among the approximately 40 species of seagrasses in the world, five are found in the Mediterranean, of which three are cosmopolitan, *Zostera marina*, *Zostera nana*, and *Cymodocea nodosa*, one is endemic, the Neptune grass, and one is a Lessepsian species, *Halophila stipulacea*.

The three cosmopolitan species of Mediterranean seagrasses have some features in common, including a single rhizome that is always horizontal, running along the sea bottom, and long, narrow,

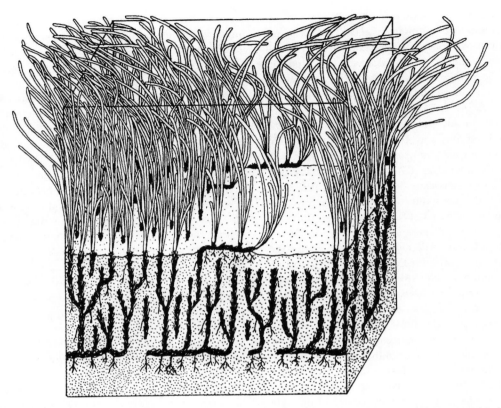

Figure 4.2 Schematic representation of a matte in a Neptune grass meadow, showing the position of rhizomes and leaves. Reproduced with permission from Boudouresque and Meinesz (1982).

flexible striped leaves. *Zostera marina* is much larger than the two other species with leaves over 1 m long. This species is abundant in the north Pacific and north Atlantic, but in the Mediterranean it is localized on the shallow bottoms of lagoons or near the estuaries. *Zostera noltii* and *Cymodocea nodosa* are much smaller plants, with leaves only 3–4 mm wide and up to 30 cm long. They look similar but can be distinguished by the leaves of *Cymodocea* which are finely dentate. Both are found growing in sheltered places such as the lagoons and the upper level of the high-light zone, often between the Neptune grass meadows and the shore because they are more tolerant to low-nutrient substrates.

The Lessepsian species, *Halophila stipulacea*, is different from the four other Mediterranean seaweeds in that it has oblong leaves, about 8 mm wide and 6 cm long, that always occur in pairs. This case will also be described in Chapter 12.

The endemic Neptune grass is the most important and emblematic Mediterranean phanerogam in terms of actual area it occupies, biomass, broad distribution, and habitat for many animal species. It is likely that prior to human impact along the coasts, Neptune grass meadows were much more extensive than they are presently. Notwithstanding, today they are present along all the coasts of the Mediterranean Sea and cover at least 38 000 km^2, corresponding to 3% of the basin's surface (see Plate 7b). Their habitat extends vertically from sea level down to the lower edge of the high-light zone (see Chapter 6). Neptune grasses' ecological amplitude and 'success' is due to the properties of its very robust rhizome, which can grow horizontally or vertically according to local circumstances. As it grows, it builds a kind of terrace, called a matte (Fig. 4.2), consisting of a tangling of roots, sediment, and various dead organisms. This structure grows continuously towards the surface because the rhizome extends from the bottom to the top of the matte where it bears the leaves which constitute the visible meadow. A 2000-year-old Roman wreck was found under a matte 3 m thick (Boudouresque and Meinesz 1982), and the broken terraces over 6 m thick, found offshore of Bandol (France), are probably much older. This plant could be among the oldest living organisms on the planet.

The leaves of the Neptune grass are arranged in sheaves of four to eight at intervals along the rhizome and reach 1 m in length and 1 cm in width. Each leaf can survive a full year, which is particularly important for sessile animals which can therefore complete an annual life cycle. The average density of leaves in a healthy Neptune grass meadow decreases gradually with depth, from approximately 7000 to about 2400 per m^2. They constitute an important concentration of biomass estimated between 3000 and 20 000 kg ha^{-1} according to the richness of the meadow, with the highest known values occurring in Malta and the bay of Naples. Their primary production is the most important of all the Mediterranean benthic biotopes, estimated at 12 000 kg ha^{-1} year^{-1} in the Port Cros Bay, France. For an average biomass including the rhizome of 35 t ha^{-1}, this rate of annual productivity is surprisingly similar to that of temperate forests, which have ten times more biomass (Boudouresque and Meinesz 1982). In sum, Neptune grass meadows constitute one of the fundamental ecosystems in the biology of the Mediterranean Sea. Its presence is particularly important for the populations of fish which can spawn safely here, sheltered by the leaves that will also hide their offspring from predators.

4.2 Invertebrates

The marine environment is home to members of all 34 or so phyla of invertebrates, 14 of which are exclusively marine. During their evolutionary development, they never have been able to leave the marine medium. It was said that the insects were absent of the marine environment, but thanks to the particular case of the trottoir which belongs to the marine domain, insects (Thysanura), myriapods, and spiders are also present.

The two major types of habitat of the marine environment, benthos and pelagos (see Chapter 6), impose quite different constraints, and offer contrasting prey for foraging and feeding inhabitants. Three main groups may be recognized: (1) holopelagic organisms, which spend all their lives in the pelagic domain, near the sea floor; (2) holobenthic organisms, which never leave the benthic domain; and (3) those that begin their life cycle (eggs and larvae) in the pelagic domain and then live as adults on the sea bottoms after a benthic 'recruitment'.

The holopelagic invertebrates are planktonic animals. As the constraints they have to overcome are associated to the necessities of floating and escaping predators, the best compromise is to be small, transparent, and lightweight. As a result of their fitness advantage to be light and highly mobile, shells and other heavy forms of protection gradually disappeared in the course of evolution. The Mediterranean plankton shows a pronounced annual cycle: low densities in winter, then proliferating in spring, leading to a 'bloom' of primary production, and reaching maximum numbers in May/June (Margalef 1985).

The Straits of Gibraltar, where water exchanges are considerable, are not a barrier for planktonic species of the euphotic zone. The plankton of the Mediterranean Sea is, for the most part, just a subset of Atlantic plankton, with no endemic species. Planktonic copepods are the best represented group, with approximately 500 species in the Mediterranean, which represents one quarter of the world total. There are 224 and 282 species in the western basin and in the Adriatic Sea respectively, 170 and 113 respectively in the north and the south of the Ionian Sea, 175 in the Aegean Sea, and 288 in the south-eastern Levantine Basin. This area was considered as the poorest area of the sea 50 years ago (Razouls *et al.* 2005–9), whereas today its high extant specific diversity is explained by the invasion and establishment of Lessepsian species, the number of which is currently increasing

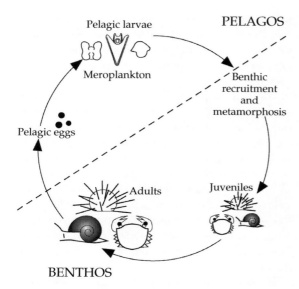

Figure 4.3 Bentho-pelagic life cycle of marine macrofauna. See text for details.

among all types of living organisms, especially the holoplanktonic species, for whom migration into the Mediterranean is particularly easy (see Chapter 12).

Based on body size, there are two kinds of benthic invertebrate: the **macrobenthos**, defined as being retained on a 1 mm mesh sieve, and the **meiobenthos**, which can pass through the sieve. This pragmatic distinction was biologically confirmed by Warwick *et al.* (1986), who recognized two distinct size groups, in a large number of benthic samples, using as sole criterion the capacity to be retained or not in the mesh. The meiobenthos is **holobenthic**, whereas the macrobenthos divides its life cycle into a pelagic stage as larva and a benthic adult stage (Fig. 4.3). Paradoxically the meiobenthos, which is narrowly linked to the sea bottom, does not present special or unique Mediterranean characteristics. For instance, some species of benthic copepods found in the sublittoral fine sands in the Gulf of Lion, France (Bodiou 1975), can be found on the coasts of Tunisia (Monard 1935) and Israel (Por 1964), in the Marmara Sea (Noodt 1955), and on the French Atlantic coast (Bodin 1977), always on a similar type of sea bottom. The meiofauna characterizes the substratum where it lives and could be a good ecological indicator of the quality and richness of the bottom, which is important in this context because of its contribution to the transfer of organic matter. The collective action of all these tiny animals is like a factory transforming the indigestible organic detritus of plants into digestible animal organic matter. Like the unicellular planktonic diatoms discussed above, the productivity of benthic diatoms, which are multicellular, is also exceedingly vast; as a result, the meiofauna is an essential link in the marine bottom's food chain.

As opposed to meiofauna that always remain in roughly the same habitat, macrobenthic animal species use the action of the currents for the dispersal of their larvae and the colonization of new substrates. But the search for favourable sites for colonization is of course unpredictable and, according to the success of recruitment, the same species can exhibit in the same location great differences in population size between years. Moreover, there are limiting factors imposed by the characteristics of the Mediterranean Sea, since the rocky shores are often steeply sloping and this decreases the size of available habitats. In addition, the limited number and width of continental shelves reduce the possibilities for soft bottom-dwelling species. Many larvae arrive in water that is too deep, and as they never find suitable sea bottoms to colonize they die. Thus, with a few rare exceptions, the active recruitment zone is limited to the proximity of the coast. It is the area of the primary production, with a maximum of detritic deposits from the continent, while the waves and the currents stir the water which is better oxygenated and richer in suspended particles. That promotes floral and fauna proliferation, particularly among certain groups, such as suspensivorous and microphageous organisms.

The planktonic stage had fundamental importance after the Messinian Salinity Crisis when the Mediterranean Sea filled with water from the Atlantic Ocean. The pelagic **meroplankton** larvae, having survived to the maelstrom of the gigantic waterfall, could again stock the deserted underwater areas without competition. That explains the dominance of Atlantic species in the benthic Mediterranean fauna and the number of neoendemic species which evolved after this event.

A peculiarity of this fauna is that many species have a smaller size than the same species occurring in the boreal Atlantic. The explanation seems to come from the higher mean temperatures of Mediterranean waters, which accelerates the onset of sexual maturity, by blocking growth in body size. The question of size and larval dispersal will be crucial in the context of global warming, since warmer water reduces the duration of larval stages and thus the distances covered by the larvae (see Chapter 13).

Benthic species settle down according to the characteristics of space and food availability. At the bottom of the high-light and low-light zones, they take advantage, as already said, of the primary production and also of the proximity of the land and of its deposits, as well as the wave action, which recycles the particles in suspension and oxygenates the water. The carnivores and active suspension feeders can live anywhere, the former because they are mobile and the latter because they can filter large quantities of seawater according to their needs. That is not true for passive suspension feeders, deposit feeders, and sediment feeders, among which there are many sessile or sedentary species that must collect their food *in situ*. For them, the ecological conditions of coastal areas are favourable because food particles are not swept away by tidal currents, but instead are regularly transported by the seawater. In this type of environment, the quantity of food is not a limiting factor and the species may appear in the form of colonies. First, it is not necessary to find a new place to colonize in each and every generation. Second, it helps save space, when the available surface area to settle on is limited. The protection is more efficient, the fecundation is easier, and the colonial life increases the foraging possibilities by incrusting forms, which create larger surfaces of predation (bryozoans, *Zoantharia*, etc.) or upright forms that catch prey or organic particles, higher up in the water column (e.g. gorgonians and other bryozoans). The food caught by one polyp feeds the whole colony. The best-known colonial species of the Mediterranean Sea is of course the red coral (*Corallium rubrum*; Cnidaria, Octocorallia) (see Box 4.3 and Plate 8a), which has been used by artisans since antiquity for carving jewellery and artwork.

Sessile species, whether colonial or isolated, are very exposed to predation. They cannot move, and their trophic traits oblige them to stay exposed without shelter while they take their food. In such a case the best defence is chemical and they develop this practice against predators by the presence of toxins, retained in their bodies, or else released into the immediate surroundings. Many bioactive molecules, each of them specific to one species, protect them against grazing. However, at the same time, a group of predators has developed individual **resistance** to these protections in the way that a given predator is insensible to the toxin of a given prey. They are essentially carnivorous gastropods which belong to this group and particularly beautiful sea slugs called **nudibranchs**, which feed on only one species of sponge, hydroid, bryozoan, or gorgon. Prey and predator constitute a kind of association in which the first one can forage without competition, while the second undergoes predation which remains limited, and thus acceptable. The best example of this kind of association is observed in the genus *Flabellina*, which feeds on hydroids and recuperates the intact cnidocysts in dorsal cerata (singular: **ceras**) to protect themselves from predation. No fewer than 35 species of nudibranch appear to be endemic (Tortonese 1985) of a total of 111 known to occur in the Mediterranean Sea (Schmekel and Portmann 1982). Another feature of note in these sessile species is that many have developed species-specific defence compounds from which chemists have isolated and purified very useful pharmacological molecules (Banaigs and Kornprobst 2007).

Concerning endemism, Pérès (1967) mentions the status of four groups (hydroids, crustacean decapods, echinoderms, and ascidians) in 1940 before the increasing immigrations of the Red Sea and notes that the proportion of endemic species is stronger in the groups of sessile species. Tortonese (1985) confirms these data with 18% for the crustacean decapoda, 22% for the echinoderms, 24% for the annelid polychaetes, 42% for the sponges, and 50% for the ascidians. Bellan-Santini (1985) notes that no **pelecypods**—a very important group in the soft bottoms—seem to be endemic and notes the presence of some endemic bathyal polychaetes and amphipods, but concludes that most species are of

> **Box 4.3. The red coral: Mediterranean jewel of the deep**
>
> The 'coralligenous' owes its name to the presence of red coral. In fact, this uniquely Mediterranean species is present on all hard bottoms where light is attenuated. This species occurs in the form of arborescent colonies, rarely exceeding 25 cm in height, but exceptionally attaining 50 cm. They are composed of a hard and calcareous skeleton covered with a matt red organic crust in which white eight-tentacled polyps are arranged in a regular fashion. The skeleton is built up by accumulation of calcareous spicules forming a hard, bright red, and very compact substance that retains its colour almost indefinitely. Very recently, a study carried out in Monaco demonstrated that this red colour is due to astaxanthin, the classical carotenoid best known in arthropods (Tsounis 2009). This longevity of the colour explains why red coral has been prized by artisans throughout the Old World since prehistoric times. Pieces of coral jewellery found in Egyptian tombs have still not faded in colour.
>
> Sadly, the same fine art objects and jewellery that have made red coral world famous may also one day be the cause of its disappearance. Currently, between 25 and 30 t of coral are harvested each year in the Mediterranean sea (Tsounis *et al.* 2007), including 4–5 t from Spain, especially along the Costa Brava (Cap de Creus and Medas Islands). One problem is that selective fishery pressure on the large colonies influences the size/age distribution. Even in deep waters where scuba divers can harvest coral—both law-respecting fishermen as well as poachers—91% of the colonies are smaller than 5 cm in height (Tsounis *et al.* 2006). As it appears that only large colonies are able to produce large quantities of larvae, there is cause for concern as to whether the reproductive potential of small colonies is sufficient to ensure their survival. If we add that the coral releases its larvae in July while the harvesting season begins in May, with the catching of the older colonies, and that even the small colonies are endangered because jewellery makers can now use pulverized fragments, it is to be feared that, due to overfishing, the red coral is steadily declining in shallow waters down to 100 m. Fortunately this species can live up to 300 m depth, but the rare colonies found in deep waters will never be sufficient to repopulate the shallow water and near-coastal zones.

Atlanto-Mediterranean origin. The biodiversity of the upper levels is remarkable and from the some 3000 species of the benthos only 700 have been collected at least once below 200 m (Fredj and Laubier 1985).

Due to its hydrology, the Mediterranean Sea is particularly rich in sessile invertebrates on its coastal rocky bottoms in both the high-light and low-light zones, and this faunal richness is the source of the beauty of its submarine seascapes (see Plates 7 and 8). By contrast, the coastal sandy bottoms and the deeper muds are more homogeneous, with a high percentage of faunal elements shared with the Atlantic.

Two zoological curiosities found in the Mediterranean Sea must be mentioned. First, an interstitial crustacean found on some shores of the western basin in France, Italy, Algeria, and Tunisia, the mystacocarid *Derocheilocaris remanei*. It is a neotenic form whose cephalic appendices have remained locomotive. It belongs to a rare group with few known populations in the world. It presents unusually narrow **granulometric** (see Chapter 9) requirements and lives only in well-sorted sands of 0.2 mm average grain size (Delamare-Deboutteville 1960). The second oddity is the fish-like cephalochordate *Branchiostoma lanceolatum*, which is also remarkable because of its slender, flattened body shape. This creature, a member of the group called lancelets, is common on the sandy shallow beaches of the western Mediterranean and is particularly well studied. Now that its full genome has been sequenced, this lower chordate, which shows many similarities to the vertebrates, is a very useful species for basic biology studies in evolution and developmental biology (Bertrand *et al.* 2007).

4.3 Fish

The Mediterranean Sea hosts at least 650 species of bony fish (Briand 2002), 750 including all 'fish' (*sensu lato*), but this number is no doubt rising, thanks to the species continually invading the Mediterranean Sea through the Suez Canal and gradually spreading in the eastern half of the basin and to other recent immigrants arriving through Gibraltar, taking advantage of global warming. In this section, we consider both the cartilaginous and the bony fish.

4.3.1 Cartilaginous fish

Known as elasmobranchs, this closely related group includes sharks, sawfish, rays, and skates; c.80 species occur in the Mediterranean. They differ from bony fish by having cartilaginous skeletons and five or more gill slits on each side of the head, while bony fish have bony skeletons and a single gill cover.

Both the bluntnose six-gill shark (*Hexanchus griseus*), the largest in its group, which is a widespread species able to reach 700 kg and averages nearly 4 m in length, and the smaller, rarer seven-gill shark (*Heptranchias perlo*) (up to 200 kg and 2 m) are present in the Mediterranean, as is the uncommon porbeagle (*Lamna nasus*), which can reach 4 m in length. In general, these sharks present little danger for humans. In contrast, the great white shark (*Carcharodon carcharias*) is also regularly sighted in the Mediterranean, and some attacks on humans have been reported in the Adriatic Sea. The mako *Isurus oxyrhinchus*, able to reach 4 m and 500 kg, and the thresher *Alopias vulpinus* are also frequent. The huge basking shark (*Cetorhinus maximus*), able to reach 15 m and 8 t, is quite common along the coasts during the early summer. Its behaviour is similar to that of whales, lazing and resting near the surface, swimming with its mouth open, and bolting down between 1000 and 2000 t of water per h, with all the food that comes with it. It spends the winter in deep waters and now requires protection because it is overhunted in the north for its liver, which is huge (Bauchot and Pras 1980). The black-mouthed dogfish (*Galeus melastomus*), the dogfish *Scyliorhinus canicula* and *Scyliorhinus stellaris*, and the smooth-hounds *Mustelus asterias* and *Mustelus mustelus* are also common. The black-spotted smooth-hound (*Mustelus punctulatus*) is only found in the Mediterranean, where it may reach 1.6 m in length. The tope *Galeorhinus galeus* and blue shark *Prionace glauca* are common as well. Three species of hammerhead (*Sphyrna*) are known to occur but they are uncommon. Nine species of other small sharks may live in the Mediterranean, including the aggressive tiger shark (*Galeocerdo cuvieri*), which has only been sighted twice, once near Malaga and once near Sicily (Celona 2000). Presumably it comes to 'visit', passing through Gibraltar.

No fewer than 17 kinds of skates live in the Mediterranean (Bauchot and Pras 1980), several species having migrating from the north, and others coming from atlantinc coast of North Africa and entering through Gibraltar. Unfortunately, the Mediterranean Sea is a more and more dangerous place for sharks and rays. The bottom-dwelling species appear to be at especially great risk due mainly to bottom-trawl overfishing. The honeycomb stingray (*Himantura uarnak*) is the only ray species noted as an invader from the Red Sea, first signalled in Israel in 1955 (Golani *et al.* 2002).

4.3.2 Bony fish

A first category of bony fish is that of swimmers who pass their entire life cycles in open marine waters, without ever seeing or approaching the sea bottom or the coasts. They all have a dark back and a clear belly to be less easily localized at the same moment from above and below. Their morphologies are adapted to permanent swimming. Among these pelagic fish, one group of smaller size lives in big shoals and feeds by filtration of water, while those of larger size are predators. The filter feeders belong to the families Clupeidae (sardines) and Engraulidae (anchovies). They have an elongated and spindle-shaped body, with low pectoral fins and their pelvian fins inserted behind and under the dorsal fin. They are very common throughout the Mediterranean and they have long been much hunted by fishermen. The outstanding characteristic of Mediterranean sardines and anchovies is their size: they are significantly smaller than those of the open oceans, and so very appreciated by fishermen for the canning industry (see Chapter 11).

Almost all the predatory pelagic species of bony fish belong to the family Scombridae. They also live in shoals made up of individuals of roughly uniform size and they are fast swimmers with a strong head, a sharp-pointed muzzle, dorsal and ventral pinnules, and a small featherlike fin located behind the body and hulls on the caudal peduncle, which improves circulation of the water flowing around the body. The smaller species, like mackerel, have a regional distribution and breed at the end of the winter in the shallow waters above the continental shelves. In contrast, the much larger tuna can accomplish migrations of very great amplitude in all the Mediterranean Sea and beyond. All the Scombridae constitute an important food supply for humans and are intensively fished. A blatant victim of insufficiently regulated industrial fishery, the bluefin tuna (*Thunnus thynnus*) is highly threatened (see Box 11.4). It breeds in early summer around the Balearic Islands, Malta, Sicily, Libya, and in the Tyrrhenian Sea, and fishing at that time is particularly destructive.

Another family of pelagic fish, the Xiphiidae, is represented by the swordfish or broadbill (*Xiphias gladius*), a large predatory fish that lives alone or in small groups. It is a migratory and very mobile species which travels between cold waters, where it feeds, and hot waters, where it breeds. Reproduction takes place in summer in the Mediterranean Sea between southern Italy, Sicily, and the Strait of Messina. It is also very popular as a gamefish and universally appreciated in cuisine for its delicate flesh.

The second category of bony fish corresponds to the benthic forms. All the 'commercial' fish (Gadidae, Sparidae, Mullidae, flatfish, etc.) are common also in the Atlantic Ocean, but one family noteworthy for the number of its Mediterranean representatives is the Sparidae or porgies. They are very common and not fewer than 17 species can be present at the same site. Porgies are emblematic of the Mediterranean and some species are clearly overexploited, making them at risk. This has triggered the domestication of the sea bream, *Sparus aurata*, which today reaches 100 000 t in production. Sea bass (*Dicentrarchus labrax*, Serranidae family) is also a formerly common species that is now increasingly reared in mariculture farms (see Chapter 13).

The most interesting ichthyologic groups of the Mediterranean Sea concern the coastal species. The abundance of the seaweeds on the hard bottoms of the upper layers of the high-light zone enable the establishment of large populations of small invertebrates, which in turn attracts predators, and particularly small fish. The main families of littoral Mediterranean fish are the Labridae, Gobiidae, Blenniidae, Pomacentridae, Tripterygiidae, and Syngnathidae. Many of them are endemic or nearly so endemic in cases where they transit Gibraltar toward the coasts of Morocco and Portugal. There are also many juveniles of other families of bigger size, including Sparidae, Mugilidae, and Maenidae.

The wrasses (Fig. 4.4a) belong to the genera *Labrus* and *Symphodus* (ex *Crenilabrus*). They live in proximity to rocks covered with banks of seaweed or Neptune grass meadows. The males are territorial and build nests of seaweed, where the wandering females can lay their eggs. With the young Sparidae (porgies) and the damselfish, the Labridae constitute the majority of the permanent swimming fish of the rocky shores or sublittoral meadows. The damselfish *Chromis chromis* is the only Mediterranean species of the tropical family Pomacentridae, which reaches its northern limit in the Mediterranean.

There are three Labridae species ecologically linked to the Neptune grass meadow, *Labrus viridis*, *Labrus merula*, and *Symphodus rostratus*, all of which present an impressive camouflage by bearing the same shade of green as the leaves of the *Posidonia* among which they live. Another example of such homochromy is found in *Opeatogenys gracilis*, a tiny fish of the Gobiesocidae family, characterized by a strong ventral sucker and living exclusively on Neptune grass leaves, of which it possesses the same green colour.

The other families are demersal, which means they dwell at or near the bottom of a body of water. The gobies (Fig. 4.4b) possess a ventral sucker, made of the two joined pelvic fins, which characterizes their relationship with the substratum. There are very numerous, with about 30 genera and 60 species in the Mediterranean Sea and the Black Sea. The male demarcates a territory where several females will come to lay eggs under a shell or a stone. The male watches the eggs and then the larvae have

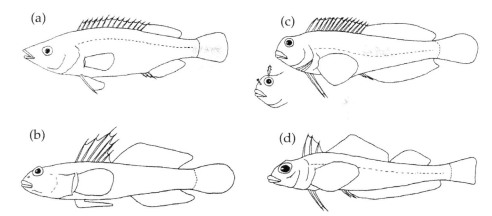

Figure 4.4 The main families of coastal Mediterranean fishes. (a) The wrasses (Labridae): a long unique dorsal fin with spines at its anterior half; anal fin with three anterior spines; caudal fin convex; terminal mouth. (b) The gobies (Gobiidae): two dorsal fins, the first with soft spines; sub-dorsal eyes; pelvian fins gathered into a sucker. (c) The blennies (Blenniidae): a long unique dorsal fin with spines on its anterior half; pelvian fins in jugular position with only two rays; short muzzle; sometimes small ramified appendices over the eyes or near the rostrils. (d) The triplefins (Tripterygiidae): very similar to the blennies, but with three dorsal fins; head always without appendices, and black in the males.

a brief pelagic stage before recruiting near the sea bottom.

The blennies (Fig. 4.4c) are diminutive benthic fish not more than 15 cm in length. They are recognizable by short snout, single dorsal fin, and thread-like pelvic fins. Some species may bear small tentacles over the eyes and/or near the nostrils. They live in shallow waters near the rocks and the seagrass meadows. The males are territorial, making nests in small holes and sometimes in the shell of a big gastropod. The males protect the eggs till hatching.

The Tripterygiidae (Fig. 4.4d) are closely related to the blennies, but with three dorsal fins. During the reproduction period, the males are very aggressive and have a black head with a yellow body, *Tripterygion delaisi*, or a red body, *Tripterygion tripteronotus*.

The sea horses (Syngnathidae) are also very present in the shallow parts of the Mediterranean Sea, with two genera, *Hippocampus* and *Syngnathus*. The short-snouted sea horse (*Hippocampus hippocampus*) is of subtropical affinity and is common along the western African coast as well. The other, *Hippocampus ramulosus*, is easy to distinguish with hair-like filaments on the head and the beginning of the back. It is found as far north as the shores of Great Britain. The pipe-fish include four endemic species of the eight species present in the Mediterranean, two of them found only in the Adriatic Sea. They are highly mimetic and remain immobile among the algae of the Neptune grass leaves to catch small prey passing nearby.

These coastal families are widely distributed in the Mediterranean, but have only limited commercial interest because of their small size. Linked with the superficial seaweeds and the Neptune grass meadows, their abundance is dependent on the microtidal system. The large number of species seems to indicate the existence of an active speciation and adaptive radiation. On the 21 species of Mediterranean wrasses, for example, five are endemic and six sub-endemic (localized also in the north and the south of Gibraltar in the Atlantic Ocean). Similarly, of the 40 species of gobies, 24 are strictly endemic, and of the 19 species of blennies, eight are endemic and six are sub-endemic (Louisy 2002).

These families occur in more limited densities in the deeper hard bottoms of the low-light zone, which also shelter some characteristic species, such as the swallowtail seaperch (*Anthias anthias*) and the cardinal fish (*Apogon imberbis*) both of which are beautiful red species.

Many fish species in the Mediterranean are now threatened by human activities, including overfishing and the re-engineering for human needs of the lagoons. Even the common sole (*Solea vulgaris*) is endangered, as the juveniles of this species live in estuaries or brackish lagoons where the food is abundant. The urbanization of the coastal areas for housing and tourism is a severe threat for this species (see Chapters 12 and 13). Another endangered species of the lagoons is the eel, especially by overfishing of young glass eels (*Anguilla anguilla*), which decreases dangerously the stocks. Such species, formerly so common in all the Mediterranean, are today becoming rare and the present population is estimated as less than only 5% of the populations present in the 1960s. A genetics study demonstrated the Atlantic origin of the Mediterranean glass eels (Maes and Volckaert 2002). By contrast, the Mediterranean does not host migrating populations of the Atlantic salmon (*Salmo salar*), the salinity being probably too high, but a few strains of brown trout are sometimes able to go to the sea for local migrations. Notably a species of sturgeon, *Acipenser sturio*, was, in the past, common in several rivers of the north-western Mediterranean, including the Adriatic area, but today this species is generally considered as extinct in the entire region, as a result of overfishing and destruction of its natural spawning areas in rivers. It was overexploited for its eggs, the famous caviar, with the fishing effort increasing drastically in France and Italy after the Russian Revolution. Reports suggest that there may remain a few individuals in south-western France (Atlantic area), but we have no proof of this. This is an unmitigated disaster, as this was one of the very rare sturgeon species able to live in seas of high salinity.

Finally, there is the group of marine animals formerly considered as fish and now classified with the more ancestral vertebrates, the lampreys and the hagfish. Two lampreys, *Petromyzon marinus* and *Lampetra fluviatilis*, live in the Mediterranean and breed in fresh water, but they become rarer and rarer as a result of freshwater degradation, both in quality and quantity. The hagfish, *Myxine glutinosa*, is especially rare and localized in the Mediterranean Sea along the North African coasts. It lives between 20 and 800 m and attacks other fish during the night in poor light conditions, so that they are not able to escape.

We now turn to the regional avifauna, considering the birds that are closely associated with the sea and excluding those species that characterize Mediterranean lagoons, even if several of them sometimes move far away from the coast for feeding at sea, such as, for example, the sandwich tern (*Thalasseus sandvicensis*) (see Chapter 3).

4.4 Marine birds

Compared to the avifauna of the Atlantic Ocean, especially its northern part, bird life in the Mediterranean is very poor, both in terms of species diversity and in the abundance of populations. Two main arguments can be advanced to explain this. First, as noted already, the productivity of the sea is low, hence offering little food for most avian species. Second, the continental shelf is rather narrow so that deep waters near to the coast only provide a thin coastal strip for foraging birds.

True seabirds in the region include species from three main groups: shearwaters, cormorants, and gulls. In addition, several raptors are closely but not exclusively associated with seawaters. The most emblematic is certainly Eleonora's falcon (*Falco eleonorae*), with a total population of approximately 4000 pairs scattered among many islands of the Mediterranean, but the peregrine falcon (*Falco peregrinus*) and the osprey (*Pandion haliaetus*) are also typical inhabitants of cliffs overlooking the sea, especially on islands of the western half of the basin (see Chapter 6). All the 25 or so breeding pairs of osprey currently breeding on Corsica established their nests on the top of rocky pillars along the coast as close as possible to water where they prey upon a variety of fish species.

As highly pelagic birds, shearwaters spend most of their time in the open sea, feeding on fish and going on land only for breeding. Shearwaters breed in colonies and lay their eggs in burrows, crevices, and caves. Several of them are threatened by predation from introduced mammals. Rats are the most severe threat for most of shearwaters. Touristic encroachments are also a severe threat in many insular coastal areas. Mediterranean shearwaters include several endemic species in the Mediterranean Sea itself, but many more species occur around archipelagos of the Macaronesian realm (Table 4.2). Some of them are quite rare and endangered; for example, the rarest seabird in Europe is the Zino's petrel (*Pterodroma madeira*),

Table 4.2 Sea birds of Macaronesia and Mediterranean Sea relatives

Species	Macaronesia	Mediterranean Sea
Shearwaters	Pterodroma feae*	–
	Pterodroma madeira*	–
	Bulweria bulwerii*	–
	Calonectris diomedea	Calonectris diomedea
	Puffinus assimilis	–
	–	Puffinus mauretanicus*
	Hydrobates pelagicus	Hydrobates pelagicus
	Oceanodroma castro*	–
	Pelagodroma marina*	–
Cormorants	–	Phalacrocorax aristotelis*
		Phalacrocorax carbo
Gulls	–	Larus audouinii*
	Larus cachinnans	Larus cachinnans

Endemic species are marked with an asterisk.

with a population not exceeding 20–30 pairs (Hagemeijer and Blair 1997). Among cormorants, only the shag (*Phalacrocorax aristotelis*) can be considered as a seabird, as it feeds in seawaters and breeds on cliffs and craggy areas. The cormorant *Phalacrocorax carbo* and the rare pygmy cormorant (*Phalacrocorax pygmaeus*) are mostly found in inland water bodies. Several species of gulls also are common in Mediterranean Sea waters. The most emblematic is the Audouin's gull (*Larus audouinii*), a beautiful gull which was dangerously declining some decades ago with a population estimated at approximately 1000 pairs in 1965, but which has since recovered with a total population today of more than 20 000 pairs (de Juana 1997). Colonies of Audouin's gull are found mainly on islets or small rocky islands not far from the coast. Some small colonies are scattered in the Aegean archipelago, but the bulk of the population is in the western part of the basin, especially in Spain (Ebro delta, and Columbrete, Balearic, Grosa, Alboran and Chafarinas Islands), in Corsica, Italy, and the Maghreb.

It could be that the spectacular and rather sudden increase of this population results from birds succeeding in learning to exploit discarded waste from fishing vessels, as did the yellow-legged gull (*Larus cachinnans*). The latter is an opportunistic species, which feeds on a wide range of prey, both terrestrial and marine. Exploitation of terrestrial and marine refuse and waste resulted in a dramatic increase in population size since the 1930s.

The ocean around Macaronesia is regularly visited by a great number of bird species in winter. Some of them more or less regularly enter the Mediterranean Sea in winter, especially when severe storms make food hard to find. These winter visitors include divers, especially the red-throated diver (*Gavia stellata*) and the black-throated diver (*Gavia arctica*), gannets (*Sula bassana*), guillemot (*Uria aalge*), razorbill (*Alca torda*), puffin (*Fratercula arctica*), the two scoters (*Melanitta nigra* and *Melanitta fusca*), and sometimes the rare long-tailed duck (*Clangula hyemalis*).

4.5 Whales

Cetaceans represent one of the most spectacular groups of mammals on Earth. They include whales, dolphins, and porpoises. It is not widely known today that they are abundant and surprisingly diverse in the Mediterranean Sea, partly because their numbers and species richness are undoubtedly much reduced compared to a few thousand years ago when the ancient Greek, Egyptian, Roman, and Phoenician mariners were plying the sea. For example, the great right whale (*Balaena glacialis*) has become rare (Cañadas *et al.* 2004). Among the 80 known species of cetaceans, no fewer than 18 visit or live full time in the Mediterranean (Watson and Ritchie 1985; Raga and Pantoja 2004), including six from the suborder Mysticetes—commonly known as baleen, whalebone, or great whales—and 12 Odontocetes, the so-called toothed whales. This means that 22% of all known whales, dolphins, and porpoises are found in a seascape that represents only 0.82% in surface and 0.32% in volume of all the world's seas and oceans. In fact, 10 are more or less common (IUCN 2003).

Members of the *Mysticeti* are very large (>10 m long) and their feeding apparatus consists of fringed plates of keratin or baleen used to filter organisms they find in the water, such as plankton or small fish. In contrast, members of the *Odontoceti* are usually less than 10 m in length, with the sperm whale (*Physeter macrocephalus*) being a notable exception. Instead of baleen, they have real teeth, as their common name suggests, in jaws that often extend as a beak-like snout behind which the forehead rises in a rounded curve or 'melon'. Whereas Mysticetes have a symmetrical skull with two external nostrils or blow holes, Odontocetes have only one blow hole, as the two nasal passages join below the body surface. Let us now review a little biology and provide lists of the two suborders of cetaceans found in the Mediterranean Sea (Watson and Ritchie 1985; Raga and Pantoja 2004).

4.5.1 Suborder *Mysticeti*

The great right whale is very rare now, only entering the western Mediterranean, coming from Madeira. It feeds at or just below the surface, largely on shoaling planktonic crustaceans while moving at about 2 knots (4 km h^{-1}) with their mouths open, closing the lips every few minutes to begin the process of straining through their copious fringes of fine baleen, and then swallowing. Today their numbers worldwide are estimated at less than 3000 individuals in all, but figures vary widely among authors.

The blue whale (*Balaenoptera musculus*) is also today rare in the Mediterranean, occuring only in the western part of the basin. It mainly eats small crustaceans, which are concentrated in cold, brightly lit shoals, less than 40 m deep. It is estimated there are less than 15 000 individuals in the world ocean today; even in the best Antartica areas there are no more than one animal per 50 km^2.

The minke or piked whale (*Balaenoptera acutorostrata*) is a species eating much more fish (herring, anchovy, capelin, etc.) than any other filter-feeding whale, also eating squids, sometimes competing with large pelagic fish. The world population seems well reconstituted, with an estimation of about 200 000 individuals, but it remains a rare species in the Mediterranean.

The sei whale (*Balaenoptera borealis*) is specialized in eating small organisms, plankton and small crustaceans, but also occasionally hunting fish as sardine, capelin, and anchovy, absorbing 900 kg of assorting food each day, feeding most actively around dawn and dusk. The world population is estimated at 80 000 individuals.

The fin whale (*Balaenoptera physalus*) is increasing its population (approximately 120 000 individuals worldwide) and is the most common whale in the Mediterranean. It is a predator with good vision, preferring fish to plankton. It mainly lives in the western Mediterranean, where it is frequently observed during spring and summer, sometimes autumn, often in groups of two to seven individuals. The Mediterranean population is estimated at 10 000 mature individuals (IUCN 2003).

The humpback whale (*Megaptera novaeangliae*) is coarse and stiff, very active in feeding larger forms of plankton and fish, normally feeding by lunging forward at the surface or by rushing on its prey from below, surfacing through the school with its mouth open. The world population is estimated at 10 000 individuals, and it is now very rare in the Mediterranean.

4.5.2 Suborder *Odontoceti*

The sperm whale world population is estimated at over 1 million. It is an extraordinary diver, reaching more than 3000 m in depth. These mammoth creatures are able to remain under water for more than 2 h, eating large prey, mainly giant squid *Moroteuthis robusta*, but also various other cephalopods, skates, sharks, etc. Sperm whales reach puberty when they are 10 years old and 12 m in length for males and 9 m for females, and they may live 70 years. They were formerly very abundant in the Straits of Gibraltar and offshore from Almeria and environs, in south-eastern Spain. Today, the Mediterranean population is less than 1000 individuals, and the species is classified as vulnerable. Finally, the long-finned pilot whale (*Globicephala melas*), in which the males can attain 8.5 m length and 4 t in weight, are very common in the Mediterranean, eating squids and fish, often in large schools.

Several species of dolphin inhabit the Mediterranean, including Risso's dolphin (*Grampus griseus*), which is relatively abundant, reaching 4.25 m and 680 kg, often travelling in cohesive groups of between three and 30, sometimes more, as they hunt mainly squids. The common dolphin (*Delphinus delphis*) attains 2.6 m in length and about 136 kg and dives down through the deep, scattering layers of plankton, 40–200 m below sea level, to feed on lantern fish and squids. They travel widely and may be very common in some areas one year while vanishing altogether in the next. Despite its name, the common dolphin is now declared endangered (ACCOBAMS 2006) in the Mediterranean Sea. Even more rare is the rough-toothed dolphin (*Steno bredanensis*), which can reach 2.4 m in length and 160 kg in size. It eats fish and squid. The most common Mediterranean species is the striped dolphin (*Stenella coeruleoalba*), occurring in all parts of the basin, west and east. In 1990, however, it was struck by a very serious viral disease (Aguilar and Raga 1993). Fortunately, populations have significantly recovered and increased since then. They live in groups as large as 500 individuals, and are able to dive to depths of 700 m to catch small prey, squid, and demersal fish. This species may reach 3 m and 160 kg.

The bottlenose dolphin (*Tursiops truncatus*) is still common in the Mediterranean but with decreasing populations, so that it is now declared vulnerable there (ACCOBAMS 2006), as indeed it is worldwide. This charismatic animal seems to enjoy the shallow waters around Mediterranean islands where it is generally seen in groups of two to 25 individuals, sometimes more, diving 5–10 min at a time to depths of 50–200 m, eating benthonic prey, cephalopods, and fish. It may reach 4.2 m for 500 kg. The much smaller harbour porpoise (*Phocoena phocoena*) is also common in the area but becoming rarer. It reaches 1.8 m for 90 kg, mainly eating fish of 3–5 kg and squid.

The so-called killer whale (*Orcinus orca*) generally lives in groups of five to 20 individuals that feed on a large variety of prey, including squid, fish, skates, sharks, sea lions, seals, seabirds, and walrus (*Odobenus rosmarus*). They also attack juveniles of other large whale species. They may reach 9.75 m in length and weigh up to 8 t. They may be observed in the Straights of Gibraltar and in the Alboran Sea. The IUCN has declared them critically endangered in the Mediterranean. The false killer whale (*Pseudorca crassidens*) is only rarely seen in the western Mediterranean. It can reach 6 m and 2.5 t, thanks to its diet of squid, tuna, and mahi-mahi (the dolphin fish *Coryphaena hippurus*). Finally, the northern bottlenose whale (*Hyperoodon ampullatus*) is quite similar in appearance to Cuvier's beaked whale (*Ziphius cavirostris*), but the former is larger, reaching 10 m and 5.4 t, is only observed at the western portal of the Mediterranean, and eats only fish. Reaching 8.5 m and 4.5 t, the latter is very rare today, with behaviour quite similar to that of sperm whales, since it enjoys deep waters, where it occurs in groups of two to six individuals, where it preys on squid, crabs, and starfish.

Both anatomically and physiologically, cetaceans show extraordinarily sophisticated adaptations for living in an aquatic environment. They have streamlined, torpedo, or spindle-shaped bodies and reduced appendages (no external ears, reproductive organs, or hind limbs). They are also large (1.2 m) to very large (up to 30 m). The largest animal that ever lived on our planet is the great blue whale that sometimes lives for part of the year in the Mediterranean Sea. Their large body size is made possible by the buoyancy provided by the aqueous medium in which they live, and their great size in turn allows them to enjoy remarkable thermoregulatory facilities.

Mysticetes eat planktonic invertebrates which are often unpredictably dispersed in patchy clumps in the Mediterranean. They are of intermediate caloric value (more than squid but less than fish) and are generally more abundant in temperate waters than in warm waters. They are the primary food of the baleen whales and long migrations appear to be based in part on patchy variation in plankton concentrations. Different types of prey selection are suggested by the feeding behaviour of baleen whales, such as skimming, gulping, or bottom-feeding. Odontocetes, by contrast, hunt actively, mainly fish and squid, sometimes seals. All cetaceans are very active swimmers and easily migrate into and out of the Mediterranean Sea. Fish feeding affects several physiological and behavioural patterns, including short gestation and lactation times. Squid eaters may be bigger. Several species reproduce in the Mediterranean.

Summary

In this chapter we highlight the biodiversity found in the Mediterranean Sea, including seaweeds, seagrasses, invertebrates, fish, birds, and cetaceans: whales and their relatives, the dolphins and porpoises. Outstanding points that emerge are as follows: (1) the dominance of plant and animal species of Atlantic origin that arrived as part of the repopulation of the basin after the Messinian Salinity Crisis; (2) the emergence of neoendemic benthic species, seaweeds, invertebrates, and coastal fish in relation to the acclimatization of formerly oceanic species to the different ecological conditions in the Mediterranean; these species are especially localized in the upper levels where the Mediterranean particularities linked to the quasi-absence of tide are the most influential; (3) as a result, there are many unique seascapes and biotopes such as the trottoir, coralligenous concretions, and Neptune grass meadows, which constitute exceptional habitats for both flora and fauna; (4) the influence of the absence of tidal currents which increase the particles deposits in the coastal areas (the coastal sedimentary bottoms of the Mediterranean are much muddier than the oceanic ones, favouring the suspension and deposit feeders and induces the abundance of sessile and colonial species); (5) the relatively high temperature prevailing at great depths, for example 13°C at 5000 m, which helps explain the presence of very particular fauna; (6) bird life in the Mediterranean is poor, both in terms of biodiversity and abundance, which may be explained by the low productivity of the sea and the reduction of the continental shelves; and (7) the abundance of cetacean species, whose numbers are rapidly decreasing. Many other species of fish are also now endangered, such as bluefin tuna and eel. Others have disappeared altogether, including the sturgeon.

CHAPTER 5

Scales of Observation

At almost any spatial scale, from satellite images taken 200 km above the surface of the Earth to a single square metre of ground viewed from a standing position, the Mediterranean Basin shows striking mosaic patterns and biological diversity that reflect its topographic, climatic, geological, and **edaphic** heterogeneity at macro-, meso-, and micro-scales of resolution. Both the patchiness and the species richness derive from a combination of ecological, geophysical, and historical factors, as well as from the profound 'tinkering' with biota and landscapes that farmers, herders, and woodcutters have practised over the past ten millennia. To understand all of this complexity, we must consider various scales of observation, in both time and space. In this chapter, we will discuss how habitats, ecosystems, and communities in the Mediterranean Basin are organized into 'life zones' that occur in recurrent and recognizable patterns along gradients of altitude that change progressively with latitude as a direct response to climatic variation. Next, to provide examples of the spatial turnover of habitats and communities that change with altitude at any given latitude, we present three 100-km-long transects, one in southern France, one in Lebanon, and one in Israel and Jordan, considering the biological turnover in each case.

Then, we will explore the dynamics of biological diversity at a much finer scale of resolution, considering factors, such as selection by herbivory, fire, and, for example, the existence of natural grazing refuges in Mediterranean landscapes, that help maintain high species diversity despite millennia of intensive livestock grazing pressure. We will also consider what are called disturbance gradients, focusing on the impact of grazing plus fire. These last two issues reveal that temporal scales of observation are as important as spatial scales.

5.1 A succession of life zones

In the Mediterranean Basin where terrain is almost always hilly or mountainous, reading a landscape can be complicated. To make it easier, it is helpful to look for the life zones that succeed each other along elevational gradients, such as can be seen by hiking up the side of a mountain, where these zones occur as more or less horizontal bands of vegetation. In each band or life zone are found characteristic assemblages of plants and animals that tend to share ecological affinities and to develop together (Holdridge 1947; Ozenda 1975; Quézel 1985). Although such communities or 'associations' are not sharply defined in nature, the life-zone concept is quite powerful for purposes of field orientation and analysis of biodiversity. However, microclimatic conditions can 'pull' certain species or even assemblages far from the expected limits of their distribution. Furthermore, as one would expect, the upper limits of each life zone shifts downwards as one travels northwards (Fig. 5.1).

The most convenient method to characterize life zones and to recognize them in the field is to note the two or three dominant tree and shrub species, which, in turn, serve as bioindicators of their particular life zone. In our region, eight readily recognizable life zones replace each other along both altitudinal and latitudinal gradients as one moves up mountain slopes or traverses the basin from south to north. In both cases, average of the minima of the coldest month (m on Fig. 5.1) drops steadily. The

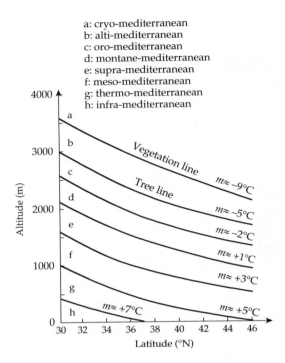

Figure 5.1 Altitudinal/latitudinal gradients showing the zonation of the various vegetation belts, or life zones, in the western Mediterranean. Note that with greater distance from the equator, the altitude at which a given life zone occurs declines steadily. m, average of the minima of the coldest month. Modified after Le Houérou (1990) and Blondel and Aronson (1995).

Figure 5.2 The Mediterranean dwarf palm in southern Spain, where it occurs on rocky hillsides near the coast and is often dug out for use in gardens. In Morocco, it is often considered a troublesome weed in cultivated fields (R. Ferris).

first band (h), occurring at the lowest altitudes and only in the warmest parts of the region, is called the infra-mediterranean life zone. It is found only in frost-free parts of south-western Morocco and in Macaronesia (see Table 6.1). In south-western Morocco, its indicator species are the argan tree (see Chapter 10), of the tropical Sapotaceae family, and an endemic acacia, *Acacia gummifera*. This is only one of the very few acacias of the 120 or so species found in Africa that is considered native to the Mediterranean region.

The second life zone, called thermo-mediterranean, is found at low altitudes in all the warmer parts of the basin, especially in North Africa and the Near East, where it is characterized by dense coastal woodlands of the wild olive tree (*Olea europaea* subsp. *oleaster*) and carob or Saint John's bread tree (*Ceratonia siliqua*). Other common components are the lentisk, false olive (*Phillyrea*), laurel, and Barbary thuja, which is a North African and southern Spanish relative of the cypress. In some parts of the western Mediterranean, this life zone also contains the cork oak, cluster or maritime pine (*Pinus pinaster*), and the Mediterranean dwarf palm (*Chamaerops humilis*; Fig. 5.2). Not surprisingly, the plant communities found in this life zone are all heavily imprinted with all the changes wrought by humans over the centuries. Many areas have been planted with stone pine (*Pinus pinea*) or cluster pine, in the past century. One prominent feature of this life zone is that nearly all woody plant species are evergreen and sclerophyllous (see Chapters 3 and 8 for discussion of this feature). Many, including the dwarf palm, are fire-resistant and vertebrate-dispersed.

Next, with increasing altitude, we encounter the meso-mediterranean life zone, which may be the most familiar one. Here, one or sometimes two species of evergreen oaks dominate a wide variety of woodlands and shrublands. The evergreen holm oak is the dominant tree species in the western and central parts of the basin, and its vicariant, *Quercus calliprinos*, dominates in the eastern part. Aleppo pine (*Pinus halepensis*) and Calabrian pine

(*Pinus brutia*) occupy large parts of this life zone in the western and eastern parts of the basin, respectively. In many regions, formerly oak-dominated formations have been more or less replaced by artificial plantations of the fast-growing Aleppo and Calabrian pines over several millions of hectares. In other areas, one finds remarkably monotonous anthropogenic communities dominated by a dwarf, prickly form of the kermes oak (*Quercus coccifera*) in the western Mediterranean, or *Sarcopoterium spinosum*, and/or various spiny legume shrubs, such as *Calycotome villosa* and *Genista acanthoclada* in the eastern Mediterranean.

The fourth life zone is the supra-mediterranean, which is the domain of deciduous oak forests that occur almost all around the basin. On each mountain range of the area, a distinctive and yet readily recognizable medley of deciduous and some evergreen tree species occur in a broad belt from about 500 to 1000 m, depending on latitude and slope. No other mediterranean-climate area outside the basin has a deciduous oak belt of this kind. **Palynological** studies indicate that many of the dominant deciduous oak species found in this life zone here today formerly occurred widely in the meso-mediterranean life zone as well. In France and northern Spain, the downy oak dominates the supra-mediterranean life zone, but other deciduous oaks do so elsewhere. Thus, from Italy to Turkey and the Levant are found the Turkey oak (*Quercus cerris*) and several other species, including *Quercus boissieri*, *Quercus infectoria*, *Quercus frainetto*, *Quercus trojana*, and *Quercus macedonia*. In this same belt, many other broad-leaved deciduous tree species also occur, such as hornbeams (*Ostrya carpinifolia* and *Carpinus orientalis*), hazelnuts, European ash (*Fraxinus ornus*), several apple family members, including mountain ash, and several small-leaved Mediterranean maples (*Acer monspeliensis*, *Acer campestre*, and *Acer opalus*). In the north-eastern quadrant, this life zone also harbours two evergreen maples, *Acer sempervirens* in Crete and *Acer obtusifolium* in the eastern Mediterranean mountains. Apparently these are the only two evergreen species of the more than 100 species of maples spread throughout the northern hemisphere.

In North Africa and the southernmost parts of Europe, broad-leaved or semi-deciduous oak stands of this life zone are dominated by the Spanish oak (*Quercus faginea*) and the zeen oak (*Quercus afares*). They have been so heavily exploited for timber, however, that only isolated relict patches subsist. Many of the formerly co-dominant tree and shrub species have been altogether eliminated by humans. In the supra-mediterranean life zone, various cold-sensitive Mediterranean plant species gradually disappear, first from north-facing slopes and then, higher up, from south-facing ones as well. These include lentisk, olive tree, kermes oak, honeysuckle (*Lonicera implexa*), rosemary (*Rosmarinus officinalis*), and buckthorn (*Rhamnus alaternus*). The heat-loving pines of the lower life zone are gradually replaced here by other pines, especially southern varieties of the Scots pine (*Pinus sylvestris*) and the numerous subspecies of black pine (*Pinus nigra*).

The montane-mediterranean life zone replaces the supra-mediterranean life zone with increasing altitude. It is dominated either by beech trees or by conifers—pines, cedars, and firs—although most of these forests have been so radically transformed that only a few intact remnants remain. In most countries around the basin, a few venerable trees of great age can be seen, accompanied by an understory sadly lacking in seedlings or saplings of the canopy species. In these montane forests, scattered centuries-old junipers and firs occur well above 3000 m in the Atlas Mountains and up to 2600 m in the Taurus ranges of southern Turkey, bearing testimony to the once majestic forests that grew there formerly.

The sixth life zone is the oro-mediterranean life zone, which is characterized by pines in the northern part of the basin, with montane pine (*Pinus uncinata*) being the most common species, along with the related mugo pine (*Pinus mugo*) of central Europe. The former is usually a much bigger tree, growing up to 25 m tall, while the latter rarely exceeds 4 m in height. In some parts of this life zone are patches of white fir (*Abies cephalonica*) and Norway spruce (*Picea abies*) that generally grow further north in Europe. The rather rare Bosnian pine (*Pinus heldreichii*) also occurs in high mountains on either side of the Adriatic. In peninsular Greece, it is sometimes accompanied by the Macedonian or Balkan pine (*Pinus peuce*) (see Fig. 5.3) and an eastern beech, *Fagus moesiaca*.

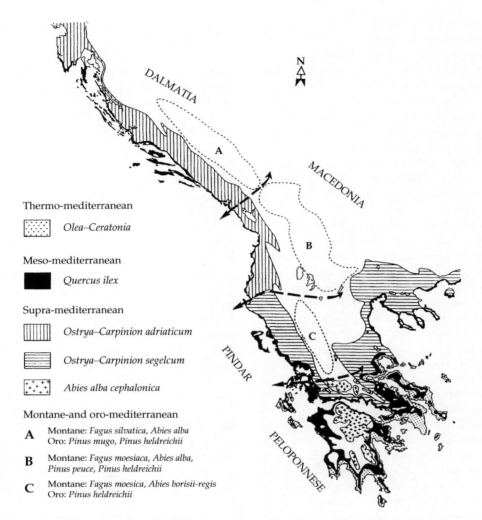

Figure 5.3 Schematic distribution of five life zones in peninsular Greece, showing regional variation especially in high mountains. After Ozenda (1975).

Still higher in altitude, the seventh alti-mediterranean or subalpine life zone usually includes dwarf junipers (*Juniperus*) mixed with diverse grasslands of *Bromus*, *Festuca*, *Poa*, *Phleum*, and other perennial and annual grasses. A large range of herbaceous perennials also occurs here, many of which are endemic owing to the geographical isolation of Mediterranean mountaintops (see Chapters 2 and 3). Indeed, biogeographical features are often highly pronounced over relatively short distances in higher altitude Mediterranean life zones. This is well illustrated in southern Greece (Fig. 5.3), where vegetation in the thermo- and meso-life zones is rather uniform throughout the peninsula, but plant assemblages in higher life zones around each of the separate mountain ranges all differ considerably. Frequently included are various representatives of widespread alpine genera like *Viola*, *Androsace*, *Saxifraga*, *Linaria*, *Arenaria*, *Primula*, and *Vicia*. In the Mediterranean mountains of France, Italy, and along the eastern Adriatic coast, yet another perennial legume–grass association occurs, consisting of *Onobrychis* and *Sesleria*. In southern Greece and western Turkey, at these altitudes, there is a very particular formation of daphnes (*Daphnes oleiodes* and *Daphnes pontica*), accompanied by a range of fescues (*Festuca*), *Primula*, *Soldanella*, and other widespread alpine plants. By contrast, in the Taurus Mountains of southern Turkey and elsewhere in the Middle East, as well as in the High Atlas of Morocco and the Sierra Nevada of southern Spain, the subalpine grasslands are treeless but have instead a multitude of dwarf, spiny shrubs with evocative names like 'goat's thorn' or 'hedgehog', all of which are legumes (e.g. *Astragalus*, *Erinacea*, and *Genista*), and a rich variety of bunch grasses of diverse genera.

Finally, the eighth cryo-mediterranean life zone, at the top of the mountains, is almost entirely devoid of vegetation, except among rocks, scree, and gravel, where a range of widespread, more or less cospomolitan alpine plants in the genera *Saxifraga*, *Androsace*, and *Aubretia* may be found. There are also many endemics as well, especially in bogs, marshes, and fens (ephemeral bogs), showing a clear Mediterranean 'signature'. An example is the forget-me-not (*Myosotis stolonifera*) communities and other assemblages of spring, bog, and rivulet communities of the high mountains of southern Spain.

Such orophilous communities in the Baetic Mountains and elsewhere in the basin are increasingly threatened by a combination of human activities—especially overgrazing by domestic livestock—and by global climate change (Lorite *et al.* 2007; see Chapter 12). This constitutes a major conservation challenge for the future (see Chapter 13).

5.2 Transects

As explained in Chapter 1, Mediterranean landscapes are often squeezed between the sea and the mountain ranges that encircle the basin and divide it into more or less isolated catchments and valleys. One convenient way to discover the extraordinary diversity of any one of the regions within the basin is simply to follow a straight line—a transect—from the coastline to the top of the nearest mountain range, 100–150 km inland. Higher elevations and distance from the coast almost always bring a higher rate of precipitation and more even distribution of rainfall throughout the year.

Having introduced the eight Mediterranean life zones in the previous section, we will now briefly describe the succession or turnover of habitats encountered along three transects in two highly contrasted parts of the Mediterranean: one in the north-western quadrant, with its subhumid climate, and two in the eastern Mediterranean, with its semi-arid climate. Precipitation gradients contribute to differentiate habitats between the two regions. For example, habitats in the north-western quadrant receive nearly 10% of their annual precipitation during the summer, a situation which contrasts sharply with the weather patterns found in the eastern Mediterranean, where no more than 2 or 3% of annual rainfall occurs during summer.

5.2.1 North-western quadrant

In the north-western quadrant, we trace a transect from the Camargue delta in Provence, southern France, to the top of the Mont-Ventoux (1912 m) that marks the limits of the Mediterranean bioclimate in this area (see Fig. 5.4). Five of the eight life zones are represented here; the infra- and thermo-mediterranean are excluded because of the persistent cooling of the region by the mistral and tramontane winds, and the cryo-mediterranean is absent for lack of sufficient altitude. In a day's drive, one can see samples of the most typical habitats of the Mediterranean from coastal brackish lagoons near the sea to lush beech-fir forests in the Mont-Ventoux (see Chapter 6).

Until the first half of the twentieth century, the coastal plain and foothills of southern France were still largely undeveloped, but right up to the second third of the century, in fact, the coastal marshes and brackish lagoons from Marseilles to Valencia, Spain, were considered unusable for agriculture and unsafe for human habitation because of malaria. After World War II, however, large drainage operations were undertaken to allow development, and the vast marine and brackish biota that once occurred here began to decline.

Just behind the coastal lagoons and marshes stretches a strip of arable lands that, in Provence, is highly variable in width due to the presence of

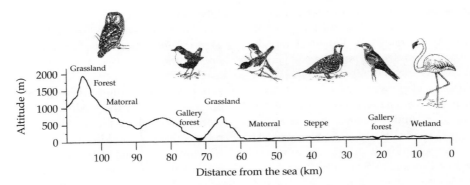

Figure 5.4 Transect through Provence, from the Camargue to Mont-Ventoux, showing bioclimatic zones, vegetation life zones, and prominent bird species of each. Birds depicted are, from left to right, Tengmalm's owl (*Aegolius funereus*), dipper (*Cinclus cinclus*), Sardinian warbler (*Sylvia melanocephala*), pin-tailed sandgrouse, roller, and flamingo.

several large river deltas. In this life zone there is a complex mosaic of wheat fields, orchards, vineyards, hay meadows, rice fields, and early fruit and vegetable plots. Each piece of land is separated from its neighbours by tall hedges of cypress trees, whose function is to break the force of the mistral wind. In other nearby regions, such as Languedoc-Roussillon, the great majority of cultivated lands have been planted as vineyards since the end of the nineteenth century.

Here and there one encounters dry and flat steppes. For example, the former delta of the Durance River in Provence, La Crau, is an area of some $50\,km^2$ in size which is covered by large round stones. There occur here several species of animals that are typical of the semi-arid habitats of central Spain and North Africa, of which the most spectacular is the pin-tailed sandgrouse (*Pterocles alchata*). La Crau is the only breeding place in France for this species, which, with no more than 100–150 breeding pairs at most, is among the most threatened bird species of the country. Other rare breeding birds found in this coastal strip include the lesser kestrel (*Falco naumanni*), the little bustard (*Otis tetrax*), and several steppe species that hardly occur elsewhere in France, such as larks (*Melanocorypha calandra*, *Calandrella brachydactyla*) and the black-eared wheatear (*Oenanthe hispanica*).

When leaving the coastal plains, one frequently encounters narrow riverine forests. Several times along the transects, both in Provence and in Languedoc, a number of rivers cut through this hilly region. Riparian forests (see Chapter 6) used to be rich and varied ecosystems, punctuating landscapes and regions and supporting a high diversity of plants and animals. Many species of plants and animals from non-Mediterranean biota penetrate the basin in these particular habitats, thanks to the abundance of standing water present all year round and also the fact that these habitats function as corridors connecting habitats that are far removed from one another. A Mediterranean note is brought to the bird assemblages of this habitat by species such as the colourful European roller (*Coracias garrulus*), whose relatives are mostly found in tropical Africa, Australia, and Asia.

In the coastal hills just above the arable plain, one enters the realm of evergreen shrublands, or matorrals (see Chapter 6), locally called **garrigues** or **maquis**. From sea level to about 200 m in altitude, a highly striking shrubland is the dwarf, highly monotonous kermes oak shrubland that result from millennia of exploitation and transformation of former woodlands (see Box 10.4). A different group of matorrals is also found, especially as we move inland and upwards in elevation. These are the garrigues, which are generally dense formations from 4 to 8 m tall (see Chapter 6 for more detail and discussion). The holm oak is the dominant tree here, but it sometimes mingles with larger deciduous downy oaks, preserved by people to provide shade for livestock. In fact, the very large extension of holm oak formations is mostly a secondary feature because before large-scale human-induced changes in landscapes, the deciduous downy oak was the dominant tree species wherever soils were deep enough to sustain woodlands (Pons and Quézel 1985). Fire and charcoal production from holm oak and downy oak coppices have for centuries been the major use of this type of vegetation. Climbing up the hillslopes, one inevitably encounters some kind of cliff or cave with very particular assemblages of plants and animals that will be described in the next chapter.

One remarkable feature of the garrigues is their abundance of invertebrates, quite different from those found in northern Europe. By turning over rocks and logs, one can sometimes see the Languedocien scorpion (*Buthus occitanus*), with its large, yellow body, or the yellow-tailed black scorpion (*Euscorpius flavicaudis*). Among insects, some outstanding groups are the cicadas (e.g. the large *Tibicen plebejus*, see Fig. 5.5 and Box 5.1), various mantises, several species of grasshoppers (e.g. the *Saga pedo*, see Box 5.2), and the two-tailed pasha (*Charaxes jasius*).

The spiked magician (see Box 5.2) is also unusual for being tetraploid and **parthenogenetic**, which means that females lay fertile eggs without sexual reproduction. In fact no male of the species has ever been seen definitely, despite one putative sighting reported in Switzerland in 2005 (see www.saga.onem-france.org). The furthest known sightings north to date are in the Czech Republic. Primarily found in the Mediterranean region, it may be moving further north with climate change.

The showy two-tailed pasha is an unusually large butterfly, with a 75 mm wingspan (Fig. 5.6). It is the sole Mediterranean representative of an otherwise tropical family; its closest relatives are found in Ethiopia and equatorial Africa. The two-tailed pasha's distribution is intimately linked to that of the strawberry tree, on whose leaves it lays its eggs to ensure a food source for the caterpillars. There is, however, one population living on Gibraltar, which evolved adaptations to live on plant species growing on limestone. With two generations produced each year, adults may be seen on the wing in May–June, and again in August–September. Adult butterflies sometimes feed on overripe figs in summer and can get completely 'drunk' from their juice!

Figure 5.5 A young cicada emerging from its nymphal sheath. Photo kindly supplied by P. Geniez.

In contrast to insects, the bird fauna of the garrigues is not very rich, since very few species have speciated in this kind of habitat (see Chapters 2 and 3). One interesting exception is the Mediterranean warblers (see Fig. 2.6). These small birds are difficult to watch for more than a few seconds at a time because they are secretive and move rapidly among the thick bushes. They are segregated according to the overall structure of vegetation with a regular ordination of the species from the small spectacled warbler (*Sylvia conspicillata*), which lives in the lowest scrub, to the larger orphean warbler (*Sylvia hortensis*), which is typical of the highest shrublands, and then the blackcap, which is widespread everywhere in various habitats in the western Palaearctic region (Fig. 5.7). However, as many as four species may occur together in the same habitat. Interestingly, not a single case of hybridization has been reported within

Box 5.1. Cicada life history

Cicadas are certainly the most popular insects in the Mediterranean because of their unmistakable stridulent singing in the hottest parts of the summer. There are 16 species of cicada in France alone and the largest and most common is certainly the plebeian cicada (*Tibicen plebejus*), much studied by Jean-Henri Fabre (see Box 8.3). Few people, however, are familiar with the extraordinary life cycle of these animals, which spend 95% of their lives underground. In July, females insert their long abdominal 'drills' into the bark of a tree where they lay as many as 300–400 eggs. Three months later, minute larvae hatch, pierce the bark to get out, and then undergo their first moult. They next drop from the tree, suspended by a delicate silk thread, and when they reach the ground, they dig a hole with their powerful forelegs, which resemble those of a tiny mole cricket. They then spend 4 years in the soil, feeding on roots. Between the egg stage and the adult stage, nearly 99% of the cicada larvae will die from predation or parasitism. On warm evenings in June, the few fortunate survivors emerge from the soil through a small vertical tunnel they construct for this purpose. These nymphs then climb up the nearest tree and seek a safe place, where they can shed their skins and emerge as light green adults (Fig. 5.5). Their chitin quickly hardens and turns to a blackish colour. Some hours later, the full-grown imago takes off for its short adult lifetime that lasts only a few weeks. This is the period when males 'sing', using the 'sawing organs' located at the base of their abdomen between the two hind legs. While singing the males suck the sap of pines, olive, almond, or other species of tree using their specialized mouth parts, which they insert deeply in the tree's bark.

Box 5.2. A giant insect and formidable predator: *Saga pedo*

The predatory bush cricket *S. pedo* is the largest grasshopper—an orthopteran of the Tettigoniidae family—and, in fact, the largest insect in Europe. It measures 15–17 cm from the tip of its antennae to the tip of its ovipositor (or sabre), the organ used by the insect to bury her eggs deep into the soil (it is parthenogenetic; see text). There are spiky teeth all over the animal's body, and the head and legs are extremely long. This secretive animal, which lives hidden in the vegetation, is not easy to find and it used to be considered rare. In fact it is rather common in grasslands and matorrals, particularly in dry and sunny habitats. It is mostly active during the night, which explains that it is usually overlooked. *S. pedo* is a formidable predator, mostly preying upon other species of grasshopper. The way it intimidates its prey is at the origin of its unusual name in French, *Magicienne dentelée* or spiked magician. Also known as the lobster of Provence or Huntress cricket, this formidable insect rises up on its hind and middle legs when approaching a prey, and waves it front legs slowly back and forth, just like a sorcerer or black magician throwing a spell.

this group of closely related species, which evolved recently; that is, during the Pleistocene (Blondel *et al.* 1996).

Leaving the garrigues, we climb to the supra-mediterranean life zone, where mixed evergreen/deciduous forests dominate and where an increasing proportion of plant and animal species have temperate, non-Mediterranean affinities. Dominant trees include several species of deciduous oak, maple, beech, Scots pine, and several species of mountain ash. Among plants, the most noticeable changes occurring with altitude is the appearance of evergreen plant species, more common in northern forests than in the Mediterranean

Figure 5.6 The two-tailed pasha adult and its caterpillar on a stem of its host plant, the strawberry tree (R. Ferris).

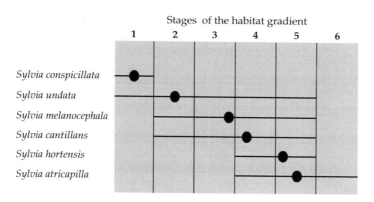

Figure 5.7 Ordination of six species of warblers in the six stages of a habitat gradient ranging from low scrubland (stage 1) to a mature forest of holm oak (stage 6). Horizontal lines indicate the number of habitats occupied by each species and the black dots are the barycentre of their distribution across the gradient (R. Ferris).

Basin, such as yew, holly, *Arctostaphyllos*, and *Pyrola*. Song thrush (*Turdus philomelos*), dunnock (*Prunella modularis*), tree pipit (*Anthus trivialis*), and several species of tits (*Parus*) indicate the close affinities of species assemblages in this life zone with those that are typical of temperate forests further north. In some places with mature mixed forests of beech and pines, the call of the black woodpecker

(*Dryocopus martius*) may strike the ear of the naturalist. Abandoned breeding sites of this woodpecker in deep holes in the trunks of large trees are sometimes colonized by Tengmalm's owl, which is a typical representative of the bird fauna of the large forest blocks of the boreal belt of conifers.

In this life zone, intricate networks of smaller rivers stemming from the high mountains collect waters falling on the huge karstic plateaux stretching northwards. These cold and oxygenated waters are the home of the brown trout and the grayling (*Thymallus thymallus*). Bird species not yet encountered along the transect include the dipper and the grey wagtail (*Motacilla cinerea*). Both these species are strictly associated with clear waters with a strong current and indicate that we are not far from the northern limits of the Mediterranean area.

Upon approaching the summit of these mountains, in the oro- and alti-mediterranean life zones, a very particular habitat consists of stony ground with some interspersed montane pines (*Pinus uncinata*) and low wind-adapted bushes of common juniper (*Juniperus communis*). Several plant species are typical of this habitat and evolved local adaptations to thrive in harsh environments that are characterized by cold and windy conditions. Examples are *Iberis candolleana*, *Crepis pygmaea*, lesser butter and eggs (*Linaria supina*), two saxifrages (*Saxifraga exarata* and *Saxifraga oppositifolia*), and the emblematic guild poppy (*Papaver aurantiacum*, syn. *Papaver rhaeticum* belonging to the *Papaver alpinum* group of species), which was called 'hairy poppy of Greenland' by Jean-Henri Fabre, although it does not occur in Greenland (see Plate 1a). In addition to very few spots in the southern Alps, this poppy is otherwise found only in alpine regions and further north in Eurasia. This oro-mediterranean habitat is also home to a very particular set of bird species like the wheatear (*Oenanthe oenanthe*), which breeds in holes under big stones, the citril finch (*Serinus citrinella*), and the common crossbill (*Loxia curvirostra*), which lives in conifers and shows striking variation in the shape and size of its bill, which is adapted to handle the seeds which are removed from the cones of a great variety of conifers (see Box 7.3).

5.2.2 South-eastern quadrant

Compared to southern France, the south-eastern quadrant of the Mediterranean is far drier and the vegetation is also more degraded. This quadrant consists of approximately 40 000 km^2 and is shared mostly among Lebanon, Israel, Palestine, and north-western Jordan, with a disjunct enclave of great interest in northern Libya (see Fig. 1.2). This region forms the bridge between the Mediterranean region and the Asian continent, and human impact and remodelling of landscapes and biota have been exceedingly important here, since at least 10 000 years BP (Naveh and Dan 1973; Le Houérou 1991; see Chapter 10), as compared with c.6000 years in Greece, and 2500–3500 years in the north-western quadrant.

Figure 5.8 provides a highly simplified transect that proceeds from the eastern Mediterranean coast near Beirut, 100 km inland at 33° N. It rapidly climbs to the top of the Anti-Lebanon mountain range a mere 40 km from the coast. The transect then descends to the Beqa'a Valley, before climbing again to the edge of the vast Syrian desert. A similarly spectacular range of landscapes and ecosystems, but almost completely different in composition, can be observed a mere 200 km further south, along a 100 km west–east transect through Israel and Palestine, then crossing the Dead Sea valley, the lowest place on Earth at −400 m, and finally climbing up the top of the Mt. Edom range of central Jordan (see Blondel and Aronson 1999:102). We will now describe those two transects in some detail.

5.2.2.1 FROM THE COAST TO MT. LEBANON

Coastal dune areas support tall tufted grasses, such as *Ammophila arenaria* and *Stipagrostis lanatus*, while cliffs and rocks near the shore harbour the silvery-leaved *Otanthus maritima* and the semi-succulent sea samphire (*Crithmum maritimum*), whose fleshy leaves can be pickled to make a pungent condiment. Other remarkable plants in the so-called spray belt are the sea daffodil (*Pancratium maritimum*) and *Eryngium maritimum* (see Plate 1b), with large white 'trumpet' flowers in early summer. This relative of *Amaryllis* was formerly common along sandy seashores of the Mediterranean, the Black Sea, and

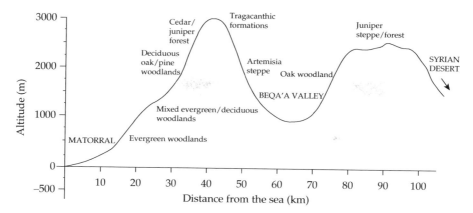

Figure 5.8 Lebanon transect from Mediterranean coast to the Anti-Lebanon range and the Syrian desert. Indications of bioindicator plant species are given for each life zone.

the Caspian Sea, but is now quite scarce due to overharvesting of the flowers and bulbs.

No fewer than 47 species and subspecies of endemic plants in 17 families are found in the coastal sand dunes of Israel (Auerbach and Shmida 1985). Recalling the crossroads-like character of the eastern Mediterranean, it is noteworthy that 29 of these taxa have strongest associations with the Mediterranean flora, whereas 14 have strongest affinities with the Saharo-Arabian desert flora, and one grass, *Aristida sieberiana*, is a palaeorelict of tropical origin (see Chapter 2).

A large number of snakes and lizards are found in the coastal area, including the fringe-toed sand lizard (*Acanthodactylus schreiberi*), which abounds on stabilized sands. Both this species and its relatives, *Acanthodactylus pardalis* and *Acanthodactylus scutellatus*, of inland desert areas, have exceptionally long, agile toes, which allow them to run about during daytime on extremely hot sands in search of their prey. Like most members of their family (Lacertidae), these lizards are endowed with remarkable camouflage equipment in their versatile skin colouring.

At low elevations near the coast, a typical thermo-mediterranean woodland once occurred here, very similar to what is found in warm coastal areas of southern Europe and western Turkey. Only fragments survive today with species like the lentisk, carob, wild olive tree, and the Aleppo, stone, and Calabrian pines. Among the oaks, *Quercus calliprinos*, the eastern vicariant of the evergreen holm oak, is a dominant species, and it occurs with wild cypress (*Cupressus sempervirens*), which derives its scientific name from the Greek word for Cyprus, an island where this evergreen tree was once extremely abundant. Indeed, this tree, which was formerly widespread throughout the Middle East (Zohary 1973), is well represented in archaeological finds in Egypt as well. But along with fir and cedar wood (see below), cypress wood, which is hard and durable, was especially prized for doors and statues in ancient times. Thus it was an important commercial product and may have reached Egypt as cut timber, rather than growing there wild (Meigs 1982; Musselman 2007). Today little of the formerly extensive populations are left anywhere. To see large populations of wild cypress today, one has to go to Crete or else to Mt. Elburz in northern Iran. In addition, the tree is cultivated throughout its natural range and indeed throughout much of the world.

Many wild and planted carob trees (Fig. 5.9) are also found in Israel, Crete, and Cyprus, from the days when animal husbandry was an important component of local agriculture and carob pods were important sources of forage and fodder for sheep, cows, pigs, and goats. Today the tree is cultivated in several mediterranean-climate regions, such as central Chile and the western Cape region of South Africa. Its seeds are used as a food additive under the name of locust bean gum, and the pods provide

Figure 5.9 Branch of a carob tree, showing primitive inflorescences growing directly from the trunk of the tree, and the large fleshy pods which are of great value as animal fodder and as a low-fat sweetener. From: Michael Zohary, Flora Palaestina, Part Two, Jerusalem: The Israel Academy of Sciences and Humanities, 1972, Plates Volume, Plate 45: Ceratonia siliqua L. ©The Israel Academy of Sciences and Humanities. Reproduced by permission.

a 'healthfood' substitute for chocolate. As carob seeds are unusually uniform in weight, they were used by the ancient Greeks in the gold trade, hence the word 'carat', still used as the standard measure of gold, silver, and diamonds.

Around 400–500 m elevation in the south-eastern quadrant, especially in Lebanon, the evergreen *Quercus calliprinos* and other typical thermo-mediterranean elements gradually give way to a series of mixed deciduous/coniferous woodlands typical of the meso-mediterranean life zone. The Turkey oak also occurs in Lebanon, as in Italy, the Balkans, and western Turkey, along with the Lebanon oak (*Quercus libani*). This latter species is recognizable from its very typical leaf traits, with serrate margins and coarse hairs on both sides of the leaves. Further south, Boissier's oak (*Q. boissieri*) and Tabor oak (*Quercus ithaburensis*) are the two dominant deciduous oak species.

At these altitudes and in close ecological association with the oaks, we also find *Pistacia palaestina*, the vicariant of terebinth (*Pistacia terebinthus*), and storax (*Styrax officinalis*), a deciduous tree of tropical origin (Styracaceae) with large, white, fragrant flowers. This tree is the source of a valuable gum obtained from wounds inflicted with a knife on the trunk of the tree. It was widely used, for incense and perfumes, in ancient and medieval times throughout the eastern Mediterranean. The sacred incense of Hebrew rites (Exodus 30:34) may have been derived from storax (Hepper 1981), and it seems likely that overexploitation of the trees for this purpose led to its becoming rare or absent in much of its natural distribution area. Concurrently, it may also have been intentionally introduced to the western Mediterranean, in the Var, southern France, by returning Crusaders in late Middle Ages (see Chapter 10).

Even more than storax, the pistachio trees (*Pistacia*) were vitally important to people in everyday life, since the hard, durable wood was valued for building and carpentry, and the small hard fruits were and still are used as a condiment (see Chapter 10). Because of their large size and great age, pistachio trees often became the object of idolotry (Musselman 2007). Additionally, as throughout southern Europe and Turkey, rockroses form a dramatic part of the shrublands in the coastal foothill areas in Lebanon and Israel, especially where grazing and fire-setting have been important in the recent past (see Plate 2a).

Dozens of wild lily and iris family (Liliaceae and Iridaceae) members occur here, especially among rocks. These include autumn crocus and saffron (*Colchicum* and *Sternbergia*), tulips (*Tulipa*), black iris (*Iris chrysographes*), gladiolus (*Gladiolus*), and *Bellevalia*, which is a relative of hyacinth, and a remarkable number of annual species. Several species of lizards, snakes, and turtles also abound in this zone, such as the land turtle *Testudo graeca* and the highly venomous and viviparous Palestine viper (*Vipera palestina*).

Climbing higher on the flanks of Mt. Lebanon in the supra-mediterranean life zone, we meet several new coniferous species, including four junipers (*Juniperus drupacea*, *Juniperus oxycedrus*, *Juniperus phoenicea*, and *Juniperus excelsa*) and the black pine. A number of apple family members also occur in the premontane and alpine zones, for example wild pear (*Pyrus syriaca*), three species of wild almond (*Amygdalus*), and three types of hawthorn (*Crataegus*). Other deciduous elements of Central Asian affinities include the deciduous gall oak (*Quercus infectoria*) and, at higher altitudes, both *Quercus brantii* subsp. *look* and *Quercus cedororum*, along with hornbeam and European ash.

In the montane-mediterranean life zone of Lebanon, from 1500 to 1900 m, there are scattered stands of Cilician fir (*Abies cilicica*), the only true fir in the eastern Mediterranean, and of course the Lebanon cedar (*Cedrus libani*), the so-called glory of Lebanon. These handsome conifers, which can reach 30 m in height, once covered the snow-clad mountain tops all the way from southern Lebanon to southern Turkey. In Lebanon alone, the area was estimated at 500 000 ha (Sattout *et al.* 2007). But from King Solomon's day to the present, these towering trees have been prized and exploited for timber. Today, only two large stands of Lebanon cedar survive, in the Mt. Troödos range of eastern Cyprus, where an endemic subspecies occurs on ultrabasic rock, and especially in the eastern Taurus Mountains of southern Turkey, where over 100 000 ha still occur thanks largely to difficulty of access (Boydak 2003). However, over 600 000 ha of Turkish cedar forest have been destroyed. Similarly, in Syria, very few cedars are left and, in Lebanon itself, only some 1135 ha (0.86% of the total forest cover) of this emblematic tree survive, scattered over 12 sites above 1400 m on the western side of the Lebanon range (Sattout *et al.* 2007). The species apparently only occurs in spots where fog and cloudiness compensate evaporation during the summer months. The absence of such conditions on the eastern flanks probably explains the absence of cedar groves there (Al Hallani *et al.* 1995).

Yet another stately and highly prized conifer occurring at high altitudes in Lebanon is the Greek juniper or eastern savin (*Juniperus excelsa*) (Savin was the ancient name for Mt. Hermon). In Biblical and Greco-Roman times these alpine conifers were prized and fought over by royal houses of the entire Near East and eastern Mediterranean. Today, on Mt. Lebanon and Mt. Hermon, *Quercus libani* subsp. *look* (from the vernacular *lik* of local sheperds) forms an open shrubby 'pygmy forest' mixed with isolated individuals of Greek juniper.

In the dry eastern slopes of Mt. Lebanon's oro-mediterranean life zone, as mentioned above, there occurs a distinctive hedgehog or thorncushion vegetation type, called tragacanthic vegetation, after a series of densely spiny *Astragalus* species that occur in the high mountains of Lebanon, Syria, Iraq, Turkey, and Iran, and which yield the valuable gum called tragacanth, from their stems and roots. The name tragacanth comes from the Greek *tragos* (goat) and *akantha* (horn), and apparently refers to the curved shape of the ribbons in which the best grade of commercial gum tragacanth is normally found. This natural water soluble gum has been used since very ancient times in pharmaceutical and food products. The millennia-old exploitation of these wild bushes has no doubt been the principal determinant of this plant community's structure and composition, just as the kermes oak formations mentioned above, and large stands of rockrose and quite a few other shrubs in the region, undoubtedly have a semi-cultural origin. We will return to this subject in Chapter 10.

Next on the Lebanese transect the eastern slopes drop off precipitously to a dry rangeland area in the rainshadow of the mountains, where the semi-desert or steppe vegetation is dominated by grasses and shrubs used for grazing sheep and goats. The flora here is rich in Central Asian shrubs, **hemicryptophytes,** bulbs, and annuals adapted to life under constant grazing pressure. The most prominent shrubs are wormwoods (*Artemisia herba-alba* complex), several of which yield a bitter juice referred to as wormwood in the Bible, la'ana in the Middle East and North Africa, and sagebrush in the western United States, where related species occur. As elsewhere in the Near and Middle East, in patches where these shrublands are badly degraded, various grasses (*Poa*, *Stipa*, etc.), saltbushes, and thistles predominate.

By contrast, the fertile Beqa'a Valley is almost entirely cultivated, with a range of crops irrigated with water pumped from the Litani River. Climbing up the slopes of the Anti-Lebanon range, amidst fields and pastures, one once again finds woodland fragments and matorral, mixed with **bath'a** (see Chapter 6). The Anti-Lebanon is almost entirely deforested, except for fragments of woodland dominated by Boissier's oak found at mid-altitudes where conifers are curiously absent. In the montane belt, occasional vestiges of a 'steppe-forest' are found, consisting of a bizarre mixture of eastern savin and other trees, along with low tragacanthic elements of the legume genus *Astragalus*, mentioned previously. Middle-sized shrub layers are altogether missing, most probably as a result of culling and overharvesting by people over past centuries. Descending to the semi-arid inland valleys, one finally reaches the arid Syrian desert, where no further trace of Mediterranean biota is found.

5.2.2.2 AROUND JERUSALEM: WHERE THE MEDITERRANEAN AREA MEETS THE DESERT AND THE TROPICS

A severe rainshadow desert also occurs on the eastern slopes of the Jerusalem hills. In this transition zone between Mediterranean, Irano-Turanian, and Saharo-Arabian biota, there is a remarkable peak of species richness in many groups of plants and animals, related in part to high inter-annual rainfall variations and particularly varied human activities over the past millennia (Zohary 1962, 1971; Aronson and Shmida 1992). Striking and abundant reptiles are the Sinai agame (*Agama sinaita*) and an endemic snake, *Ophisaurus apodus*, the only representative in Israel of the large Anguidae family.

The Dead Sea depression is part of the Pliocene Afro-Syrian Rift valley, which extends from East Africa northward to Turkey and includes the Sea of Galilee (Tiberias Sea) and the Red Sea, as well as the Beqa'a and Litani River valleys of Lebanon. Along both shores of the Dead Sea are scattered oases, where freshwater springs permit the cultivation of speciality crops of tropical origins that cannot be grown elsewhere in Israel or Jordan.

In the uncultivated parts of oases and in the canyons and gorges of dry river beds, there are arid tropical trees of 'Sudanian' affinities, such as wild date palm, horseradish tree (*Moringa aptera*), and the depauperate remnants of an *Acacia-Ziziphus* woodland very similar in physiognomy to those found in the Rift Valley in Kenya or Tanzania. Outside these specialized habitats, a sparse desert vegetation, composed mostly of Saharo-Arabian elements, is found concentrated in the dry watercourses, called **wadis**, whereas 90% or more of the area is bare of all vegetation, except annuals that appear immediately after big rains.

On the Jordanian side of the Dead Sea, the valley slopes are much steeper than on the Israeli side. Along with desert vegetation on the Jordan slopes, with their numerous cliffs, canyons, and sandstone areas, there is a gradual transition into upland Mediterranean woodland, with numerous remnants from the once abundant juniper/pistachio/oak formations (see Davies and Fall 2001 for evidence and an excellent discussion of the relationships between modern and palaeofloras in this region). These abrupt slopes are cut by 15 deep, rather inaccessible canyons, where a surprising collection of elements from different phytogeographical regions intermingle. The Mediterranean elements include sclerophylls, such as oleander (*Nerium oleander*), *Thymelaea hirsuta*, *Quercus calliprinos*, and red juniper (*Juniperus phoenicea*), and deciduous elements such as Syrian ash (*Fraxinus syriaca*) and various Rosaceae. Species from the Anacardiaceae, like *Pistacia atlantica*, *Pistacia palaestina*, and *Rhus tripartita*, along with shrubs like *Noea mucronata* (Asteraceae), and the leguminous bladder senna (*Colutea istria*), represent the Irano-Turanian contingent, while *Retama raetam*, *Ochradenus baccatus*, *Calligonum comosum*, *Anabasis articulata*, *Zygophyllum dumosum*, and tamarisk (*Tamarix*) are some of the more common Saharo-Arabian elements. Finally, *Acacia raddiana*, the succulent *Caralluma*, and jujube (*Zizyphus*) are signature of tropical 'Sudanian' elements in this crossroads region of four distinct floral regions.

5.3 Small-scale, within-landscape diversity

We will now look briefly at species diversity and turnover at the spatial scale relevant to a bird, an insect, or a mouse; that is, between a few square metres and a small watershed or landscape. One of the simplest statistics used to make comparisons among sites in this size range is species richness in a given group of organisms within standardized plots of logarithmically increasing size. We adopt a popular and useful tool to measure diversity and to investigate its changes in relation to the spatial dimension, which is to recognize several nested levels of diversity, as depicted in Fig. 5.10.

Distinguishing the components of diversity is closely associated with quantifying the local distribution of species, similarity and dissimilarity among local species assemblages, and rate of change in species composition with respect to ecological conditions or distance. The division of diversity into alpha (α), beta (β), and gamma (γ) components, as proposed by Whittaker (1972), characterizes patterns of diversity and turnover on different scales. Alpha and gamma diversities pertain to the number of taxa at the local and regional scales respectively. Gamma, or regional diversity, is an expression of the pool of species that occur at the scale of a series of habitats; that is, a landscape. It is within this pool of species that each local habitat will 'select' those species that will constitute each separate habitat-specific species assemblage. Beta diversity measures the rate of change, or turnover in diversity, between two habitats. In other words, the total, or gamma diversity of a landscape, consists of two additive elements, the average diversity within landscape units (alpha diversity) and the diversity among landscape units (beta diversity). Finally, the term delta (δ) diversity is sometimes used to describe changes in diversity patterns among larger units of space, such as landscapes or regions (Whittaker 1972; Blondel 1995; Huston 1999; Loreau 2000). A recurrent pattern in most Mediterranean ecosystems and landscapes is that alpha diversities are not always impressive but beta and gamma diversities tend to be very high indeed,

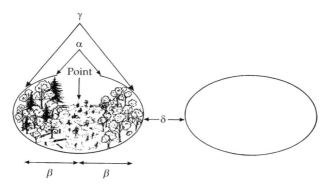

Figure 5.10 The different components of diversity within and between communities in a landscape. See text for details. After Blondel (1995).

because of the physical and biological heterogeneity of the systems (see Chapters 1, 2, and 10).

5.3.1 Disturbances and plant species diversity

All disturbance events, such as drought, fire, grazing, and cutting, have strong effects on the various components of diversity. There are interactive effects as well among the various types of disturbance that vary depending on the components, as well as the spatial and temporal scales scrutinized. In this section, we discuss a few examples concerning vascular plants, which lend themselves particularly well to an analysis of the components of diversity at varying scales. In Chapter 7, we will discuss the effects of recurrent fires on the dynamics of bird populations and discuss the role of predatory ants at the landscape scale.

Most of the variation and turnover discussed thus far in this chapter can be understood in terms of events and trends linked to 'long history'—that is, over recent evolutionary time—even if some examples of human-mediated change in vegetation structure were noted. In what follows, we will pay more attention to the impacts of humans and their livestock, on the scale of short history (ecological time), over the last several hundred or few thousand years.

As Naveh and Whittaker (1979) and many subsequent authors have pointed out, moderately grazed woodlands and all the various shrublands in the eastern Mediterranean region show some of the highest alpha diversities for plants in the world. It is generally argued that the unusual species richness of these landscapes, especially in annual plants and geophytes, is the product of relatively rapid evolution under conditions influenced by drought, fire, cutting, grazing, and other disturbances by humans and other animals (see Chapter 2). Naveh and Whittaker (1979) suggested that the longer the period of human perturbations, the greater the possibility for differentiation to occur (see also Chapter 10). Other things being equal, this would suggest that higher alpha diversity would be found in the eastern Mediterranean as compared to the western part.

In Table 5.1 are shown vascular plant species richness data from five sites in northern Israel and five in southern France. Vascular plant species richness at all three spatial scales considered is much higher in the eastern Mediterranean sites than those in France, across a range of vegetation structures. Yet in each area, recent land use history greatly affects plant species diversity. Age since fire or other major distubance events have a clear effect on floristic diversity. Thus, grazed oak woodland at Allonim in northern Israel had two- to three-fold greater plant species diversity than the ungrazed woodland site at Neve Ya'ar. Disturbance-adapted annuals contributed 97 of 135 total species in this site and half to two-thirds of species richness in the Mt. Gilboa and Mt. Carmel sites as well. Similarly, data from a study site in southern France (Cazarils) revealed that moderate grazing over 4 years had a significantly positive effect on plant species diversity as compared to fenced, ungrazed plots in the same landscape units (Le Floc'h *et al.* 1998). These data corroborate the widely held view that *moderate* disturbance regimes result in higher species diversities than very heavy disturbance, or the absence of disturbance (Huston 1994; Seligman and Perevolotsky 1994; see also Chapter 7).

Data presented in Table 5.1 for the French sites of Puéchabon and St. Clément show that with increasing age following a major fire or clear-cutting, species diversity declines gradually. In addition to changes in species richness, the botanical composition and life-form spectra change in response to the frequency and intensity of disturbance events and/or to decreasing or increasing intervention by humans.

At much smaller scales, disturbance events, such as ant hills, the holes dug by small mammals, and the grazing of plants by snails and insects are sources of spatial and temporal variability that contribute to maintain species richness of plants and animals. The role of these small-scale disturbances in the dynamics of plant species diversity has been studied in abandoned vineyards near Montpellier, southern France. Abandoned farmlands, known as 'old fields' throughout the Mediterranean, are known to be rich in plant species (Escarré *et al.* 1983; see Chapter 6), and have a high degree of dynamism (Papanastasis 2007; Marty *et al.* 2007). This may be especially so in the first years after field abandonment (Bonet and Pausas 2007), due to high immigration and extinction processes in the early

Table 5.1 Species richness of vascular plant communities in 1, 100, and 1000 m² plots

Locality	Vegetation structure	Disturbance regime or age since fire	Number of vascular plants in plots		
			1 m²	100 m²	1000 m²
Israel					
Mt. Gilboa	Open grassland	Lightly grazed	29	105	179
Mt. Carmel	Open shrubland	Disturbed	20	75	119
Allonim	Dwarf shrubland	Grazed/burnt	21	88	135
Mt. Carmel	Oak woodland	Ungrazed	14	48	65
Neve Ya'ar	Oak woodland	Grazed	10	36	47
France					
Puéchabon	Holm oak coppice	1 year after cutting	9	45	64
Grabels/Bel Air	Kermes oak shrubland	10+ years after fire	12	40	54
Puéchabon	Holm oak coppice	10+ years after cutting	6	33	51
St. Clément	Open pine forest	1 year after fire	7	29	30
St. Clément	Kermes oak shrubland	5 years after fire	8	23	28

Source: From Naveh and Whittaker (1979), reprinted in Westman (1988); F. Romane and A. Shmida, unpublished results (last line).

stages of secondary succession. Thereafter, a steady turnover sets in, with perennial herbs and woody species gradually replacing annuals and biannuals and a trend to greater stability and lower plant diversity. This trend is well illustrated in the above-cited study by Le Floc'h *et al.* (1998) and still better in the work by Lavorel *et al.* (1994), who created artificial disturbances in three different 'old fields' that had been abandoned for 1, 7, and 15 years.

5.3.2 Habitat and plant species turnover at the landscape scale

We have already mentioned that many Mediterranean plants are highly adapted to frequent and varied disturbance events. Multiple effects must be considered as well, as for example, the effects on plant diversity of the combination of overgrazing and burning (Papanastasis *et al.* 2002). Studies of artificial exclosures—such as on either side of fences—have demonstrated the signficant impact on plant diversity apparent to the naked eye (e.g. Noy-Meir *et al.* 1989).

There are, in addition, natural or 'geological' grazing refuges where goats and sheep cannot graze, farmers cannot wield their ploughs, and fires are infrequent. The presence of these natural refuges in a landscape contribute to plant species richness across the spectrum of life forms present and contribute to the **resilience** of plant communities at the landscape level. In a grassland area of northern Israel, with a grazing history several millennia old, Shitzer *et al.* (2008) studied plant assemblages inside and near to the small (<5 m²) natural grazing refuges created by outcrops of basaltic rock. In particular, they examined the differences of impact on very small-scale species richness (1 m²) across a range of plant life forms (Table 5.2).

Although species richness was significantly greater in the near-refuge than in the refuge quadrats, three different groups of species were found to be present: those that were only very rarely found in refuges (53 species of 103), mostly small and medium-height annuals with low capacity for competition, or perennial plants; a small number found primarily inside the refuges (12/103), including tall perennial grasses, tall annuals, climbers, and a shrub; and a third, intermediate group (38/103) found both inside and outside refuges, most of which were dominant tall grasses. These detailed results show that small natural 'geological' refuges are effective in maintaining grazing-susceptible species in a landscape. More generally,

Table 5.2 Small-scale species richness (mean number of species in 1 m^2), total and by groups, in refuge and near-refuge quadrats, the mean difference between them (N–R), and the significance of the difference

Species group	Near refuge (N)	Refuge (R)	Difference (N–R)	P value (SR)
All species	18.64	8.48	10.15	<0.001
Annual grasses	4.83	1.63	3.20	<0.001
Annual legumes	2.58	0.67	1.91	<0.001
Annual forbs	6.70	2.44	4.25	<0.001
Geophytes	0.58	0.38	0.20	0.003
Perennial dicotyledons	2.23	1.41	0.82	<0.001
Perennial grasses	1.59	1.65	−0.06	0.163
Shrubs	0.12	0.29	−0.17	<0.001

Significance calculated by the Wilcoxon signed rank test (SR).
N = 63 sites.
Source: Shitzer *et al.* (2008).

the contribution of such refuges to species richness and life-form diversity at the landscape scale is much greater than their small size, relative to the total study area (Shitzer *et al.* 2008).

Ortega *et al.* (2004) also found positive correlations between plant diversity and structural complexity, described as habitat patches in human-modified Mediterranean landscapes in Spain. Del Barrio *et al.* (2006) conducted a follow-up study focusing on the influence of linear landscape elements in particular, whether of human or non-human origin, on plant species diversity. Linear elements, such as hedges and tree plantings along field edges and roadways or railways, as well as rivers and streams, were studied in five different Mediterranean landscapes, in terms of land-cover patch mosaics, presence of unique species, overall plant species richness, and alpha diversity in 1000 m^2 plots. 'Core habitats' were compared with the various linear elements. Notably, species density, percentage of unique species, and alpha diversity per plot were significantly higher ($P < 0.05$) in or along the linear elements than in the 'core' habitats. This result not only highlights the influence of linear structural elements in augmenting species diversity across landscapes, but also shows how linear elements—whether hedgerows or rivers—provide refuge areas and ecotones where 'species flow' and turnover are high. Thus the spatial heterogeneity of a landscape is accentuated and contributes to biodiversity at all spatial scales.

Finally, what we referred to at the beginning of this chapter as disturbance gradients have considerable impact on species diversity as well. Such gradients are often set in place and in motion by spatially and temporally variable factors, such as grazing (Garnier *et al.* 2007) and fire (Diaz-Delgado *et al.* 2002; Mouillot *et al.* 2005). Easier access to time series satellite imagery and new Geographic Information Systems (GIS) software is now making it easier to study the effects of recovery from fire over large regions and long time periods. At the landscape and regional scales, fire history, climate change, topography, and dominant type of vegetation clearly affect postfire response of vegetation and entire ecosystems. In particular, increased fire frequency, as is occurring in many parts of the Mediterranean Basin, may reduce ecosystem resilience; that is, the ability to recover to a pre-disturbance state, as defined, for example, in terms of standing biomass of plant species richness (see Chapter 11). Using Landsat imagery to monitor vegetation recovery after successive fires in a 32 000 km^2 area of Catalonia, north-eastern Spain, between 1975 and 1993, Diaz-Delgado *et al.* (2002) found trends that were observable several years after burning, but not immediately following a fire. Among many other significant results, they found a positive correlation between mean annual rainfall and post-fire ecosystem resilience. They also found that woodlands dominated by resprouting evergreen holm or cork oaks were highly resilient to an individual fire, but showed a larger decrease in resilience, over a period of many years, under conditions of recurrent fire than did woodlands dominated by pines that regenerate from seed and harbour a fast-growing herbacous understory. These results have clear implications for management in areas throughout southern Europe where reducing the risk of wildfire is a major concern for private and public land managers.

Mouillot *et al.* (2005) went even further in this direction, studying 200 years of landscape changes on a 3760 ha area of central Corsica, using various maps dating from 1774, statistics on land cover from 1848 and 1913, and a complete 40-year fire history from 1957 to 1997. In this study, the border effect between forest and shrubland, the relative flammability of the two, the combination of past landscape patterns and the variable colonization abilities of differing forest species were all used to interpret patterns of fire at the landscape scale.

The advantages of studying correlationships of species diversity and ecosystem response to disturbance along gradients are also now being used to help predict and inform management decisions in the face of desertification (Kéfi *et al.* 2007; see Chapter 13) and other trends and disruptions linked to global climate change (Kazakis *et al.* 2007; see Chapter 12).

In the next chapter, we will enter into the details of living systems in the sets of habitats, which characterize the various life zones.

Summary

The aim of this chapter was to show that the most striking feature of the Mediterranean region is the diversity of biota, at whatever scale of observation. From moist beech-fir forests in the northern fringe of the Mediterranean to tropical-like oases near the Dead Sea, from one square metre to the next, from one landscape or island to the next, the diversity of life forms in the region is truly striking. We show that from sea level to the top of the mountain ridges that encircle the basin, a series of vegetation belts or life zones has been defined. They are characterized by a few dominant tree or shrub species. Then we illustrate these scales effects on the diversity of life along three 100 km long transects from the seashore to the top of one mountain range in southern France and two on the eastern Mediterranean shores. This entailed crossing a succession of life zones characterized by specific sets of dominant tree or shrub species and associated assemblages of plants and animals. Within each life zone, local heterogeneity of habitats provides opportunities for a large variety of populations and species to find conditions that are suitable for them, thus adding some component of diversity in the systems. Then we examined a finer scale of resolution and discussed a series of micro-scale disturbance events and their impacts on plant diversity.

CHAPTER 6

A Patchwork of Habitats

The goal of this chapter is to present the major habitat types encountered in the Mediterranean area and to organize them into broad categories. The vegetation types described in the previous chapter can assume a surprising number of variations on a theme, depending on local conditions and human land-use histories. The combination of a great many adjacent habitats in ecosystems with a large number of possible pathways of degradation, transformation, and secondary succession, even within a single micro-region, gives Mediterranean landscapes a distinctive mosaic quality that sets them apart from central and northern Europe to the north and the arid and humid tropical zones further south. For plant and animal populations, this pattern of landscape-scale 'patchiness' has profound consequences.

Simplifying somewhat, the north–south bioclimatic gradient discussed in Chapter 1 is reflected by forest and shrubland formations. In the humid and sub-humid parts of the north-western and north-eastern quadrants of the basin (see Fig. 1.2), dense woodlands and true forests are growing back since widespread rural exodus and agricultural abandonment began at the beginning of this century (see Chapter 10). In contrast, on the southern and eastern shores of the basin, where rainfall is lower and where human population pressure is growing rapidly, ecosystems in all life zones are often badly depleted and degraded.

Of course, there are many exceptions to this schematic picture. There are degraded shrublands and 'badlands' in almost all parts of the north-western quadrant (i.e. south-western Europe), and beautiful stands of nearly undisturbed fir or oak/beech forests in a few remote mountain ranges in North Africa, Greece, Turkey, and on several of the larger islands.

Let us now review the main habitat types, starting with forests and woodlands. For each habitat type, we will mention some of the most important or characteristic plant species and discuss the range of plant life forms found there. We will also introduce some of the 'special' habitats in the Mediterranean, including matorrals, steppes and grasslands, 'old fields', cliffs and caves, riverine forests, various continental wetlands, microtidal habitats, and any other marine habitats. We leave aside those areas that are so intensively cultivated or urbanized that there is little or nothing left that is 'Mediterranean' about them. Managed woodlands in cultural landscapes will be discussed in Chapter 10.

6.1 Forests and woodlands

In the Mediterranean region, the term 'forest' should not bring to mind an image of high, dense stands of trees with a closed canopy. Few such forests exist in the region today and probably have never been common in the past (Quézel 2004). Recent palaeoecological studies (De Beaulieu *et al.* 2005) suggest that forest cover was not as dense and uniform as formerly thought, even during the so-called Holocene optimum (roughly 9000–5000 years BP). Instead, throughout the Holocene, Mediterranean landscapes have tended to be rather open and heterogeneous. Climatic and edaphic factors favour instead a more open formation, sometimes called woodland or park woodland. In some areas, however, dense forests do occur where rainfall is sufficient and deforestation has ceased for enough time to allow the forests to grow. **Riparian** forests are a special case we will discuss below.

By one estimate, Mediterranean forests and woodlands cover approximately 73 million ha or about 8.5% of the region's land area (Merlo and Paiero 2005; but see Chapter 10). Their distribution is highly uneven, however, with 77% on the northern shores, 15% in the east (including Turkey), and only 8% in the south; that is, North Africa (Merlo and Paiero 2005). These differences are mostly due to differences in climate, which is hotter and drier in the southern and eastern parts of the Mediterranean than in the north. However, a surprising 27% (21.2 million ha) of Turkey is covered with forests, some of which are quite lush (Kücük 2008).

Only 11% of the forest area in the Mediterranean consists of artificial plantations, mostly of pines and eucalypts. (In many areas, people manage the woodlands as agro-forests or as *dehesas*, as will be described in detail in Chapter 10.) Eucalypt forests cover around 800 000 ha in Portugal, mostly used for paper. Forests are highly diverse in their architecture, appearance, and woody plant species composition, since, as mentioned in Chapter 3, there are approximately 290 species of trees in the region, of which more than 200 are endemics (Quézel and Médail 2003). By contrast, there are only 135 species in all of central and northern Europe (Médail 2008a). Table 6.1 indicates some of the most common tree species in the four Mediterranean quadrants and the various life zones. The distribution of the oaks and pines is only partially indicated, as additional information on these crucial groups will be provided in Chapter 7. Note that most species are noted for a single life zone: the one where they occur most commonly, even though in fact they can occur in three, four, or even five in the varying quadrants and peninsulas of the region. Some species, for example stone pine (see Plate 9a), cypress, chestnut (*Castanea sativa*), olive, and cork oak, have been so widely dispersed and planted by people that their natural area of distribution is now difficult or impossible to determine.

Mediterranean forests are also highly varied in the growth forms, morphology, physiology, and phenology of the dominant trees in each region. For example, four leaf types occur in various combinations (see Chapter 8). Tree leaves may be sclerophyllous and evergreen, leathery in texture, and often spiny or prickly. A second group has laurel-like leaves (Fig. 6.1a) that are somewhat softer and shiny but still evergreen, like the foliage in many tropical forest trees. Thirdly, they may be 'semi-deciduous' and remain on their stems over winter, with reduced or terminated growth and photosynthetic function. Leaves are not shed until spring, when they are replaced by a new crop of leaves. Examples of such species are Spanish oak, gall oak, and downy oak. The fourth group has typically deciduous leaves, such as predominate in northern temperate forest trees. Many examples are found in the north-eastern quadrant, such as *Carpinus*, *Corylus*, *Ostrya*, and *Zelkova* species, occurring primarily in higher-rainfall areas and higher-altitude zones. One particularly interesting tree of this group is the ironwood tree. This majestic tree is restricted to a small part of the north-eastern quadrant and represents the sole member of the witch-hazel family (Hamamelidaceae) in the Mediterranean flora, but it was once widespread throughout the region, during the Miocene. Among other, more widespread deciduous species in the region, many show reduced leaf size as well as other adaptations to a warm, summer-dry climate, when they are compared to congeneric species found further north. An example is the small-leaved Montpellier maple (*Acer monspessulanum*) (Fig. 6.1b), whose leaves are much reduced in size compared to those of the maple species found in northern boreal forests. Similar trends are found in oak, ash, mountain ash, hornbeam, etc.

The largest and most diverse evergreen sclerophyllous forests in the Mediterranean area today are the 'laurophyllous' forests found on the wetter, northern coasts of the larger Canary Islands: La Palma, Gran Canaria, and Tenerife. These forests are relicts of a now virtually extinct Tertiary flora that was widespread in southern Europe and northern Africa about 15–40 mya (see Chapter 3). This forest type is named laurisilva because it includes no fewer than four species and subspecies of the tropical Lauraceae family, as well as several endemic broad-leaved evergreen trees, such as

Table 6.1 Some of the most abundant and characteristic canopy tree species in six life zones of the four quadrants of the Mediterranean Basin (excluding riverine or riparian forests, which are discussed later in the chapter)

Life zone	Geographical extension in the four quadrants	Dominant tree species	Elevation
Infra-mediterranean	SW Morocco, Macaronesia	*Argania spinosa, Acacia gummifera* various Lauraceae, *Pinus canariensis*	<250 m
Thermo-mediterranean	SW, SE, NW	*Olea, Ceratonia, Pistacia lentiscus*	<500 m
	SE, NE	*Zelkova sicula*	
	SE, NW	*Styrax officinalis*	
	SW, NW	*Pinus halepensis*	
	SW, NW	*Pinus pinea* subsp. *mesogeensis*	
	SW, NW	*Quercus suber*	
	SW	*Tetraclinis articulata*	
	SW, SE	*Pistacia atlantica*	
	NE	*Pinus brutia*	
Meso-mediterranean	NW	*Quercus faginea*	0–600 m in the northern quadrants;
	NW, SW	*Acer monspessulanum,*	
	NW, SW	*Quercus ilex, Quercus coccifera, Quercus suber*	500–1000 m in the southern ones
	NW, SW	*Juniperus oxycedrus*	
	NW, SW, NE	*Pinus pinea, Pinus*	
	NW, NE	*Celtis australis*	
	NW, NE, SE NE	*Laurus nobilis*	
	NE, SE	*Celtis tournefortii, Quercus infectoria, Quercus calliprinos*	
	SE	*Cedrus brevifolia* (Cyprus)	
	SE	*Quercus alnifolia* (Cyprus)	
	SE, NE	*Quercus aegilops*	
Supra-mediterranean	NW, NE	*Quercus humilis, Quercus frainetto, Quercus cerris*	600–1200 m in the northern quadrants;
	NW, NE	*Cupressus sempervirens*	
	NW, SW	*Castanea sativa*	800–2000 m in the southern ones
	NE	*Quercus macedonia, Quercus trojana*	
	NE	*Abies cephalonica*	
	NE	*Ostrya carpinifolia*	
	NE	*Carpinus orientalis*	
	SE	*Quercus afares*	
	SE (Crete)	*Zelkova abelicea*	
	SE, NE	*Quercus infectoria*	
	SE, NE	*Acer sempervirens*	
	SW	*Abies pinsapo,*	
	SW	*Abies maroccana* (very scarce)	
	SW	*Quercus faginea, Quercus canariensis*	
Montane-mediterranean	NW	*Fagus sylvatica*	>1000 m
	NW	*Pinus nigra* subsp. *nigra*	
	NW	*Pinus nigra* subsp. *clusiana*	
	NW	*Pinus nigra* subsp. *laricio*	
	NW	*Pinus nigra* subsp. *salzmannii*	

Table 6.1 (*Continued*)

Life zone	Geographical extension in the four quadrants	Dominant tree species	Elevation
	NW	*Pinus sylvestris*	
	NW, NE	*Carpinus betulus*	
	NW, SW	*Cedrus atlantica*	
	NE	*Cedrus libani*	
	NE	*Abies cephalonica, Abies cilicica,*	
	NE	*Abies nebrodensis*	
	NE	*Juniperus foetidissima*	
	SE	*Pinus nigra* subsp. *pallasiana*	
	SE	*Pinus heldreichii*	
	SE	*Abies numidica*	
	SW	*Pinus nigra* subsp. *mauretanica*	
	SW, NW, NE	*Cupressus sempervirens*	
Oro-mediterranean	NW	*Pinus uncinata*	>2000 m
	NW	*Juniperus excelsa*	
	NW, SW	*Juniperus communis*	
	NW, SW	*Arceuthos drupacea*	
	NE	*Pinus mugo*	
	SW	*Juniperus thurifera, Juniperus turbinata*	

NE, north-east; NW, north-west; SE, south-east; SW, south-west.
Source: After Quézel (1976b) and P. Quézel, F. Romane, M. Barbéro, and A Shmida, personal communications.

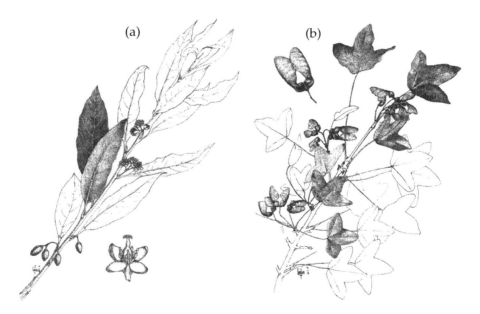

Figure 6.1 (a) The evergreen sclerophyllous bay tree, and (b) the small-leaved, deciduous Montpellier maple (R. Ferris).

Canary Island strawberry tree (*Arbutus canariensis*), *Myrica faya*, and *Visnea mocanera*. All of these trees share their broad-leaved, sclerophyllous leaf shape with the widespread bay tree, which still occurs widely throughout the Mediterranean Basin. The highest concentration of tropical relict species is found in the wild olive/carob/holm oak (or the related *Quercus calliprinos*) associations that dominate in foothills and uncultivated slopes at lower altitudes throughout the basin. These associations can be considered as a 'laurophyllous' vegetation type, which survived since the Tertiary, albeit with continuously changing and regionally variable botanical composition.

At low altitudes, especially in parts of southern Iberia, Turkey, and North Africa, open woodlands or park-like glades alternate with very dense and much lower stature vegetation types. These formations are usually the result of human management schemes of some sort, involving livestock raising, coppicing or woodcutting, and some regime of proscribed fire. At higher altitudes, however, open formations of conifers (e.g. *Pinus*, *Abies*, and *Cedrus*) are found with an understorey of spiny shrubs (e.g. *Astragalus* and *Genista*) on rocky outcrops. Taller, denser forests of varying composition are found in the supra-, montane-, and oro-mediterranean life zones (see Table 6.1). Although heavily influenced by humans, these forests tend to reflect more closely the natural potential for vegetation given local soils and climate. They often show a thorough mixture of evergreen trees, conifers, and winter-deciduous trees, shrubs, and vines. Where mean annual rainfall is greater than 500 mm, sclerophyllous elements tend to diminish in numbers, while deciduous ones gain in number and ecological importance.

Mediterranean forests and woodlands contain a surprising number of conifers, including pines, junipers, cypress, cedars, firs, and the Barbary thuja. But broad-leaved trees are much more important than in northern and central European forests (Scarascia-Mugnozza *et al.* 2000). Like the ironwood, the Barbary thuja is a **monotypic** palaeorelict restricted to North Africa with some remnants in southern Spain (see Chapter 3). Conifers are still more abundant in Mediterranean mountains than in the lowlands, and their contribution to overall vegetation cover is higher at higher altitudes. The 'primeval' forests at higher elevations in much of the area probably combined conifers and broad-leaved species in intricate mixtures with many species. The segregation often seen in forest canopies today in the basin, whereby pines or evergreen oaks can form nearly pure stands, is almost always a product of human interventions and does not reflect the natural dynamics of these forests.

6.2 Matorrals

We will use the Spanish term *matorrals* to designate all the kinds of shrubby vegetation (shrublands) that occur in the basin and that have been given various names, in various languages. Throughout history, factors including thin soils, forest clearance for timber, grazing by livestock, and/or repeated fires have produced low-, medium-, and tall-stature matorrals, more or less dense or open, and many of them appear to be natural. However, dozens of distinctive shrubland formations that occur around the Mediterranean Basin are clearly secondary; that is, they develop as the direct result of some combination of human activities.

Given the wide range of substrates, microclimates, and local land-use histories to be found in the basin, it is not surprising that matorrals show a wide range of structural forms and floristic composition (Tomaselli 1981). There are different local names to designate the diverse local forms of these natural and semi-natural shrublands and the landscapes they dominate: *garrigue* and *gariga* and *maquis* or *macchia*, in France and Italy; *xerovuni* and *phrygana*, in Greece; *matorral* and *tomillares*, in Spain; *choresh* or *maquis*, in Israel; and *bath'a* throughout the Near East are just some of the collection. Comparable terms used in other mediterranean-climate regions, where similar vegetation types are common, include: *chaparral* and *coastal sage*, in California; *matorral* and *jaral*, in Chile; *fynbos*, *renosterveld*, *karroid shrubland*, and *strandveld*, in South Africa; and *kwongan* and *mallee* in southern Australia.

In the Mediterranean region, as elsewhere, the names for the various matorrals are often poorly defined and their usage varies from one region and language to another. In many countries, *maquis* (or *macchia*) is considered as the first major stage in forest or woodland degradation, followed by *garrigue*, *phrygana*, or *bath'a*, which are all of still lower stature and complexity than *maquis*. In France, however, the distinction between *garrigue* and *maquis* is usually made on the basis of substrate, such that *garrigues* are said to occur primarily on limestone substrates (see Plate 10a), whereas *maquis* is reserved for those formations occurring on acid, silicaceous soils. In addition to holm oak and the cohort of associated species found in nearby *garrigues*, in the French *maquis* (e.g. in Corsica, the Maures massif near Marseille, and the Albères mountains in the eastern Pyrenees) are found such **calciphobe** species as strawberry tree, heaths, and heather (*Erica* and *Calluna*), as well as certain rockroses (e.g. *Cistus ladaniferus*), lavenders (*Lavandula*), and other shrubs. Yet this dichotomy has only limited value and a third matorral type also occurs, which is somewhat intermediate in floristic terms and is found on dolomite substrates. In Spain, this distinction between *garrigue* (or *gariga*) and *maquis* (*machia*) is not made. Similarly, Zohary (1962) designated as matorrals any sclerophyllous evergreen vegetation type that is dense and capable of attaining 4–6 m in height, but usually much lower. That includes woodland and shrubland.

Matorrals are dominated by shrubs with evergreen, broad, small, stiff, and thick sclerophyllous leaves, and depending on how open the formation is, a more or less rich understorey of annuals and herbaceous perennials (di Castri 1981). Thus, the most characteristic feature of matorrals is that they consist of a fine-grained mosaic of almost all the growth forms recognized by plant ecologists, including the full range described long ago by Raunkiaer (1934) in his attempt to propose a universal system of classification based on the position of leaf- and stem-renewal buds relative to ground level (Table 6.2).

Prominent examples of matorral shrubs that occasionally grow to be trees are the various evergreen oaks, carob, dwarf palm, bay, and lentisk, but also various species of *Arbutus*, *Daphne*, *Laurus*, *Phillyrea*, *Myrtus*, *Rhamnus*, and *Viburnum*, all of which are sclerophyllous evergreens. The low

Table 6.2 Main growth forms recognized by Raunkiaer (1934)

Growth form	Definition	Example
Therophyte	Annual	Poppy
Cryptophyte	Bulbous plant	Crocus, tulips
Hemicryptophyte	Perennial herb	Alfalfa, rhubarb
Chamaephyte	Shrub	Thyme, lavender
Phanerophyte	Tree	Kermes oak, laurel

For more details, see Orshan (1989).

and medium shrub layers of matorrals, especially in the western Mediterranean, include a number of familiar mint family members, such as lavender, rosemary, thyme, and others, as well as the bright-flowering rockroses (*Cistus*) and their relatives (*Fumana* and *Helianthemum*). In the Near East and North Africa, where human pressure is higher, matorral understoreys tend to have fewer shrubs and more hemicryptophytes.

Even if matorrals appear predominantly evergreen, about half of their woody species are in fact winter-deciduous. Examples are the maples, most *Pistacia* species, smoke bush (*Cotinus*), storax (*Styrax*), and *Rhus*, as well as numerous deciduous oaks. The brooms *Cytisus*, *Genista*, *Spartium*, and *Teline* are notably common and rich in species and subspecies in Iberia and Morocco and have the distinction of bearing evergreen stems that are photosynthetically active all year round. Most species in this so-called **retamoid** group have small deciduous leaves that fall off during droughts.

A great many **geophytes** also occur in matorrals, in a number of monocotyledon plant families. Among these, the Orchidaceae is represented by over a hundred species, but unlike most tropical orchids, here the family has only terrestrial species. As many as 50 species of orchid have been found to co-exist in a single matorral area of 100 ha in Greece. Many of them are common in frequently disturbed habitats, which are also rich in solitary bees that pollinate them. It has been suggested that chances for successful cross-pollination in terrestrial orchids increase when they are visible to their insect pollinators from great distances. This reliance on optical clues for specialized pollinators may be one of the selective pressures determining the distribution of most terrestrial orchid species to open habitats and also contributing to the great evolutionary success of this group of plants in matorrals (Dafni and Bernhardt 1990) (see also Chapter 8). Moreover, a remarkable range of strategies have evolved in the orchid family to attract pollinators, including various forms of deception (Cozzolino and Widmer 2005; Pellegrino et al. 2007). Recent studies have also shown that natural hybridization is much more common than was formerly thought (Pillon et al. 2006), and this has serious consequences for conservation science and practice (Cozzolino et al. 2006).

Much work has been devoted to deciphering the pathways of degradation of former Mediterranean forests and woodlands and their transformation into matorrals. Although most matorrals appear to retain the ability to recover spontaneously if human intervention ceases, some of them are so badly degraded that, without long-lasting intervention, they remain blocked in stunted, species-poor formations. A large number of possible **ecosystem trajectories** exist in Mediterranean ecosystems, both towards degradation and recovery. The history of local land use plays a critical role and many matorrals are so far removed from the primaeval forests that they never return spontaneously to the stature and structural complexity of true forests, unless there is massive human intervention. Instead, they are 'locked' or 'blocked' in self-perpetuating systems that can endure for centuries with little visible change. Two examples mentioned in Chapter 5 were the low-altitude formations dominated by the dwarf kermes oak, which extend unbroken over hundreds of hectares at a stretch, in the north-western quadrant of the basin, and the tragacanthic formations in the eastern Mediterranean mountains, especially in Turkey and Lebanon. At mid-altitudes as well, spiny legume shrubs, including *Calycotome spinosa*, brooms, and gorse (*Ulex europaeus*), can form large stands, sometimes described as 'leopard's skin' because patches of low green 'brush' or scrub alternate with patches of bare ground and rocks. In many such cases, former uses by people have lead to these formations, arising and persisting even long after the use by people has ceased, as was mentioned in Chapter 5 for tragacanth, kermes oak, and rockrose formations. In other examples, coppicing for wood production can be sustained over centuries, also without visible changes (see Chapter 10).

The Spanish term *tomillares*, the Arabic and Hebrew word *bath'a*, and the Greek term *phrygana* all refer to much lower plant formations than the matorrals, but that are closely related to them floristically and historically. The *tomillares* of Spain are shrublands characterized by a large number of thyme (*Thymus*) species (*tomillo* in Spanish). *Phrygana* in Greece and Turkey also consists of formations dominated by dwarf shrubs typically 20–70 cm tall. Structurally similar to these formations, *bath'a* formations are the Near Eastern equivalent and occur interspersed with higher matorral types, in the higher rainfall areas near the coast, as well as at middle altitudes on mountain slopes. More commonly, *bath'a* occupies large stretches in semi-arid, Mediterranean-desert border regions further inland. Common species here include *Calycotome villosa*, *Genista acanthoclada*, *Eryngium*, *Satureja*, and *Sarcopoterium spinosum*, as well as several bulb species, some rockroses, and hundreds of species of annuals.

Describing these dwarf formations, Shmida (1981) noted that canopy cover may be as extensive as in matorral, but total biomass is much lower. Leaves of dominant shrubs tend to be small, soft, and pungent, emitting volatile terpenes when touched. Winter foliage is regularly replaced by even smaller, more sclerophyllous foliage during summer. Leaves and shoots die back in part or in full in a surprisingly flexible fashion that is determined by both rainfall distribution and temperature extremes (Orshan 1972). Such facultative drought-deciduous species are often spiny and hemispherical shrubs that appear to be resistant to herbivory, fire, and prolonged drought. They occasionally include stem- or leaf-succulents, such as the spurges (*Euphorbia*; these are highly toxic to animals). This botanical composition and life form spectrum is highly suggestive that these formations are the result of long-standing and deep ecological footprint of humans and their domestic livestock, often at the fluctuating transition zone between the mediterranean-climate zone and the desert, where in human geography terms, there is an age-old struggle between 'the desert and the sown' (Reifenberg 1955).

6.3 Steppes and grasslands

The history and phylogeny of a great many plants of the Mediterranean Basin (especially in the eastern parts) can be traced to the arid and semi-arid steppes of central Asia, the so-called Irano-Turanian floristic region described in Chapter 2 and referred to in Chapter 5. Recall that Zohary's law suggests that disturbed areas are generally invaded by colonizing elements coming from the drier habitats in the vicinity rather than from the wetter ones. This tendency is reflected at another spatio-temporal scale by an apparent 'intrusion' of Irano-Turanian elements in adjacent Mediterranean vegetation zones, as we saw in Chapters 2 and 5. A great many Irano-Turanian steppe elements have also colonized disturbed sites in the western Mediterranean, especially North Africa (see Le Houérou 2001), and have also been introduced—inadvertently in most cases—to most of the other climatically similar areas around the world. This biogeographical success is due to their exceptionally long period of pre-adaptation to frequently and variously disturbed sites.

In southern Mediterranean regions, particularly at higher altitudes, and in localities receiving less than 300 mm mean annual precipitation, natural vegetation is mostly made up of mixed annual and perennial grasslands, with only scattered bulbs, shrubs, and trees. The most extensive of these are the vast 'alfa steppes' of North Africa and parts of south-eastern Spain, with their dominant bunch grass, alfa (*Stipa tenacissima*), and the shrubs wormwood and *Rhanterium* (Asteraceae) (Le Houérou 2001; Maestre and Cortina 2002). In certain areas in North Africa, very large and old betoum trees (*Pistacia atlantica*) are found, no doubt because they were considered sacred and therefore off-limits for cutting by local people and nomads (see Chapter 10 for more discussion of sacred trees). Where aridity combined with a high water table creates saline or alkaline pans punctuating the steppes, salt-tolerant and semi-succulent saltbushes of the Saharo-Arabian floristic group, primarily *Atriplex*, *Haloxylon*, *Salsola*, and *Suaeda* (all Chenopodiaceae), and saltgrasses such as *Aelopus*, constitute extensive formations.

6.4 Old fields

As noted in Chapter 2, there are between 1500 and 2500 species of annuals, biennials, and bulbous plant species in the Mediterranean flora that mostly occur in early stages of succession in cultivated, fallowed, or abandoned fields, terraces, or pastures. Abandoned farmlands are also known as old fields, and they are drawing increasing attention in the Mediterranean region. Here, arable lands and pastures were historically opened up for temporary cultivation in the full range of forests, woodlands, and shrublands, as well as wetlands, only to be abandoned subsequently. In a Mediterranean landscape mosaic, this creates a dynamic set of interactions in old fields driven by re-colonization of herbaceous and woody plants from varying habitats, as well as by the widespread annuals and biennials that invaded during the period of cultivation. This later group of sun-tolerant and highly competitive annual or biennial species are called 'ruderal' and 'segetal' species, depending on whether they grow in uncultivated sites like roadsides, or else in sporadically cultivated fields (see Chapter 8). Numerous ecological studies have been and are being conducted on biota in these habitats, which will no doubt expand in the future, especially on the northern shores of the Mediterranean, as a result of changing agricultural policies and socio-economic conditions (Marty *et al.* 2007; Papanastasis 2007; Bonet and Pausas 2007).

6.5 Cliffs and caves

Many limestone-dominated landscapes in the Mediterranean contain cliffs, caves, and escarpments, which provide highly specialized habitats for a number of plants and animals. In addition, there are tens of thousands of caves, shafts, and 'avens' among the labyrinth of the hills and mountains. The larger cliffs and virtually all of the caves are made of compact limestone and often harbour unusual micro-habitats and rare species. In southern France, only one species, the rare Petrarch's fern (*Asplenium petrarchae*), is endemic to cliffs on south-facing slopes near the sea coast.

Cliff habitats include step-crevices, vertical faces, overhangs, pavements, and sloping cliffs. Solar insulation and soil moisture conditions are often limiting factors to plant growth. Ecological adaptations related to a perennial life form, prolonged and conspicuous flowering, high germinability of seeds, and various long-distance seed dispersal mechanisms also characterize **chasmophytes**. Prominent among these species are an eastern Mediterranean contingent, including *Varthemia iphionoides* and *Phagnalon rupestre* (Asteraceae), *Rosularia lineata* (Crassulaceae), and several species of *Micromeria*, *Stachys*, and *Teucrium* (Lamiaceae). A second group has been identified as a 'middle Mediterranean' element, containing many mustard family (Brassicaceae) members, such as the showy wallflowers (*Erysimum*) and the *Brassica cretica* species group. The spectacular flowering capers (*Capparis spinosa*) are visible from far off on cliffs of warmer areas.

Chamaephytes are by far the most common life form found in cliffs but geophytes also occur, including cyclamens (*Cyclamen*), ferns, succulents (*Sedum*, *Cotyledon*, and *Caralluma*), wild snapdragon (*Antirrhinum majus*), and sticky-weed (*Parietaria judaica*). In some groups where most species are annuals, the cliff-dwelling species are all shrubs. Cliff faces and overhangs also harbour a number of tree species with a dwarfish appearance because of water and nutrient limitation. Some of these appear to be clinging to the cliffs like twisted survivors of a thousand storms. Among the most spectacular are Phoenician juniper (*Juniperus phoenicea*), holm oak, and fig tree (*Ficus carica*). It has been suggested by Snogerup (1971) that eastern Mediterranean cliffs may have served as refugia for various matorral and phrygana species of trees or shrubs during cold periods of the Pleistocene.

This highly specialized habitat offers breeding sites for one of the most unique group of birds in the Mediterranean. Several large raptors use ledges to breed in colonies (Fig. 6.2). A ballet of griffon vultures (*Gyps fulvus*) is certainly one of the most unforgettable sights for a bird watcher. A number of more secretive cliff-dwelling species include Bonelli's eagle (*Hieraaetus fasciatus*), eagle owl, Egyptian vulture (*Neophron percnopterus*), raven (*Corvus corax*), peregrine falcon, kestrel (*Falco tinnunculus*), stock dove (*Columba oenas*), and, in the southern Mediterranean, colonies of the lesser kestrel. In some parts of the basin, notably in the south and east, cliffs dominating the sea are inhabited by large colonies

Figure 6.2 A typical Mediterranean cliff with some of its inhabitants. From left to right: lesser kestrel, Bonelli's eagle, Egyptian vulture, and eagle owl.

of Eleonoras's falcon. This species has a fascinating biology since it breeds very late in the season—in July and August—taking advantage of the autumnal migration of passerine birds upon which it feeds (Walter 1979). Quite surprisingly, the only winter ground of this falcon is Madagascar.

Looking in more detail at the cliff habitat, the observer finds many other birds as well, such as the shy blue rock thrush that builds its nest in the darkest parts of overhangs or the black redstart (*Phoenicurus ochruros*). Colonies of alpine swifts (*Apus melba*), rock swallows (*Riparia rupestris*), and the rare pallid swift (*Apus pallidus*) occur near the coast, not to mention various species of bats, which breed in deeply fissured vertical rock faces. In winter, some birds, which breed further north in the Alps, overwinter in cliffs or on their ridge. This is the case of the alpine accentor (*Prunella collaris*), the wallcreeper (*Tichodroma muraria*), and sometimes small flocks of snow finches (*Montifringilla nivalis*).

The life of these 'cliff birds' is not always peaceful. Ravens have been observed robbing recently captured prey from nests of eagle owls, while those same owls themselves sometimes catch and eat peregrine falcons, which explains why these two species never breed together on the same cliff!

6.6 Riverine or riparian forests

Mediterranean riverine forests were once complex, biologically varied ecosystems, extending over more than 2000 km² in the basin, punctuating landscapes and entire regions with a high diversity of plants and animals. Unfortunately, most of them have been definitively removed and replaced by agriculture. Only some remnants are left, for example along the Morača River in Montenegro, along the Strymon River (Lake Kerkini, Greece), in the middle Po valley of northern Italy, and in the lower Rhône River valley of southern France. The few remaining patches that have been preserved give an idea of what such habitats must have once looked like. In the Nestos delta of eastern Greece, for example, about 60 ha of intermittently flooded riparian forest with poplars (*Populus alba*), alders (*Alnus*), and willows remain today in a nature preserve. Thanks to particularly favourable edaphic conditions, these forests are dominated by deciduous trees, such as oaks, poplars, elms, alders, and willows (Pérez-Corona *et al.* 2006), upon which climb luxurious vines of wild grape (*Vitis silvestris*), hops (*Humulus lupulus*), and various species of clematis (*Clematis*). In the Balkans, a dominant tree species of this habitat is Oriental plane. Ribbon-like, these forests may penetrate from temperate areas into the warmer Mediterranean ecosystems, bringing with them a series of species and life forms that would not survive outside the shadow of the trees and moist microclimate.

6.7 Wetlands

The Mediterranean Basin includes a variety of wetlands, from large inland lakes to small temporary ponds and extensive coastal lagoons. Wetlands occur in all parts of the basin but they are more diversified and cover larger areas on the European shores than in North Africa and the Middle East. However, with approximatively 250 bodies of water totalling 1.3 million ha, Turkey encompasses a wide array of wetlands that are highly diversified in terms of hydrological regimes. In summertime, wetlands often stand out as oases of greenery and moisture among the dry and yellowish countryside. Britton and Crivelli (1993) estimated at 21 000 km² the area covered by wetlands in the basin, of which 4700 km² are coastal lagoons, 2800 km² are freshwater lakes and marshes, and 11 600 km² are temporary salt lakes, found mostly in North Africa. Except for wetlands that are connected to large permanent rivers and some inland freshwater lakes, the main characteristic of Mediterranean wetlands is the fluctuations in water levels and salinity, which reflects the large variation in rainfall both within and between years. Five main categories of wetlands can be recognized in the Mediterranean, as will be described below.

6.7.1 Freshwater lakes

Except for the many human-made reservoirs built to increase water supply, most inland freshwater lakes in the Mediterranean are of glacial origin. These are limited to high altitudes in the mountain ranges that were affected by Pleistocene glaciations—the Sierra Nevada, Pyrenees, Alps, Apennines, Dinaric Alps, and the Atlas ranges (see Chapters 1 and 2). Most of them are nutrient-poor, steep-sloping, deep, and have little or no emergent vegetation. Some inland lakes in Italy, however, are of volcanic origin and occupy ancient calderas. A series of inland lakes in the Balkan Peninsula are parts of karstic formations resulting from fracturing of limestone blocks. The most famous of these are the inter-connected Greek lakes Megali Prespa and Mikri Prespa and also Lake Vegoritis, which are among the richest and most productive aquatic ecosystems of the Mediterranean (e.g. Catsadorakis 1997). They are rich in both emergent plants, such as common reed (*Phragmites australis*), bulrush (*Typha latifolia* and *Scirpus lacustris*), and yellow iris (*Iris pseudacorus*), and submerged vegetation, such as nenuphar (*Nymphaea alba*), *Ranunculus*, fringed water-lily (*Nymphoides peltata*), and whorl-leaf water milfoil (*Myriophyllum verticillatum*), and are also biological hotspots for endemic fish species and wildlife. A large diversity of birds, including the white pelican (*Pelecanus onocrotalus*), the dalmatian pelican (*Pelecanus crispus*), as well as several herons, spoonbills, cormorants—including the rare pygmy cormorant—and ibises, breed there in large colonies (Catsadorakis 1997; Catsadorakis and Crivelli 2001).

6.7.2 Deltas and coastal lagoons

The most extensive wetlands in the Mediterranean are alluvial flood plains, coastal lagoons, and deltas of the main rivers flowing down from nearby high mountain ranges: Guadalquivir and Ebro in Spain, Rhône in France, Po in Italy, and Axios and Evros in Greece. The most important lagoon systems occur over more than 200 km from the mouth of the Rhône to the French/Spanish border and from Venice to Trieste along northern Italy's Adriatic coast. Large deltaic systems do not occur in the **Maghreb** part of western North Africa, since the short, highly seasonal rivers found there do not provide enough sedimental material for delta formation. The only large delta in North Africa is the Nile delta, whose waters flow down from tropical Africa.

The formation of extensive deltaic systems is favoured in the northern Mediterranean by the small amplitude of tides, which allows the development of offshore sand banks inside which alluvial sediments deposit in shallow waters. Coastal lagoons are produced by the accumulation in coastal waters of sand and silt deposits that are brought by rivers and continuously reshaped by marine currents and wind. This results in the building of offshore bars that are more or less parallel to the coast and encircle inland lagoons. When powerful rivers empty into the sea, they form deltas with numerous channels continually changing course. This gives rise to a maze of marshes and lagoons interspersed by sand dunes or mud flats, arising from the river's meanderings and oxbows over the course of centuries and millennia. The ongoing changes of river arm configuration within a delta result in a perpetual upheaval in habitats. Thus, a large Mediterranean delta is a moving mosaic of wetlands with contrasting salinity and seasonal water levels, usually no more than 2 m deep. Salt concentration varies widely in space and time from fresh water to hypersaline waters (up to $40 \, \text{g} \, \text{l}^{-1}$), in relation to rainfall and seasonal water levels. Coastal lagoons are typically isolated from the sea by sand dunes, which are open here and there, allowing connections with the sea. Coastal dune vegetation varies in importance from a narrow spit with marram grasses (*Ammophila*), to extensive woods of Phoenician juniper and stone pine.

A recurrent feature in all coastal Mediterranean aquatic systems is their huge variation in biologically important factors, such as flooding periodicity, water salinity, and soil salinity, which all have profound influence in the structure and dynamics of plant and animal communities. Many of these factors vary enormously during the course of the year, from year to year or over even longer periods of time. As a result, plant and animal communities are highly dynamic and do not exhibit long-term predictable successional changes in species composition, except when lagoons are managed for the production of salt. When this is the case, a beautiful pinkish colour is characteristic of this type of habitat (Box 6.1). Wherever water levels widely fluctuate, current assemblages of plant species usually reflect recent past events.

> **Box 6.1. Pink lagoons**
>
> Flying over Mediterranean lagoons and salt pans in a plane, passengers will inevitably be surprised and delighted by the variety of colours that characterize brackish and saline waters of Mediterranean coastal marshes and lagoons. From whitish to pale pink to plainly red, depending on the salt concentration of the water, there is an infinite range of delicate colour shades. These colours are due to minute algae, the pink-coloured *Dunaliella salina*. In spite of their colour, these algae belong to a group of green algae, the class Chlorophyceae (family Dunaliellaceae). These halophile micro-algae are particularly resistant to salt thanks to their high concentration of β-carotene, which protect them against intense light and high concentrations of glycerol. These properties make them used in cosmetics and dietetic foodstuffs. Very few organisms can thrive in hypersaline waters, but low species richness is often compensated by extremely high population densities. *D. salina* transmits its pinkish colour to the whole food chain of the lagoon, namely tiny crustaceans which feed upon it, especially the brine shrimp *Artemia salina* (branchiopod). This small animal, which can remain metabolically inactive in total stasis for several years, has a biological life of 1 year during which it develops to a mature length of 1cm on average. Brine shrimp is among the very few species which can tolerate extremely high levels of salinity and unusually high water temperatures. Because they are rich in lipids and unsaturated fatty acids, they constitute an important food supply for several species of fish and birds, which feed upon them regularly. The pink or red colour of the emblematic flamingo derives from the concentration of β-carotene in its food (see Box 11.8). Other species that are more or less pinkish thanks to this food supply are the beautiful slender-billed gull and the shelduck (*Tadorna*).

In areas that are flooded for a few months and where salt concentration remains high, vegetation is mostly composed of **halophytes** in the Chenopodiaceae family, especially several species of *Arthrocnemum* and saltgrasses, such as *Aeluropus* and *Paspalum*. In these flatlands, which often cover large areas, the only tree species are tamarisks, which thrive under a wide range of salinity and water levels (e.g. *Tamarix africana* and *Tamarix canariensis* in the Iberian Peninsula, *Tamarix gallica* in France, and *Tamarix tetranda* in the Balkan Peninsula). They can survive flooding for up to 6 months or more at a water depth of 1 m. Where salt concentration in the soil is lower, wet grasslands, including many papilionoid chamaephytes, may extend over huge areas. These shrubs include *Dorycnium jordani* and, in the eastern Mediterranean, legume shrubs *Prosopis farcta* and wild licorice (*Glycytthiza glabra*).

Aquatic vegetation in permanent water bodies also varies according to salinity. It includes the ditch grasses *Ruppia* and various algae (e.g. *Ulva*, *Chaetomorpha linum*; see Box 6.2) in saline waters, passing to freshwater plants (e.g. sago pondweed (*Potamogeton pectinatus*), brackish water-crowfoot (*Ranunculus baudotii*), water milfoil (*Myriophyllum*), and *Zannichellia*) and large reed beds as salinity decreases.

Mediterranean wetlands are rich in fish species, which occupy the various habitats according to their tolerance of highly saline, brackish, or fresh water. Sea fish include sea bass, gilt-head (*Sparus aurata*), and common sole, whereas carp (*Cyprinus carpio*), pike (*Esox lucius*), and pike-perch (*Sander lucioperca*) live in fresh water. Species which tolerate the large fluctuations in salinity levels found in brackish lagoons include eels (*Anguilla anguilla*), several species of mullet (Mugilidae), sand-smelt (*Atherina boyeri*), and a tiny Mediterranean seahorse, *Syngnathus abaster*. Many fish species, called **diadromous** and euryhaline, depend completely on brackish lagoons for spawning so that large migrations of these fish occur between the sea and these inland bodies of water.

The periodic but unpredictable drying out of many Mediterranean wetlands has resulted in the evolution of several strategies that allow animals

> **Box 6.2. Characeae as breeding and foraging algae for animals**
>
> In many bodies of water, including those that dry up for several months in summer, plant communities of shallow water include Characeae, an important group of freshwater green algae that resemble vascular plants. Characeae have the exceptional ability to fix lime in their tissues (up to 70% of the plant) so that they become brittle and coarse to the touch. These submerged plants are of primary importance as food for the hundreds of thousands of ducks that overwinter in the Camargue. Thick carpets of Characeae are also excellent breeding sites for many fish, amphibians, and aquatic insects, while **oogons** of these algae constitute an important part of the diet of teal *Anas crecca* in winter (Tamisier 1971).

to escape the effects of drought and desiccation. For example, several bird species, such as the flamingo (see Plate 10b), marbled teal (*Anas angustirostris*), stilt (*Himantopus himantopus*), and ruddy shelduck (*Tadorna ferruginea*), are peripatetic and opportunistic, taking advantage of temporarily favourable conditions wherever they occur. These birds can breed in quite different areas from one year to the next, depending on water levels. Similar flexibility in behaviour is found in other animals as well. In many invertebrates, especially small crustaceans, such as amphipods, ostracods, and copepods, reproduction and growth occur in winter and early spring. These animals await the return of favourable conditions by spending the hot dry summer in quiescent stages such as eggs. Thus their life cycles are often seasonally reversed as compared to those of similar organisms in central Europe.

Productivity of coastal lagoons is exceptionally high, having been estimated to be eight to 10 times greater than that of the sea. The economic value of lagoon fisheries along the French Mediterranean coast exceeds that of the Mediterranean trawling fleet (Britton and Crivelli 1993), and coastal lagoons yield 10–30% of the total Mediterranean production of fish. In the saline lagoons of the Camargue (see Plate 10b), the fauna is often reduced to a few highly adapted species (Britton and Johnson 1987), notably the brine shrimp *Artemia salina* (see Box 6.1), which can survive salinity levels up to $300\,g\,l^{-1}$, larvae of the dipterans *Ephydra*, a copepod, *Cletocamptus retrogresses*, a few microturbellaria, and nematodes. It has been estimated that the biomass of invertebrates may reach $500-1000\,g\,m^{-2}$ in late spring, when water temperature is rising and most organisms are at their peak of annual growth. There may be as many as 30 000–50 000 brine shrimps per square metre at this time or year! Such high concentrations constitute the main food of flamingos, shelducks, avocets (*Recurvirostra avosetta*), and Kentish plovers (*Charadrius alexandrinus*), as well as swarms of migrating waders.

Taking advantage of the exceptional diversity and productivity of Mediterranean wetlands, many species of birds breed together in mixed colonies. Extending over 145 000 ha in southern France, the Camargue is one of the most famous and best preserved wetland areas of the basin (see Box 13.4).

More than 10 species of terns, gulls, and waders breed together on small islets scattered in the large lagoons of the Camargue. Colonies may include up to several thousand breeding birds and attract secretive species, such as ducks (*Netta rufina*, *Tadorna tardorna*) and redshank (*Tringa totanus*), which benefit from the protection provided by other species. From the core breeding area, these birds then disperse to a variety of feeding grounds, according to species-specific habitat preferences and foraging techniques (Fig. 6.3).

Some species, for example avocet, stilt, and Kentish plover, do not leave the lagoon and forage only in shallow water at a depth proportional to the length of their legs. They mostly feed on small crustaceans, including phyllopods (*Artemia*), amphipods (*Gammarus locusta*), and on a large variety of aquatic insects (Ephydridae, Syrphidae, Chironomidae, and Dolichopodidae). They also occasionally eat molluscs (*Hydrobia*) and marine worms (*Nereis diversicolor*). The slender-billed gull is highly specialized to Mediterranean lagoons, where it catches flat fish using the unusual technique—for a gull—of plunging its long neck into shallow water. The oystercatcher (*Haematopus ostralegus*), which is not very common in the area, regularly one of them, the sandwich tern (*Sterna sandvicensis*) is restricted to this feeding area. The common tern (*Sterna hirundo*) and the smaller little tern (*Sterna albifrons*) forage both at sea and in lagoons and canals. The rare gull-billed tern (*Sterna nilotica*) is more terrestrial and feeds mostly inland on large insects, such as mole crickets, grasshoppers, and dragonflies, and amphibians, lizards, and crustaceans (*Triops, Branchipus*), which it finds in rice fields, sometimes far from the breeding colony.

Finally, the two most common gulls, the black-headed gull and the yellow-legged gull, are highly eclectic in their feeding habits and foraging habitats (Duhem *et al.* 2008). As a consequence they benefit from human-induced changes of landscapes, as they make use of a large variety of food resources found in marshes, farmland, and rubbish tips. Accordingly, their population sizes have increased tremendously during the past century, from some few hundred pairs of black-headed gulls in the 1930s to nearly 10 000 breeding pairs in the 1960s, and from a handful of yellow-legged gulls to more than 4000 pairs today (the annual rate of increase is 9%). Lebreton and Isenmann (1976) demonstrated that population increase of the black-headed gull was mainly a result of a higher winter survival rate, due to the use of predictable food resource provided by rubbish tips and boat trawling. The yellow-legged gull represents a threat for the breeding colonies of other gulls, terns, avocets, and even flamingos, because it robs their eggs and eats their young fledglings. Entire colonies would be completely destroyed by yellow-legged gulls if control measures were not repeatedly taken to limit their numbers.

6.7.3 Temporary marshes

Not all bodies of water in the region are connected to the main rivers, to the sea, or to large permanent lagoons. One very interesting example is the temporary or **endoreic marshes**, occurring in natural depressions with no outlet. They are completely dependent on rainfall and therefore dry up completely for several months each year. These temporary marshes also occur in other parts of the world, including mediterranean-type climate

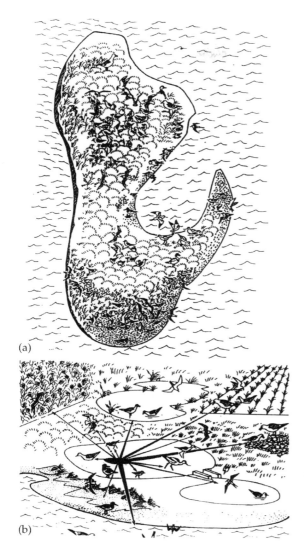

Figure 6.3 Organization of a typical breeding colony of birds in the Camargue. (a) As many as 14 species breed together on an islet and benefit from mutual protection from predators. There are often several subcolonies, two of which are shown: in upper part of drawing, a flock of black-headed gulls (*Larus ribidunbus*) with several pairs of slender-billed gull, ducks, and redshank; in lower part, flocks of common tern, sandwich tern, and other species. (b) Each species leaves the islet to forage in species-specific habitats.

forages along the coasts of lagoons and the sea, searching for large molluscs (*Cardium*) and beetles (tenebrionids and carabids). Several species of fish-eating terns regularly forage at sea, but only areas, for example Chile and California, where they are known as vernal pools, and often show high biodiversity of plants and animals with fascinating life-history adaptations (Thompson 2005: 117).

In our region, temporary marshes are not only variable over time, they are also highly variable in size, from a few hectares to many square kilometres, as well as having variable duration of inundation, geological substrate, and levels of salinity (Grillas and Roché 1997). In spite of this large diversity, they share several common characters, as far as living organisms are concerned, because of the alternation of wet and dry phases. This alternation favours species with short life cycles and strongly decreases interspecific interactions, such as competition and predation. Their biodiversity is often exceptionally high, especially for annual plants (Bonis *et al.* 1995), amphibians, and crustaceans. They harbour rare and often threatened plant species, such as the water fern *Marsilea strigosa*, *Pilularia*, loosestrife (*Lythrum*), rushes (*Juncus*), quillwort (*Isoetes setacea*), and the thrumwort (*Damasonium stellatum*), as well as a buttercup, *Ranunculus laterifolius*. Some of these species are known from a handful of sites at most. For example, the very rare *Teucrium aristatum* is only known from two sites, one in France (Crau) and one in Spain. As many as 4% of the plant species that are considered as being threatened in France occur in these temporary marshes (Grillas and Roché 1997). Because of the strong constraints linked to unpredictable water supply, plant and animal species specialized in this type of habitat have evolved remarkable growth forms and life-history traits, especially with regards to the production and dispersion of seeds, eggs, spores, and various diapause forms. Most plant species found here are annuals or biennials, with a life cycle lasting only a few weeks (Grillas and Roché 1997). They produce large quantities of seeds that may remain in the ground for a very long time, until favourable conditions for germination occur, which is an insurance against local extinction. As many as 320 000–530 000 seeds m^{-2} have been reported in the Doñana marshes, Spain (Grillas and Roché 1997).

The very distinctive habitat of temporary marshes also harbours a large number of invertebrates, such as crustaceans adapted to prolonged droughts (copepods, phyllopods, ostracods), which often occur in incredibly large numbers since fish predators are absent (Brucet *et al.* 2006). The unmistakable *Triops cancriformis*, which resembles a small limule, lays eggs that, after drying up of the marsh, may encyst for several months before hatching when the marsh fills up again. Temporary marshes are also favourable breeding sites for many insects whose life cycle includes an aquatic stage, such as dragonflies and many beetles.

These marshes are also of paramount importance as breeding sites for amphibians. In March and April, marbled newt, common newt, as well as several species of frogs and toads, especially the Iberian spadefoot, the parsley frog (*Pelodytes punctatus*), and several others (*Alytes obstetricans*, *Bufo calamita*, and *Rana perezi*), lay their eggs here over a period of several weeks. As a response to the short period of flooding that occurs mostly in winter, several species of amphibians have evolved a winter breeding season so that metamorphosis of the tadpoles occurs before drying up of the breeding site (Morand 2001). For example, the Iberian spadefoot spreads its breeding season from October to February in Andalucía, thus increasing its chances for successful reproduction. Of course, there are high extinction risks for these small populations tightly linked to fugitive habitats. Most of them presumably function as parts of a larger **metapopulation** with exchanges of individuals among subpopulations breeding in discrete habitat patches scattered over short distances within a landscape. The more small ponds occur in a landscape, the lower the extinction risks for populations of these of amphibians. Unfortunately, as much as 30–50% of these habitats has been destroyed throughout the basin, which renders many species characteristic of these habitats vulnerable to local extinction (Morand 2001). Global changes, especially climate warming and an expected decrease of rainfall in the Mediterranean Basin (IPCC 2007), represent serious threats for many of these fragile habitats (see Chapter 12). In addition, many of the temporary marshes that are located near the coast will be threatened by an increase in salinity as a result of rising sea level. Others may be subject to shrub encroachment (Médail *et al.* 1998; Rhazi *et al.* 2004).

6.7.4 Chotts, sebkhas, dayas, and gueltas

In the most arid parts of the basin, notably in North Africa, central Turkey, and in some parts of the Iberian plateau (Laguna de Gallocanta), where annual rainfall does not exceed 400 mm, there occur large endoreic temporary wetlands too dry and salty to be included in the previous section. The largest of these are called *chotts*. Many of the depressions, where they occur in North Africa, were once extensive freshwater lakes when the climate was more humid than today. Some of them, such as Chott Djerid in Tunisia, are among the largest wetlands in the Mediterranean. A large chain of endoreic drainage basins also occurs in Algeria, at high altitudes (approximately 1000 m) on the Plain of Chotts between the two main ranges of the Atlas Mountains. These temporary wetlands are usually devoid of aquatic vegetation and have a crust of halite or anhydrite covering the lake floor. Their margins are covered by a scattered vegetation mostly consisting of halophytic bushes (e.g. *Salicornia*, *Arthrocnemum*). Isolation from permanent water bodies, long periods of complete desiccation, as well as large seasonal variations in salinity make these habitats inhospitable to most species and life forms. Those species that are able to colonize them are either highly resistant to desiccation (e.g. the Cladocera), or else colonize readily over long distances (Corixidae), which allows them to occupy these ephemeral habitats for short periods before moving on to other sites (Boix *et al.* 2004).

In the arid zones of North Africa and the Near East, smaller depressions, called **sebkhas**, are occasionally filled following heavy rainfall. Since evaporation is about 10 times higher than atmospheric precipitation in these areas, surface water remains no longer than a few weeks. This may suffice, however, to provide breeding grounds for large colonies of nomadic birds, such as flamingos and stilts, as well as stop-over sites for thousands of migrating birds. A large portion of central Turkey, where rainfall is usually less than 400 mm year^{-1}, is drained by the Lake Tuz, a *chott* 90 km long and 32 km wide, but no more than 1.5 m deep. Although used as a salt pan yielding two-thirds of Turkey's industrial salt production, a colony of flamingos and several tens of thousands of wintering geese regularly visit this wildlife hotspot.

Some even smaller bodies of water in North Africa are also of great importance for wildlife, in arid regions where water is scarce. Examples are *dayas* and *gueltas*. The former are small endoreic temporary ponds where water occurs for some weeks or months after large rainfall; the latter are deep holes in the bed of rocky wadis. *Gueltas* usually retain permanent water and are important spots for several species of plants, fish, amphibians, and sometimes breeding birds, for example the ruddy shelduck.

6.7.5 Intertidal mudflats

Since there are practically no tides in the Mediterranean Sea (see Chapters 4 and 9), there are also virtually no intertidal mudflats except in the Gulf of Gabès of the southern Tunisian coast, around the nearby Kneïs Islands, and to a much lesser extent at the head of the Adriatic Sea near Trieste. In the Gulf of Gabès, a tidal amplitude of 3 m creates nearly 200 km^2 of mudflats, which support seagrasses, such as *Zostera noltii* and *Zostera nana*, and the saltgrass *Spartina maritima*. These habitats are used as stop-over places for thousands of migrating waders in both spring and autumn.

6.8 Diversity of marine habitats

As described in Chapter 1, the Mediterranean Sea is a huge mass of water occupying two large basins separated from each other and from the Atlantic Ocean by shallow straits, one of which is also very narrow. This water mass is called the pelagic domain or pelagos, an ancient term referring to the open ocean or sea. The ecological realm at the lowest level, or just below the pelagos, which includes the sea bottom and some superficial subsurface sediment layers, is called the benthic domain or benthos.

The benthos corresponds to a boundary between the water and the continent, called the water/substratum interface. Another interface, between the air and water, is of course situated at the surface of the sea. They are the most important regions of the seas and oceans, regrouping a very high percentage of the forms and abundance of sea life. The air/water interface receives the solar energy transformed by the photosynthesis of the phytoplankton, and this so-called primary production is the starting point of the marine food chain. At the other end of the water column, the water/substratum interface receives all the particles that fall down through the water. The closer these two interfaces are to one another, the richer the range of life forms present. This situation corresponds of course to the shallow coastal waters. We will briefly review the pelagos, then the benthos, and then a special habitat near the coasts, the river estuaries.

6.8.1 Pelagos

While the benthos is stationary, the pelagos is in constant motion. Sea currents renew and oxygenate the water, carry living organisms from place to place, and transport dissolved and particulate organic matter that is necessary for marine animals and plants. Mediterranean currents also create linkages to the Atlantic and the atmosphere. The loss of water by evaporation is at the source of a very specific thermohaline circulation which differs from that of the world oceans. The water deficit is compensated by an inflow from Atlantic Ocean through the Straits of Gibraltar (see Chapter 1).

Contrary to the atmosphere, the pelagos is a genuine habitat in the sense that some of its inhabitants spend their entire live here, including the various kinds of passively floating, drifting, or somewhat motile organisms called plankton that have been discussed in Chapter 4, as well as the **nekton**, which is the technical term for active swimmers among marine fauna. During its entire life, for example, a tuna may never see the sea bottom or any coast. In the pelagos, there is a vertical distribution of habitats determined by the penetration of sunlight and topography of the deep bottoms (Fig. 6.4). The uppermost, sunlit layer is called euphotic and is home to the majority of the living biomass in the sea with the exception of a thin upper part, where solar energy is intense such that ultraviolet rays are too aggressive for life. Very few organisms survive there. Divided into an epipelagic zone, which is the habitat for the **photophilous** species, and a mesopelagic zone, habitat for the **sciaphilous** ones, the euphotic layer extends over the entire Mediterranean Sea with a vertical extension in relation to clarity of the water; it is deeper in the eastern basin and in the open sea.

In fact, the Mediterranean Sea is blue because of the scarcity of phytoplankton and the generally low density of suspended particles. Surface waters entering from the Atlantic, which are already stripped of a part of their nutrients by the oceanic phytoplankton, lose even more of their nutrient content as they move eastwards, as a result of feeding by phytoplankton. Finally, the *mare nostrum* is 'underfed' because of the lack of vertical exchanges between the water masses which, in the world's oceans, continuously fertilize the surface layers from the bottom upwards. Large rivers like the Ebro (Spain), Rhône (France), and Po (Italy), several large rivers outflowing from the Balkans, and the outflows of the Black sea are also sources of nitrates and phosphates, but they only fertilize near-coastal waters. The Nile River lost its influence on the sea after the Aswan Dam was built in 1960, and there are no other important rivers traversing the southern Mediterranean shores (see Chapter 1). Atmospheric inputs of nutrients may, however, be important on the African coast, via dust deposition, which also increases along a gradient from west to east. Aeolian inputs of iron, an important trace element for phytoplankton growth, are among the highest recorded in the region, and can exceed the river inputs (Turley 1999). Even if these windblown particles only arrive episodically, they can still be very important sources of nutrients for oligotrophic waters like those of the Mediterranean.

Additionally, coastal regions permanently receive continental contributions of nutrients as waves and currents renew the particles and the oxygen in the water. A portion of the microphytobenthos may also be put into suspension by water turbulence and thereby further enrich coastal waters (see Chapter 4). Consequently, the coastal areas are not only biologically rich, as mentioned above, but also always relatively productive. In other words, in addition to increasing oligotrophy along a macrogradient from west to east (Chapter 4), there is also a marked

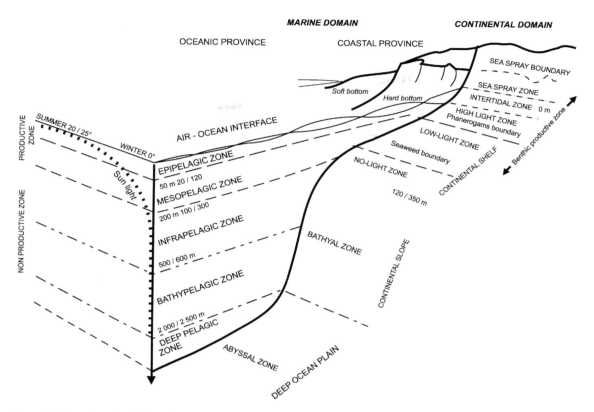

Figure 6.4 The main marine habitats. After H. Boutière, unpublished work, with modification.

gradient of decreasing nutrient supply from the Mediterranean shores towards the open sea.

6.8.2 Benthos

Unlike the pelagos, the benthos is not a homogeneous, mobile environment or habitat. It is the layered and immobile component which corresponds to the interface between the sea water and the continental substratum. It extends from the shorelines to the deepest depths of the sea and can be divided into horizontal zones, each of which corresponds to a set of ecological conditions providing habitat for specific communities.

From the top downward the characteristics of benthic habitats include their levels of moisture and sunlight, and their topography. The wetting factor concerns the two upper levels; that is, the sea-spray zone and the intertidal zone. The sunlight factor concerns the next two zones, which are always submerged, the so-called high-light zone and the low-light zone. The two deepest zones, which receive no sunlight, are only distinguished by their topography. They are the bathyal zone, which corresponds to the continental shelf, and the abyssal zone, which includes the large plains of the basin bottoms (Fig. 6.4). In the oceans, there is a seventh zone, lying under 6000 m of seawater. This zone, called the deep trough, does not exist in the Mediterranean Sea.

In each of the six benthic zones in the Mediterranean, the most important conditions are (1) the nature of the sea bottom, which may be hard or soft, sandy or muddy, (2) the strength of the currents, (3) exposition to waves, and (4) nutrient supplies. Biotic factors include competition among organisms for a specific niche or site, or a specific food source. The variability of environmental factors leads to the existence of many communities defined by Pérès and Picard (1964). In Chapter 9, we will discuss the evolutionary ecology of some of the most notable members of these different communities.

6.8.3 Other marine habitats

Coming back towards the shores, there is a coastal habitat worth mentioning, to round out our discussion of the patchwork of marine habitats, namely the estuaries. The inputs of the rivers bring large quantities of mixed sediment which accumulate in the shallow muds at close proximity to the river mouth before to be exported toward the open sea (see above). On the higher levels of the estuary, *Salicornia* can grow when the salt content of the mud is enough. On the lower bottoms, *Nostoc* and other **Cyanophyceae** are often very abundant forming thick layers on the surface. All the components of the fauna are highly euryhaline and eurythermal. Monospecific populations of polychaetes (Nereidae) burrow in the mud, a few species of bivalve molluscs are also frequently encountered, especially *Cardium lamarcki* and *Abra alba*, as well as the green crab *Carcinus mediterraneus*, which is never found in the sea itself. Isopods and amphipods are abundant in the plant deposits. All species of flatfish tolerate the brackish waters of the estuaries quite well, as do sea breams, sea bass, and mullets. In the low-light zone in front of the estuaries, the bottoms experiencing inputs of rivers show a very high rate of sedimentation where sessile species cannot survive. The gastropod *Turritella communis* can be exceptionally abundant in these areas, constituting up to 95% of the local macrofaunal population (Pérès 1985).

The harbours are also specific habitats with areas of deposition and pollution, but they result of the human action and can no more be considered as natural biotopes. In the most polluted areas, there is an intense development of very small polychaetes, essentially *Capitella capitata*, which are the first colonizers after catastrophic episodes of eutrophication. The vegetation is limited to the presence of green algae.

In the next chapter, we will use case studies to give some idea of how organisms became adapted to and thrive within the mosaic structure of Mediterranean habitats, which tend to present varying degrees of ecological and geographical isolation.

Summary

The aim of this brief review of the main Mediterranean habitat types was to give an idea of the diversity of habitat present in the basin. We did not go into detail about **ecotones**, which constitute an entire array of habitats by themselves. In a sense, the mosaic pattern of most habitats and landscapes makes the Mediterranean as a whole a huge ecotone. Scaling habitats across the three dimensions of space provides a picture of the diversity of life zones and ecological opportunities organisms have at their disposal to choose suitable conditions for settling and evolving.

CHAPTER 7

Populations, Species, and Community Variations

One basin-wide 'constant' we saw in Chapter 2 is the exceptional degree of environmental heterogeneity in regions and landscapes, field-size plots, or even at the scale of 1 m² of soil. In other words, Mediterranean landscapes are patchy across many spatial scales, and this favours, over evolutionary time, various processes of adaptation of populations, such as local differentiation, **phenotypic plasticity**, and subtle mechanisms of habitat selection in animals. In addition, the very high levels of endemism in many groups of plants and animals (Chapter 3) and the emergence of habitat-specific species assemblages (Chapter 6) add to the huge variation of living biotas in the Mediterranean region.

To study biota in such a patchy spatial and temporal environment, populations and communities should not be approached as independent units, but rather as interactive ones, developing within a landscape of other populations and communities. Indeed, the realization that regional dynamics can have very marked influence on ecological and evolutionary processes at many hierarchical levels, including those of species and populations, has been a major development in the field of ecology over the last few decades (Gotelli 2002).

In this chapter, we examine some of the processes of differentiation occurring over evolutionary time in the Mediterranean region, at the levels of populations, species, and communities, within the prevailing context of a 'moving mosaic' of habitats. As emphasized in Chapters 1 and 2, the many geographic discontinuities in the basin linked to the abundance of islands, peninsulas, and a dozen mountain chains in close proximity to the sea, as well as the highly dissected structure of almost all landscapes between the mountains and the coasts, have all influenced ecological and evolutionary processes in the region. There also is clearly an important historical discontinuity between the western and eastern halves of the basin, as reflected by the presence of many pairs of vicariant species (Dallman 1998; Quézel and Médail 2003; Thompson 2005). In the first section of this chapter we explore this issue in some detail for trees and birds. Next we consider life on Mediterranean islands and the so-called insular syndrome which operates at the scales of populations, species, and communities, among both plants and animals. Then we discuss various examples of genotypic variations that allow organisms to adapt to local conditions in a heterogeneous environment, as well as the phenomenon of phenotypic plasticity whereby a single genotype may express itself in differing **phenotypes** across a range of environments. Detailed examples of genetic and phenotypic variation as a response to habitat and landscape heterogeneity are devoted to sexual and chemical polymorphism in a plant, the thyme, and to how a small bird, the blue tit, copes with habitat heterogeneity. Finally we consider seasonal effects and the impact of migration on birds, one of the best-studied groups of Mediterranean organisms.

7.1 East–west vicariance patterns

A remarkable number of east–west pairs of vicariant species have been identified among the dominant tree genera of the Mediterranean region, such as the Aleppo pine in the west and Calabrian pine in the east (Fig. 7.1a). These two species only co-exist in parts of Greece, Turkey, and Lebanon, where they also produce natural hybrids (Barbéro et al. 1998). Similarly, Spanish juniper (*Juniperus thurifera*) in the western Mediterranean is replaced by Greek juniper in the east (Barbéro et al. 1992).

Among the evergreen Mediterranean oaks, the dominant western holm oak is replaced by *Quercus calliprinos* as the dominant oak in the eastern half of the basin (Fig. 7.1b). There is much complexity in the holm oak complex: *Quercus ilex* subsp. *ilex* in the eastern part of the range of the species in Mediterranean Italy and *Q. ilex* subsp. *rotundifolia* which is widespread in the Iberian Peninsula and North

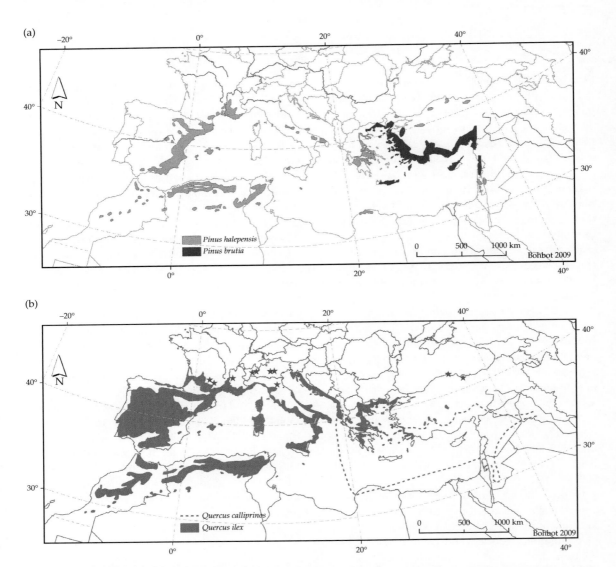

Figure 7.1 Distribution of the most widespread pine and oak species in the Mediterranean Basin showing west–east disjunctions between (a) Aleppo pine in the west and Calabrian pine in the east and (b) holm oak (sensu lato) in the west and *Quercus calliprinos* in east. After Quézel (1985) and Barbéro et al. (1998). For discussion of differing taxonomic treatments of holm oak, kermes oak, and *Quercus calliprinos*, see text.

> **Box 7.1. Consequences for genetic variability of human selection on oaks**
>
> By comparing many genetic loci of individual trees of holm and cork oaks in a mixed population, Lumaret and co-workers have shown that genotypes of the cork oak are consistently clustered while those of holm oak are scattered and show little similarity among them (Fig. 7.2). What emerges most clearly from these patterns is not habitat selection or adaptation on the part of ecotypes within species, but rather effects clearly related to human selection processes. Whereas the holm oak has been managed by people throughout its distribution range for many millennia, it has never undergone specific selection for any trait (except for isolated cases in central Spain where it was considered as a fruit tree, and selection was carried out for sweet acorn production to feed pigs) (Lumaret *et al.* 2002). In contrast, cork oak has been subjected to ongoing selection for improving its highly useful outer bark everywhere it occurs in the western Mediterranean. Hence there is much lower genetic diversity in this species. Several alternative hypotheses, such as recent expansion from glacial refugia and the short time available for differentiation to have taken place subsequently, have also been suggested to explain the low nuclear genetic diversity of cork oak (Jiménez *et al.* 1999).
>
>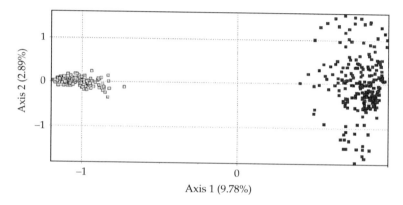
>
> **Figure 7.2** Positions obtained from correspondence analysis (CA) of the multilocus genotypes of 220 holm oak trees (black squares) and of 171 cork oak trees (grey squares) sampled in nine and eight populations located throughout the western Mediterranean respectively, and analysed according to polymorphism at eight nuclear microsatellites loci (Lumaret and Jabbour-Sahab 2009). The percentage of **inertia** is indicated in parentheses for each axis.

Africa (Lumaret *et al.* 2002) and a zone of natural hybridization between them in southern France, west of the Rhône (Michaud *et al.* 1995). Analyses of chloroplast DNA by Lumaret *et al.* (1991, 2002) and Toumi and Lumaret (2001) have shown that gene flow between the two forms is restricted to this zone. Kermes oak is also part of the group, and, following Zohary (1961), some authors consider *Q. calliprinos* to be an eastern vicariant of *Quercus coccifera* (see Table 7.3 below), while others consider that there is a continuum between the two taxa and that they form part of a swarm of interfertile subspecies, which also include *Quercus aucheri* in Turkey and eastern Greece. In fact all three may be considered as a single species (R. Lumaret, personal communication). The endemic species *Quercus alnifolia*, found only on ultrabasic substrates in Cyprus (Barbéro *et al.* 1992), is also relatively close to these three taxa (Toumi and Lumaret 2001). Let us recall that there may be an historic role of humans in the distribution of these oaks (see Box 7.1), as indeed is known to be the case for many economically important plants throughout the basin (see Chapter 10).

Among the deciduous Mediterranean oaks the situation is still more complex, with a large series of vicariant species that will be described later in this chapter. Similar patterns also occur in deciduous plant species, as for example in the terebinth (west) and its eastern 'cousin' *Pistacia palaestina*, which is a co-dominant with *Q. calliprinos* in the meso-mediterranean life zone, just as the terebinth co-dominates with holm oak in the same strata (*étage*) in the western and central part of the basin.

Examples of vicariant pairs among animals are common as well. Among birds, they include the black-eared wheatear to the west and the pied wheatear (*Oenanthe pleschanka*) to the east, the Neumayer's rock nuthatch (*Sitta neumayer*) to the west and the eastern rock nuthatch (*Sitta tephronota*), the peregrine falcon to the north and the Barbary falcon (*Falco pelegrinoides*) to the south. Many other examples (Vuilleumier 1977; Haffer 1977) support the view that processes of allopatric speciation occurred through vicariance, thanks to isolation of biota at different epochs in the past, especially in the eastern part of the basin. In some of these examples, hybridization between and among species suggests that they are closely related and recently descended from a common ancestor.

An important additional framework for differentiation of organisms is provided by the extensive archipelago of small and large islands, as will be discussed in the next section.

7.2 Life on islands

Biologists have always been fascinated by the evolutionary biology of island biota, particularly in relation to factors determining species diversity, adaptive radiation, and evolutionary changes within and between populations. Much work has been devoted to island biology since the first seminal book of Wallace (1880). More recently, the theory of island biogeography of MacArthur and Wilson (1967) shed a new light on island biology, which, in turn, has been rejuvenated by a series of recent studies, for example those of Grant (1998), Lomolino (2000), Heaney (2007), and Whittaker and Fernandez-Palacios (2007), to cite just a few. An island is a self-contained region whose species originate either from ancestral species and populations when the island was connected to adjacent mainland areas or from immigration and colonization events from outside the region. The notion of 'self-contained' signifies that, within the island, each species has an average net growth rate sufficient to maintain a viable population size. However, if populations cannot be saved by some rescue effect on an ecological scale of time, species are necessarily sinks over 'long' evolutionary time, because once the island has been colonized by a **propagule**, evolutionary changes make the new incipient species so narrowly specialised to its new environment that it cannot re-colonize the mainland from which it originated (Blondel 2000). In this context, what makes island life in the Mediterranean region particularly interesting is that each island shows a unique array of bioclimatic and biological features and its own unique set of native plants and animals. This is especially true of the larger ones, which have been entirely disconnected from any continent since at least the Messinian Salinity Crisis (see Chapter 1). This means that plant and animal species had to colonize them from nearby mainland areas.

7.2.1 Differentiation

When a propagule of a species succeeds in immigrating and then colonizing an island, it is confronted with new sets of environmental factors, both biotic and abiotic. Moreover its genetic background is different from that of the mother population, because of the severely reduced diversity in the colonist population resulting from one (or a few) immigrants, which normally carries with it only a fraction of the genetic diversity of the mainland source population. As a result, potential difficulties may arise from inbreeding effects, resulting in the so-called founder effect. These new ecological and genetic conditions constitute new selection regimes that inevitably lead the founding population to diverge from its mainland mother population. Divergence may eventually lead to speciation if reproductive isolation is attained before new individuals of the same species colonize the island. Two examples of such processes in Mediterranean islands are the co-occurrence of the blue chaffinch and the chaffinch in the Canary Islands, and that of the Corsican swallowtail butterfly

(*Papilio hospiton*) and the European swallowtail (*Papilio machaon*) in Corsica. In both cases, the former species succeeded in colonizing the islands and to differentiate from its mother mainland population or from an ancestral taxon into a full species *before* a second colonization event of individuals of the same sister species occurred. Complete genetic isolation between species of the two pairs allowed them to co-exist in **sympatry**. Since rates of immigration, eventually followed by successful colonization and differentiation, are much lower in geographically isolated islands than in those that are near the source mainland, extinction rates are also lower, with the result that the overall rate of turnover is lower on remote islands and the degree of differentiation is higher. In addition, endemism rates are lower in groups with high dispersal aptitude than for those with no adaptations for long-distance dispersal. Indeed, endemism rates are higher in plants than in most vertebrates and also much higher in less mobile animals—such as reptiles, amphibians, or non-flying mammals—than in birds or bats (see Chapter 3).

Nonetheless, overall levels of endemism are usually quite high on Mediterranean islands. For example, on some of the larger islands, endemic species of insects may account for 15–20% of the insect fauna (Blondel and Cheylan 2008). Amphibians include 12 endemic species in the Mediterranean archipelago, equivalent to 41% endemism (see Chapter 3). In contrast, average endemism rates are lower in reptiles, reaching only 10%. But some lineages are specific to the Mediterranean, such as the genus *Archaeolacerta*, in Corsica and Sardinia, and *Euleptes* (formerly *Phyllodactylus*), on several Tyrrhenian Islands. In birds, there are few endemic species in the islands considered here, except in the Canary Islands, because islands within the Mediterranean Basin itself are too close to the nearest mainland for differentiation to occur between one colonization event and the next (Table 7.1).

However, at the subspecies level, morphological changes on islands have led taxonomists to recognize many subspecies of birds on Mediterranean islands. For instance, on Corsica, more than half of native bird species are considered to be taxonomic subspecies. This does not necessarily mean, however, that morphological changes are associated

Table 7.1 Bird species endemic to islands in the Mediterranean biogeographical region

Island	Endemic species
Canary Islands	*Columba bollii*
	Columba junionae
	Apus unicolor
	Anthus berthelotii
	Saxicola dacotiae
	Fringilla teydea
	Serinus canaria
	Cyanistes teneriffae
Cyprus	*Oenanthe cypriaca*
	Sylvia melanothorax
Western islands (Balearic Islands, Sardinia, Corsica, and nearby islets)	*Sylvia sarda*[1]
Corsica alone	*Sitta whiteheadi*

[1] Small populations of this species also breed near the coast in south-eastern Spain.

with large genetic changes. For example, the citril finch (*Carduelis citrinella*) occurs on both the European mainland and the island of Corsica. Although this species is restricted to the subalpine life zone on the mainland, it is widespread from sea level up to high mountains in Corsica, where its ecology and behaviour much differ from those on the mainland. However, there is a surprisingly low genetic divergence between the citril finch population of Corsica and those of the Alps and the Pyrenees, despite the Corsican population having been isolated from the mainland populations for a long time (Pasquet and Thibault 1997).

7.2.2 Depauperate biota

Evolution and genetic differentiation of endemic forms on islands is only one aspect of the story of island life; the differentiation of endemic species is often just the tip of the iceberg. Comparing island patterns with mainland patterns can help in understanding ecological and evolutionary processes that occur as a result of isolation. Area is an especially important determinant of the insular faunas for two reasons. First, as the size of the island decreases, so does the size of a species' populations. Thus the probability of extinction increases as the area of an island decreases. Second, as the size of an island increases, it contains an increasing number of different habitats, suitable for colonization by a wider range of species than a small island. Altitude is an additional factor increasing the diversity of habitats and hence the opportunities for species to colonize an island.

The most obvious character of island communities is that they are impoverished in comparison with communities occupying areas of similar size on the nearby mainland. For example, 108 species of birds regularly breed on Corsica, an island of 8680 km², as compared to 170–3 species found breeding in three areas of similar size in continental France (Blondel 1995). The relationship between species richness and area is formalized by the equation $S = CA^z$, where S is the number of species found on the island, A is the size of the island, C is a constant which depends on the taxon involved, and z is the slope of a double logarithmic plot of species against area. z can be considered as an estimation of the dispersal power of the organisms. The smaller the value of z, the poorer the dispersal power. As expected from this relationship, the smaller the island, the lower the number of species. In butterflies, Hockin (1980) reported that 75% of the variation in richness on Mediterranean islands is accounted for by area, and 93% if the distance to the nearby mainland is included in the analysis. The species/area relationship is illustrated for birds in Fig. 7.3.

As a general rule, rates of species impoverishment are function of dispersal abilities of organisms. As compared to species richness on the mainland, impoverishment in Corsica is 38% in birds, 43% in reptiles, and 68% in non-flying mammals. Similarly, in butterflies, most of the variation in species richness on Mediterranean islands can be explained by differences in the islands' surface area. For example, there are only 24 species of butterflies in one of the smallest islands, Formentera (115 km²), as opposed to as many as 89 species on Sicily, the largest of all the Mediterranean islands (25 700 km²), with intermediate values of 27 species in Minorca (694 km²) and 49 species on Corsica (8680 km²) (Hockin 1980). However, there are some exceptions to the regular trend of decreased

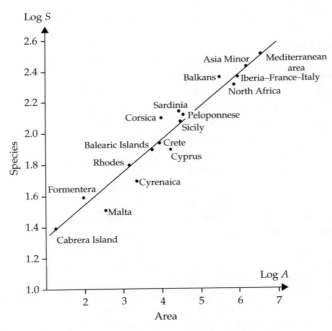

Figure 7.3 Double-logarithmic species/area relationship of birds in some islands and mainland regions of the Mediterranean area (after Blondel 1986).

species richness on islands. For example, insectivorous groups of vertebrates, which feed upon small flying insects, are much less impoverished than most other groups. There are as many species of swifts and swallows in Mediterranean islands as in the nearby mainland, and 26 species of bats occur in Corsica, which is exactly the same number as on the mainland.

7.2.3 Changes in ecology and morphology: the insular syndrome

Species impoverishment on islands results in a cascade of changes in fundamental processes at the levels of communities, species, and populations. Studies conducted on Mediterranean islands, among others, provided good examples of these changes in birds. They may be summarized in six main points that are all components of the so-called insular syndrome (Blondel 2000). These components include (1) sorting processes, (2) population sizes, (3) niche enlargement, (4) territorial behaviour and aggressiveness (in animals), (5) body size, and (6) mobility and dispersal.

First, not all the species of the source mainland have the same chances of becoming successful colonists on an island. There is a kind of sorting process for island-colonization candidates. There is a large literature indicating some kind of **disharmony** for many taxa: plants, ferns, ants, butterflies, mammals, and birds (Blondel 2000; Whittaker and Fernandez-Palacios 2007). In birds, the most successful colonists are species that are rather small, widespread, and abundant on the mainland, and rather flexible in habitat selection and foraging habits (Blondel 1991). They are, so to speak, pre-adapted to cope with the new environmental conditions they will find in their new habitats. In other words, the small generalist species will be favoured. Similar strategies in plants are those of small, generalist **dioecious** flowers. In birds, the smaller the island, the lower the number of species and the higher the proportion of resident species that are widespread in the overall region. Thus, island communities are not a random subset of the communities on the mainland. For example, the breeding bird fauna of Malta includes only 28 species of land birds, examples of which are the house sparrow (*Passer domesticus*), the kestrel, the turtle dove (*Streptopelia turtur*), the barn swallow (*Hirundo rustica*), the nightingale, or the goldfinch (*Carduelis carduelis*). All of them are extremely common and abundant throughout the entire Mediterranean Basin. To take another example, the woodpecker family (Picidae) is represented by nine species in the Mediterranean, but only the great spotted woodpecker (*Dendrocopos major*) and the wryneck (*Jynx torquilla*) regularly occur in large islands, such as Corsica, Sicily, and Sardinia. As a rule, large species, including carnivorous species, which are usually bigger than their prey, are under-represented on islands. This makes island communities disharmonic in terms of taxonomic composition and trophic structure.

Second, population sizes of the species that occur on islands are often much larger than those of their mainland counterparts. This process, described as 'density compensation' by MacArthur *et al.* (1972), has been interpreted as resulting from a release from interspecific competition; that is, the ecological space that would have been occupied by the missing species being actually filled by the smaller number of species present in insular communities (see Table 7.2). In fact, since the very same mechanism that produce larger populations on islands have not been experimentally demonstrated, it is better to use a neutral term, such as density inflation (Blondel *et al.* 1988).

Third, many species on islands often occupy more habitats, and forage over a larger spectrum of microhabitats, than on mainlands, and prey upon a larger range of food items. This is a process called niche enlargement. As a consequence, species impoverishment at the scale of a whole island is not necessarily reflected at the scale of local habitats. For example, some habitat types on Corsica actually have more species than their mainland counterparts, as shown by changes in species diversities along two habitat gradients that match each other on the basis of vegetation structure (Table 7.2). The higher number of bird species in Corsican matorrals is a result of niche enlargement whereby species 'spill over' from forests to matorrals, hence increasing local species diversities (Blondel *et al.* 1988). All these processes result in larger population sizes since populations are more flexible in habitat

Table 7.2 Habitat niche enlargement and density inflation in tit species in matching habitat gradients on Corsica as compared to the nearby mainland

Species	Population density in each habitat (breeding pairs per 10 ha)					
	1	2	3	4	5	6
Mainland						
Great tit				2.2	3.1	3.2
Blue tit						11.5
Coal tit						0.2
Crested tit						1.8
Total				2.2	3.1	16.7
Corsica						
Great tit	1.6	1.7	2.5	3.6	2.6	4.7
Blue tit		0.5	0.2	3.3	7.9	14.2
Coal tit				1.2	2.1	4.1
Total	1.6	2.2	2.7	8.1	12.6	23.0

Habitats range from low matorral to mature oak forest (ranked 1–6). Each species occupies on average more habitats and has higher population sizes (breeding pairs per 10 ha) on Corsica than on the nearby mainland.

Source: After Blondel (1985).

selection, hence reducing extinction risks. This process of niche enlargement has been observed in many insular species of both animals and plants, as perceived when comparing them to closely related mainland species. An excellent example is the endemic Balearic cyclamen (*Cyclamen balearicum*, Primulaceae) as compared to the widespread *Cyclamen repandum* (Debussche and Thompson 2003). We will come back to this well-studied genus below.

A classic explanation of higher population densities and niche enlargement in species-poor islands is that extra resources become available because of a reduced number of competitors. As a consequence, it is argued that island habitats include similar number of individual birds as mainland habitats, but with fewer species.

Fourth, changes in territorial behaviour and aggressiveness are classical components of the insular syndrome in island populations of vertebrates. The social behaviour of many reptiles, mammals, and birds reveals remarkable shifts, such as reduced territory size, increased territory overlap, acceptance of subordinates, reduced situation-specific aggressiveness, and abandonment of territorial defence, as compared to mainland populations.

These changes are often associated with unusually high densities, niche enlargement, low fecundity, and the production of a few competitive offspring. One explanation of these shifts in behaviour is the 'defence hypothesis' (Stamps and Buechner 1985), which suggests that a release in aggressiveness and territorial defence may occur as a result of trade-offs between defence costs of the territory in crowded populations and costs of reproduction. If defence costs become exaggerated, animals would benefit by expending less energy in territory defence and reallocating their resources in breeding activities so as to produce young that are more competitive. One example of such shifts is that of the black rat on a very small islet of 6.4 ha off the coast of Corsica, where as many as 141 black rats lived in close vicinity and did not severely compete for resources. As a result of a stranger–neighbour effect and alleviation of aggressiveness, they knew each other and lived together in 'good company'. However, these animals have not lost their aptitude to be aggressive towards unknown intruders. Granjon and Cheylan (1989) introduced five rats (four males and one female) from the 'mainland' of Corsica onto this small islet, after having equipped them with radio-tags. Within a few hours, all five animals were recovered dead, killed by resident rats. The introduced rats were unknown and therefore treated as enemies and exterminated by the 'legitimate' territory owners.

In birds on Mediterranean islands, there is also evidence that territory owners are often much less aggressive than on the mainland, accepting subordinates in their territories and even allowing intruders to breed between adjacent territories. Compared to their mainland conspecifics, for example, Corsican blue tits are much less aggressive and readily share their territory with other pairs.

Fifth, changes in body size are known as the Foster's rule (Foster 1964), according to which large species tend to become smaller and small species tend to become larger; hence the reduced variation of body sizes on islands (van Valen 1973). Disentangling which factors determine changes in body size is difficult because direct experimentation is not possible. In fact, there are so many exceptions to this trend that the rule has been questioned by many scientists. For example, among birds, one obvious exception to the rule is that of the large raptors. It has often been claimed that there should be fewer raptors on islands because their large size prevents them from constructing large viable populations and also because of a reduction of the diversity of prey. Notwithstanding, there are in fact many large raptors on most Mediterranean islands, especially vultures, eagles, and falcons. Crete, for example, harbours as many as 11 species of raptor: lammergeyer, Egyptian vulture, griffon vulture, black vulture (*Aegypius monachus*), buzzard (*Buteo*), golden eagle, lesser kestrel, kestrel, Eleonora's falcon, lanner falcon (*Falco biarmicus*), and peregrine falcon! However, the apparent over-representation of large raptors in many Mediterranean islands is simply due to the size of the islands and their proximity to the mainland in relation to the dispersal range of the various species. In any case, for large birds, islands that are not far from the mainland are clearly not necessarily biological islands if the birds which breed there are part of a single larger population that extends over large mainland and island areas. In that case, repeated exchanges and gene flow of individuals between the mainland and the island make the latter but a part of a larger range of distribution. A similar situation had been described in butterflies of large islands of the western Mediterranean (see Chapter 3).

One famous example of changes in body size of islands is that of the dwarf hippos and elephants, which inhabited Mediterranean islands before their extermination by humans (see Chapter 11). In such relatively small mountainous areas as Cyprus, Malta, and many islands of the Aegean Sea, a small animal body size should logically be selected for, if it allows these species to make a better use of the food supply and to construct larger populations that are less vulnerable to extinction. After crossing the body of water separating this island from Turkey, some 100 000 years ago, the size of the dwarf hippo of Cyprus (*Phanourios minutus*) apparently reduced from that of a normal hippo to that of a pig (Diamond 1992)! Besides size reduction, the insular syndrome shown by these animals on Cyprus also included anatomical changes in their legs, which became shorter, more robust, and without fingerwebs. Presumably these adaptations allowed them to walk on the tips of their fingers, enabling them to walk and climb in the mountainous terrain of Cyprus, which is typical of Mediterranean islands in this regard.

There has been much debate on the selection pressures which lead to dwarfism or gigantism (e.g. Lawlor 1982; Angerbjörn 1985). It has been proposed that a release from predation and/or from interspecific competition favours gigantism in small species, especially rodents. In this context, one interesting observation reported by Vigne (1990) is the reverse trend of decrease in the size of endemic rodents (*Prolagus* and *Rhagamys*) during the Holocene in Corsica, presumably as a response to increased predation pressures from predators introduced by humans. However, this 'response' did not prevent these species from going extinct as well.

Other morphological changes on islands involve colour and camouflage. For example, many lizard species on small islands are darker than their mainland counterparts, or closely related species; insular forms of the Lilford's wall lizard (*Podarcis lilfordi*) of the Balearic Islands are completely black, even though they live on a white limestone substrate! Another common adaptation is a trend towards herbivory in many insular species. For example, adult individuals of the lizard *Gallotia* from the Canary Islands are almost completely herbivorous, a feature that is highly unusual in the Lacertid family.

Finally, a commonly encountered response of both plants and animals to new selective pressures in insular environments is a reduction in morphological traits that allow dispersal over long distances (Carlquist 1974). This is very apparent, for example, in the Balearic lizards mentioned above, which have shorter legs and tails than their mainland counterparts, which limits their ability to travel. Hence, we are faced with an apparent paradox: the best long-distance immigrants are organisms with powerful dispersal abilities, whereas the most efficient colonists of islands are poor dispersers once established within the island's set of ecosystems. The mechanisms responsible for this are a combination of ecological and evolutionary responses. At the ecological level, bird communities in continental islands quite often include a disproportionate proportion of sedentary species or species that become sedentary, which is repeatedly the case in remote oceanic islands (McNab 2002). At the evolutionary level, a repeated trend is the reduction of morphological traits such as powerful wings in animals. Flightlessness is common in island populations of birds and insects, because the advantages of sedentariness increase as the advantages of dispersal decrease, including the danger for winged organisms to be accidentally blown off islands while flying before strong winds, which is something that often occurs on islands.

Flightlessness is also an energy-saving mechanism in birds whenever a permanent habitat with a local year-round food supply and the absence of predation favour strong habitat fidelity (McNab 1994). Walking while exploiting resources of a given habitat, in the manner of herbivorous non-arboreal animals, led to the progressive atrophy of wings. In other words, wings lost their usefulness because of the small distances to be covered and the absence of predator in insular environments.

7.3 Community dynamics in heterogeneous landscapes

Turning now to communities in complex habitat mosaics, we will examine how these communities are organized at smaller scales of space and how their dynamics are driven by factors which make the mosaic of habitats 'move' in space and time. This process involves repeated colonization and extinction events that operate at the landscape scale. As noted already, a landscape is made up of many local habitats that continuously change in time so that species and communities inhabiting them must also change continuously. We will illustrate these processes using the turnover of habitats and species assemblages as a response to recurrent fires.

7.3.1 Dynamics of bird communities after fire

Almost all contemporary fires in the Mediterranean region are anthropogenic in origin, be it intentionally or accidentally. At first sight, high fire-return rates have destructive effects on ecosystems, as we will discuss in Chapter 11. But natural fires play such an important role in the dynamics of Mediterranean landscapes and communities that they must be considered as a driving force in the moving mosaic of habitats within a landscape. The type, extent, size, frequency, season of occurrence, and behaviour of fire are attributes that characterize the fire regime of a region and to which organisms are adapted. However, current fire regimes in Mediterranean ecosystems have been shown to vary in relation to land use by humans and fragmentation (Mouillot et al. 2002, 2003; Pausas 2004, 2006). Provided their return rate is not too short, fire is a natural component of the dynamics of most ecosystems. Fire does not seem to alter the physical structure of soils; nor does it destroy the organic matter in the soil unless recently burnt areas are too rapidly and intensively grazed by sheep and goats. Changes in the local fire regime may lead to severe landscape changes, sometimes exceeding ecosystem resilience (Diaz-Delgado et al. 2002).

Fire provides an excellent illustration of the role of natural disturbance in the functioning of ecosystems at the scale of landscapes, the so-called patch dynamics first explored by Pickett and White (1985) and Huston (1994). Indeed, fire fits the definition of a disturbance as 'any relative discrete event in time that disrupts ecosystem, community, or population structure and changes resources, substrate availability, or the physical environment' (Pickett and White 1985:7). Schematically, Mediterranean landscapes that are periodically and frequently submitted to fires are characterized by a turnover of four habitat types, which replace each other in space and time, creating a moving mosaic. These are grasslands, low matorrals, high matorrals, and forests with, of course, a huge range of local variants depending on substrates, periodicity of fires, history of land use, dominant plant species, and other factors. In his studies in the Mt. Carmel region of northern Israel, Naveh (1999) demonstrated the crucial role of fire as an evolutionary and ecological factor in shaping landscapes and vegetation. Many communities and species of open habitats, including xero-thermophilous species, are narrowly dependent on such disturbance events.

Using birds as a model, a study of post-fire successional processes was carried out by Prodon et al. (1987) and the data have been reanalysed by Jacquet (2006). The experimental design involved 186 study plots evenly distributed among a series of 11 habitats, ranging from grasslands to mature forests in the holm oak series. At the scale of this landscape, 51 bird species were censused. They ranged from species of open vegetation, for example the woodlark (*Lullula arborea*), the corn bunting (*Miliaria calandra*), and the linnet (*Carduelis cannabina*), to forest species, for example the European robin (*Erithacus rubecula*), the blackcap, and the chaffinch, with matorral species, such as several species of warbler, and the nightingale, between them. At each census spot, the dominant plant species were recorded, as well as vegetation profiles. The data were used to model the relationships between vegetation structure and bird communities and illustrate the turnover of species in the 11 habitats (Prodon and Lebreton 1981; Jacquet 2006).

At the scale of the whole range of habitats within the mosaic, all 51 species found in the post-fire gradient were already present somewhere within the landscape. Accordingly, the whole gradient could be considered as a closed system, within which processes of local extinction and re-colonization operated. Species of open vegetation colonized habitats immediately after fire and then were replaced by matorral species, such as warblers and the nightingale, as the resprouting vegetation grew taller. In turn, matorral species were replaced by forest species until the recovery process was completed (return to a forest stage). Recovery time, or resilience, has been estimated at approximately 50 years for vegetation but only 35 years for the bird fauna. Resilience time is longer as the vegetation becomes taller and more complex, and it differs according to the oak species involved. It is on average longer in holm oak-than in cork oak-dominated woodlands. The fact that recovery time is on average shorter for birds than the vegetation is at least partly due to the fact that several bird species exhibit a strong site fidelity and tend to remain in habitats that are suboptimal for them (Jacquet 2006).

The turnover of bird communities as a response to fire-induced habitat changes highlights the importance of scale effects in community investigation, as pointed out repeatedly by Wiens (1989). Because environmental heterogeneity plays a prominent role in structuring communities, ecological processes operate on spatial and temporal scales far larger than those usually used in field studies. At the scale of a landscape and over long periods of time, the survival of all the species of a mosaic of habitats involves the existence of a disturbance regime that is unpredictable in time and space in the short term but predictable in the long term. This regime periodically moves up and down the position of any given habitat patch within a landscape, including patches of grasslands, matorrals, and forests. At a broader geographical scale, the combination of a disturbance regime and community dynamics, resilience, and **inertia** results in a dynamic equilibrium that may be fairly stable in the long term. Such a system is characterized by (1) a given pool of species that is a legacy of history and (2) a regime of disturbance that is specific to each region. To be sustainable in the long term, such a system requires areas large enough for the spontaneous occurrence of stochastic and chance events. Blondel (1987) coined the term **metaclimax** to define both the spatial scale required for maintaining a self-sustaining system and the disturbance regime that guarantees all the habitat patches required for the regional survival of all the species legated by history.

Communities of small mammals have also been shown to recover quickly after fire with a similar sequence of local colonization and extinction events that match successional changes in the structure of vegetation (Torre and Díaz 2004). The process is similar to that reported above for birds, but with a smaller number of species. More recently, Sarà et al. (2006) used null models to investigate patterns of species co-occurrence of terrestrial vertebrates (reptiles, birds, and mammals) in habitats that had been burnt at different time intervals in northern Sicily. They demonstrated that fire disrupted patterns of community organization soon after fire, with species colonizing randomly the new habitat, but that communities progressively re-organized themselves through sorting processes. As a result, 50 years after fire, there was a non-random co-occurrence pattern of vertebrates in the old mature woodlands. These studies provide an idea of the speed of the moving mosaic in Mediterranean forest habitats periodically disturbed by fire events.

7.3.2 The spatial dynamics of predatory ants

Ants contribute a major part of the overall insect biomass in most terrestrial ecosystems and may include more than half of the individual insects (Wilson 1992). Therefore, their role in ecosystems at both scales of species-specific interactions and functional groups must be important. Some important functions of ants are seed dispersal (myrmecochory), soil **bioturbation**, and predation, to say nothing of some extraordinary species-specific interactions with butterflies (*Maculinea*). For example, the caterpillars of the large blue butterfly (*Maculinea arion*) are delicately transported by ants in their nests where the latter carefully raise them because they produce rewarding sweet secretions that the ants are fond of (Thomas *et al.* 1989; Anton *et al.* 2008). Caterpillars will stay in ant nests until they pupate and become full-grown butterflies.

In a typical Mediterranean landscape, including several life zones along an altitudinal gradient (420–1880 m) on the slope of the Mont-Ventoux in southern France (Chapter 5), ant species diversity amounted to 64 species (du Merle *et al.* 1978). A careful analysis in 58 study sites of ant species richness and species-specific distributional patterns revealed a significant discrimination of community structure in relation to vegetation belts and vegetation units. The richest communities occurred in the thermo-, meso-, and supra-mediterranean life zones, and then declined with increasing altitude. Several ant species are strongly associated with particular plant species; for example *Aphaenogaster gibbosa* with thyme and *Formica gagates* with downy oak.

An experiment designed to assess one important function of ants, namely their predation pressure on eggs of other insects, revealed that many species move seasonally across different habitat patches, resulting in different ant assemblages over time. Predation has been studied in a small mosaic of three habitat patches, a clearing (942 m^2), a forest edge (478 m^2), and a forest patch (952 m^2). At the scale of the three habitats combined, 14 species have been recorded (13 in the clearing, 10 in the forest edge, and six in the forest) and 175 ant nests have been found. More than half of these species (nine) are predators of insect eggs. The egg-eating activity by ants has been measured using traps supplied with eggs of the Mediterranean flour moth (*Anagasta kuehniella*). Results showed that a very high proportion of the moth's eggs were eaten by ants in the three habitats (up to 76%) and that important between-habitat exchanges of ants occurred. The relatively low number of predatory ants resident in the forest was compensated by high seasonal migration rates from the two other habitats, especially the clearings. In particular, there was an intense exploitation, especially in mid-summer, of the forest habitat by ant colonies invading it from colonies established in the two nearby habitats. Species involved in this dispersal were *Leptothorax unifasciatus*, *Pheidole pallidula*, and *Myrmica specioides*. Such between-habitat migrations of predatory ants suggest that clearings contribute to controlling populations of insects including foliage-eating insects harmful for tree foliage.

Thus, the patchy geographical configuration of the landscapes plays an important role in the dynamics of insect populations at the scale of the ecotone between forest and open habitats. This is because many forest insects and their predators carry out parts of their life cycle in habitats other than forests. This mosaic structure allows some highly mobile species (e.g. *L. unifasciatus*) to find, at any time of the year, optimal environmental conditions for breeding. Predator species that exploit forest resources from neighbouring habitats may reach higher population sizes, and hence achieve higher controlling effects on insect populations at the scale of a combination of several habitats than at the scale of only one habitat. This example illustrates the importance of neighbouring effects in some important ecosystem processes at the scale of a mosaic of habitat patches.

7.4 Adaptation, local differentiation, and polymorphism

Mediterranean biotas are particularly rich in disjunct distributions of closely related species, subspecies, or populations (Verlaque *et al.* 1991; Thompson 1999), as are floras of other mediterranean-type climate regions in the world (see Chapter 3). Many of these disjunct distributions may reflect the geological and tectonic complexity of the Mediterranean Basin or movements of tectonic microplates, as discussed in Chapter 2. But others may result from recent processes, such as island isolation, dispersal, or human-induced habitat fragmentation. In addition, the rich diversity of habitats described in Chapter 6 is conducive to a wide range of selection regimes that shape life-history traits of populations depending on local conditions. Species and populations may respond to these selection pressures in a number of ways. They may evolve **local specialization**, whereby local specific combinations of genes adapt organisms to local conditions. They may also evolve phenotypic plasticity, whereby the same genotypes may be expressed in different phenotypes according to environmental conditions.

Many studies in recent years have demonstrated that life-history traits, which have important fitness consequences, may evolve quite rapidly, even within a few generations, provided they are submitted to strong directional selection pressures. There is in fact a growing body of evidence showing that plants and animals exhibit a high amount of intraspecific variation, both genetic and phenotypic, at different spatial scales, among individuals within populations, among subpopulations within a landscape, and among populations across a species' range. In particular, the study of the spatial structure of genetic markers, notably maternally inherited chloroplast DNA in plants, has an important tool for studying evolutionary change associated to episodes of colonization and isolation (Thompson 1999). The spatial genetic structure of populations may provide insights into the evolutionary significance of such events as historical associations among populations or how isolation shapes patterns of disjunct distributions of related organisms. In this context, the study of the role of selection, gene flow, and genetic drift in shaping character variation and evolution may be particularly insightful. For example, the existence of several tectonic microplates, progressively squeezed over eons between the main African and Eurasian plates (see Chapters 1 and 2), caused spatial isolation events by splitting species into two or more disjunct populations, which subsequently evolved in isolation (e.g. Verlaque *et al.* 1991). Then human influence on the distribution, size, and abundance of natural habitats presumably greatly affected the spatial structure of local populations.

7.4.1 Polyploidy in orchard grass

Polyploidy is an important evolutionary factor influencing the evolutionary biology of plant populations, and some Mediterranean groups clearly illustrate the range of evolutionary processes operating on genetic differentiation in **diploid**/polyploid taxa (see Thompson 2005). In this context, **ploidy** levels in plants is an important question to address. Stebbins (1971) noticed that many polyploid plant species have spread over a wide range of environmental conditions, whereas closely related diploids have remained restricted to much smaller areas. A related question is whether the extraordinary environmental heterogeneity of the Mediterranean Basin resulted in a higher **ecotypic variation** of local populations than that found in non-Mediterranean parts of the species' ranges. Lumaret (1988) has shown that the perennial orchard grass (*Dactylis glomerata*) consists of a complex of no fewer than nine distinct subspecies, with several insular endemics, and as many as 15 diploid types, three tetraploid, and one hexaploid, the latter being confined to North Africa. Several tetraploid forms (e.g. the subspecies *glomerata*, *hispanica*, and *marina*) all have wide distribution ranges and exhibit morphological and physiological variations. Stebbins and Zohary (1959) pointed out that the differentiation of closely related tetraploid forms of *D. glomerata* subsp. *glomerata* resulted from **autopolyploidy** of diploids derived from both temperate zone and Mediterranean groups, thus providing the plants with ecological attributes adapted to both climatic regions. Polyploidy appears to be an evolutionary response to the wide range of ecological conditions within the Mediterranean region itself. In fact, almost all forms of *D. glomerata* in the Mediterranean group are to some extent adapted to drought conditions. Representatives of this complex of taxa range from sea level to well above the tree limit in high Mediterranean mountains, with a predominance of tetraploids in more severe ecological conditions. Thus, at least in some cases, tetraploidy may widen the habitat range of this complex at both ends of the various ecological gradients along which the species occur and disperse. In the hyper-arid regions of Libya, hexapolyploidy may be a further form of genetic adaptation to extreme ecological conditions. In addition to providing a common means whereby plant taxa may diversify ecologically and genetically (Petit and Thompson 1999), polyploidy may confer enhanced resistance to attack by pathogens, insects, and nematodes.

The mechanisms allowing polyploids to adapt to a wider range of environments than related diploid species are associated with biochemical, physiological, and developmental changes (Bretagnolle 1993). For example, cell size, which determines photosynthetic rates and DNA content, is higher in tetraploids, which provide them with a greater photosynthetic capacity than diploids. Moreover, **autotetraploids** exhibit greater mean performances in fitness-related traits over a range of environments as compared to their related diploids, as shown not only in the orchard grass (Bretagnolle 1993), but also in another perennial grass, *Arrhenatherum elatius* (Petit *et al.* 1996). In the latter species, the diploid subspecies *sardoum* is endemic to pine forests and open scree slopes in mountains, whereas the tetraploid subspecies *elatius* occurs in a wide range of habitats, such as road sides, open fields, waste ground, and woodlands, all over Europe. Petit and co-workers experimentally demonstrated that the tetraploid *elatius* has better performance than the diploid in all environments, as indicated by significantly higher values for stem height, leaf surface area, total seed number, and number of spikelets per flowering tiller. Interestingly, these studies have shown that the greater variation in the ecology and geographical distribution of tetraploids does not result from greater phenotypic plasticity but rather from between-population genetic differentiation (Petit and Thompson 1998). Thus, greater vegetative stature and inflorescence size of the tetraploid suggest that their more widespread distribution is due to their better performances rather than to a greater capacity to buffer environmental variation. Indeed, differences in vegetative stature were observed between open habitats and woodland in tetraploids, but not in diploids.

The highly dynamic evolution of polyploidy in many groups and greater genetic differentiation may be critical components of their success in spatially variable environments such as those found in the Mediterranean area. Adaptations to Mediterranean conditions in polyploid groups also include morphological traits allowing water-saving mechanisms and seed retention throughout the summer drought. For example, the seeds of the tetraploid *Dactylis glomerata* subsp. *hispanica* are not shed until autumn, when conditions become suitable for germination (Lumaret 1988).

7.4.2 Continental and island populations of cyclamen

Plant species which have isolated populations on both Mediterranean islands and the nearby mainland provide opportunities to investigate evolutionary processes associated with the genetic differentiation of populations. Balearic cyclamen, one of the 20 species of this genus occurring around the Mediterranean, is endemic to five geographically isolated sites in southern France and to the Balearic Islands of Mallorca, Menorca, Ibiza, and Cabrera (Debussche *et al.* 1996). On the mainland, this geophyte lives in the understorey of Mediterranean evergreen oak and pine woodlands, usually on north-facing slopes or gorges in shady rocky forests or else on limestone outcrops (Debussche *et al.* 1995). In contrast, on the Balearic Islands, it occurs in a wider range of habitats on a larger altitudinal gradient. The presence of this species with poor long-distance dispersal ability suggests that it was already present in all these regions by the time the islands became isolated from one another, probably in the Pliocene (5 mya; see Yesson *et al.* 2009). Therefore, they have been genetically isolated for several million years and, subsequently, their populations were further fragmented through climatic changes and human activities.

The effects of habitat isolation on levels of genetic diversity and patterns of genetic differentiation in this species were investigated by Affre *et al.* (1997), using biochemical techniques. Their study produces two interesting results. First, as island biogeography would predict, island populations contained less diversity (fewer alleles, less **heterozygosity**) than continental populations, suggesting that the colonization of the Balearics Islands by this cyclamen led to a loss in genetic diversity, probably through increase selfing. Second, and rather unexpectedly, the authors found that the genetic differentiation among 'terrestrial' island (i.e. highly disjunct) populations was greater than that among 'true' island populations (as measured by Wright's fixation index F_{ST}). This greater differentiation among isolated continental populations may be due to the fact that they have undergone more severe isolation effects, either due to glaciations or human-induced habitat fragmentation, or probably both. In summary, this endemic species of cyclamen is characterized by a marked population structure, high effects of inbreeding on genetic diversity, and much lower levels of genetic diversity within populations than in widespread congeneric species (Thompson 2005).

7.4.3 Sexual and chemical polymorphism: the story of thyme

Some fascinating examples of intraspecific variation in reproductive systems and essential oil contents have been studied in many Mediterranean plants, such as rosemary, mint (*Mentha*), oregano (*Origanum*), and rue (*Ruta*), and also among closely related species, such as true lavender (*Lavandula angustifolia*) and the closely related *Lavandula stoechas* and *Lavandula latifolia*. In particular, an enormous body of work has been conducted on such variations in an emblematic aromatic shrub of the north-western quadrant, the wild thyme (*Thymus vulgaris*), which forms a dominant component of open vegetation in eastern Spain, southern France, Italy, and the western Mediterranean islands (Thompson *et al.* 1998; Thompson 2002). Over the last half century, more than 25 scientists have endeavoured to decipher the genetic and ecological factors that determine the evolution and maintenance of polymorphic variation in the reproductive system and chemical compounds present in this species, and the work goes on.

Let's start with the sexual polymorphism. Thyme is a gynodioecious species, which means that plants bear either hermaphroditic or exclusively female flowers. Sex determination is controlled by the interaction of cytoplasmic genes that produce female phenotypes and nuclear genes that restore male function and thus the hermaphrodite phenotype. When the inheritance of sex is cytoplasmic; that is, in the absence of genes that restore male function, females produce only females because the male function is sterile. The frequency of females in natural populations of thyme averages around 60%, varying from 5% to more than 90%, depending on populations. What maintains gynodioecy and such high average frequencies of females in thyme and why is there such marked variation in female frequencies? Since female phenotypes reproduce only via ovules, they suffer a genetic cost relative to hermaphrodites. What advantage compensating for the lack of pollen production allows them to persist in populations? Research has shown that if sex inheritance is cytoplasmic, females need only a slight advantage over hermaphrodites, in terms of seed production, to be maintained and invade the populations (Gouyon and Couvet 1987; Couvet *et al.* 1990). In addition, the genetic cost for females may be offset if hermaphrodite progeny suffer inbreeding depression (thyme is self-pollinating), or else if plants re-allocate to seed production the resources they save by not producing pollen. Indeed, female thyme plants, as in other gynodioecious plants, generally produce two to three times more seed than hermaphrodites, and their offspring are more vigorous (Thompson and Tarayre 2000). However, the maintenance of gynodioecy requires that hermaphrodites also occur in the population, which implies that the restoration of male function must sometimes occur. Sexual phenotypes are thus determined by the interplay of cytoplasmic male sterility genes and nuclear genes that restore male fertility. The explanation for variable female frequencies is that populations are not at equilibrium; that is, there is a high variation in nuclear and cytoplasmic gene frequencies. Female frequencies depend on the age of populations, with high female frequencies in very young populations that have recently invaded abandoned fields or sites that have recently been burnt. Founder effects during early colonization cause a lack of variation in cytoplasmic feminizing genes (Manicacci *et al.* 1996; Tarayre *et al.* 1997) in association with an absence of appropriate nuclear restorer alleles, resulting in cytoplasmic sex determination and the exclusive production of females. Later, the arrival of nuclear restorer alleles via seed or pollen dispersal causes female frequency to decline. Thus, there is in thyme a 'moving mosaic'— of mating systems, this time—that is driven by the occurrence of repeated extinction/re-colonization events resulting from ecological disturbances, such as fire, that cause spatial variation in the genes that determine sexual phenotype. This is a fine example of metapopulation dynamics, with non-equilibrium dynamics prevailing in young populations where female frequencies may be as high as 90%, and equilibrium being maintained across suites of interacting populations (Olivieri et al. 1990).

The second part of the thyme story concerns polymorphism in the composition of essential oils in plant cells (Thompson 2005:144–164). Mediterranean plants are renowned for their fragrances, which give an exquisite smell to matorrals and landscapes in the region, and many of these plants are widely used, fresh or dried, for cooking in cuisines from the entire world. A study by Ross and Sombrero (1991) of plant species in the Mediterranean Basin has shown that a disproportionate number—approximately 90 of the 153 genera, in 50 families—of plants that produce and accumulate aromatic volatile oils in their cells occur in the Mediterranean region.

In thyme, six genetically determined forms of the compound monoterpene have been recognized on the basis of the dominant forms produced in glands on the surface of the leaves (Vernet et al. 1986; Thompson 2002). One could expect that such polymorphic variation is maintained because the different forms differ in fitness, according to spatial or temporal variation of abiotic and biotic factors. It appears in fact that phenolic compounds dominate in dry hot sites, whereas non-phenolic chemotypes dominate in inland and hilly sites that are often colder and moister (Vernet et al. 1977; Thompson 2002). Spatial differentiation in the distribution of chemotypes also occurs at much smaller spatial scales (Gouyon et al. 1986), sometimes even at the scale of a few metres. Such patterns clearly suggest that chemotype variation has something to do with local adaptation. For example, Amiot et al. (2005) and Thompson et al. (2007) have demonstrated that the adaptation of phenolic thyme chemotypes to hot dry conditions in summer and the resistance of non-phenolic chemotypes to early winter freezing are the keys to polymorphism in this species.

There is also evidence that, in thyme, secondary compounds are closely involved in biotic interactions that may influence the relative fitness between thyme chemotypes. For example, there is a marked variation between chemotypes in their palatability to slugs, and the snail *Helix aspera* definitely prefers non-phenolic phenotypes (Linhart and Thompson 1999). Interestingly, phenolic compounds may differ within the same genotype, depending on the age of the plant. Seedlings that are expected to be highly vulnerable to herbivory have a phenolic phenotype which protects them, but this phenotype will be replaced by a non-phenolic one as the plant gets older and becomes less vulnerable to herbivores. As explained by Linhart and Thompson (1995), the phenotype most preferred by snails may 'hide' behind a less palatable phenotype during early seedling development. These authors have subsequently found that different herbivores prefer different chemotypes, and that no one chemotype provides the best defence against all potential herbivores and pathogens (Linhart and Thompson 1999; Thompson 2005). Hence variation in parasites, as well as herbivore preference and abundance, may all have an influence on the maintenance of polymorphism in essential oils.

Inhibition of the germination of other plants may also be one of the functions of essential oils produced by thyme and other plants. Tarayre et al. (1995) demonstrated that phenolic chemotypes of thyme have a greater inhibitory effect than non-phenolic chemotypes on the germination of an associated grass species, *Brachypodium pinnatum*. Ehlers and Thompson (2004) reported similar results for the effects of thyme on another perennial grass, *Bromus erectus*, which responded differentially to different chemotypes of thyme. Thus, variation in chemical compounds in a species may partially determine and maintain local variation in the composition of plant communities. Similarly, secondary compounds such as aromatic oils can confer adaptive mechanisms in this widespread Mediterranean shrub, and environmental heterogeneity plays a key role in the determination and maintenance of chemical and reproductive diversity at the level of local populations (Keefover-Ring et al. 2009). Are there similar examples among Mediterranean trees?

7.4.4 Oaks, pistachio trees, maples, and pines

Mediterranean oaks, pines, and pistachio trees play important roles in almost all ecosystems of the basin and are all rich in systematic complexity and ecological diversity. Much information is available on the genetic structure and breeding systems of these species, which are dioecious or **monoecious** and highly outbred. In species that have been of great importance for humans over the centuries, the genetic structure of populations may have been modified to some extent through selection pressure from human activity, thus obliterating, or at least superseding, the expression of natural processes (see Box 7.1). Yet it is worth considering whether any of the trends we have examined thus far in this chapter can be discerned operating at the species level among these long-lived species.

Variable life-history traits of clear adaptive significance in Mediterranean oaks, pistachios, maples, and pines include growth form, leaf and fruit longevity, sexual reproductive, and seed-dispersal systems. Among Mediterranean oaks and pistachio trees, for example, we find species contrasting in one important life-history trait, which is **evergreeness** versus summergreenness (see Chapter 8). Six species of Mediterranean oak are evergreen but the other 35–40 species are all summergreen or deciduous. Among Pistacia species, the ratio is greater still: one evergreen species for about 10 deciduous ones. In both groups, vicariant species and interfertility among sympatric species are the norm (see Table 7.3).

In fact, there exists a continuum between evergreen and deciduous oaks and pistachios in the Mediterranean. At one extreme, some deciduous species have no overlap between leaf generations punctuated by cold winter seasons, whereas at the other extreme leaves of evergreen species of the previous season fall in late spring, after the current year's new leaves have emerged. In this situation, there is thus a nearly total overlap in annual leaf 'crops' on a given tree. An intermediate 'semi-deciduous' group also occurs among oaks, as shown in Table 7.3. One similar case among pistachios, Pistacia saportae appears to be a natural hybrid between a deciduous and an evergreen species.

Table 7.3 Vicariant series among Mediterranean oaks

Leaf form	Group[1]	North Africa	Western Mediterranean	Eastern Mediterranean	Zagros[2]
Evergreen					
Q. coccifera	B	Q. coccifera	Q. coccifera	??	
Q. ilex	A	Q. ilex[3]	Q. ilex	Q. calliprinos	Q. baloot
				Q. aucheri[4]	
Q. suber	B	Q. suber	Q. suber	Q. alnifolia[5]	
Deciduous					
Q. ithaburensis	B			Q. ithaburensis	Q. brantii
Q. cerris	B		Q cerris	Q. hartwissiana	
Q. infectoria[6]	A	Q. faginea	Q. canariensis[7]	Q. boissieri	Q. boissieri
Q. libani	B	Q. afares	Q. pyrenaica	Q. libani	Q. libani
Q. robur	A		Q. humilis[8]	Q. anatolica	

[1] A, acorns fall 6 months after ripening; B, acorns usually fall 18 months after ripening.
[2] Zagros, Caucausus, or Himalaya.
[3] Includes Q. rotundifolia of Spain.
[4] Endemic in Greece and Aegean Islands.
[5] Endemic in Cyprus.
[6] Special group of semi-deciduous species which lose only some of their leaves in autumn.
[7] Occurs only in a restricted area of southern Spain, near Gibraltar.
[8] Extends into parts of central Europe.

Sources: M. Barbéro, P. Quézel, F. Romane, and A. Shmida, personal communication.

In general terms, evergreen oaks and pistachios can be considered as old, 'primitive' taxa, with tropical or subtropical affinities, while deciduous species may be more 'recent', having resulted from adaptive radiation into colder, extra-tropical latitudes in more recent times (see Chapter 2). Notably, the evergreen species of both groups are restricted to warmer life zones of a given region and, within landscapes, to the warmer drier habitats.

These observations have prompted several researchers to ask whether a clear difference between related deciduous and evergreen species could be determined in terms of ecophysiological water-use efficiency or, conversely, water-loss prevention under conditions of water stress. In a detailed comparison of several evergreen and deciduous oaks of the same age growing in a controlled uniform environment, Achérar and Rambal (1992) found no clear difference in water-use strategy, except among very young seedlings.

Contrary to the distribution pattern described above for the oaks, a very different pattern occurs in the Mediterranean maples: the deciduous Montpellier maple is widespread and occurs in many life zones, whereas the two evergreen ones are more narrowly distributed (*Acer sempervirens* in Crete and *Acer obtusifolium* subsp. *syriaca* in the eastern Mediterranean), and both are restricted to cool, moist mountain zones. A third group includes a number of deciduous species, for example *Acer campestre*, *Acer hyrcanum*, *Acer platanus*, and *Acer tataricum*. This group is primarily restricted to cool, mixed deciduous forests of premontane formations (meso-, supra-, and oro-mediterranean life zones).

The 16 species and subspecies of Mediterranean pines (Table 7.4) are also quite varied in their life-history traits, climatic and soil requirements, and biochemical composition (Barbéro *et al.* 1998). Considerable variation occurs in their reproductive biology. Generally, when passing from subtropical zones towards the temperate zones, there is a change in both pines and oaks from biannual to annual fruit maturation, which is presumably a response to increasingly cold winters. By reducing fruit maturation time, these trees no longer need to maintain viable fruits on the tree in a period when they would be subject to freezing, but rather 'release' them for germination in the autumn of the same year of their development.

Some Mediterranean pines are widespread and expanding their ranges (e.g. Calabrian and Aleppo pines), while others are highly sensitive to fire, long-lived, and also limited to particular soil types. In terms of area of distribution, the same pattern prevails as among oaks and pistachios, namely that the widespread generalists are found in lower life zones (most affected by humans), while restricted taxa and endemics are most common either in mountainous areas (e.g. Bosnian and montane pines) or else on islands (e.g. Canary Island pine, *Pinus canariensis*). A good example of this group is provided by Bosnian pine in Greece and the closely related *Pinus leucodermis* in the Calabrian Mountains in southernmost Italy. This is an ancient (Miocene) species restricted to dolomite substrates, which does not survive fire, invade, or regenerate readily, but can live for very long periods, even as much as 3000 years (M. Barbéro, personal communication). In the highly variable black pine, however, and in maritime or cluster pine, biochemical studies indicate that more successful colonizers among subspecies (e.g. *Pinus nigra* subsp. *clusiana*) are characterized by greater intrapopulation variability of certain chemical markers (e.g. flavanoids and protoanthocyanids) than edaphically restricted or otherwise narrowly distributed subspecies, such as *Pinus nigra* subsp. *salzmanii* in southern France or *Pinus nigra* subsp. *mauretanica* in North Africa, both of which are restricted to dolomite substrates (Barbéro *et al.* 1998). In pines, as for the evergreen oaks, the long history of human manipulation of genotypes must be taken into account when attempting to interpret the intricate tapestry of their taxonomy and distribution within the Mediterranean region.

In Chapter 6, we noted that Mediterranean forest cover in the past was not as dense and uniform as formerly thought, and that many landscapes were originally open and heterogeneous. However, it also appears that in the past, some 5000 years ago, which corresponds to the post-glacial 'climatic optimum', Mediterranean forests showed a much denser, more intricate mixture of oaks, pines, junipers, and deciduous trees than today. Thanks to fossil pollen and charcoal data, forest reconstruction or landscape archaeology are increasingly

Table 7.4 Distribution of the Mediterranean pines with respect to substrates, climatic zones, and life zones

Species	Substrates[1]						Climatic zones[2]				Life zones[3]				
	Ma	Ca	D	Si	Ub	Ss	H	SH	SA	A	TM	MM	SM	MtM	OM
Pinus halepensis	+++	++	++	+		+	++	+++	+++	++	++	+++	++		
Pinus brutia	+++	++	++	+		+		+++	+++	+	+	+++	++		
Pinus pinaster			+	+++	+++	+++	++	+++			+	+++	+		
Pinus hamiltonii			+	+++		+++	+++				+	+++			
Pinus maghrebiana	+++	+	++	++	+	++		+++	+		+	+++	++		
Pinus pinea			+			++	++	+			+	+++	+		
Pinus canariensis				+++		+++		+++			+	+++	+		
Pinus nigra subsp. salzmannii		++	+++	++			+++	++					++	+++	
Pinus nigra subsp. laricio			+++	+++			+++	++					++	+++	
Pinus nigra subsp. nigra		+++	+++	+++	++		+++	++					++	+++	
Pinus nigra subsp. mauretanica			+++				+++						++	++	
Pinus nigra subsp. dalmatica			+++				++	+++				+	+		
Pinus nigra subsp. pallasiana	+++	++	++				+++	++					++	+++	
Pinus sylvestris	+++	++	+++	++			+++	++					++	+++	
Pinus heldreichii		++	++	++	+++								++	+++	++
Pinus uncinata														+++	+

[1] Ma, marls; Ca, limestone; D, dolomite; Si, acidic sands; Ub, ultrabasic volcanic rocks; Ss, sandstone.
[2] H, humid; SH, sub-humid; SA, semi-arid; A, arid.
[3] TM, thermo-mediterranean; MM, meso-mediterranean; SM, supra-mediterranean; MtM, montane-mediterranean; OM, oro-mediterranean.

Source: Barbéro et al. (1998).

feasible, and one essential theme stands out clearly: especially at lower altitudes, each species of Mediterranean pine was associated to one or more broad-leaved and sclerophyllous oak formations. For example, even today, Aleppo pine occurs consistently with holm oak, while Calabrian pine occurs with *Quercus calliprinos*. *Pinus pinaster* subsp. *hamiltonii* occurs with cork oak, and Scots pine with downy oak (Barbéro *et al.* 1998). The other conifers, such as junipers, cypress, cedars, thuja, and firs, that were formerly part of the conifer canopy layer in different life zones, have been altogether removed or drastically reduced by humans. Pines recover better and spread faster than junipers and far more than most other conifers, and they are also far more frequently planted than any conifer or other kind of Mediterranean tree. In Chapter 10, and again in Chapter 13, we will come back to pines and oaks and other conifers in the context of cultural landscapes.

7.4.5 Blue tits in a habitat mosaic: coping with habitat heterogeneity

The blue tit is a small passerine bird common in broad-leaved and mixed forests at low and mid-altitudes. Its distributional range covers the whole Mediterranean region, including the Canary Islands. The five subspecies of this oceanic archipelago show considerable variation in both systematics and ecology (Garcia-del-Rey *et al.* 2006). This complex differs so much from all other blue tit forms that it has been proposed by Kvist *et al.* (2005) to constitute a distinct species, *Cyanistes teneriffae*. In most parts of the Mediterranean Basin, forested landscapes are mosaics of habitats with patches where evergreen holm oaks are dominant and patches where deciduous downy oaks are dominant. The spring development of new leaves occurs 4 weeks later in holm oak than in the deciduous downy oak, which results in a corresponding 4-week interval in the development of leaf-eating caterpillars that the tits use as prey for themselves and their young. To what extent are blue tits able to adjust their breeding time to this huge seasonal variation of food availability between the two types of oak woodland patches? To address this question, two landscapes, including several habitat patches dominated by the two types of oaks, have been selected, one on the mainland near Montpellier and the second in the island of Corsica. The mainland landscape is relatively close to the large deciduous forest blocks of central France. By contrast, in many parts of Corsica, most of the habitats suitable for tits are evergreen oak forests and matorrals, interspersed with small patches of deciduous trees. Birds which have the best breeding success are those that best synchronize the period when they raise their nestlings with the brief 2–3 week window of caterpillar availability in spring. Indeed, both in deciduous oak habitats on the mainland and evergreen oak ones on Corsica, blue tits start to breed at a time that allows fairly good synchronicity of nestling period and the period of caterpillar availability (Fig. 7.4a). This means that they start

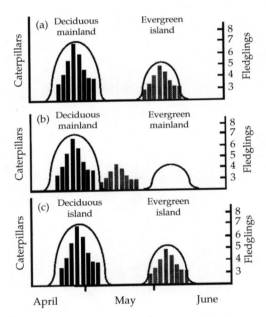

Figure 7.4 The breeding schedule of blue tits in Mediterranean habitats dominated by the deciduous downy oak on the mainland near Montpellier and in habitats dominated by the evergreen holm oak in Corsica. On each graph, the curves indicate caterpillar abundance and histograms the number of fledglings. (a) Between-landscape scale (i.e. mainland and Corsica). (b) Within-landscape scale on the mainland. (c) Within-landscape scale in Corsica. Note (1) the delay of 4 weeks in the timing of caterpillar abundance and nestling period in evergreen habitats compared to deciduous habitats, and (2) the mismatch between the nestling stage and food availability in evergreen habitat patches of the mainland landscape (b) (after Blondel 2007).

to breed 4 weeks later in the evergreen habitats of Corsica than in the deciduous habitats of the mainland. These two populations, which are isolated by the sea and thus deprived of all contact between them, are assumed to be adapted to their local habitats (Blondel *et al.* 1993). Indeed, experiments in aviaries have demonstrated that the difference in laying date between these two populations of tits has a genetic basis (Lambrechts *et al.* 1997). Hence they are locally genetically specialized to their habitats.

From these observations on the mainland and assuming that tits evolved similar adaptations to local food resources *within* each of the two habitat mosaics (mainland and Corsica), as they did *between* them (Fig. 7.4a), one may predict that there would be a similar difference of 4 weeks in the start of breeding depending on whether they breed in deciduous or in evergreen oak habitat patches, whatever the landscape is located on the mainland or in Corsica. Observations do not necessarily fit these predictions. Tits in the patches which are the less common in each landscape (evergreen on the mainland, deciduous on the island) may start to breed approximately at the same date as those in the most common patches. They breed too early in evergreen mainland patches and too late in deciduous Corsican patches in relation to the peak of food abundance (Figs 7.4a and 7.5). This mismatching between the time of breeding of the birds and food availability results in a lower breeding success than in the commoner habitats. The explanation for the weak inter-habitat differentiation of blue tit life-history traits in each region is that gene flow across the different habitat patches within each landscape prevents birds from evolving adaptation to the mosaic character of their local environment (Dias *et al.* 1996; Lambrechts *et al.* 1997).

However, in the Corsican landscape, a small isolated catchment which is isolated by mountain ranges, the dominant tree species is the deciduous downy oak. The two study populations, one in evergreen oakwood and the other in deciduous oakwood, were rightly timed on food availability (Fig. 7.4c). In other words, breeding patterns at the within-landscape scale in Corsica were similar to those observed at the between-landscape scale between the mainland and Corsica

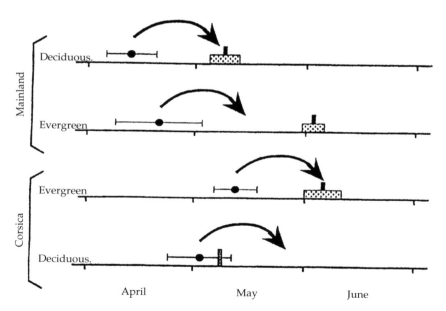

Figure 7.5 Schematic representation of the matching between breeding time of blue tits (black dots represent laying date, ±SD) and the time of maximum abundance of caterpillars (black bars) in the commoner habitats (deciduous on the mainland, evergreen on Corsica), and the mismatching in the less common habitat. The black arrows indicate the mean date of maximal food demand, when the young birds in the nest are approximately 10 days old. Closeness between arrow and black bars indicates good match (after Dias and Blondel 1996).

(Fig. 7.4a) (Blondel *et al.* 1999). Birds from the habitat with deciduous oaks started egg laying on average 1 month earlier, laid on average 1.5 more eggs, and fledged about 35% more young than those from the large evergreen holm oak forest. Thus, blue tits in the two habitats in Corsica—which are only 25 km apart—are as well synchronized to the short window of food availability as those which breed in the far distant habitats of downy oaks on the mainland and holm oak on Corsica. Why such a difference, which is at first sight unexpected?

This case study is an illustration of the insular syndrome discussed above, providing one example of the consequences on phenotypic variation of geographic isolation. At first sight, the similar geographic configuration of habitat patches in the mainland and insular landscapes, where blue tits were studied, was expected to produce similar patterns of mismatching between laying date and optimal breeding time in these two landscapes (see Blondel *et al.* 2006). Extending these studies to five populations in each landscape, three in evergreen oakwoods and two in summergreen oakwoods, the phenotypic variation of laying date was much larger on the island than on the mainland. In the five habitats of Corsica, tits started egg laying within a range of 29 days as compared to only 9 days on the mainland (Blondel *et al.* 2006) (Fig. 7.6). Thus populations of blue tits only a few kilometres apart have evolved specialization to local habitats, which is a demonstration that adaptive response to habitat-specific selection regimes may operate on a scale which is much smaller than the scale of potential dispersal and gene flow.

Interestingly, what makes Corsican populations differ so much from those on the mainland and thus explains a larger phenotypic variation of traits with higher habitat-specific adaptations at a small spatial scale is reduced dispersal in island birds, which, as explained above, is a component of the insular syndrome (Blondel 2000). This has been shown to occur in Corsican birds (Blondel *et al.* 1988). In gene flow/selection dynamics, stronger habitat fidelity combined with habitat-specific assortative mating results in lower dispersal rates and hence lower gene flow in Corsica than on the mainland. These patterns are conducive to local specialization and population structuring, whereas high dispersal on the mainland leads to phenotypic plasticity and, possibly, local maladaptation due to gene flow.

This study of Mediterranean blue tits shows that population processes operate at the scale of a landscape, not just at that of habitats. In habitats that are clearly isolated, selection regimes may result in a fine local specialization of life-history traits. But wherever the dispersal range of a species by far exceeds the size of local habitat patches, differences

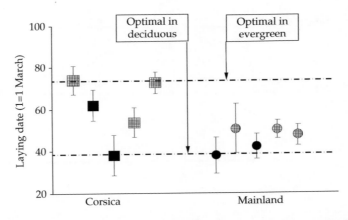

Figure 7.6 Phenotypic variation of laying date for blue tits (±1 SD). Horizontal dashed lines denote the best date for starting breeding relative to food availability in deciduous and evergreen oakwood. Black squares, deciduous habitats; hatched squares evergreen habitats. Note the much higher variation of laying date in Corsica than on the mainland (modified from Blondel *et al.* 2006).

> **Box 7.2. Mediterranean constraints on breeding in insectivorous birds**
>
> The mosaic of habitat patches dominated either by deciduous trees or by evergreen sclerophyllous trees provides a quasi-experimental opportunity for investigating the ecological and evolutionary responses of birds to habitat-specific features and constraints. The large differences in the tempo and mode of food resources between evergreen and deciduous oaks make the relationship between food and breeding performance of tits quite different in the two habitat types. In evergreen oaks, leaf-eating caterpillars, which are the main prey of nestling tits, are four to five times less abundant than in the rich deciduous oakwoods. As a result, the total biomass of chicks in the nest is on average 59 g in poor evergreen habitats (6.3 chicks × 9.3 g) compared to 86 g in rich deciduous habitats (8.3 chicks × 10.3 g). The poor body condition of young tits raised in evergreen habitats make them unlikely to be recruited in the population, hence source–sink population structuring between deciduous and evergreen habitats in many heterogeneous landscapes of the Mediterranean. In the evergreen, sclerophyllous habitats where the spring development of the foliage and associated leaf-eating insects occurs late in spring when temperatures are high, various combinations of constraints including food shortage, high parasitic loads, and high temperatures make breeding conditions much variable and often extremely poor (Tremblay *et al.* 2003; Simon *et al.* 2004; Blondel *et al.* 2006).

in selection regimes may result in local maladaptation of populations due to gene flow. Thus, most life-history traits in Mediterranean species that are confronted to high environmental heterogeneity must result from a fine balance between local specialization and phenotypic plasticity.

Finally, long-term studies, such as that on blue tits, provide opportunities to go in the details of ecological features and constraints that are typical of the Mediterranean bioclimate. Two of them are summarized in Box 7.2. The study of Mediterranean blue tits provides a pertinent example of phenotypic differentiation at small spatial scales. Many other case studies could be described in detail. One of them is that of the common crossbill, which is summarized in Box 7.3.

7.5 Species turnover in time: migrating birds

In his seminal book, *The Palaearctic-African Bird Migration System*, published in 1972, Reginald Moreau described and analysed the magnitude of the task that faces the estimated 5 billion birds leaving the Palaearctic region each autumn to spend the winter in more favourable areas further south. If we are to dare a comparison, the seasonal gigantic shift back and forth of birds between Eurasia and Africa each year is, *mutatis mutandis*, similar to the series of contraction/expansion cycles of biota across Europe as a result of glacial episodes during the Pleistocene. Keeping apart the genetic consequences of these huge movements, the most obvious difference is in time scales, one year in the first case, tens of thousands years in the second. Given its strategic location at the border between the two major continental land masses of the Old World, the Mediterranean region, with a moist and mild climate from autumn to late spring (see Chapter 1), plays a key role in this system. Thus, on the kaleidoscopic picture of plant and animal communities in space is superimposed a kaleidoscope of bird communities which seasonally change in time. As soon as the breeding season is completed and sometimes even before, many Mediterranean habitats are invaded by migrating birds, which either will stay in the region for some time before going further south in tropical Africa, or will stay for the whole winter. Surviving individuals of long-distance migratory species will pass again through the Mediterranean to reach their northern breeding grounds in spring. Moreau calculated that, for bringing back to their breeding grounds the surviving individuals the next spring, some 1200 birds must cross the sea each day per kilometre width for a full 3 months!

Box 7.3. Morphological variation in crossbills

The common crossbill is widespread in Europe, from northern Europe to North Africa, and in many of the larger Mediterranean islands. This bird occurs in many habitats dominated by different species of conifers. Depending on which species of conifer the birds eat the seeds, the size and shape of the beak strongly differ. In Europe, the morphological differentiation of crossbills feeding on the seeds of pines, spruces, and larches is a classical case study in evolution. Faced with the high diversity of native species of conifers in the Mediterranean Basin, the crossbill responded by developing a remarkable range of bill shapes. Comparing bill morphology of crossbills inhabiting montane and black pines in the same geographic area of north-eastern Spain, Borras *et al.* (2008) showed that the large and thick spines of montane pine cones have selected for a larger and more robust beak than did those of the black pine. Another pattern of variation of bill morphology has been described between crossbills feeding on the seeds of Aleppo pine and those of the Scots pine between continental Spain and the Balearic islands (Alonso *et al.* 2006). The bills of birds inhabiting montane pines are more robust and have a greater depth than those inhabiting black pines. Thus, the difference between the bill shapes of Mediterranean populations mirrors the conifer diversity in the Mediterranean and is similar to those found between various populations reproductively isolated in Central Europe and North America. This strongly supports an adaptation of crossbills to the cones and seeds of the various species of conifers. Interestingly, Questiau (1999) has shown from molecular phylogenetics that the various crossbill lineages over Europe must be included in only two species, the two-barred, well-differentiated crossbill *Loxia leucoptera* and the common crossbill, which encompasses in a single **polytypic** species all the populations occurring across Europe. The various populations of this species are linked by gene flow, but they retain phenotypic differences in bill size, no doubt as a response to persistent directional selection pressures from the size and hardiness of the seeds of the differing conifers the various crossbill populations feed upon.

7.5.1 Seasonal aspects

Thus, in the Mediterranean region, there is a permanent turnover of birds and four main categories to consider: residents, which remain in the region all the year round; long-distance trans-Saharan migrants that occur in the region only for brief periods in autumn and spring; summer visitors, which invade the region in spring from their African winter quarters for breeding; and winter visitors that do not breed in the region but invade it for spending there the whole winter (Table 7.5). From year-round bird censuses carried out over four successive years, in a sample of habitats in the Camargue, Blondel (1969) showed that from a total of several thousands individual birds censused, only 13% were long-distance migrants, 24% were breeding birds, either resident or summer visitors, and no less than 63% were winter visitors. This latter category also included the multitude of ducks and waders which overwinter in the Camargue (Box 7.4). A schematic picture of this seasonal ballet is given in Fig. 7.7. Of course, there is much overlap between these categories. Many resident species start to breed in the heart of winter, in December/January, for example the griffon vulture or the eagle owl, while the first autumn migrants, notably the waders, which breed in the far north, appear in Mediterranean wetlands already in late June. Among the earliest autumn migrants are wood sandpipers (*Tringa glareola*), which leave their breeding grounds soon after their young emancipated and arrive in Mediterranean wetlands at the very beginning of summer. At this epoch, many species in the Mediterranean—bee-eaters, rollers, herons—still rear their young. In fact, migrating or wandering birds may be observed all the year round in any part of the Mediterranean.

Between 230 and 250 species of birds leave their Eurasian breeding grounds in autumn and winter in the Mediterranean or further south.

Table 7.5 Examples of bird species of the four main categories that occur in Mediterranean habitats according to season

Residents	Summer visitors	Long-distance migrants	Winter visitors
Green woodpecker	Scops owl	Tree pipit	Wren
Crested lark	Bee-eater	Garden warbler	Meadow pipit
Fan-tailed warbler	Roller	Willow warbler	Water pipit
Cetti's warbler	Tawny pipit	Wood warbler	Song thrush
Moustached warbler	Melodious warbler	Pied flycatcher	Dunnock
Serin	Subalpine warbler	Wheatear	Chaffinch
Goldfinch	Nightingale	Rock thrush	Robin
Mallard	Turtle dove	Redstart	Chiffchaff
Kestrel	Lesser kestrel	Icterine warbler	Teal

Box 7.4. Mediterranean wetlands as winter quarters for wildfowl

Mediterranean wetlands are of prime importance for migratory birds fleeing cold weather further north, because they rarely freeze. They are wintering grounds for a large variety of ducks, as well as coots (*Fulica*), geese, cranes, and sometimes swans (*Cygnus columbianus*) that leave their breeding grounds in central, eastern, and northern Eurasia in late summer. Some wintering species such as the wigeon (*Anas penelope*) breed as far east as eastern Siberia. In some large wetlands, such as the Camargue, the Guadalquivir delta, or Lake Ichkeul (Tunisia), more than 150 000 individuals may overwinter. But there are regional differences in the composition of local assemblages. For example, Lake Ichkeul is very important for greylag goose (*Anser anser*) (8% of the overwintering waterfowl), wigeon (60%), and pochard (*Aythia ferina*) (30%), whereas the Camargue is a winter ground of primary importance for shoverler *Anas clypeata*, teal *Anas crecca*, red-crested pochard (*Netta rufina*), and gadwall (*Anas strepera*).

Wintering ducks belong to two main categories. The first consists of granivorous species (e.g. mallard (*Anas platyrhynchos*) and teal) that feed on the protein-rich seeds of a variety of submerged macrophytes, such as pondweeds (Potamogetonaceae) and the cosmopolitan algae Characeae. The second category includes herbivorous species that exploit vegetative tissues, such as leaves and stems (e.g. gadwall, wigeon). Ducks spend the daytime in large concentrations in large bodies of water without vegetation. These are used as day roosts where they rest, groom, and, at the end of the wintering season, start to display before leaving the site. At dusk, they leave the pond and disperse to a variety of habitats where they will forage in farmland and shallow waters of seasonal marshes during most of the night. Since the main diet of the second group is less rich than that of the first, herbivorous species are obliged to extend their feeding time into the daylight hours. Large concentrations of wintering ducks inevitably draw in their wake large raptors, such as spotted eagles (*Aquila clanga*) and white-tailed eagles (*Haliaeetus albicilla*), which survey the ducks' wintering places from nearby trees and take every opportunity to feed upon them for the greater pleasure of bird-watchers!

Migrants passing through the western half of the Mediterranean are mainly drawn from the western Palaearctic, Scandinavia, Central Europe, and western Russia, while those passing through the eastern half come from further east. Some amazing travellers, such as the wheatear and the willow warbler (*Phylloscopus trochilus*), come from as far east as eastern Siberia, while others, for example the barn

Figure 7.7 The Mediterranean as a region of choice for migratory and wintering birds. Some overwinter in tropical regions of southern Asia. The figure also shows (bottom right) the turnover in time of breeding, migrating, and wintering birds. Note the overwhelming importance of autumn migrants and wintering birds, and the length of the breeding season which largely overlaps with the spring migration.

swallow, move as far south as South Africa. About one-third of these migrants do not go further south than the Mediterranean and stay in the basin for the whole winter. The moist and frost free winter climate in lowlands and coastal habitats around the Mediterranean Basin is particularly favourable for birds that forage on the ground and in shallow water. Large numbers of wintering passerines, such as thrushes, dunnocks, wrens, and kinglets (*Regulus*), benefit from the diversity of insects, snails, earthworms, and other invertebrates that they find on the ground, as well as from the fruits that are plentifully produced by a wide range of bushes and trees (see Chapter 8). Contrary to the picture in central and northern Europe where birds are much more abundant in summer than in winter, the opposite is true in the Mediterranean (see Fig. 7.7).

7.5.2 Long-distance strategies

The second category, long-distance migrants, including two-thirds of the Palaearctic migrants, consists of insectivorous birds, which depend on small flying insects for food. From late October until the end of March, these birds are unable to sustain themselves in Europe and therefore migrate to the tropics. Over 130 species of land and freshwater birds winter mainly south of the Sahara. This means that, each autumn, they have to negotiate the crossing of the Mediterranean Sea and the Sahara desert. Only soaring raptors and storks avoid crossing the sea, using flyways at the two extremities of the basin, the Bosphorus-Dardanelles to the east and Gibraltar to the west. Thousands of birds may be observed in a single day. But nearly all small species cross the sea on a broad front. As the Admiral Lynes wrote in 1910, 'there is not a single hectare of land or sea that is not overflied by migrant birds in the Mediterranean region'. This observation was confirmed by radar observations in the 1960s. In spring time, when weather is favourable, it is an amazing experience to sit on a beach and to search through a telescope tiny black spots moving some metres above the sea, approaching the coastline. When close enough, it is always a surprise to see that these spots are a pied flycatcher (*Ficedula hypoleuca*), a nightingale, a roller, a rock thrush, or any other of the dozens of species that had left the coast of Algeria the previous evening. All of them seem exhausted when they land. Quite often, alas, they will end their journey in the beak of a wandering yellow-legged gull that purposefully lingers along the coast in wait of its prey. Sometimes, when a weak tailwind and a clear sky make flying conditions especially good, an avalanche of birds descends upon the first bushes on the coastline. Then, urged by the necessity to reach their breeding grounds as soon as possible, they will continue their journey, 'gliding' from bush to bush in a north-eastern direction while feeding as much as possible. It is only at dusk that they will take off again for a non-stop nocturnal flight.

7.5.3 Refuelling

In autumn Mediterranean habitats are of paramount importance for all these migrants by providing stop-over feeding sites, which allows the birds to build up the energy reserves they will need to cross the sea and beyond that the daunting stretches of Saharan desert. The genetics, ecology, morphology, and physiological adaptations of small, long-distance migrant birds are quite astonishing. They have been beautifully studied by a handful of enthusiastic scientists and summarized in a special issue of the ornithological journal *Ibis* (Crick and Jones 1992; Berthold 1993).

Moreau (1972) thought that avian migrants crossed the Sahara in a single non-stop flight of at least 40–60 h because conditions in the Sahara were too difficult to allow them to find resting and feeding places. However, Bairlein (1992) showed that some trans-Saharan migrants regularly stop-over in suitable desert habitats where they rest and 'refuel'. This suggests that the fat load the birds accumulate in the Mediterranean prior to migration is not sufficient for a non-stop crossing of the Sahara.

During their Mediterranean stop-overs, which may last a few days to several weeks, long-distance migrants prefer open or semi-open habitats, such as matorrals and various types of ecotones, where they find a large supply of food. There they accumulate a considerable amount of fat, which will be used as fuel for their journey. Birds gorge themselves with food, mainly fruits, which are rich in sugar and carbohydrates that can be readily and quickly transformed into glycogen and fat. Nearly all species, including those that are mainly insectivorous during the breeding season, such as warblers and flycatchers, eat fruits to fatten themselves. It has been calculated that premigratory fat deposition can be very rapid. Within one week, a bird of 20 g may increase its body mass of 40–50%. When flying, small birds expend about four times more energy than when they are resting. Body-weight decrease, as body reserves are consumed during migratory flight, is on the order of $0.8\% \, h^{-1}$ in a bird weighing 10–20 g. Thus, since birds can accumulate up to 50% of their fat-free weight in fat reserves, they can make a non-stop flight of 30–50 h, which allows them to cross the Mediterranean Sea and the Sahara desert without refuelling. Birds leave the shores of the sea in autumn when they are fat enough to risk the crossing of the sea and then the desert. In contrast, the amount of energy spent for crossing back the large inhospitable barriers in spring makes the birds particularly lean when they reach the northern side of the sea. For example, in southern France, redstarts weighed on average 13.8 g in spring, but as much as 17.5 g in autumn (Blondel 1969).

Having now surveyed some components and several examples of the evolutionary and ecological variability of Mediterranean populations, species, and communities as a response to environmental heterogeneity, we move on, in Chapter 8, to consider life-history traits of selected organisms *vis-à-vis* the functioning of Mediterranean ecosystems.

Summary

The aim of this chapter was to demonstrate that from vicariance effects at large scales of space and time to the evolution of life-history traits at the scale of local habitats and landscapes, the range of genetic and ecological variations of species and populations in the region's plants and animals is enormous. On islands, communities are highly integrated constructions of interacting populations of relatively small numbers of species frequently characterized by adaptations, forming part of an insular syndrome, six components of which are described in detail.

Several examples in plants and animals show how population responses to habitat heterogeneity in space and time lead to a wide range of adaptations, including local specialization, phenotypic plasticity, polymorphism, and polyploidy in plants. In some cases, long-lasting human selection had strong effects on genetic variability as illustrated by the holm oak and the cork oak.

At the scale of landscapes, a variety of disturbance events, including fires, play a pivotal role in the dynamics of species and communities and contribute to the moving mosaic of habitats of Mediterranean landscapes. Because the Mediterranean Basin is a meeting place between Eurasia and Africa, where billions of birds spend the winter, the basin is also visited by a succession of birds, which migrate, overwinter, or breed in the region. The variety of migratory behaviours adds to the diversity of communities and ecological processes.

CHAPTER 8

Life Histories and Terrestrial Ecosystem Functioning

We would need an entire book to describe what is known of the evolutionary consequences of 'mediterraneity' on living systems and organisms. Here we highlight some recent developments and discoveries in the immense field of evolutionary ecology of Mediterranean biota in the context of their terrestrial ecosystems and landscapes. The examples and discussions given are but a small part of this fascinating field. Two key questions are: first, what are the evolutionary responses of organisms to Mediterranean-specific selection pressures? Second, to which extent do the biological traits that we see in the region's plants and animals result from factors that are not Mediterranean-specific, such as phylogenetic constraints or historical effects not related to the Mediterranean bioclimate specifically? An additional question we address is the extent to which the suite of life-history traits and the 'assemblage' of species into biotic communities confers enhanced resistance and resilience to ecosystems in Mediterranean environments where strong spatial and temporal heterogeneity influence populations and communities.

8.1 Evergreenness and sclerophylly

Evergreenness is a recurrent and striking feature in plant assemblages of the Mediterranean area and indeed in all five of the mediterranean-climate regions (see Chapter 3). But, in fact, evergreen broad-leaved foliage is one of the most ancient features of vascular plants, as shown by fossil remains of primitive gymnosperms and angiosperms. It is thus from evergreen ancestors that most modern floras have evolved (Raven 1973; De Lillis 1991), and evergreenness may have been an adaptation to the uniform and warm climate of the Cretaceous and a large part of the Tertiary. In contrast, the deciduous leaf habit, or summergreenness, apparently evolved later, possibly as a response to the appearance of climatic seasonality in the early Oligocene, about 35 mya. Seasonality may logically have been conducive to the evolution of changes in the phenological cycles of leaves, leaf longevity, and changes in the energetic balance and maintenance costs of photosynthetic tissues, which all differ sharply between the various kinds of deciduous and evergreen leaves. Yet evergreen broad-leaved plants survive and co-exist with deciduous species. Indeed, they dominate vegetation not only in the five mediterranean-climate regions, but also in the majority of vegetation types in the tropics, as we saw in Chapter 2. What then is special about presence of broad-leaved evergreen vegetation that is apparently typical of the Mediterranean? For one thing, it is mostly sclerophyllous, a trait we discuss next by asking whether sclerophyllous leaves in Mediterranean plants represent an adaptive life-history trait, an artefact from the past, or both.

8.1.1 Is sclerophylly a Mediterranean adaptation?

The term sclerophylly designates rigid, more or less leathery leaves that present a low surface-to-volume ratio, a large development of veins per unit of leaf area, a thick cuticule, and a tendency to brittleness when hit or folded. Some Mediterranean sclerophylls have relatively large leaves, such as laurel, lauristinus (*Viburnum tinus*), and ivy (*Hedera helix*), or else pinnately compound, as in carob, terebenth, and lentisk. But most sclerophyllous leaves in the region's flora are small and undivided, as in the holm oak, boxwood, myrtle (*Myrtus communis*; see Plate 2b), and smilax (*Smilax aspera*). All of them, however, are relatively long-lived and resistant, which no doubt explains why the ancient Greeks and Romans considered such evergreen plants as myrtle and laurel to be symbols of love, strength, and eternity.

Five lines of evidence attest to the adaptive value of sclerophylly in Mediterranean ecosystems.

1. A classical view of evergreenness in general, and sclerophylly in particular, is to consider this feature as a straightforward adaptation to environmental conditions, especially the prolonged drought in the hottest part of the year, which is typical of all mediterranean-climate regions (Mooney and Dunn 1970). Indeed, sclerophyllous leaves allow plants to control their transpiration by closing stomata during periods of water stress, which in turn enables them to survive periods of summer drought (Schiller et al. 2007). In all five of the mediterranean-climate regions, the large number of shrubs and trees that exhibit sclerophylly are thus considered as having independently evolved a 'water-saving' strategy, resulting from evolutionary convergence among the floras of these five regions with the same bimodal climate regime (see Chapter 3). The rationale is that if the survival of long-lived species depends on their ability to maintain a positive carbon balance over the year, then evergreen, sclerophyllous foliage presents the advantage—as compared to deciduous plants—of allowing a longer period of active photosynthesis despite the stressful hot-season drought.
2. Additional support for an adaptationist view of sclerophylly related to drought comes from quantitative studies showing that plant associations in progressively higher life zones—that is, those with milder, wetter climates—show a declining percentage of sclerophyllous species in six different Mediterranean mountain ranges studied along altitudinal transects (Barbéro et al. 1991). But water-saving is not the only issue; there are complex feedback cycles at work as well: even in evergreen plants, rates of transpiration and carbon gain vary in response to fluctuations in soil moisture and hence soil–plant–water relations (Zavala 2004). Tolerance to and recovery from water loss in plants also have anatomical consequences in perennial plants, since debilitating xylem cavitation can occur. Sclerophylls appear to recover better than non-sclerophylls in this specific respect (Salleo et al. 1997).
3. Sclerophylly also appears to be associated with the chronic state of nutrient deficiencies found in the soils of most mediterranean-climate regions (Specht and Rundel 1990), especially those of the western Cape Province and Australia (Lamont 1982). Indeed, evergreen species also tend to predominate in nutrient-poor habitats of the mediterranean-climate regions and elsewhere (Monk 1966), since plants growing there must develop mechanisms for efficient nutrient uptake and/or to tap widely spaced pulses of nutrients related to unusual climatic events. Clearly, in the Mediterranean region, sites and habitats with particularly nutrient-poor soils, such as gypsum or bauxite badlands or dolomitic sands, show a preponderance of evergreen sclerophyllous shrubs from genera such as *Achillea, Cneorum, Coriaria, Daphne, Genista, Globularia, Helianthemum, Helichrysum, Juniperus, Lavandula, Rosmarinus, Thymelaea*, and *Thymus*, representing a wide range of phylogenetically unrelated families. The overall rationale then, with regards alleged adaptation to summer drought and nutrient-poor soils in sclerophyllous trees and shrubs, is that a balance between photosynthetic gain and maintenance costs of non-photosynthesizing tissues can only be achieved, under limiting environmental conditions, by reducing the allocation of photosynthates to leaf growth and by prolonging leaf longevity. Indeed, the lifespan of most sclerophyllous leaves exceeds 2 years, averaging 3 years in holm oak and the olive and 5–6 years in kermes oak. Being 'costly' to produce, a plant

with sclerophyllous leaves tends to keep them a long time.
4. A different line of argument is that sclerophyllous leaves offer little reward to herbivorous animals and thus can be expected to suffer less from the strong and sustained herbivory pressure found in the mediterranean-climate regions than softer, broader, and relatively more sweet-leaved trees and shrubs. This 'anti-herbivore' effect is achieved partly by virtue of unpalatable and bad-tasting compounds, such as monoterpenes and tannins, which deter many plant-grazing and browsing animals (Herms and Mattson 1992).
5. One additional explanation for sclerophylly being so widespread in the Mediterranean flora, and indeed in other mediterranean-climate regions as well, may be related to the overall resistance and resilience of shrubs and trees sharing this trait in the face of disturbances or perturbations, such as fire and cutting, and browsing by large domestic animals. For example, holm and kermes oak, boxwood, and cade juniper (*Juniperus oxycedrus*) all resprout vigorously from the stump when cut down or burned. In support of this interpretation, human impact in the Mediterranean area as a whole resulted in a significant desiccation of the basin and an increasing importance of sclerophyllous plant species at the expense of deciduous trees (see Chapter 2). Deciduous oaks and other conifers, by contrast, generally die following such harsh treatment. Note that even on 'normal' calcareous and silicaceous soils in the Mediterranean region, sclerophyllous taxa are common as well, including the boxwood, buckthorn (*Rhamnus*), heathers (*Erica* and *Calluna*), carob, oleander, olive, and, of course, the conifers and the evergreen oaks.

However, despite many efforts to understand its ecophysiological function, sclerophylly is still a rather empirical term and the function of sclerophylly presumably differs between Mediterranean plant species and tropical plants. The precise selective forces contributing to the evolution, geographical distribution, and functional role in ecosystem dynamics of sclerophylly and some of its correlates (e.g. leaf longevity and evergreenness) are still poorly understood (Aerts 1995; Read and Sanson 2003). Let us return now to our original question and consider the arguments *against* there being any straightforward adaptive value of sclerophylly.

8.1.2 Is sclerophylly an artefact of the past?

Studies by De Lillis (1991) and others that focus on the water and gas-exchange responses to summer drought by sclerophyllous and deciduous species suggest that evergreen species are in fact *not* better adapted to water stress than co-occurring deciduous species! Detailed physiological studies of two co-occurring species of Mediterranean oak—one sclerophyllous and one deciduous—revealed that despite differences in biochemical composition, size, and mass per unit area, the leaves of the two species responded similarly to water limitation (Damesin *et al.* 1998). Similar results were obtained in a comparison of evergreen and deciduous oaks of the same age growing in a controlled environment (Achérar and Rambal 1992). It may be, however, that differences in function between the two groups become more apparent at the ecosystem level than at the leaf scale because of differences in leaf area indices between the two leaf types (Schulze 1982). A high total leaf area per unit of soil area could allow the evergreen species to maximize their winter and spring carbon gain at a time when the water supply is not limiting. In addition, differences between the two morphotypes in the timing of leaf fall and leaf-litter decomposition may result in functional differences in the economy of nutrients favouring evergreens at the ecosystem level.

Going even further, Schulze (1982) suggested that sclerophylly should be considered as an 'epiphenomenon' of the water-stress adaptation; that is, a by-product rather than a response to water stress. The same could be said of the argument that sclerophylly is an adaptation to nutrient-poor conditions. Indeed, experimental additions of phosphorus and nitrogen have been shown to alter the structure of leaves on adult sclerophyllous shrubs, not only in the mediterranean-type ecosystems, but also in the floras of upper montane forests in temperate zones and in nutrient-poor, tropical lowland forests (Beadle 1966)!

An additional perspective on sclerophylly can be gained by considering phylogenetic constraints (Herrera 1992). The Mediterranean vegetation originated from tropical and temperate sclerophyllous ancestors of the late Cretaceous, which were uniformly distributed over the Laurasian continent (Raven 1973; Axelrod 1975; see Chapters 1 and 2). The Eurasian flora originally evolved from that flora, which was rich in Ginkoaceae and other hard-leaved plant families during the Mesozoic. Later, during the Tertiary, so-called laurisilva forests (Chapters 2 and 3) spread along the continental borders of the Tethys Sea and gave rise to the extant flora, which is in large part—but not exclusively—evergreen and sclerophyllous (Axelrod 1975). Indeed, it is relevant to recall that many other leaf types and growth forms co-occur with sclerophylls in the Mediterranean flora, as described in the following sections.

8.1.3 Other leaf types

In addition to sclerophyllous leaves, many other leaf types occur in the Mediterranean flora, which also provide protection against desiccation while still allowing photosynthesis. These include woolly leaves (e.g. *Phlomis*, some rockroses, and, of course, lavender), succulent leaves (*Euphorbia*, *Sedum*), and tiny, short-lived leaves, such as are found on the various thymes and also the numerous legume shrubs commonly called 'broom' (e.g. *Genista*, *Lygos*, *Retama*, *Spartium*, and *Teline*). All are technically **leptophylls** or slender-leaved. Several groups even bear evergreen stems that are photosynthetically active all year round. These include the so-called retamoid legumes, named after the above-mentioned *Retama* and its relatives (see also Chapter 6), some of which have small deciduous leaves, which fall during drought periods and may not appear at all in dry years. In the drier areas particularly, many species in this group are generally **aphyllous**, or nearly so, and confine their photosynthetic activity to their stiff, evergreen stems. The same thing occurs in *Osyris alba*, sole Mediterranean member of the sandalwood family, which combines facultative leaflessness with an ability to parasitize the roots of many trees and shrubs.

An interesting additional example of leaflessness is the attractive flowering aphyllanthe (*Aphyllanthes monspeliensis*), a monospecific endemic lily family member of the western Mediterranean. Unlike its well-known relatives, the bulbous asphodels (*Asphodelus*) and the star-of-Bethlehem (*Ornithogallum*), the aphyllanthe has specialized rhizomatous roots and rush-like stems, with no leaves at all. Mediterranean species of *Asparagus*—which is also close to the lilies—also bear very reduced leaves or none at all. Finally, a few species such as the widespread Butcher's broom (*Ruscus aculeatus*), a prickly perennial herb of forest and matorral understoreys in the Mediterranean region and elsewhere in Europe, produce long-lived sclerophyllous **phyllodes**, as well as evergreen photosynthesizing stems.

Thus, Mediterranean plants from many families and life forms show a remarkably wide range of leaf types. What is more, some of them bear two different types of leaves at different times of year! This is a foliar strategy of particular relevance here since is found in many small shrubs ('subshrubs' of *maquis*, *bath'a*, and *phrygana*; see Chapter 6), particularly in the eastern half of the basin. These plants show what is called seasonal dimorphism in their leaves (Orshan 1964; Margaris 1981; Palacio et al. 2006). In winter and spring, they bear relatively thin, large leaves with high photosynthetic rates, which fall then from the plant and are replaced by reduced, thickened ones of the typical sclerophyllous type during summer and autumn. The stems on which the two types of leaves occur are also different: so-called dolichoblasts for the winter leaves and brachyblasts for the smaller summer leaves. Examples are common in the mint family such as Jerusalem sage (*Phlomis fruticosa*) and the remarkable, cushion-like *Sarcopoterium spinosum*, of the rose family (Christodoulakis et al. 1990), both of which are widespread in the eastern Mediterranean. Various other mint family members in the region show seasonal dimorphism, including several *Salvia* and *Satureja* species (Aronne and De Micco 2001). Similar examples are known from other mediterranean-climate regions, for example in *Ceanothus* (Rhamnaceae) from the California chaparral (Comstock 1985).

A particularly detailed study carried out in southern Italy by Aronne and De Micco (2001), on the phenology, morphology, and leaf anatomy of *Cistus incanus* subsp. *incanus*, indicates that seasonal dimorphism in the leaves of this rockrose effectively produces 'seasonally different plants' from the same individual, which helps the plant to withstand the great range of growing conditions found in the Mediterranean region in the course of a year and between years.

Summing up what we have seen thus far, phylogenetic origins and relationships *and* adaptive, ecophysiological traits, and those which provide defence against herbivores, fire, and cutting can both help explain the widespread occurrence of sclerophylly in the Mediterranean flora and vegetation today. The co-existence of many other adaptations—such as seasonal dimorphism in leaves, the widespread symbiosis of Mediterranean plants with root fungi and bacteria of various sorts, as well as specialized roots (*Aphyllanthes*), root-parasitism (*Osyris*), etc.—further supports the view that *both* adaptive and non-adaptive components are at work in the frequency of sclerophylly in Mediterranean flora and vegetation. We also begin to see that there is a remarkable diversity of life-history traits that probably have great bearing on the resistance and the resilience of Mediterranean ecosystems. Let us now consider in more detail other common and characteristic plant types of the region in relation to ecosystem functioning.

8.2 Autumn-flowering geophytes: a strategy for surviving competition and drought

In addition to sclerophylly and evergreenness, one highly distinctive and recurrent trait of the Mediterranean flora is the bulbous life form. These so-called geophytes have a fleshy, subterranean storage organ, which is usually the only portion of the plant that survives the extended period of summer dormancy. This life form also provides protection from periods of very high or very low temperatures in many regions with marked seasonality (e.g. Al-Tardeh *et al.* 2008). It also has obvious advantages for plants occurring in regions with many herbivores. In terms of evolutionary ecology, we can say that underground storage organs correspond well to the 'storage-pulse' model of resource supply—that is, water and nutrients common to most arid and Mediterranean ecosystems (Noy-Meir 1973)—and also help the plants cope with the seasonality of mediterranean-climate regions.

A great number of unrelated species in many different families have evolved the bulbous strategy, or something similar, throughout the basin. For example, there are 217 species of geophytes in the Israeli flora, which represents about 10% of the total (Fragman and Shmida 1998). In addition to avoiding drought by shedding their above-ground parts when water is limiting, geophytes gain a head start over the annual plants with which they compete for resources. In early springtime, the subterranean bulbs, corms, or rhizomes of geophytes allow them to begin growth as soon as temperatures rise above a certain threshold. Annuals, by contrast, need 2 or more weeks to germinate and establish roots, during which time they are especially vulnerable to grazing and trampling by animals (see below).

Well-known geophytes are found in small number of dicotyledon groups, including cyclamen (Primulaceae; see Chapter 7), and a great many monocotyledons, such as the lily, crocus, tulip, iris, amaryllis, sea squill and, above all, orchid families (see Chapter 6). With about 250 species of orchids, the Euro-Mediterranean flora is surprisingly rich in orchids although it includes only terrestrial species. The largest orchid genera in the region are *Ophrys* (20–50 species) (see Plate 3), *Orchis* (25–40 species), *Dactylorhiza* (5–28 species), and *Cephalanthera* (2–10 species) (Dafni and Bernhardt 1990). While some orchid species occur as far north as Scandinavia (e.g. 20 species in Norway), at least two subfamilies are mainly eastern Asian (Raven and Axelrod 1974) and only differentiated extensively in the Mediterranean area upon arrival there. Species richness may be very high in some parts of the basin, for instance 100 species in Greece, but declines to the north or towards the arid zones, with 86 species occurring in Turkey, 40 in Syria, 29 in Israel, and only one in Egypt (Dafni and Bernhardt 1990). Orchid species richness also declines with increasing altitude in mountainous areas and with aridity. There are many species of terrestrial orchids in North Africa that have never been studied at all, and this clearly represents a high priority for research and conservation.

One typical and unusual feature of Mediterranean geophytes is their wide range of flowering times when compared to that of the overall Mediterranean flora and to geophytes in other parts of the world. Whereas most Mediterranean plants flower in spring, many of the geophytes are **hysteranthous**; that is, they flower during autumn, completely independently of leaf development. In Israel, 10% of the native flora (2241 species in 130 families) flowers in autumn and this group consists mostly of geophytes in the lily group (25 species in 10 families) (Shmida and Dafni 1989). Two different types of hysteranthous geophytes have been identified in the Israeli flora, an *Urginea*-type and a *Crocus*-type. In the former, flowering was progressively delayed in the course of evolution until an autumn-flowering strategy was established. In the latter, the opposite trend occurred (Fig. 8.1).

The hysteranthous habit appears to confer several adaptive advantages, both in terms of optimizing water and nutrient uptake and in terms of attracting potential pollinators. Thus, flower size, plant height, and seed-dispersal mechanisms have all been found to have significant correlations with the atypical flowering time of these autumn-flowering geophytes.

Similar complexity occurs in the cyclamens. In a study combining morphological and molecular phylogenetic analyses, Debussche *et al.* (2004) showed there are four distinct subgenera in *Cyclamen* with differing phenological patterns in each. Anciently evolved phenological traits are very often conserved within lineages. The hysteranthous flowering habit occurs in three of the four subgenera, and the same is true of synanthous species—whose leaves and flowers appear at the same time—which also appear in three subgenera. Only one of the four groups is always synanthous and only one of the four is always hysteranthous (see Thompson 2005:124).

While hysteranthous flowering—a life-history strategy wherein the plant flowers when nothing

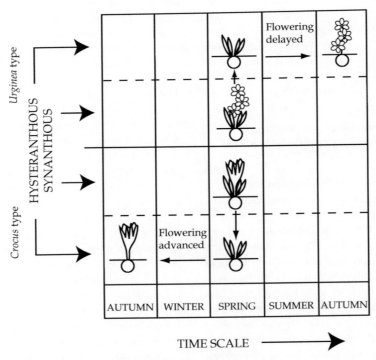

Figure 8.1 Probable evolution of hysteranthous flowering among Mediterranean geophytes. The initial condition where flowers and leaves appear together is termed synanthous flowering. (After Dafni *et al.* (1990); reproduced with the permission of L.P.P. Ltd.)

else at all of the plant is present—is more or less limited to geophytes, a strategy called 'precocious' flowering does occur in a few herbaceous perennials and woody plants in the Mediterranean region. An example is the male dogwood, *Cornus mas*, whose bright yellow flowers appear on bare stems of this small tree in early to mid-February, even before the almond tree (*Amygdalus dulcis*), the traditional harbinger of spring. By flowering long before leafing out in the spring, the tree avoids or reduces the risk of losing leaves to frost, and it also may draw the attention of early pollinators before other trees and shrubs are flowering.

8.3 Annuals in highly seasonal environments

Even more abundant than geophytes, in terms of species richness and perhaps overall biomass, are the annuals and **ephemerals**, which figure so prominently in almost all the basin's regional floras and plant communities, especially in the drier, eastern half. In the hills near Jerusalem, for example, an average of 143 plant species and a maximum of 189, mostly of very small annuals, were found in 1000 m² quadrats in a particularly rainy springtime (Shmida 1981). This is probably a world record in terms of alpha-level diversity for vascular plants at this spatial scale. Indeed, annual plants may well be considered one of the specialities of the Mediterranean Basin, often constituting half of the dominant vegetation present, whereas they rarely amount to more than one-tenth in other biomes of the world (Raven 1973).

Annuals are also far more adaptable to changing conditions in space and time than geophytes or woody plants. For one thing, to cope with strongly seasonal environments, being an annual plant is a remarkable successful strategy (Grime 1979). Many annuals have a relatively small number of large seeds that are highly sophisticated storage organs allowing survival during the unfavourable dry season. Seeds of such annuals, or therophytes (see Table 6.2), are relatively short-lived, lasting only a single drought season. Many other annuals produce long-lived **diaspores** that protect seeds over several years until an appropriate season for germination presents itself. Annuals of both types are found in large numbers in a variety of Mediterranean habitats. Following agricultural abandonment, they are the first colonizers, only to be replaced by perennial grasses, shrubs, and trees as succession progresses. In many heavily grazed areas, ecosystems are artificially maintained in an early successional stage, and this has possibly resulted in the recent evolution of many grazing-tolerant annuals from ancestral herbaceous perennials (Pignatti 1978). As mentioned in Chapter 2, there are over 1500 species of ruderals and segetals that are subtly adapted to one or more situations in the Mediterranean and have latterly become highly successful 'weeds' in many other parts of the world. In the Mediterranean, where grazing ruminants have long been ubiquitous, desert and desert-fringe ephemerals should be considered part of the same guild (Sami 2004).

Adaptive shifts in the timing of phenological phases (phenophases), such as germination and the onset of flowering and seed production, are also common in Mediterranean annuals, as well as morphological variability in regards to height, shape, and overall biomass. To a large extent they share these features with annuals of deserts but differences do occur between the two groups at both species and population levels (Aronson *et al.* 1992, 1993; Sher *et al.* 2004). Indeed, observations of annual plants show that particularly at the soil surface level, Mediterranean and arid environments share a high degree of unpredictability of resource availability, both in space and in time. This problem is somewhat mitigated for annual plants when their seeds germinate under the canopy of trees, shrubs, or even subshrubs (Tielbörger and Kadmon 2000; Gilaldi *et al.* 2008), a fact which helps explain frequently observed patterns of vegetation in all parts of the region and is of course a major factor in ecosystem functioning, both in managed and unmanaged landscapes (see Chapters 10 and 13).

Common adaptations of annuals to disturbance in the Mediterranean are delayed germination via hardseededness and other mechanisms (e.g. in *Medicago*), and both genotypic and phenotypic responses to interactions between temperature and moisture occur (Groves 1986). Another common adaptation among Mediterranean annuals is the possession of morphological features predisposing fruits to dispersal by grazing animals and to burial in the soil. Genera such as *Bromus, Erodium, Emex, Hordeum, Medicago, Stipa,* and *Trifolium* all provide examples of such adaptations (Shmida and Ellner 1983), which appear to be adaptations to the spatial heterogeneity and unpredictability mentioned above, since they help spread the risk of offspring mortality in space, just as delayed germination is a means of spreading the risk in time. Given the panoply of adaptations that Mediterranean annuals have acquired, both for seed dispersal and for surviving in frequently disturbed sites, it is not surprising that they are among the most widely dispersed plants in the world today.

8.4 Herbivory and plant defences

In its simplest form, herbivory is the removal of plant parts, especially young leaves, by grazing or browsing animals, from very small to very large. Outbreaks of tiny chewing insects may result in the complete defoliation of trees, as is sometimes the case of the 'bag worm' larvae of the processionary moth (*Thaumetopoea pinivora*) in plantations and natural stands of various pine species, or that of the related moth (*Thaumetopoea pityocampa*) in Mediterranean oaks, as well as pine woodlands. Sometimes such insects may cause the death of trees. One example is that of the bug *Matsucoccus feytaudi*, which resulted in the near total eradication of the cluster pine in huge plantations of this tree in southern Europe in the 1960s. Global warming is likely to increase phenomena of this sort in the region, as elsewhere, in the decades to come (Hódar and Zamora 2004).

8.4.1 Herbivory and the structure of plant communities

Since grazing by animals is usually selective, floristic composition and the relative abundance of species change under sustained grazing—and browsing—pressure. Indeed, this may be a powerful force in shaping the structure and species composition of plant communities and influencing the direction of ecosystem trajectories. In particular, seedlings of many species may be vulnerable to predation by animals because they lack physical and chemical defences, which develop only later. A recent study in Spain (Manzaneda *et al.* 2005) provides insight into the effects of herbivory on the frequency and distribution of the foetid hellebore (*Helleborus foetidus*), a common understory plant throughout the western Mediterranean. During the first growing season after fire, when seedlings of many species develop together, preferential consumption of one species over others by herbivores may shift the outcome of interspecific competition and hence of biotic communities that dominate a site for decades to come. For example, annual composites, crucifers, umbellifers, grasses, and legumes all tend to increase under moderate grazing, while perennial forbs and grasses, and tall annual grasses, tend to decrease (Noy-Meir *et al.* 1989). If shifts occur among shrub species that may become dominant, this can lead to profound and long-lasting changes in the species composition and hence the functioning of communities (Focardi and Tinelli 2005).

Noy-Meir (1988) showed that normally dominant grasses can be replaced in large degree by ruderal forbs (dicotyledon annuals) in a 'vole year' in Mediterranean grasslands. A vole year occurs when there is an unusual outbreak of large populations of the social vole (*Microtus socialis*), a small but potent rodent that wreaks such havoc in grassland swards that opportunistic ruderals are able to invade and take over the terrain. Similar events have been observed in North African steppes, where 'eruptions' and abrupt crashes of populations of rodents such as sand rat (*Psammomys obesus* and *Psammomys meriones*) can have tremendous impact on grassland and shrubland ecosystems (E. Le Floc'h, personal communication). In an evolutionary perspective, such vole years and similarly 'catastrophic' years of other causes can explain in part how annual forb species that now occur in grazed grasslands could have existed before the coming of domestic grazers, 'when pressure from large wild grazers was probably too light to open gaps in the dense tall grassland' (Noy-Meir 1988). In the Mediterranean area, and especially the eastern and southern parts, the long-standing practice of permitting overgrazing and overbrowsing by domestic animals has resulted in plant communities that are now weighted, so to speak, in favour of plant species that are avoided by livestock, such as asphodels and brooms, not to mention various foetid and even toxic plants, referred to above as anti-herbivores. However, we defer further discussion of domestic herbivores until Chapter 10.

8.4.2 Defence against herbivory

Plants have evolved several types of defence—both physical and chemical—against herbivory. In addition to sclerophylly, the most common 'anti-herbivore' defence mechanisms include the production of secondary metabolites, especially tannins and terpenes, and physical defence mechanisms, such as long and sharp thorns, which limit accessibility or availability to foliage. As part of their protection against herbivory, many Mediterranean bulbous plants are rich in bitter or toxic secondary compounds, which help deter grazing animals and insects. These alkaloids and other compounds accumulate not only in leaves but also in underground storage organs, which are especially attractive to rodents, boars, and other burrowing animals, since they contain moisture, as well as carbohydrates, even in the dry summertime. One example is the red squill, or sea onion (*Urginea maritima*, Liliaceae), which is found in sandy coastal areas throughout the Mediterranean region and a few areas of the Middle East and Europe. Its bulbs contain a series of cardiac glycosides, flavonoids, and many other chemicals (Al-Tardeh *et al.* 2008) which have stimulated great interest for potential use in medicine and various industries (Gentry *et al.* 1987).

In addition to the well-defended bulbs of red squill, and other geophytes, many evergreen Mediterranean shrubs contain volatile essential oils in their leaves which appear to play a role in deterring herbivores (e.g. Papachristou *et al.* 2003). Examples include myrtle, thyme, mint, sage (*Salvia*), basil (*Ocimum basilicum*), lavender, coriander (*Coriandrum sativum*), dill (*Anethum graveolens*), oregano, rue, laurel, rosemary, and fennel (*Foeniculum vulgare*). The great majority of Mediterranean taxa producing these aromatic compounds are found primarily in the mint family (Lamiaceae or Labiatae), parsley (*Petroselinum crispum*), and, to a lesser extent, sunflower family (Asteraceae).

The ecological role of biogenic volatile organic compounds (BVOCs) in leaves and other plant parts is complex and even more frequent among Mediterranean plants than is commonly realized. These oils are highly flammable, which has often led to the suggestion that they are involved in fire return feedback dynamics in conjunction to the high carbon/nitrogen ratio in the leaves of Mediterranean and other mediterranean-climate regions sclerophylls. Together these traits may inhibit or retard decomposition once the leaves finally fall off the plant and reach the ground (Bond and Keeley 2005). This in turn may have significance in setting up feedback cycles wherein low-nitrogen soils and leaves of low nitrogen content are favoured at the ecosystem level in some mediterranean-climate regions and some savannas as well (Bond *et al.* 2003).

It has also been suggested they may inhibit germination of the seedlings of competitor species, a process called allelopathy, mimic insect pheromones as a means of attracting pollinators, and even reduce water stress by providing anti-transpirant action (Margaris and Vokou 1982). Further, BVOCs, of which the most common are isoprenes and monoterpenes, may be involved in the adaptive responses of Aleppo pine to drought and variation in available soil nutrients (Blanch *et al.* 2007).

Not surprisingly, however, there is not only significant among-species variation within a community, but also genetically controlled within-species variation in BVOC content of many aromatic Mediterranean plants (Thompson *et al.* 1998; Thompson 2005). Building on a large body of previous work, Gouyon *et al.* (1986) concluded that the distribution of intraspecific variability (chemotypes) in thyme (e.g. Thompson 2002) appears to be heavily influenced by the environment (see Chapter 7). There are also cases known where an endemic species of a group has a very different BVOC make-up, or signature, than more widespread congeneric species. An example is the Corsica-Sardinia endemic *Ruta corsica*, within the widespread western Mediterranean genus *Ruta* (e.g. Bertrand *et al.* 2003). Many more examples of regional variations are discussed by Thompson (2005:144–167). Finally, in the future we will be living in a warmer world with higher BVOCs emissions (see Chapter 12). As Peñuelas (2008) put it, it will be 'an increasingly scented world'.

8.4.3 The formation of galls

One last, very special case of 'herbivory' we will mention is that of the production of galls. The relationship between a gall-forming insect and its host plant is considered parasitic. The insect oviposits eggs in a host plant, which induces growth deformities in the plant, and these are then used as food resource by the young larvae upon hatching. In thyme, there are variations in rates of infestation by the parasitic fly *Janetiella thymicola*, in relation to chemotype (Thompson 2005:162–163). In false olive, the gall-inducing insect is a cecidomyiid fly, *Schizoma phillyreae*. Adult flies emerge from the galls during the flowering period of the shrub and seek suitable breeding sites. Females oviposit one egg in the ovary of open flowers. The gall begins to develop 6–8 weeks after flower fertilization. Then the larvae grow inside the ovary and remain in a larval stage for at least 3 years (Traveset 1992). The cost of this to the plant is a drastic reduction in the number of viable seeds because up to 97% of the initiated fruits of a plant may become galls.

Herbivory may have important effects on a series of processes, such as plant dispersal, growth patterns, reproductive success, and plant forms (Ginocchio and Montenegro 1992). By removing only parts of the plant, herbivores may leave other parts capable of regeneration through the iteration of new modules. Hence, galls may have important effects on the structure and shape of vegetation. They were also of special interest to Theophrastus, father of botany (Box 8.1).

Box 8.1. Theophrastus (371–c.287 BC), father of botany

Perhaps the most fascinating adaptations to Mediterranean ecosystems are those that associate plants and animals in close interactions. The study of plant–animal interactions began in the Mediterranean area with the Greek philosopher and naturalist Theophrastus (Thanos 1992), who was a close friend of Aristotle (see Box 4.1). After the death of Aristotle, Theophrastus succeeded him as head of the Peripatetic School in Athens. His two surviving botanical works, *Historia Plantarum* (Enquiry into Plants) and *De causis Plantarum* (On the Causes of Plants), were major influences on medieval science and have led to his being considered as the founder or the 'father' of botany. Theophrastus paid considerable attention to seeds that are consumed by the larvae of beetles (thought by him to be produced by the seed itself), and he was intrigued by plant galls as a source of tannin, especially the galls produced by certain oaks and pistachio trees (for example, gall oak and terebinth).

Theophrastus also recognized the repellent function of plants producing toxic compounds, which may cause poisoning or death to animals that eat them. As examples, he cited the black hellebore (*Helleborus cyclophyllus*), which is lethal to horses and cattle, the deadly root of nightshade *Aconitum*, which is avoided by sheep, and spindle bush *Euonymus*, whose fruits and leaves are lethal to sheep and goats if eaten in large quantities. Theophrastus was the first to observe the role of animals in seed dispersal, citing the cormlets of the corn-flag (*Gladiolus segetum*) as being dispersed by moles, the caching of acorns by jays, and the mechanisms of mistletoe fruit dispersal by mistle thrush (*Turdus viscivorus*). He identified the two species of mistletoe occurring in Greece (*Loranthus europaeus* and *Viscum album*) and wondered how it was that these curious plants grow only on specific host trees. He came to the right conclusion: birds consume mistletoe berries, the seeds of which pass unharmed through their digestive track. They then establish their seedlings from bird droppings, which happen to fall in a good place on a suitable host tree.

8.5 Pollination

Many plant species need a vector to transport pollen from one individual to another. The two most important vectors are wind and insects. In the Mediterranean area, other animals, such as mammals, including bats, and many birds, are not involved in the pollination process, in contrast to what happens in tropical regions. Mutualistic plant–animal systems involving insect pollinators, also called plant–pollinator networks for the generalist systems (see below), are those in which both participants gain some reward from the association. Insects transport the pollen of the host plant, and the plant offers nectar, pollen, stigmatic secretions, and, sometimes, other resources, such as edible oils and floral fragrances (Simpson and Neff 1981), to insect visitors. Indeed, it is selectively advantageous for the plant to attract insects with a suite of attractants, such as floral shape, pigmentation, scent, and edible rewards (nectar, pollen, oils), that are generally concealed in more or less long floral spurs.

8.5.1 Generalist insect-pollinated plants

A prominent feature in the Mediterranean flora is that most **entomogamous** (i.e. insect-pollinated) plant species are generalist; that is, they can be pollinated by many insect species over a wide range of families, including ants and other unexpected visitors. Gomez *et al.* (1996) have shown that several plant species in the high mountains of southern Spain, for example *Alyssum purpureum*, Spanish sandwort (*Arenaria tetraquetra*), and English stonecrop (*Sedum anglicum*), and arid lands, for example *Lepidium subulatum*, *Gypsophila struthium*, and the round-seeded broom *Retama sphaerocarpa*, are mostly pollinated by ants. This is presumably because, in the extreme environmental conditions that prevail in these habitats, ants by far outnumber other potential pollinators. However, although ants are the most abundant group of terrestrial insects, and are well known as dispersing agents for seeds, in their relationships with flowers they seem to be mostly nectar robbers. This may be correlated to the fact that most ants secrete toxic liquids, which render pollen grains unviable.

In Mediterranean habitats most pollinating insects are air-borne; for example, flies, hoverflies, bee-flies, butterflies, wasps, beetles, and especially bees of two different categories (O'Toole and Raw 1991). The first category includes small (3–7 mm) slow-flying bees with a short tongue and relatively low energetic demands. The second is dominated by medium to large (10–25 mm) fast-flying bees with long tongues and much higher energetic demands. In fact, a loose correlation between flower traits (e.g. size, tube length, and nectar production) and pollinator traits (e.g. size and tongue length of pollinators) indicates that insects of various sizes visit flowers of a large range of sizes. However, in a detailed study of the relationships between flower colours and the natural colour vision of insect pollinators in the flora of Israel, Menzel and Shmida (1993) found a general trend towards higher frequencies of ultraviolet blue and blue colours in flowers predominantly visited by bees, as compared to higher frequencies of blue-green and ultraviolet green colours in those predominantly visited by flies and beetles. In Mediterranean ecosystems, these authors argue, a highly competitive pollination 'market' exists, which is dominated by hymenopteran insects characterized by (1) a high colour-detection capacity with ultraviolet, blue, and green receptors, and (2) a strong learning capacity at the level of individual bees for floral features, which guide the pollinators, according to reward experience rather than by innate search images. Thus, floral colours and shape are adaptive 'advertising' signals recognized by fast-learning pollinators (see Thompson 2001).

8.5.2 Competition among plants

The spring peak of flowering in Mediterranean landscapes produces a surplus of flowers relative to the number of potential pollinators. This may result in sharp competition among plants for pollinators, especially because most plants are generalist *vis-à-vis* their insect pollinators. Simplifying greatly, early-flowering plants invest largely in rewards (nectar and pollen) and 'advertisement' (large colourful flowers) (Cohen and Shmida 1993). Later in the year, there is a surplus of insects over flowers and, by flowering in mid-season, some plants try to 'sell' their rewards 'cheaply' rather than competing for the services of pollinating insects earlier in the year (Dafni and O'Toole 1994). The fact that, as compared to spring-flowering plants, those that flower in summer tend to reduce their investment in flower size and rewards could be a response to a large number of non-competing potential pollinators.

One interesting example of competition among plants for pollination in the Mediterranean area is that of the almond tree. It blooms in January/February, which is the wettest and coldest period of the year, with the lowest activity of pollinating bees. At that time, when almond trees are in full bloom, the potential pollinating honey bees are mostly inactive, remaining in their hives even on sunny days, unless ambient temperature rises above 15° C. Why did the almond tree not evolve a more 'appropriate' blooming time? Based on a study of the foraging behaviour of honey bees on almond trees and on competing flowering plants, Eisikowitch et al. (1992) suggests that almond trees have a very low competitive ability. Trees would not achieve pollination unless they shift their blooming time toward the lowest period of competition with other flowering plants. As a result their potential pollination period lasts only 18 days, a period which is nevertheless sufficient for survival of the species. This is a typical case of avoidance of competition whereby a specific blooming period results from a trade-off between two selection pressures, one towards early flowering to avoid competition among plants for pollinators and the other towards more favourable climatic conditions later in spring.

8.5.3 A plethora of strategies

Although most plant species are generalist insect-pollinators in the Mediterranean area, there are some notable exceptions. The best known of these is the fig (Box 8.2; Fig. 8.2a) and the many orchid species.

Another surprising feature of figs that has recently come to light is called 'intersex **mimicry** of floral odour' (Grison-Pigé et al. 2001), since the flowers on female trees, in some cases, mimic the odour of the flowers on male trees to attract pollinators. Thus, the pollination of fig trees occurs, to a certain extent, by deceit!

Sophisticated pollination systems, many of which involve deceit, are also found among the orchids, as first described and interpreted in evolutionary terms by Darwin (1862) in his seminal book *The Various Contrivances by which Orchids are Fertilised*. Although several orchid species are **autogamous** (e.g. *Ophrys apifera*; see Plate 3a), a great many species are **allogamous** and insect-pollinated by one or several species. Some groups are pollinated by hawk moths or butterflies (e.g. *Platanthera* and *Anacamptis*), and others by parasitic dipterans (*Herminium*). But the great majority are pollinated by a large number of different species, including sawflies, carpenter bees, flies, syrphids, or honey bees (e.g. *Epipactis* and *Himantoglossum*). However, solitary bees are by far the dominant pollinators of most orchid flowers. Usually a small number of pollinating species are involved. For example, although *Platanthera chlorantha* is visited by 28 species of insects, only six of them are responsible for 97% of the total pollination events (Nilsson 1978).

It is well known that, in many more or less species-specific systems, flowers resemble the target insect species that is 'supposed' to pollinate it. The lower petal (labellum) of the flower has a certain shape and bears certain spots or ornaments that are 'designed' to attract insects. In this case, one can speak of 'legitimate pollinators', and the patterns and shape of the flower probably result from an adaptation to mimic those of the insect. Well-known examples are found among the *Ophrys*, some species of which have been given the name of an invertebrate which resembles the design of the flower (e.g. *Ophrys fusca*, *Ophrys tenthredinifera*, and *Ophrys araneifera*; see Plates 3b and 3h). Some Mediterranean orchids may flower as early as late November, but the peak flowering season is March/April, a period when the density and diversity of actively pollinating solitary bees and wasps coincide with the peak flowering periods of orchids. A close matching between the activity of pollinating insects and the flowering period of orchids may differ among regions of the basin, depending on altitude and latitude (Borg-Karlson 1990).

Box 8.2. The sophisticated fig–wasp system

Some sophisticated mutualistic associations between insect-pollinated plants and pollinating insects justify using the term 'coevolution' as defined by Janzen (1980): 'an evolutionary change in a trait of the individuals in one population in response to a trait of the individuals of a second population, followed by an evolutionary response by the second population to the change in the first'. One example is that of the fig tree, which has been a companion of humans in the Mediterranean area for at least 100 000 years (Khadari *et al.* 2005a) and domesticated since 11 500 years BP (see Chapter 10). The fig tree is allogamous, and its flowers are hidden within closed inflorescences and hence they must be pollinated by a specialized insect. All species of fig (over 700, mainly in the tropics) are pollinated by species-specific chalcid wasps of the family Agaonideae. The species that pollinates the Mediterranean fig is *Blastophaga psenes*, and the story of pollination of this fig–wasp mutualism has been nicely described by Kjellberg *et al.* (1987) and Anstett *et al.* (1995). It is a story of great importance for humans since unpollinated figs abort, except in parthenocarpic cultivars. Dried figs, which constitute the great bulk of the world market, must be pollinated by suspending male figs in female trees to ensure adequate sugar content and good flavour. This practice is called caprification.

Over a thousand tiny flowers line the inside of the inflorescence of the fig tree, which is a compact infolded structure called 'a fig' (Fig. 8.2b). A female wasp loaded with pollen enters the fig; while penetrating the opening at the top of the fig, she may lose part of her antennae and wings. If she enters a fig on a male tree she will oviposit in the flowers and then she promptly dies. After the eggs hatch, the young larvae feed inside the developing ovules of the fig. Several weeks later, when the fruit is almost ripe, the young wasps emerge and the wingless males fertilize the females. This allows the young winged female loaded with pollen to leave the fig and to venture a trip to another fig-host plant. A new cycle will thus begin. If the wasp visits a female tree, she enters the fig, pollinates in passing the stigmas thus ensuring seed production, but she cannot oviposit because she does not get access to the ovules. This system is a strict mutualism, because each species benefits from the other and, indeed, is necessary for its survival: the plant for pollination and the insect for food, oviposition sites, and shelter. Thus the male trees produce inedible figs containing insect: these are called caprifigs. The female figs produce tasty figs and are therefore called domestic figs, even if they are wild (Kjellberg *et al.* 1987). In some countries, cultivated figs are caprified by hanging male figs on cultivated fig trees or by planting male trees in elevated sites upwind of the fig orchard, in such a way as to ensure the pollinating insects an easy flight downwind. This enables the farmer to plant only one male for every 20 female (fruit-producing) trees. In some districts such as the Meander valleys in Turkey, there is also a very active market for male figs for use in caprification.

In many cases, plant–insect interactions may be much more complicated than a simple symmetric reward for the two partners. In some cases, orchid flowers offer insects a greenhouse-like shelter in unfavourable weather conditions as much as 3°C warmer than ambient temperatures (Pellegrino *et al.* 2007). In the case of *Serapias*, the flowers mimic an invitingly warm and safe sleeping hole for solitary bees, which are the main pollinators of the orchid (see Plate 3i). This provides an energy gain and a safe nocturnal lodging for the insect (Dafni and Bernhardt 1990). It has also been noted that insects sleep in the tunnel-shaped flowers of various *Cephalantera* species, which results in pollination.

Some other orchids not only offer no material award to pollinators, but have even evolved a 'cheater' or 'trickster' behaviour of 'deceiving' their pollinators by exploiting naive insects that are conditioned to forage among certain floral models.

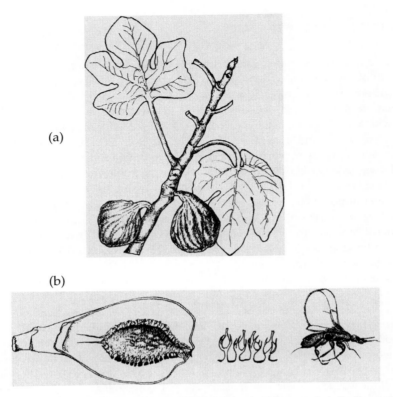

Figure 8.2 (a) Figs on a stem, and (b) an open fig showing the continuous line of enclosed flowers and a pollinating wasp (R. Ferris).

These 'fraudulent' orchids imitate the suite of cues that normally attract pollinators, but without giving them any reward (Bernhardt and Burns-Balogh 1986). This trickery may be particularly rewarding because nectar and carbohydrate resources are costly for the plant to produce. This strategy is selectively advantageous in perennial plants that must withstand seasonal aridity. In orchids, it may have played an important role in the extraordinary speciation and adaptive radiation of the family (Cozzolino and Widmer 2005).

Another strategy for plants is to mimic other plant species that do offer rewards to pollinators. The most widespread form of this kind of trickery is called Batesian mimicry, wherein a low-density non-rewarding species mimics the flowers of a specific species offering rewards at a much higher level (Schemske 1981). Examples in the Mediterranean area are the orchid *Cephalantera longifolia* that mimics the rockrose *Cistus salviifolius*, or the related *Cephalantera rubra*, which mimics several species of bellflowers (*Campanula*). Sometimes, however, the mimic species lacks a specific model and flowers early in the season when there are large numbers of inexperienced generalist pollinators on the wing. This form of Batesian mimicry, which is based on naiveté (Little 1983), is common in *Orchis*, *Ophrys*, and *Dactylorhiza*, three genera common in the Mediterranean Basin. Of course, such systems can only have evolved in situations where plants occur in the same habitats as other species which bloom at the same time, bear similar attractive cues, and are truly rewarding.

The second kind of system is called Müllerian mimicry and occurs in cases where at least three species bloom at the same time and offer similar floral attractants. At least two of the three species mimic each other and offer some version of the standard edible reward. In one known case in the Mediterranean region, *Orchis caspia* offers no reward, yet attracts four species of solitary bees that pollinate the orchid along with co-occurring species that flower at the same time and offer tangible rewards, such as the summer asphodel (*Asphodelus aestivus*) (Liliaceae) and Greek oregano (*Salvia fruticosa*) (Lamiaceae). *O. caspia* 'takes advantage' of naive bees that are unable to discriminate between flowers of plants which offer a reward and the non-rewarding orchid. There are still other unusual and intriguing systems among Mediterranean orchids, for example the brood-site deception whereby flowers simulate the oviposition site of a pollinator species and thereby attract a female insect ready to lay eggs. For example, *Epipactis consimilis* attracts female hoverflies (Syrphidae) to lay their eggs on the labellum of the flower by mimicking an aphid which is the normal oviposition target for these flies (Kullenberg 1961, 1973; Kullenberg and Bergström 1976).

Finally, there are examples of sexual deceit. This system involves 'pseudo-copulation' since it relies on inducing male pollinators to attempt copulation with parts of the flower that resemble the female insect. Similarly, a substitute of an edible reward can be a provocative olfactory stimulus for insects. For example, flowers of *Orchis galilea* emit a strong, musky smell, which acts as a species-specific sexual attractant for the male of the pollinating bee, *Lasioglossum marginatum* (Bino *et al.* 1982). A study of the chemical composition of the odour of *Ophrys* species has shown a strong chemical similarity with that of the male insect pollinators (Borg-Karlson and Tengoe 1986). These males attempt to copulate with the flower but only receive depositions of pollinia (pollen sacs) for their efforts. Orchids with a pseudo-copulatory syndrome have a labellum which resembles in shape and colour the dark body of female insects (Borg-Karlson 1990; Dafni and Bernhardt 1990). The shiny spots that decorate the labellum of several *Ophrys* species (e.g. *Ophrys speculum*) are similar to those that appear on the back of a female wasp when she crosses her wings.

To conclude this section on pollination, we note a remarkable long-term study underway in a nature reserve near Athens by Theodora Petanidou and co-workers (Petanidou *et al.* 1995, 2008; Potts *et al.* 2006). In terms of ecosystem functioning, few interactions are as important as plant–pollinator interactions, except perhaps plant–disperser networks, the subject of the next section.

8.6 Fruit dispersal by birds

If the animals most frequently responsible for plant pollination are insects, it is vertebrates, and primarily birds, that are responsible for the seed dispersal of plants producing fleshy or otherwise edible seeds and fruits. Plants producing bird-dispersed fleshy fruits are a prominent component of most Mediterranean woodlands and shrublands (Herrera 1995). Birds eat whole fruits and then regurgitate or defecate intact seeds ready to germinate. The rewards provided by plants to the birds are, so to speak, the price they must pay for having their seeds dispersed far from the mother plant. Bird-dispersed plants are taxonomically very diverse in the Mediterranean area and include dozens of species in the pistachio, olive, myrtle, and laurel families among many others. More than half of them rely entirely on birds for their seed dispersal, but several species of mammals including foxes (Traba *et al.* 2006) partake of this feast to some extent as well.

8.6.1 Rewarding fleshy fruits

Fruit production in Mediterranean woody plants is usually high both in terms of numbers and biomass, and this is good news for the birds (Herrera 2002). For example, annual fruit production in some Spanish habitats can be as much as 1 400 000 ripe fruits ha^{-1}, corresponding to a range of 6–100 kg dry mass ha^{-1} (Herrera 1995). Nearly all ripe fruits in Spanish matorrals (90.2%) are consumed by avian dispersers (Herrera 1995; see Table 8.1). These fruits are characterized by 'flag' features, such as bright red, black, or blue colours, and their conspicuous location at the end of vertical stems makes them readily visible to birds. They are nutritionally rich, containing much fat and protein in their pulp, as well as secondary metabolites that may confer enhanced detoxification ability against toxic compounds for the birds who eat them (Herrera 1984). Moreover, there is a fairly close matching

Table 8.1 Proportions of ripe fruit crops removed by avian seed dispersers in various woody Mediterranean plant species in a wide range of families

Plant species	Percentage of fruits eaten by birds
Asparagus aphyllus	100
Pistacia lentiscus	91–100
Smilax aspera	86–100
Phillyrea angustifolia	72–99
Osyris alba	76–98
Daphne gnidium	92–97
Myrtus communis	89–95
Rhamnus alaternus	61–93
Lonicera etrusca	80–90
Viburnum tinus	51–75
Prunus mahaleb	50–68
Cornus sanguinea	36–49
Pistacia terebinthus	25–30

Source: Various sources in Herrera (1995).

between the ripening season of fruits and the seasonal patterns of occurrence and migration of avian dispersers in Mediterranean shrublands (Herrera 2002; Hampe 2003).

The functional significance of the links between seed-dispersing birds and plant-dispersed assemblages in Mediterranean woodlands and matorrals must be approached at the community level because there are hardly any species-specific mutualistic systems involved. Instead, several species of birds eat and disperse the seeds of many species of plants, such that it is unlikely the local extinction of one partner would seriously affect the local survival of the other. The only known example of a species-specific interaction of this kind occurs in the southern Alps and involves the nutcracker *Nucifraga caryocatactes* and the rock pine (*Pinus cembro*) (Crocq 1990). As much as 95% of the food of the nutcracker is provided by the large oily seeds of the rock pine. In turn, the tree relies almost completely on the bird for dispersal of its seeds. Nutcrackers harvest the entire production of seeds of this pine in autumn and then bury small clusters of 7–10 seeds in small holes scattered throughout their territory. The birds return in winter and spring to eat the seeds, but some caches are forgotten or abandoned. Seeds from these caches will ensure the regeneration of the pine.

8.6.2 Two categories of fruit-dispersers

In terms of vegetation dynamics and of the survival of billions of birds that have to leave their northern breeding grounds in winter, the process of fruit consumption by birds in the Mediterranean area is of paramount importance. As in the case of plant–insect interactions leading to flower pollination, this mutualistic system implies that both partners gain some reward from their association. Birds transport the seeds of a plant, which in exchange offers them a rich food supply. There is much evidence that Mediterranean avian seed dispersers have evolved morphological, physiological, digestive, and behavioural adaptations to take advantage of the abundant and profitable fruit supply provided by many woody plants. Morphological and physiological pre-adaptations, such as flat broad bills and various digestive adaptations, allow these birds to handle and swallow the fruits efficiently. Most of them are only seasonal frugivores, since they shift from an insect-dominated diet in their spring/summer breeding grounds to a fruit-dominated one in autumn/winter. Plant–bird associations primarily involve a large number of passerine birds of small to moderate size (10–110 g) in the families of thrushes, warblers, and flycatchers. Members of the tropical bulbul family (Pycnonotidae) are also important seed dispersers in the eastern Mediterranean and North Africa (Izhaki *et al*. 1991). These species are active during the period of fruit availability in the same habitats as the plants whose seeds they disperse. Seed dissemination is basically a within-habitat or landscape-scale process, since most seeds remain only a short time (20–30 min) in a bird's digestive tract. Thus, they travel only short distances from the mother plant to the location where birds eventually drop or eliminate them.

There are two main categories of avian fruit-dispersers in the Mediterranean area, corresponding to two distinct strategies for overwintering. The first category involves the millions of long-distance trans-Saharan migrants that transit the Mediterranean between August and late October en route to their tropical winter grounds (see Chapter 7). Examples are whitethroat (*Sylvia communis*), garden warbler, pied flycatcher, and willow warblers.

Extensive consumption of sugar-and carbohydrate-rich fruits by these birds at their Mediterranean stopover sites may play an important role in determining their migration schedules (Bairlein 1991). In fact, as pre-migratory fat deposition is a prerequisite for successful migration across the Saharan desert, the consumption of large quantities of fruits in the Mediterranean area appears to be a *sine qua non* in the Palaearctic–African migration system of these birds, as explained in Chapter 7.

The second category is that of bird species which overwinter in the Mediterranean from October to early March. As many as 12–16 species, reaching densities of up 15 individuals ha^{-1}, belong to this group with robin, blackcap, song thrush, and blackbird (*Turdus merula*) as dominant species (Debussche and Isenmann 1992; Herrera 1995). Fruits of 20–30 plant species contribute more than 75% and often more than 90% of these birds' food supply. Most birds in this category usually eat the fruits of several plant species, each for a short time. This presumably helps to compensate for the unbalanced nutritional composition of each species' fruit pulp. However, lipid-rich fruits, such as those of pistachios, olive tree, and lauristinus, invariably predominate in the diet of overwintering birds. Water and lipid content increases in the fruits of winter-ripening plant species as compared to summer- or even autumn-ripening ones. Fruits with a high content in energy-rich lipids are thus most abundant at a time of year when the energy demand of avian seed dispersers is highest. In a Spanish lowland shrubland, the three most common plant species (lentisk, myrtle, and smilax), which together make up 63% of the fruit-producing cover, all have a remarkably long fruiting period of 2.2–3.5 months, which corresponds well with the overwintering period of birds (Herrera 1984). These same winter-fruiting species also have the highest levels of digestible lipids (up to 58% of pulp dry mass) and, not surprisingly, represent the most important food plants for this second category of overwintering birds.

8.6.3 Co-evolution or serendipity?

In the Mediterranean Basin, woody plants tend to ripen their fruits much later in the year than those in temperate regions. This has sometimes been interpreted as resulting from a co-evolutionary process. However, the late-fruiting season of Mediterranean woody plants could simply be the result of physiological factors related to evergreenness and the mild, rainy winters that occur in many parts of the basin (Debussche and Isenmann 1992). Moreover, as seen in Chapter 2, the extant flora of the Mediterranean Basin includes many species of fleshy fruit-bearing plants of tropical origin that appeared long before the differentiation of most of the contemporary bird species that disseminate their seeds today. The fossil record reveals that at least 83% of genera of bird-dispersed woody plants in southern Spain were already present in western Europe before the onset of the Mediterranean climate in the Pliocene, 2.8 mya (Herrera 1995). This does not mean, however, that the system did not originally result from co-evolutionary processes. The more or less tight linkages existing between fleshy fruit-bearing plants and birds presumably evolved as far back as the early Tertiary, when both the plant and bird species present were different. Many of the lineages that initiated the process were presumably the ancestors of birds that only occur today in tropical Africa, having definitively left the Palaearctic region during the Miocene climatic deterioration. Subsequent extinction and speciation events occurring in both plant and bird groups did not necessarily cause any breakdown of the mutualistic association. The present-day system may therefore be the result of an opportunistic reshuffling over time of different sets of species.

However, current co-evolutionary processes may be occurring as well. For example, the rather close fit between fruit size and disperser gape width may result from bird selection against large fruits in the plant species they visit regularly. The gape width of bird dispersers sets an upper size limit to the fruits they can ingest because birds usually do not nibble at fruits. Hence, the fraction of the fruit crop not removed by birds significantly increases with increasing fruit diameter, both among species and among individuals within species. Similarly, migratory habits, increased food passage rates, conferred detoxification ability, and seasonality in food preferences and size of digestive organs, all are likely to reflect current adaptations to exploit the superabundant food supply provided by Mediterranean plants.

8.7 Decomposition and recomposition

In this section we briefly consider earthworms and dung beetles, two of the major functional groups involved in decomposition and recomposition of organic matter. In both cases, we will see that contrasted but complementary life-history traits occur in organisms belonging to different functional groups in these soil-borne insects that play very strong roles in Mediterranean ecosystems, driving the nutrient and carbon cycling that are just as important as photosynthesis to ecosystem functioning.

8.7.1 Earthworms

Since Darwin (1881), who was fascinated by the role of earthworms in structuring soils, the contribution of earthworms to physical, chemical, and biological soil processes has received much attention. These animals may constitute a dominant part of the life of soils, with biomass amounting to $1000\,kg\,ha^{-1}$ in rich soils of temperate Europe and up to $4000\,kg\,ha^{-1}$ in permanently grazed pastures (Lavelle 1988). The amount of soil they process may reach $230\,t\,ha^{-1}\,year^{-1}$. As in many other groups of animals, species diversity of earthworms is much higher in the Mediterranean Basin (150 species in southern France alone) than in all of northern Europe (30 species). This difference can be explained by Pleistocene glaciations, which destroyed most species occurring north of the Mediterranean area (Bouché 1972; see Chapter 2). New species are still being discovered as well; for example, 10 new species were recently described for Cyprus (Pavlícek and Csuzdi 2006).

Earthworms are generally assigned to four main functional groups on the basis of their location and activity in the soil: (1) epigeic (acid-tolerant species living in the leaf litter they feed upon), (2) **anecic** (burrowers, which also feed upon leaf litter but are sensitive to acidity), (3) endogeic (deep-dwelling species, which actively process and mix soil particles through continuous intestinal transit), and (4) epianecic (deep burrowers active in rich forest soils, being more acid-tolerant than the anecics). Each of these functional groups contributes to such major ecological processes as organic matter recycling, drainage, and bioturbation, but in different and complementary ways.

The main aspects of earthworm soil activity are (1) physical effects whereby earthworms promote soil porosity, aeration, water drainage, and bioturbation (Romanyà et al. 2000), (2) biogeochemical effects, such as litter decomposition, organo-mineral migration, phosphorus and nitrogen recycling, and shifts in pH and carbon/nitrogen equilibria (Cortez and Bouché 2001), and (3) indirect contributions to maintaining high plant diversity, by providing better germination and early growth conditions for trees and herbaceous plants (Granval and Muys 1992). Finally, earthworms are important source of proteins for many predators. Up to 200 species of birds and mammals feed on them in the Mediterranean area (Granval 1988).

8.7.2 Dung beetles

Along with herbivory by insects, grazing and browsing by wild ungulates and livestock are the main types of predation of green plants by animals. Recycling the non-digested part of this food is a major process of ecosystem function. It allows the return to the soil of organic matter and minerals, especially nitrogen and phosphorus. It also contributes to the proliferation of micro-organisms, detritivore communities, and earthworms. In addition to the enrichment of soil in organic matter, a rapid decomposition of faeces decreases the risk of disease transmission among animals, because vertebrate parasite populations that reside in dung are killed in the process, and this may in turn locally play a role in increasing areas available for plants.

Each day, an individual sheep or goat drops approximately 2 kg of faecal matter whereas a cow can deposit as much as 28 kg (Lumaret 1995). Decaying faeces are recycled by large and complex assemblages with a subtle turnover in time of various micro-organisms, flies, and dung beetles. Animal dung represents both habitat and resources for at least 135 species of dung beetles (*Aphodius, Geotrupes, Bubas, Scarabeus*, etc.) in the Mediterranean Basin. All these insects use this highly concentrated resource as food, substrate for laying eggs, and food source for their larvae. Many of them complete their life cycle in or near dung before it disappears completely. A key factor in the colonization of newly deposited dung is its odour, to which insects are highly sensitive. Dung must be moist in order to be colonized by insects and this represents a constraint in Mediterranean habitats, where high temperature and low precipitation cause dung to desiccate quickly. The summer drought period also slows down insect activity so that the recycling of dung is much slower in Mediterranean habitats than further north. Lumaret (1995) showed that the complete drying out of dung dropped in summer and is achieved within 2 or 3 weeks in Mediterranean habitats, as compared to 2 or 3 months in winter. Indeed, dung beetles are mostly active in spring and autumn and then decrease in numbers in summer. High desiccation rates associated with reduced beetle activity in summer explains that 30–50% of total summer droppings still remain unrecycled after 8 months (Lumaret and Kirk 1987). These climatic constraints explain why only a few species of dung beetles have successfully colonized the most arid parts of the Mediterranean.

Dung beetles have evolved different strategies to exploit their unpredictable and temporary food resource. Schematically, three main groups are recognized (Fig. 8.3), of which the first seems to be the least adapted to Mediterranean bioclimates. These are the dweller species (e.g. *Aphodius*, *Ontophagus*)

that live inside the dung they use as food for themselves and for their larvae. Dwellers colonize dung as soon as it has been dropped and their eggs are laid directly in it. The larvae pupate in the ground under the dung and emerge as fully developed insects through a tunnel they dig themselves. It is crucial that the dung remains moist for the entire life cycle of the insects, because digging the hard desiccated crust of dry dung is not possible for them. Females of some Mediterranean species have evolved an adaptation to cope with drought. They dig tunnels under the pad and lay eggs in small pieces of dung they put in small lodges, a behaviour very similar to that of the second category, the tunnellers.

The tunnellers are represented by species of *Ontophagus*, *Bubas*, *Copris*, and many species of *Geotrupes*. They avoid desiccation during hot Mediterranean summers by burying small pieces of dung in the ground down to a depth of 1.5 m. For example, *Bubas bubalus* individuals bury 200 g of dry faecal matter for their breeding activities. At the bottom of the tunnel, they prepare a breeding chamber filled with pellets of dung, each of which will receive an egg.

The third category is that of rollers, such as *Scarabeus* and *Sisyphus*. Like the tunnellers, they also bury dung with eggs inside, but before digging a tunnel in the ground, they prepare a large ball they will roll several metres before burying it in a suitable place. The famous nineteenth century entomologist of southern France, Jean-Henri Fabre (see Box 8.3) was the first to describe this behaviour

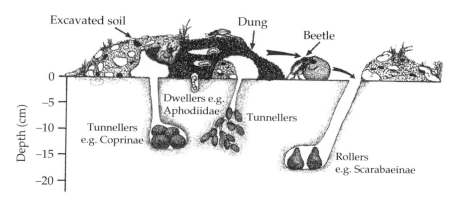

Figure 8.3 The three main categories of dung beetle: dwellers, tunnellers, and rollers (after Lumaret 1995).

> **Box 8.3. The insect world of Jean-Henri Fabre**
>
> Although he was one of the most remarkable and prolific of nineteenth century naturalists, and certainly one of the most gifted observers of Mediterranean biota, Jean-Henri Fabre (1823–1915) is poorly known today to most readers. The son of peasants, he worked for 30 years as a school teacher, being paid a mere pittance, before being dismissed from the position he had held for 20 years at the Avignon High School, when he got in trouble for teaching young ladies about the pollination of flowers and a church-based conspiracy got him fired on charges of indecency (L. Jones, personal communication)!
>
> He was saved from misery and ruin by his close friend, the economist, philosopher, and naturalist John Stuart Mill. After serving as curator of the famous Requiem Museum of Avignon, between 1866 and 1873, Fabre finally left Avignon, and after 9 years in Orange purchased some uncultivated land and a house at Sérignan, in the Vaucluse, where he lived for 37 years. There he patiently studied the behaviour of wasps, wild bees, crickets, cicadas, and other insects in his garden, and wrote numerous books and booklets about them. His writing ultimately brought him fame and financial independence, as well as the sobriquet of the 'Insects' Homer'. His 10-volume masterpiece, *Souvenirs Entomologiques*, was published over a 30-year period and translated into many languages. Charles Darwin called him an 'incomparable observer' and he was elected as an honorary member of scientific societies in London, Brussels, Stockholm, Geneva, and St Petersburg. Late in his life, he was also heralded for his writing by literary figures, such as Maurice Maeterlinck, Edmond Rostand, and Romain Rolland. Today he is best remembered in France, of course, but also in Japan, where he is something of a cult figure. The house where he lived in Sérignan is now a museum belonging to the National Museum of Natural History, and Fabre's name is indelibly associated with the nearby Mont-Ventoux, the famous Provençal mountain, which he climbed several times each year, and which had previously been immortalized by the Italian poet and humanist Petrarque in the fourteenth century. A recently discovered and thoroughly charming endemic bulb found only on Mont-Ventoux was very appropriately named in his honour: *Leucojum fabrei*.

for the sacred scarab beetle (*Scarabeus sacer*). The size of the ball is closely adapted to the size of the hind legs of the insect. A related species, *Scarabeus semipunctatus*, lives in sandy regions of the Mediterranean. Lumaret (1995) has shown that females of this species will stop rolling their balls of dung and bury them as soon as they find a suitable place. It is possible to stimulate the burying behaviour by wetting the soil just in front of an actively dung-rolling insect.

Because of the many constraints due to drought in the Mediterranean area, which make dung available for only a short period of time, tunnellers and rollers are less abundant than dwellers. The proportion of the first two categories combined is 46%, as compared to only 28% in central Europe (Lumaret 1995). Moreover several dwellers in the Mediterranean have completely reversed their life cycle, as compared to that of tunnellers, concentrating their activity in the coolest and wettest months of the year. One threat to the biological diversity and abundance of dung beetles is the generalized use of veterinary drugs, which makes domestic animal dung poisonous for these insects (Wall and Strong 1987). In addition, climate change, land-use change, vermifuge chemicals that are given to livestock to eradicate parasites, and other factors that impact habitat quality also affect the biogeography and, indeed, the critical ecological interactions of these formidable actors in Mediterranean ecosystems and landscapes (Verdú and Galante 2002; Cabrero-Sañudo and Lobo 2006).

In the next chapter, we will review some life-history traits of native organisms in the various life zones of the Mediterranean Sea.

Summary

In this chapter, we have provided some examples of life-history traits and of various terrestrial plant and animal species that, in some cases, survive as palaeorelicts and, in others, have clearly evolved quite recently. Our goal was not to make an exhaustive review, but rather, by using some selected case studies, to draw attention to the huge number of traits, or suites of traits, that allow terrestrial Mediterranean species of plants and animals to survive, adapt, compete, and cooperate with other species. These suites of life-history traits and the networks of interactions within biotic communities clearly confer enhanced resistance and resilience to ecosystems in Mediterranean environments affected by strong spatial and temporal heterogeneity and many different kinds of human and non-human disturbances. We also showed from the example of fruit-eating birds that ecosystem functions in the Mediterranean can have profound influences well beyond the limits of the basin. The abundant fruit production of many Mediterranean plant species contributes greatly to the wealth of the bird fauna of the whole Palaearctic region.

CHAPTER 9

Life in the Sea

In Chapter 4, we provided an overview of present day biodiversity in the Mediterranean Sea, including whales and birds, along with flora, invertebrates, and fish. At the end of Chapter 6, we described the array of marine habitats where suitable conditions for life occur in the sea. In this chapter, we will discuss ecological adaptations of marine organisms—except for marine birds—in relation to their respective habitats or life zones, both within the pelagos, which extends from the sea-spray zone down through the various sunlit underwater zones and the no-light zone, and in the sea bottom area, called the benthos. We begin with a brief overview of the very specific conditions that plants and animals must face in order to grow and reproduce in seawater.

9.1 Marine life specificities

The most important contributing or driving factors influencing the evolutionary ecology of marine organisms are sunlight (Boeuf and Le Bail 1999), salinity (Boeuf and Payan 2001), temperature regimes, and, of course, available dissolved oxygen.

Sunlight is fundamental for all plants, terrestrial and marine, which perform photosynthesis because it is the ultimate source of all primary production in the sea, just as it is on land. For some marine animals, life in the very deep sea is possible wherever feeding material drops from the upper level of water. For those nearer the surface, light is necessary for the development of body pigmentation and overall for the implementation of the pigmented layer of the retina in the back of the eye. Young fish larvae, which are completely transparent at their birth, can see after 10 days. This is a very important life-history trait in early development and growth, and pertinent later in development, both in predator–prey interactions and breeding behaviour (Boeuf and Le Bail 1999).

Salinity is one of the main environmental limiting factors for the geographic distribution in seas (Eckert *et al.* 1999) and hence the ecological behaviour of marine animal species. The average salinity of the Mediterranean Sea is 38–39 psu (which corresponds to 38–39 g l^{-1} and an osmolarity of 1140–1170 mOsm l^{-1}; see Chapter 1). The eastern half of the Mediterranean is more salty than the western half, but there is also a great deal of local and regional variation. For example, many intermediary or ecotonal areas are complex mixes of seawater and fresh water, producing what is called brackish water, which ranges in salinity from just a few grams per litre to nearly the salinity of seawater. Such habitats include estuaries and various kinds of coastal lagoons (see Chapter 6). They harbour a very abundant and diversified fauna which often evolved adaptations to thrive in these habitats. These naturally brackish water systems, with specific biotic communities that can tolerate large variations in key ecological factors such as salt concentrations, belong to the so-called paralic domain (N.B. the term paralic derives from the ancient Greek *para* meaning 'near' and *halos* meaning 'salt').

Lagoons in particular are important in the coastal ecology of the Mediterranean Sea and rather extensive in area, particularly in Italy, with 7.6 million ha, Egypt, which has 330 000 ha, Tunisia, with 72 000 ha, and France, with 43 000 ha (Rouzaud 1982). These lagoons also represent a huge potential for aquaculture (see Chapter 13). Bivalves and gastropods, including oysters, mussels or sea-snails, and winkles, are among the most well-adapted aquatic invertebrates, in terms of their tolerance to frequent changes in salinity; that is, from fresh water to brackish and seawater. As a result of this ecological amplitude, these creatures can develop and persist in a wide range of environments. A few crustaceans have an anisosmotic regulation (Eckert et al. 1999) and are able to develop and grow at various salinities, and the juvenile stages of many highly commercial fish also flourish in these habitats where they find an external osmotic pressure very close to their internal osmotic conditions. Larvae and juveniles in the Mediterranean Sea particularly enjoy these brackish water areas where they may develop under excellent conditions without the high energetic cost of osmoregulation (Boeuf and Payan 2001).

The temperature of the Mediterranean Sea fluctuates between 11 and 27°C at the surface, but sometimes varies less than 1°C per year in deep waters. This is a remarkably small range of temperatures as compared to that of other seas and oceans. But this leads to special constraints on animals because increasing salinity and temperature result in decreasing oxygen availability. As a result, it may be difficult for some species to breathe in shallow waters during summer, especially in the lagoons where oxygen is largely lacking. Eutrophication is a common event in many Mediterranean coastal lagoons during the summer season, and this can be catastrophic for some species' survival. High average temperatures also reduce the duration of fish larval stages and accelerate sexual maturation. Accordingly, many Mediterranean species are smaller than comparable ones in the Atlantic Ocean. In fact, many Mediterranean fish species develop faster to maturity and are able to reproduce at smaller sizes than their congeners in the Atlantic.

The influence of these factors will apply both in the benthos and the pelagos with consequences appropriate to the respective communities of these two domains. After these general comments about the specificities of life in a marine environment, we will now discuss the main biological characteristics of the two domains already briefly presented in Chapter 6.

9.2 Pelagos

The pelagic environment is a three-dimensional space where water masses move horizontally and vertically, according to temperature and salinity gradients under the influence of surface heat and freshwater fluxes. It is this so-called thermohaline circulation, mentioned briefly in Chapters 1 and 6, that drives seawater movements at the scale of the Earth and is sometimes called the ocean conveyor belt. Locally, the wind can provoke different coastal currents as it blows off the sea, towards the coast, or the opposite (see Chapter 1). The Mediterranean Sea has its own thermohaline circulation dynamics, which differ from that of the oceans. This is due to the fact that the Mediterranean Sea is an evaporative basin whose water deficit is compensated by an inflow from Atlantic Ocean through the Straits of Gibraltar (see Chapter 1).

As seen in Chapter 6, the primary structuring factor of the pelagos is the penetration of sunlight into the water which divides it into an euphotic domain and a no-light domain (see Fig. 6.4). The first one is divided in epipelagic zone and mesopelagic zone, which concentrate biological activity due to the amount of primary production by the phytoplankton, upon which depends the entire pelagic food chain. All marine flora and the great majority of the fauna live here, including passive plankton, and the actively swimming nekton which comes here to forage. The second domain, the deepest one, receives no sunlight at all. The upper part, or infrapelagic zone, has a close trophic relationship with the upper levels through migratory species, while the lower levels, right down to the deepest waters—the **bathypelagic** and **abyssopelagic** zones—are home to communities adapted to permanent night. Contrary to what one might think, these zones are not a biological desert. Individual organisms are widely scattered, but considering the importance of the volume of water they occupy, they are numerous all the same.

Having briefly redefined the spatial division of the pelagos, we can now discuss its living systems and their adaptive behaviour. There are two well-differentiated categories of pelagic organisms: (1) the plankton, which, as mentioned above, includes all the fauna and the flora that are passively transported by currents, and whose only basic requirement for survival is the ability to float; and (2) the nekton, which includes all marine animals that move independently of the currents; that is, fish, squids, whales, and other cetaceans. The nekton includes mostly social animals, which travel, eat, and breed in more or less large groups called shoals.

The phytoplankton is composed of unicellular algae with average size between 0.2 and 200 μm as explained in Chapter 4. They do not occur in the most superficial layer of the water because their specialized kind of photosynthesis is hampered by strong light. Moreover, since Mediterranean Sea water is very clear, the sunlit zone extends much deeper than in any of the oceans. The maximum of chlorophyll a, and therefore maximum productivity, occurs at depths of 20–50 m, according to the seasons and the areas, but may occur as deep as 70 m, in summertime in the eastern part of the basin. Each spring, in February and March, there is an annual 'bloom', which is the term used to describe the exponential growth of a colony of phytoplanktonic algae associated with nutrient-enriched water. Starting in April, however, colony growth rates decrease under the pressure of grazing by zooplankton; most populations are at their lowest density in July. Despite the generally low densities at that time of year, incidents of eutrophication are increasingly frequent near the urbanized coasts, a problem that is exacerbated by the above-mentioned absence of admixture of fresh waters by currents or tide.

The Mediterranean zooplankton is an impoverished subset of the oceanic microfauna and does not present any special features as compared to oceanic zooplankton. There are no structured communities in the pelagic environment because there is no shortage of available space in such a homogeneous environment. There are only two modes of foraging: filtration and predation. The thermal barrier formed by the deep sea waters that remain at 12–13° C constitutes a warm environment that may be unsuitable for species coming into the Mediterranean from the much deeper, colder Atlantic waters. From the bathypelagic and abyssopelagic species which are found at great depths in all the great oceans, only the most **eurythermic** can cross the Straits of Gibraltar and enter Mediterranean waters. Once arrived, these pre-adapted newcomers can develop a very large vertical distribution due to the homogeneous temperature of the deep waters, but this phenomenon is actually quite rare and almost accidental, as the deep current of the Straits actually flows away from the Mediterranean towards the Atlantic. Certain deep-dwelling species in the Mediterranean in fact are thought to be remnants of the last ice age when the Gibraltar currents were inversed (Razouls *et al.* 2005–9). The zooplankton moves passively on a horizontal plan, but can also migrate vertically according to a circadian rhythm. The amplitude of their migration routes ranges from a few metres to several hundreds metres depending on the species involved, but always according to the same principle. These diurnal routes are traced especially by crustaceans (copepods and **krill**), which are very much sought-after by many marine predators. During the day, they hide in the deep, no-light zone to escape their predators. Then at night, they actively rise towards the surface in search of the plentiful food source produced by the phytoplankton. When they have eaten their fill, they stop grazing and cease all movements, which allows them to return passively, to their optimal night time depths by the simple force of gravity (Collignon 1991). This circular, vertical movement pattern is crucial to ecosystem dynamics because it allows a rapid transfer of energy; that is, the biomass synthesized by the migrant crustaceans during the day, from the superficial producing layers towards the deeper sea layers. In other words, the nocturnal predation of zooplankton does not greatly affect the phytoplankton, which regenerates its biomass, the following day (Margalef 1984; Frontier and Pichod-Viale 1991). Margalef (1974) was the first to take into consideration the energy transferred in the sea, which is now a fundamental research subject in marine ecology. Specialist of phytoplankton, Ramon Margalef was also a naturalist and humanist. A notable precursor of the scientific study of biodiversity, he worked on the interspecific relationships and created what is now called Margalef's diversity index, which helps analyse marine community structure.

Mediterranean fish include large migratory predators (sharks, tunas, and swordfish), smaller predators (mackerels), and little plankton filterers (anchovies and sardines). Except for the sharks, all these fish always travel in shoals when hunting for food. The most elaborate case concerns the small filterer fish which constitute approximately 40% of the Mediterranean fish fauna. For example, both anchovies and sardines remain in shoals all their lives, have no sexual dimorphism, and provide no care at all for their eggs or offspring. The shoals of these fish are also noteworthy for the near-perfect synchronicity of their sexual coming-of-age. As a result, all the members of a shoal may participate in reproduction at once, in highly intense sessions of group breeding and genetic mixing. A shoal of these fish species may in fact be regarded as something like a 'super-organism' which eats, moves, and breeds almost as a single ecological unit or individual. No other animal groups are known that present such a highly socialized and synchronized life from birth till death (Bauchot and Bauchot 1964). Moreover, the shoal is a very energy-saving organization because individual fish are distributed in such a way that there is nearly no wave turbulence among them, allowing easier, synchronized movement of the group as a whole. Their tight swimming formations also provide a means of defence against predators because the shoal can adjust its form in three planes in order to avoid attacks, thereby preventing predators from targeting a particular fish. In the evolutionary arms race, however, marine predators, such as whales, swordfish, tunas, and sharks, have responded by grouped attack behaviour, which results in a 'blowing up' or 'atomization' of the shoal, putting all individual fish at risk to the predator. When this happens, many kinds of marine birds also swoop in to take advantage of the situation. Lastly, for better or worse, the shoal behaviour allows fishermen to take large quantities of fish at a time. A traditional fishing technique in the Mediterranean is to fish at night, from a boat, using a very strong lantern, called a *lamparo* (lamp, in Spanish), which attracts fish which then are caught in great numbers with nets. This very ancient practice apparently originated in Spain and Italy. From there it was exported to Algeria and Morocco after the Second World War and is now practised along the Atlantic coast of Morocco as far south as Agadir. But today, fisheries are usually industrial, with devastating effects on marine life (see Chapters 11 and 13).

Other members of the nekton are squids, turtles, and marine mammals, and other actively mobile, deep sea dwellers as well. They live in a very homogeneous and quiet environment and are not fast swimmers. They are specialized to hunt and capture rare, widely scattered prey. They generally have large mouths with numerous teeth, and many of the deep water fish and cephalopods also bear photophores, which are light-emitting organs which appear as luminous spots on their bodies. Population sizes are not large and these animals are never gregarious because of the rarity of their prey. There are also 'part-time' nekton species like the large benthic shrimps or 'gambas' that spend large time periods resting in the silt of the sea bottoms, but emerge to graze on plankton and other types of food in deep waters.

Below the huge volume of the pelagos, the thin benthic pellicle is a completely different environment. It is necessary to keep in mind, however, the fact that life on the sea bed is entirely dependent on what takes place in the water itself. These two entities are inseparable and complementary. The mobile pelagos is a means of transport for organic and mineral particles and organisms, but also for energy, while the stationary sea bottoms are an area of reception for all that falls through the water column, either in a single drop, or sometimes after being put back in suspension one or more times. Whatever the trajectory, however, the benthos is always at the end of the trip.

9.3 Benthos

The benthic metazoans can live on the sea bottom (epifauna) or be buried in the sediment (endofauna). Floral elements are always sessile while animals can be sessile or mobile. Among the latter, there are swimmers (fish), walkers (crabs, lobsters), and creepers (gastropods, octopus); those in perpetual movement throughout their lives (porgies); and those that can remain motionless for a time, waiting for prey, for example sea horses, or hiding from predators. The sizes, colours, morphologies, and behaviours vary, but are always adaptive with regards to the spatial distribution of a species in the benthos domain. The relation with the substratum is important, because, for many species, the first

> **Box 9.1. The different trophic groups present in the marine benthic flora**
>
> There are seven main trophic groups in the marine benthic fauna: (1) active suspension feeders, which create a water current to suck their food; (2) passive suspension feeders, which only present a trap to catch organic particles; (3) deposit feeders, which salvage the organic deposits with adapted organs (cirra, trunk); (4) sediment feeders, which swallow the sediment with no particular special feeding adaptation, since the digestive tube does the sorting; (5) herbivores of small size, like the meiobentos, which graze on the microphytobenthos or greater in size, like the seaweed grazers; (6) highly diverse, carnivorous predators; and (7) detritus feeders, which are also often carnivorous, but not always.
>
> The food supplies are divided in deposits, falling particles, suspension particles, and different types of prey. Together they form the trophic structure of the biotope and the fauna is distributed according to the relative importance and availability of each of these components. The faunal assemblage or community present in each type of biotope corresponds to the content of the trophic web that best manages the populations so that the maximum of available food can be consumed.

problem is to find a place to settle. An organism never remains accidentally in its biotope. A benthic faunal community is 'assembled' in part as a function of the range of possible relationships with the substratum and also in such a way that each species searches for the best possible conditions for foraging (Box 9.1)

The various zones in succession described below correspond to the evolution of ecological factors in the benthos. We shall focus particularly on the Mediterranean 'peculiarities' in each of the benthic biotopes, and in the composition of their corresponding communities.

9.3.1 The sea-spray zone

Ranging in depth from 1–2 to 5–6 m, this zone is a very harsh habitat, burned by the salt and the sun, whipped and scoured by the wind, washed by fresh water when it rains, and drenched with seawater during storms. Only very resistant and specialized species of plants and animals can survive here for very long. These conditions are quite similar in fact in the sea-spray zones of all the temperate bodies of waters, so that organisms found here are not very different along oceanic and Mediterranean shores. On hard bottoms, the flora is represented by lichens and endolithic Cyanophyceae. Sessile animals cannot survive here, and the fauna is represented only by vagile invertebrates, little winkles (*Melaraphe* (=*Littorina*) *neritoides*), isopods crustaceans (*Ligia italica*), and flies (*Fucelia maritima*), all of which can avoid the most exposed spaces. The isopods and flies eat small bits of organic debris while the wrinkles graze blue algae encrusted on the rocks. At the inferior limit of the zone, the presence of a large endemic barnacle, *Euraphia depressa*, signals the beginning of the intertidal zone. Fixed on places exposed to the north and generally present in small groups, in holes or crevices, this very particular crustacean is the highest perched creature in the marine domain. Remarkably, it can resist exterior temperatures above 50° C.

Barnacles (*Euraphia* in the sea-spray zone and *Chthamalus* in the intertidal zone) present very interesting life histories related to their habitat. These are crustaceans whose larvae are free-swimming pelagic organisms transported by waves. When they arrive at a favourable site, they attach themselves with the aid of their tiny antennae and immediately begin adult development. They are protected by a calcareous outer wall made of skeletal plates closed at the top by two lids or operculae. They feed exclusively on organic particles carried by sea spray and waves, which they filter from the water with specialized thoracic appendices. When resting, they are completely closed, which allows them to resist dehydration and lack of food. They can withstand an entire summer without imbibing a single drop of water, even though the conditions are much harsher than anywhere along the coasts of other temperate seas and oceans. Among Mediterranean barnacles, *Chthamalus* are among the most highly dependent on temperature and rock type when choosing where to fix themselves and develop (Herbert and Hawkins 2006). Very subtle differences in substrate exist that may not appear obvious to the casual observer. Soft substrates, for example, also occur which are never affected by the lapping of waves, and marine faunas are noticeable there by their absence. At most, some sand fleas may be found, which hide in the silt and sand during the day and forage outside at night. The presence of wet wracks or pebbles increases the faunal diversity with some insects, myriapods, and gastropods added to the small crustaceans.

9.3.2 The intertidal zone

As noted in Chapter 4, the tides of the Mediterranean Sea are not significant, except for the northern Adriatic and the eastern coast of Tunisia. However, although the intertidal zone is not large, it does exist. On hard substrata, it is revealed by the existence of characteristic species which define several levels in the zone. Indeed, biota on the intertidal rocky shores show a well structured vertical distribution. As the percentage of emerged versus submerged surface area increased from the bottom upward, organisms are distributed in horizontal belts, according to their resistance to dehydration. The upper belt corresponds to populations of small barnacles (*Chthamalus stellatus* and *Chthamalus montagui*) or pedunculate cirripeds found in Algeria on very exposed places. The lower belts include algae that form overlapping parallel strips. The calcifuge *Rissoella verruculosa*, which is an endemic red algae thought to be a palaeorelict of the Tethys Sea, marks the middle of the intertidal zone. Exposed to high temperatures, its thallium becomes dry and breakable, but regains its plasticity once moistened. Below it occur two encrusting algae, the black *Ralfsia verrucosa* and the brown *Nemoderma tingitanum*, whose short, densely packed filaments resemble a tooth brush, which is the best possible design to resist dehydration. Grazers like limpets are attracted by this food resource, and, as an adaptation against the waves, they return always to the same position (a process called homing) where they perfectly fit their shell to the surface of a rock and thereby resist being torn off their homes, even by the strongest waves. Experiments carried out on limpets in the genus *Cellana* have demonstrated that homing behaviour is an adaptation which regulates local density and dispersion to maximize utilization of food resources and thus reduce intraspecific competition for food at high density of limpets (Mackay and Underwood 1977).

Belts in the exposed sites have more vertical amplitude and present a bigger vegetal biomass than those in the sheltered ones. It is the place of the 'trottoir', a bioconstruction composed of dead thalli of a calcareous algae, at the lower level of the zone (see Chapter 4). It offers a very particular habitat in the part of dead, but not yet consolidated concretions. Occupying the cavities inside this portion of the trottoir, an abundant fauna occurs with typically marine species (annelids, bivalves, isopods, amphipods) together with typically terrestrial species (myriapods, aerial gastropods, small wingless insects, and even an endemic spider, *Desidiopsis racovitzai*). The latter can live here thanks to the presence of bubbles which remain inside the network when the water level is high.

On the intertidal soft substrata, the lapping of the waves is more or less important, but always present, and there is a permanent movement of water through the sediment. In the sandy substrates which practically always constitute the lines of coast, populations of small crustaceans and annelids shelter under the pebbles and migrate vertically with the variations of the sea level. The frequent presence of Neptune grass meadows near the Mediterranean shores induces the deposit of dead leaves floating into shore. Called 'banquettes', they provide shelter and feed many marine and terrestrial invertebrates, including macrobenthos (see Chapter 4), such as the sand fleas and annelids, but also meiobenthos, such as sowbugs, flies, entognatha (collembola), and apterygote insects. Concerning the sandy beaches, the fundamental ecological work of Delamare-Deboutteville (1960) on the environmental biology and the populations of such biotopes has revealed the variety of the associated fauna and the major ecological role of the interstitial groups, particularly mystacocarids (see Chapter 4).

Intertidal mudflats (see Chapter 6) can also be found in sheltered areas like harbours that include populations of annelids (Nereidae), which take advantage of the large quantities of organic matter found there. As is well known, polluted sites are often characterized by the presence of green algae (Pérès 1967). In the highly polluted areas, very small polychaetes of the family Capitellidae are the only faunal elements present in large quantities after a strong eutrophication.

9.3.3 The high-light zone

Beginning at the permanent immersion level and defined as the area where seagrasses and heliophilous algae can live, the high-light zone is the fringe just below the coastal line. It receives the great bulk of the sunlight which enters the sea. The constant level of the sea, due to the near-total absence of tides, is an important factor for species distribution in this zone, both on hard and soft bottoms. In contrast to the fauna and flora of megatidal oceanic shores, where the higher level of this zone is under water for approximately 8 or 9 hours, twice a day, the homologous Mediterranean organisms are continually beaten by the waves on exposed rocky shores. Such a situation is favourable for some species of seaweeds which settle here in the surf level. The genus *Cystoseira* for example is particularly well adapted to this situation. From 29 species in this genus, more than 20 are endemic to the Mediterranean (Cinelli 1985), including *Cystoseira meditarranea* which is the most common species (Boudouresque 1969). They produce dense patches of vegetation—sometimes called *Cystoseira* forests—on exposed rocky shores, to depths of about 1 m. They harbour a great abundance of epiphytic fauna and flora which are constantly grazed by meiofauna and smaller members of the macrofauna, especially crustaceans and annelids. These algal patches also attract predators which take advantage of these concentrations of food, and all together they constitute a particular habitat found only along Mediterranean shores (Vergés *et al.* 2009).

Species assemblages associated with *Cystoseira* systems vary from place to place. Some species are found only in sheltered sites, for example, *Cystoseira crinitae* and *Cystoseira zosteroides*, which occur at more than 50 m depth (Ballesteros *et al.* 2009), but, mostly, they are found in exposed or semi-exposed sites (Pérès 1967) near the shores. To survive in the habitat, organisms must firmly anchor themselves to a rock to avoid being carried away by the movements of the water. Each group of sessile organisms has evolved its own means to remain stationary, for example, the claws of the crustaceans, podia with suckers of the urchins and starfish, and the slimy 'foot' of the gastropods and sea anemones. There are also many sessile species that live their whole lives underwater, such as hydroids, bryozoans, *Balanus* (crustacean, Cirripeda), and of course the mussels and other bivalves that fix themselves to slippery rocks or other hard objects with their byssus gland that secretes an organic matter forming adhesive threads—also called byssus—that allow the animal to attach itself to a solid object or to a Neptune grass rhizome. One species of Mediterranean bivalve, the rough pen shell (*Pinna nobilis*), has a very unusual story, thanks to its byssus (see Box 9.2).

Below the zones of strong swell, the flora of the hard bottoms is less productive in terms of biomass. This is the domain of coastal fish that forage along the sea bottom in search of small prey. The wrasses are very common in the quiet rocky bottoms of the high-light zone, along with gobies, blennies, and triplefins. All these fish are territorial, at least during the breeding period, and males usually take care of the young and look for nests (see Chapter 4). Also present in the shallow hard bottoms are young porgies (sparids), damselfish (*Chromis chromis*), and mullets, who browse for small invertebrates among the seaweeds. The rainbow wrasse (*Coris julis*), which occurs only in the Mediterranean and in the Atlantic, is particularly interesting since it lives in harems (Michel *et al.* 1987). What's more, young individuals are always female and then became male later on. This same life-history strategy is also found in the dusky grouper (*Epinephelus marginatus*), which generally lives in much deeper waters, but is able to live in the high-light zone if it finds a suitable shelter,

Plate 1 (a) Guild poppy (*Papaver auranticum*); (b) *Eryngium maritimum*. Photographs: J. Blondel.

Plate 2 (a) Hairy rockrose (*Cistus corsicus*); (b) myrtle (*Myrtus communis*) branch, showing typical evergreen sclerophyllous leaves common to many woody plants in the Mediterranean region, and the fleshy fruits eaten by many birds and mammals. Photographs: J. Blondel.

Plate 3 Orchid diversity; (a) *Ophrys apifera*; (b) *O. araneifera*; (c) *O. bertolonii*; (d) *O. lutea*; (e) *O. bicolor*; (f) *O. scolopax*; (g) *O. sphegodes*; (h) *O. tenthredinifera*; (i) *Serapias cordigera*. Photographs: J. Blondel.

Plate 4 (a) Rosalia longicorn (*Rosalia alpina*); (b) chameleon (*Chamaeleo africanus*); (c) ocellated lizard (*Lacerta lepida*). Photographs: (a, c) J. Blondel; (b) M. Cheylan.

Plate 5 (a) The blind cave salamander (*Proteus anguinus*); (b) the invasive Louisiana crayfish (*Procambarus clarkii*). Photographs: (a) CNRS; (b) J. Blondel.

Plate 6 (a) Subalpine warbler (*Sylvia cantillans*); (b) blue rock thrush (*Monticola solitarius*); (c) blue chaffinch (*Fringilla teydea*). Photographs: J. Blondel.

Plate 7 (a) Coralligenous concretion; (b) Neptune grass meadow (*Posidonia oceanica*). Photographs: S. Ruitton.

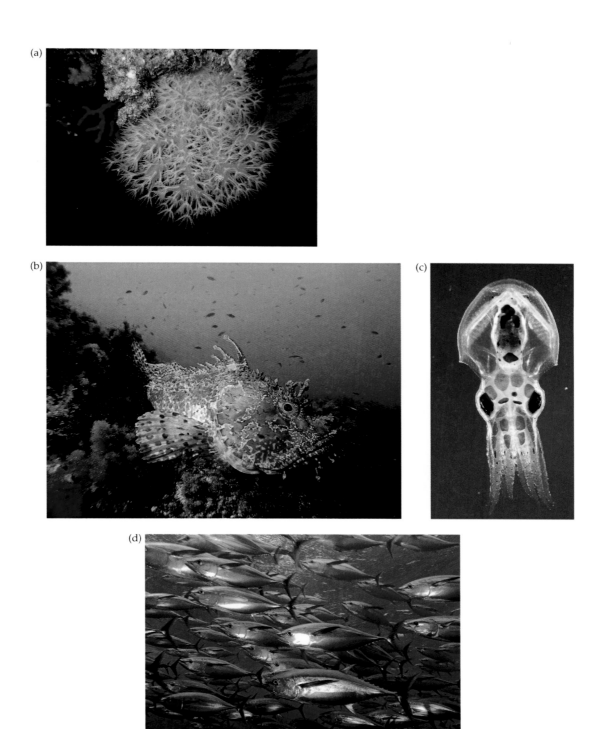

Plate 8 (a) Red coral (*Corallium rubrum*); (b) large-scaled scorpionfish (*Scorpaena scrofa*); (c) *Octopus* larva; (d) yellowfin tuna (*Thunnus albacares*). Photographs : (a, b): S. Ruitton; (c) J. Lecomte, photothèque CNRS/Laboratoire Arago; (d) Fadio/IRD-IFREMER/M. Taquet.

Plate 9 (a) Stone pine (*Pinus pinea*) woodland; (b) *montado* with cork oak (*Quercus suber*). Photographs: J. Blondel.

Plate 10 (a) *Garrigue*, an example of shrubland (matorral); (b) Camargue, a wetland in southern France. Photographs: J. Blondel.

> **Box 9.2. Silk of the sea**
>
> Mediterranean Neptune grass meadows shelter an endemic bivalve with an amazing history. This is the rough pen shell, whose species name *nobilis* is fully justified. This brownish-red, fan-shaped shell, with a gorgeous pearly interior, sometimes exceeds a metre in breadth, making it one of the largest shellfishes in the world. Highly sedentary, it remains half buried in the sediment or among the seagrass leaves throughout its life. This of course makes it easy to collect. Indeed, the species is now threatened, partly due to the decline of the Neptune grass meadows, but also as a result of pollution, trawling, and, above all, unauthorized harvesting by divers.
>
> Like the equally large Atlantic species, *Pinna rudis*, the rough pen shell sometimes produces pearls of acceptable—but not first-rate—quality that draw the attention of divers and merchants. Of course, given its size and delicate beauty, seashell collectors and connoisseurs have always admired the large nacreous shells of both species as objects for their salons. But the real treasure of the rough pen shell is its byssus filaments, which are unusually fine and strong and were used in ancient times to produce the legendary sea silk, or silk of kings.
>
> Now largely forgotten, the byssus threads produced by the rough pen shell once drew the attention of the finest textile artisans of ancient Chaldea, Persia, Egypt, and Greece. In ancient Egypt, only royal family members were allowed to wear clothing made of byssus, which cost, indeed, a king's ransom for obvious reasons. No fewer than 1000 mussels were required to assemble enough byssus to make 200–300 g of silk. Then hundreds of hours were needed to transform the silk into a garment. This ancient art was still widely practised in Italy up to the eighteenth century, but has now been lost except in Sardinia, where a few women still weave the silk of the sea. The rough pen shell is now fully protected by law but is still collected illegally for its beautiful shell. Hopefully it is not too late to save the species from extinction.

especially in protected areas where it is not hunted by fishermen. *Symphodus ocellatus*, a Mediterranean wrasse, undergoes sex change at a certain stage of development, which results in a trade-off between immediate mating success and future reproduction opportunities created by competition between males and conflict between the sexes. Such behaviour can only be understood by examining intersexual conflict and intrasexual competition simultaneously (Alonzo and Warner 1999).

If the diminutive fauna of the seaweeds attract many predators, the seaweeds themselves are not heavily attacked. Herbivorous limpets or wrinkles are generalist grazers of young sprouts and never attack mature algae. But there are two exceptions: the salema (*Sarpa salpa*), a porgie which eats algae and Neptune grass leaves, and the sea urchins, who are the main herbivorous species of Neptune grass on the rocky shores. *Paracentrotus lividus* is the most common sea urchin, travelling in troops of several dozens or more, grazing one patch of seaweed and then the next with the help of their chisel-shaped teeth enclosed in a framework known as 'Aristotle's lantern', an anatomical feature common to all discoidal sea urchins whose special habitat is the hard sea bottoms (Brusca and Brusca 2002). As pollution gets worse, especially in the vicinity of large towns and cities, it's impact is steadily getting more obvious on hard bottoms with sessile organisms. Near Banyuls, southern France, for example, *Cystoseira mediterranea* is gradually being replaced by more resistant species such as *Cystoseira compressa*. The chemical pollutions are particularly hazardous because they are not always rapidly detected.

The soft substrata of the high-light zone are generally sandy bottoms. As a consequence of the quasi-absence of tides, granulometry progresses regularly from the shores to the low-light zone; that is, from gravels and pebbles to coarse, medium, and finer sands. This succession can be explained by the fact that most of the sedimentation is of continental origin. The heaviest particles settle first and the finest ones are carried the furthest, adding to the muddy bottoms which are generally in the low-light zone.

Granulometry is also associated with currents, which are strongest in shallow waters, moderate over fine sands, and much lower over the deep mud. However, deposits of organic particles are inversely proportional to the strength of the currents: they are very low on the coarse sands and increasingly heavy with greater depth. These are the baseline conditions that define the distribution of fauna on the sea bottoms.

In general, the coarse sands—with particles under 1 mm in diameter, in a zone 1–10 m in depth—are very poor in terms of biomass and biodiversity, except for the subterranean meiofauna. Bottom currents cleanse and oxygenate the coarse sands, but those sands are quite poor in organic matter. Only highly mobile species occur here, on the surface or below. The surface fauna includes crabs, hermit crabs, shrimps, and small gastropods, all of which are ubiquitous and present also on other bottoms. Within the sediment are found the lancelets *Donax* and *Tellina*, which are bivalves with very smooth and flat shells and long siphons. These are surface predators, partly buried in the sand like sand eels, as compared to the wholly-buried predators (nemertians, many species of the family Naticidae and annelids of the genera *Glycera* or *Nephthys*) who move in the sediment without relation to the surface because the interstitial network is well oxygenated. As mentioned, there is an abundant interstitial meiofauna, consisting of many poorly studied groups, the most important being free nematods, copepods, ostracods, turbellarians, gastrotrichs, and micro-annelids (Bodiou and Boucher 2006).

Descending still further into the sea, the fine-sand communities are more important than those of the coarse sands. Generally situated between 10 and 25 m in depth, they receive significant quantities of nutrients and benefit from the primary production of the microphytobenthos. The substratum is dense and hard because, with an average grain size approximately 150 µm, there are many more friction surfaces per unit of volume than in the coarse sands. The animals must consume large quantities of energy to penetrate or to move in the substratum and a great proportion of the fauna stays just below the interface. In strategic terms against predation, it is always advantageous to remain at least partly buried, and bivalves are particularly well represented in the fine sands. These creatures are sedentary, with short siphons and bulging ornamented shells, which relates to the fact that they are not deeply buried and do not move a lot. Inside the sediment, there are also annelid polychaetes in the oxygenated layer and occasional 'irregular' sea urchins (of the 'sand dollar' group), which are deposit feeders. Contrary to the coarse sands, there is an abundant interface fauna of shrimps, crabs, brittle stars, gastropods, and large quantities of little peracarid crustaceans. This fauna takes advantage of the bottom life deposits and of the primary production. In turn, it provides food for predators including young fish, small cephalopods, and large crustaceans. Almost all of the numerous species which live on the surface have protective colouring or camouflage mimicking the sand. This is found for example in grey shrimps (family Crangonidae), crabs, brittle stars, and young flatfish. Other creatures, like the well-known hermit crabs, use a shelter like an empty gastropod shell.

These sandy bottoms are important nursery grounds for young fish because of the abundance of prey they can find here after their benthic recruitment. The fine sands are rich in biodiversity with many accessible, small prey on the surface, while the large populations of bivalves constitute an ideal food for fish such as sea breams.

Another unusual predator is a soft bottom starfish of the genus *Astropecten*, which remains buried in the sand and feeds by turning its stomach inside-out to trap and digest various small bivalves (J.-Y. Bodiou, personal observation). After finishing a meal, it turns its stomach back inside. There are not passive suspension feeders in sandy areas, except in the deeper portions of the community, where tube-dwelling polychaetes come before the muddy bottoms of the low-light zone. Between coarse and fine sands, the medium sands constitute a transition zone with biota common to one or the other of the two communities.

Generally with sandy bottoms, the Neptune grass meadow is situated in the high-light zone, playing a fundamental role as support of marine fauna and flora (see Chapter 4 and Plate 7b). The leaves of a square metre of meadow correspond to an available surface of 10–25 m^2, where small species can find a place to settle, including hydroids, bryozoans, foraminifera, tube-dwelling polychaetes, and many phytophageous invertebrates. Some of them are endemic to the Neptune grass leaves, like the bryozoan *Electra posidoniae*. Neptune grass is negatively influenced by increasing salinity, a trend occurring at present in several places in the Mediterranean (Fernandez Torquemada and Sánchez Lizaso 2005). In contrast, epiphytes are apparently not affected. Neptune grass is more resistant to decreasing salinity, possibly as a result of its terrestrial origin. The leaves are also a shelter and a hiding place for many fish and invertebrates, prey, predators, and grazers of the leaves, but often of the epiphytic algae. There are also a group of species for which the foliage of the Neptune grass is an exclusive habitat. Generally green, there are fish, such as the wrasses *Labrus viridis* and *Symphodus rostratus*, and crustaceans, such as the shrimp *Palaemon xiphias* or the isopod *Idothea hectica*. One stunning and sought-after inhabitant of the Neptune grass meadows is the sea horse, of which there are two species in the Mediterranean: the hairless *Hippocampus hippocampus* and the hirsute *Hippocampus ramulosus* (see Chapter 4). Both are frequent here but can be also found in other habitats, particularly in the lagoons. Their preference for coastal areas puts them in constant danger because they are slow swimmers vulnerable to predators. Fortunately, they are not easy to spot. Like its cousins of the family Syngnathidae (pipe-fish), their coloration provides camouflage among the old brown leaves coloured with incrusting algae. There they hide and wait for little prey swimming within their reach. An interesting biological particularity of the sea horses is that the female confides her eggs to the male, which keeps them in a kind of 'marsupial' pocket until the birth of the juveniles. The period of gestation is 3–5 weeks depending on ambient water temperature (Louisy 2002). Sadly, their populations are now very low because, until quite recently, they were sun-dried and sold as souvenirs to the tourists by fishermen who took them in their nets. Now protected, the best place for Mediterranean sea horses is among the Neptune grass meadows which are also now protected.

In 1982 an important cooperative international programme was organized, called GIS Posidonie (www.com.univ-mrs.fr/gisposi/spip.php?rubrique1). This group not only investigates Neptune grass but also the entire Mediterranean seaweed flora, focusing on distribution, biology and ecology, physiology, threats, and protection and restoration measures needed in the near future.

Under the meadow lives a sciaphilous fauna in the cavities of the 'matte' (see Chapter 4). They are normally found deeper in the low-light zone, and they take advantage of the darkness between the rhizomes to install themselves a bit higher. It is the advantage of the pelagic larval phase of the macrobenthos: the juveniles can settle everywhere and survive if the conditions are favourable. One may consider this habitat as a superposition of two ecosystems. A lower part, with sciaphilous species, is an enclave from the low-light zone in the high-light zone and an upper part, which is the genuine Neptune grass habitat with an exceptional group of endemic leaf-dependent species.

The Neptune grass meadow constitutes an important primary production stock, it shelters a numerous and diversified fauna and paradoxically only few herbivorous species have a direct trophic relation with it, such as the sea urchin *Paracentrotus lividus*, some crustaceans, gastropods, and the fish salema. This is due to phytochemical compounds which make the leaf unattractive for the grazers, which feed essentially on the epiphytic seaweeds. So the Neptune grass meadow is a multidimensional habitat for coastal organisms (Mazella *et al.* 1995), it shelters a complete food chain, which could not exist without it, but it is not concerned by the interactions between its components. While the rhizome production stays confined in the matte, the dead leaves fall on the matte between the sheaves and stay for a while because they are rich in polyphenols and slow to decay. Before bacterial attack can begin, they must be partially broken down by peracarid crustaceans (isopods and amphipods). This necromass is called 'litter' and is regularly exported by the storms. A portion of these fragments drift to the coastline and accumulate there in the shape of benches, which protect the coastal dunes from storm damages, but the greater part migrates to other sea bottoms and particularly toward the great depths of the bathyal zone. It is thought that all the sea floors of the Mediterranean Sea, even the deepest, receive a 'contribution' of Neptune grass litter at some point each year. The littoral Neptune grass beds are an important element in the global trophic budget of the Mediterranean Sea. The high-light zone shelters four other species of seagrasses (see Chapter 4), which are not nearly as important in terms of ecosystem functioning.

9.3.4 The low-light zone

Below the high-light zone and its photophilous assemblages, the low-light zone is a semi-dark domain. In the absence of tide, bottom currents are steady and the annual variations of the temperature are low (between 13 and 17° C). Living conditions are much less harsh and more constant here than in the high-light zone. Higher plants are absent, but photosynthesizing algae play a fundamental role as builders of the most important biotope of the zone (Ros *et al.* 1985), which is the coralligenous concretions of calcareous algae, as described in Chapter 4 (see Plate 7a). Covering large areas, these formations constitute one of the most important hotspots of biodiversity on the Mediterranean bottoms. Thanks to its mode of construction, it presents a large range of various substrata, surfaces to occur and volumes to occupy which offer a shelter to all sizes of benthic organisms. The most tiny crustaceans and polychaetes, as well as large fish like the dusky grouper and conger eels find niches in it. All the groups of invertebrates are represented here, all the trophic types are present, all positions in relation with the substratum are possible, the fauna can be sessile, sedentary, or vagile, and it can swim, walk, creep, or clutch with claws or suckers. The coralligenous concretion is in fact like a Tower of Babel where everyone can find a place to live and things to eat, but also can be eaten. This is reminiscent of the amazing ecology of coral reefs. But we have to clarify that the bioaccretions of the temperate areas are mainly of plant origin, while the tropical reefs are especially built from the skeletons of madrepores (Cnidaria, Scleractinia) and living algae which build the terraces that occur at the top where they receive light necessary for photosynthesis. The majority of the sessile fauna is found on the vertical walls of the blocks, under the overhangs, and in the network of the internal cavities (Fig. 9.1). Being constituted of suspension and deposit feeders, this fauna feeds on the microparticles which are continuously deposited on the bottom at this depth. Whereas many other bioaccretion systems in temperate waters, the coralligenous is mainly of plant origin despite the abundance of the fauna that it shelters.

The mobile fauna includes many small detritivores, grazers, and predators. Grazers can be herbivorous, including sea urchins and gastropods, or carnivorous, such as the nudibranchs. Predators can actively search for food, or lie in wait for prey to come near. In that case, their camouflage must be adapted to the irregular and coloured aspect of the blocks themselves. The red scorpionfish (*Scorpaena scrofa*) is particularly well camouflaged here, becoming nearly invisible when it is motionless (see Plate 8b). This species may reach 50 cm and is included in a family of 200 species distributed in all the oceans. In the red scorpionfish, distributed in the Atlantic and the Mediterranean, the anterior part of the dorsal fin presents a modified spine with a very active venom. The red scorpionfish is found in the low-light zone and is replaced by *Scorpaena elongata* in deep waters, down to 1300 m.

The biodiversity of this habitat cannot be explained only by the structure of the substratum. The general ecological conditions of the low-light zone are very favourable: temperature is optimal, water is never turbulent, and currents are slow and continuous. This means there is a continuous 'rain' of fine particles, providing abundant food for the resident fauna.

Finally, the coralligenous concretion is home to hard-substratum sessile fauna, including microphagous animals, limestone fauna (borers), grazers, microcave dwellers, and predators. Its interest lies in the number of ecological niches it can offer. In an extensive general review, Laubier (1966) identified more than 500 species in this kind of habitat. However, surprisingly, only seven species were exclusive to it. Above all, this is an ecological crossroads of great interest for the maintenance of biodiversity in the coastal areas because the mode of reproduction of the marine organisms with a pelagic larval period makes repopulation of the surrounding areas possible if and whenever necessary. That explains the presence of so many sciaphilous species in the dark environment of the Neptune grass matte. The population of many deep areas, particularly between hydrothermal vents, is a very good example to illustrate this. The question is how are wild fauna, specialized to a specific environment (hydrothermal vents or Neptune grass meadows), able to re-colonize new areas, sometimes very far away? Larval stages that are much more mobile and less specialized may be the explanation. Other types of rocky bottom of the low-light zone are occupied by sessile species assemblages dominated by populations of large sponges, cnidarians, bryozoans, and brachiopods. Here, suspension feeders predominate, along with their predators, especially the urchin *Cidaris cidaris*, which grazes primarily on sponges (Pérès 1985). Such deep-sea sea urchins are not easy to study, compared with the coastal shallow-water species. Extrapolation with shallow-water echinothuriids suggests that larval development is lecithotrophic, which means lacking a planktotrophic phase. Of the deep-sea species examined to date, only *C. cidaris* has a reproductive phase that clearly produces a larva, although the limited samples available did not permit any determination of period of reproduction (Tyler and Gage 1984).

In addition, there are three types of soft bottom in the low-light zone, consisting of fine terrigenous sediments and organogenous remnants mixed together (Pérès 1985). First, there are coastal detritic assemblages composed of broken shells and small fragments of calcified organisms—principally bryozoans and algae—with an admixture of sand and low quantities of silt in the upper part of the low-light zone, often linked with the presence of coralligenous assemblages. They present the largest faunistic diversity of the low-light soft bottoms with numerous species of molluscs, crustaceans,

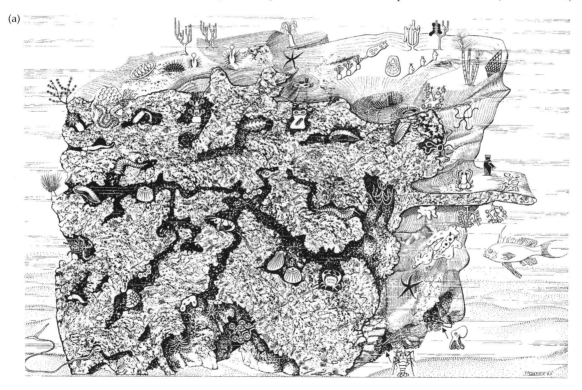

Figure 9.1 (a) A cross-section and stylized three-dimensional view of a coralligenous concretion, showing microhabitats of 50 common species, indicated by (b) drawings. Reproduced with permission from Laubier (1966).

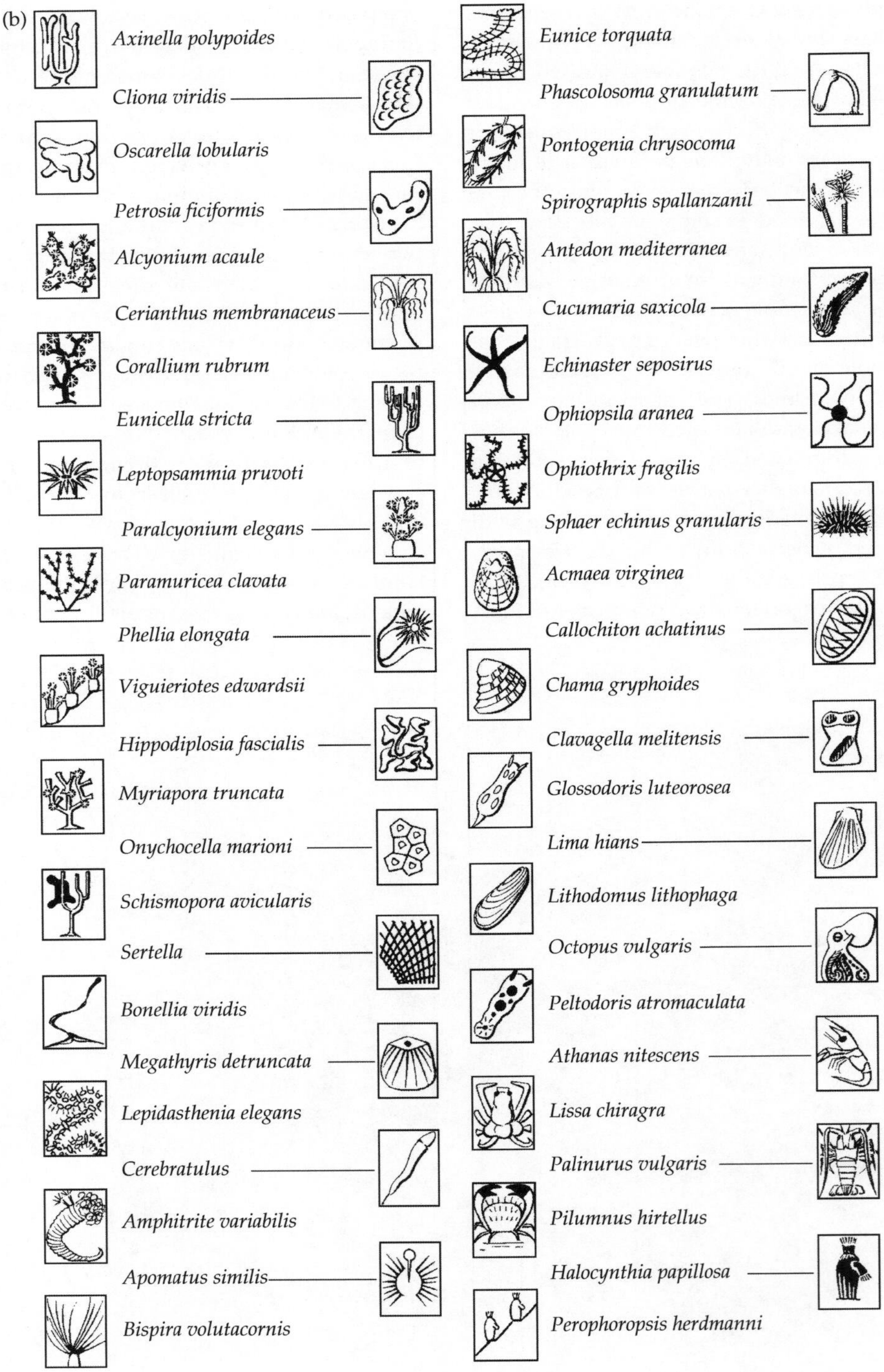

(b)

- *Axinella polypoides*
- *Cliona viridis*
- *Oscarella lobularis*
- *Petrosia ficiformis*
- *Alcyonium acaule*
- *Cerianthus membranaceus*
- *Corallium rubrum*
- *Eunicella stricta*
- *Leptopsammia pruvoti*
- *Paralcyonium elegans*
- *Paramuricea clavata*
- *Phellia elongata*
- *Viguieriotes edwardsii*
- *Hippodiplosia fascialis*
- *Myriapora truncata*
- *Onychocella marioni*
- *Schismopora avicularis*
- *Sertella*
- *Bonellia viridis*
- *Megathyris detruncata*
- *Lepidasthenia elegans*
- *Cerebratulus*
- *Amphitrite variabilis*
- *Apomatus similis*
- *Bispira volutacornis*

- *Eunice torquata*
- *Phascolosoma granulatum*
- *Pontogenia chrysocoma*
- *Spirographis spallanzanii*
- *Antedon mediterranea*
- *Cucumaria saxicola*
- *Echinaster seposirus*
- *Ophiopsila aranea*
- *Ophiothrix fragilis*
- *Sphaer echinus granularis*
- *Acmaea virginea*
- *Callochiton achatinus*
- *Chama gryphoides*
- *Clavagella melitensis*
- *Glossodoris luteorosea*
- *Lima hians*
- *Lithodomus lithophaga*
- *Octopus vulgaris*
- *Peltodoris atromaculata*
- *Athanas nitescens*
- *Lissa chiragra*
- *Palinurus vulgaris*
- *Pilumnus hirtellus*
- *Halocynthia papillosa*
- *Perophoropsis herdmanni*

Figure 9.1 *Continued*

echinoderms, and ascidians. They can present 'nullipores' facies composed of free calcareous algae settled on a sediment particle (see Chapter 4).

The muddy detritic assemblages with broken shells and gravels contain a significant admixture of fine deposits. The sedimentation rate is sufficiently low to allow the settlement of sessile species on little hard supports. In contrast with coralligenous species, they are generally not colourful. The sessile cnidarian *Alcyonium palmatum* is common on these bottoms with other mud-loving species as the polychaete *Aphrodite aculeate*, which protects its gills under dorsal scales (elytra) overlaid by a thick felt layer from which derives the common name of 'sea mouse', sediment feeders as holothurians (*Pseudothyone raphanus* or *Stichopus regalis*), and coumpound or solitary ascidians.

The terrigenous mud-shelf assemblages with a sediment relatively fluid due to a high sedimentation rate where the settling larvae of sessile species cannot survive. Tap-rooted cnidarians of the genera *Veretillium* and *Pennatula* are adapted to this kind of substratum, with burrower crustaceans (*Callianassa truncata* and *Goneplax rhomboides*) and many polychaetes.

Species adapted to the various kinds of muddy bottoms (see Chapter 4) are generally suspension and deposit feeders, but holothurian echinoderms, as well as a few other inhabitants, are able to ingest the sediment, thanks to an alimentary canal that digests organic matter while the inorganic sediments are rejected through the anus. Lastly, the mud is always relatively fluid so that it is easy for many invertebrates to find a shelter below the surface. But as the mud is not oxygenated, animals can breathe thanks to tubes in the case of polychaetes, and siphons for the bivalves. Crustaceans make burrows where they create a water current by movement of their appendages, and many of them go out only by night (i.e. mantis shrimps, *Squilla mantis*).

9.3.5 The no-light zone

The low levels of deep-water circulation and the oligotrophy of the sea make foraging rather difficult in the deep Mediterranean, which explains why the benthic fauna is impoverished there as compared to the oceans. The shallowness of the Straits of Gibraltar and the current of intermediate water towards the Atlantic prevent the deep species of the Atlantic from penetrating into the Mediterranean. As a result, bathyal assemblages mainly include eurybathic species extending from the lower part of the shelf till the great depths. Near the shelf edge and at the upper part of the slope where there are stronger currents and turbidity, hard-substrata assemblages occur which include numerous brachiopods. Deeper on the slope, two deep-coral populations can be found on rocky places, the 'yellow coral' to 300 m and two species of 'white coral' between 300 and 800 m. Some places benefit from favourable conditions when they receive the inputs of important rivers, such as the Ebro and the Rhône. The canyons which cut into the shelf canalize the falling particles, thereby inducing a greater diversity in these favourable zones.

Fish that have been observed or caught include Holocephali, Macrouridae, and Trachitchthyidae. Two species known in the Atlantic have been recently sighted in the Mediterranean. The orange roughy (*Hoplosthetus atlanticus*) is today common as a seafood dish, which is raising problems of overexploitation. This is a species living between 400 and 2000 m, but generally deeper than 900 m, and able to live for 150 years (Maul 1986). It reaches maturity only at age 20 or so, and females do not reproduce every year and do not spawn many eggs. Another species, the roundnose grenadier (*Coryphaenoides rupestris*), lives between 200 and 2200 m (Cohen et al. 1990). It may form large schools at 600–900 m and feed on a variety of fish and invertebrates, but primarily on pelagic crustaceans, such as shrimps and small peracarids (amphipods and cumaceans). Cephalopods and lantern fish constitute a lesser portion of the diet of this species. It is also landed by fishermen and clearly at risk of overexploitation. All these species, with a long life span and generation time, should be considered as biological curiosities rather than fishery stocks! The chimaeras (Holocephali) live as deep as 2600 m on ocean floors. They have elongated, soft bodies, with a bulky head and a single gill-opening. They grow up to 150 cm in adult body length, although some also have a very long tail. In many species, the snout is modified into an elongated sensory organ (Stevens and Last 1998).

Among invertebrates, an interesting species is the cosmopolitan mesobathyal lobster *Polycheles typhlops* (decapoda; Polychelidae), common in the waters around Sardinia, between 400 and 1400 m. Data related to the reproductive biology of this species show that there are seven and four stages of development for the females and males, respectively (Cabiddu et al. 2008). Monthly variations of the percentage distribution related to various stages of development of the ovary and the presence of ovigerous females indicate that the species does not follow a marked seasonal reproductive model, with a long main period for egg hatching that seems to occur between spring and autumn. The Norway lobster (*Nephrops norvegicus*) is also abundant in the Mediterranean Sea, but at much greater depths than in the Atlantic. It may be observed from 100 to 800 m. It seems that its distribution is more influenced by the structure of the floor and the temperature than the depth. It eats polychaetes, small crustaceans, molluscs, and echinoderms. Males grow more rapidly than females and moult more often (Quéro and Vayne 1998). It lives in colonies in holes dug in the mud, almost never moving far from its home. Norway lobster is also heavily exploited in the Mediterranean (such as Spain, France, and Italy), and fisheries have to be carefully organized. The common spiny lobster (*Palinurus elephas*) lives at low depths (2–150 m), the juveniles enjoying Neptune grass meadows between 15 and 25 m, and the pink spiny lobster (*Palinurus mauritanicus*) lives much deeper, from 40 to 600 m. They particularly enjoy deep faults at the border of the continental plateau (Quéro and Vayne 1998). Such areas offer good shelter, with rocks and coral massifs, and large numbers of lobsters may join together on reduced surfaces. Rates of growth are very low as is the case for all deep water species: 300 g at 4 years, 2 kg at 10 years. A typical Mediterranean species from the deep Ligurian Sea, the spider crab *Paromola cuvieri* (in fact closely related to the sponge crab *Dromia*), crossed the Straits of Gibraltar more than 50 years ago and was soon discovered in the Orkney Islands and Norwegian Sea (Gordon 1956). It is a species exhibiting marked sexual dimorphism in claw structure. In the female, the claws are very slender and only slightly longer than the shell, whereas in the male they are about three times as long as the shell and even stouter than the walking legs.

Among cephalopods, the veined squid (*Loligo forbesi*) (minimum 100 m), the elegant cuttlefish (*Sepia elegans*), the broadtail shortfin squid (*Illex coindetii*) (minimum 150 m), and the horned octopus (*Eledone cirrhosa*) live between 50 and 500 m in the Mediterranean, much deeper than in the Atlantic. Due to the temperatures prevailing at the surface and in the depth in the Mediterranean, most species common to the two bodies of water live at much lower depths in the Mediterranean.

Abyssal species are present especially in the vicinity of the Matapan trough (5121 m deep), near south-western Greece. The deep-water fauna of the Mediterranean is characterized by an absence of distinctive characteristics and by a relative biological impoverishment. Both features are the result of events that took place after the Messinian Salinity Crisis (Emig and Geitsdoerfer 2004). The main drivers involved in producing these effects are the historical sequential faunal changes during the Pliocene and thereafter, in particular those that began during the Quaternary glaciations and that are still in progress today. Thus, the existing deep Mediterranean Sea appears to be younger than any other deep-sea constituent of the World Ocean (Emig and Geitsdoerfer 2004).

The deep sea is poorly known in the Mediterranean, since many oceanographic campaigns have been undertaken by either surface and submersible means, due to the absence of hydrothermal vents. For the Mediterranean in general, however, a clear picture is emerging. As we saw in Chapter 4, the Mediterranean is rich in species close the coast and 25% of the species are endemic, mostly in ascidians (50%) and hydroids (27%), echinoderms (26%), crustacean decapoda (crabs and shrimps, sensu lato, 13%), and also many strictly coastal, small fish. However, in sharp contrast to the rich flora and fauna near the coastlines, the deep blue, open sea is biologically poor.

Summary

This chapter summarizes some prominent features of life in the Mediterranean Sea. In the pelagic domain, the plankton is not fundamentally different from that of the Atlantic Ocean, only impoverished. The nekton (fish, squids, turtles, and whales) is well represented. It presents no endemic species. Most of the fish live in shoals, the most elaborate social systems being those of the sardines and anchovies. In the benthic domain, the bottoms change regularly from the shorelines to the depths.

The rocky bottoms are particularly rich in algae in the high-light zone, offering food and shelter for many small benthic invertebrates which in turn attract many small predators. These are the most diverse and colourful of Mediterranean coastal waters, of which they are very characteristic and whose populations are very stable in relation to the microtidal fluctuations. The algae are no longer dominant here, but they are present in the low-light zone where calcareous forms create an important 'reef', the so-called coralligenous concretions, which are a true hotspot of Mediterranean marine biodiversity. Rocky bottoms are not common at this level. The soft bottoms are sandy near the coast and increasingly silty with greater depth. A characteristic fauna lives in each biotope (coarse sand, fine sand, coastal mud, and deep mud) with morphologies, colours, relations with the bottom, and foraging modes adapted to the respective habitats. On the sandy bottoms are found the abundant Neptune grass meadows, the most important biotope of the Mediterranean from a faunistic point of view and also the source of primary production exported towards the deep-sea bottoms.

The no-light zone is not a biological desert, despite its appearances, since it shelters nektonic and benthic species, sometimes of large size, but never gregarious, which makes sense given the low quantities of food available here.

CHAPTER 10

Humans as Sculptors of Mediterranean Landscapes

Nowhere else more than in the Mediterranean region has nature moulded people so much and have people in turn so deeply influenced landscapes. Human pressures on Mediterranean ecosystems have existed for so long that di Castri et al. (1981) argued that a complex 'co-evolution' has shaped the interactions between Mediterranean ecosystems and humans through long-lasting but constantly evolving land-use practices. There is perhaps some exaggeration but not much in this claim of the historian Fernand Braudel (1985) who wrote: 'What is the Mediterranean? One thousand things at a time. Not just one landscape but innumerable landscapes. Not just one sea, but a succession of seas. Not just one civilization but many civilisations packed on top of one another. The Mediterranean is a very old crossroads. Since millennia, everything converged on it.' As a result, human activity should be considered as an integral ecological feature of the region, including effects on gene flow, differentiation, and local selection pressures and constraints (see Blondel and Aronson 1999; Thompson 2005). Studies of intraspecific variation within plant and animal species, such as those we discussed in Chapter 7, show that organisms may evolve life-history traits as a response to human-induced habitat changes. The transformation of landscapes and habitats has also had profound consequences on the distribution and dynamics of species and communities. One cannot understand the components and dynamics of current biodiversity in the Mediterranean without taking into account the history of human-induced changes.

10.1 Human history and Mediterranean environment

Landscapes have been designed and re-designed by humans for almost 10 000 years in the eastern part of the basin and 8000 years in its western part (e.g. Braudel 1985; Pons and Quézel 1985; Butzer 2005). The consequences on biological diversity of this long-lasting action on landscapes and habitats will be discussed in some detail in Chapter 11. What makes any reconstruction of human history difficult is that settlement histories are linked to population growth and decline, wars, and invasions, all related to so-called 'millennial long waves' (Whitmore et al. 1990). There are intricate linkages among ecology, climate, land use, demography, and politico-economic integration at the regional level. Let us look back, first of all, to the arrival of humans in the region, in the process of moving out of Africa.

10.1.1 Early humans

From the dawn of human history, there were many populations of *Homo erectus* all around the Mediterranean. At Atles del Tell, Algeria, a jawbone of *H. erectus* was recently found. It appears to be around 700 000 years old, much more ancient than the Tautavel man (450 000 years BP), found in the Eastern Pyrenees in 1971, which was long considered to be one of the most ancient *Homo* fossils in Europe. In the Levant, Turkey, and Mesopotamia, records of permanent human settlements, including fair-sized cities, go back to the interglacial periods of the late Pleistocene, when humans lived as hunter–gathers in caves.

Little is known, in fact, of the first steps of the 'out-of-Africa' process referred to above. The main routes of the earliest colonization by humans of Eurasia probably passed through the Levantine Corridor (see Chapter 1) along the eastern edge of the Mediterranean. Recently, the Acheulean site of Gesher Benot Ya'akov in the Dead Sea Rift of Israel has provided evidence of hominid movements and technological developments dating back to 790 000 years ago! Archaeological material from this site shows that these early humans already had technical skills quite similar to those of African stone tool traditions, illustrating the corridor role of this area between Africa and Eurasia (Goren-Inbar et al. 2000). This site provided a wealth of information on palaeoenvironments and hominid adaptations. The site is near the shallow Hula Lake (whose remnants are today an important nature reserve with thousands of ducks, cranes, pelicans, and ibises). The sequence excavated represents 100 000 years of accumulation and includes an extraordinary archive of palaeobiological evidence giving many details on the changing landscapes. It includes data on the flora (pollen and spores, wood, bark, fruits, seeds), fauna (fish, turtles, birds, and mammals of various sizes). One aspect that favoured hominid adaptation and survival in this Levantine region was a very rich environment with an abundance of natural resources. The fauna included a great diversity of edible species, and the exploitation of wild game was much enhanced by the production and control of fire.

Generally speaking, the first significant impact of Palaeolithic humans was on population dynamics of many animal species well before the Neolithic revolution and the establishment of permanent settlements. For example, in two sites along the northern and eastern rims of the Mediterranean Sea (Israel and Italy), where remains of human meals span the Middle through Epi-Palaeolithic periods, beginning about 200 000 years ago in Israel and 110 000 years ago in Italy, large series of small animals, such as tortoises, shellfish, medium-sized birds, and mammals, were important to human diet throughout the Middle, Upper, and Epi-Palaeolithic periods (Stiner et al. 1999). Interestingly, the types of small animals most often consumed by prehistoric foragers—tortoises, shellfish, partridges, hares, and rabbits—shifted dramatically in both areas during the Upper and Epi-Palaeolithic from a large proportion of slow-reproducing and slow-moving easily caught animals to an increase of agile fast-reproducing very mobile animals much more difficult to catch. This shift was probably a result of human pressure. Further evidence that shifts in prey species resulted from hunting pressure and human population growth—and not just from environmental changes—is that the mean size of prey steadily decreased during this period (Stiner et al. 1999). Zooarchaeological records in the southern Levant, from the Mousterian—Middle Palaeolithic (100 000–33 000 years BP)—also reveal changes in early human hunting patterns between 70 000 and 10 000 years ago (Davis 2003a). These changes include a gradual shift from hunting very large animals, such as aurochs, rhinoceros, and horse, in the Mousterian, to medium-sized ungulates such as fallow deer in the Upper Palaeolithic, and then gazelles, birds, and fish in the **Natufian** period. These changes have been interpreted as a shift in the balance between humans and animals in the southern Levant, brought about by an increase of human populations. Demographic pressure ultimately stimulated people not only to consume smaller prey items, but also to start husbanding animals for food and also to seek new territories overseas. These changes in Palaeolithic game exploitation can be used as a barometer of human population trends, which seem to be characterized by a series of demographic pulses at the end of the Pleistocene. Then, Mesolithic cultures, which represent the transition from forager (Epi-Palaeolithic hunter–gatherers) to food producing economies, began around 10 000 years BP in Israel and 8000 years BP in the western Mediterranean.

In the western part of the basin, a very long co-existence between modern humans and Neanderthals has recently been documented. Occupation of Gorham's cave, Gibraltar, by Neanderthals has been firmly established until some 28 000 years BP (Finlayson et al. 2006). These populations survived in the southernmost part of Europe and thrived on the rich natural environment which surrounded the Gibraltar area, with a variety of communities of plants and vertebrates on the sandy plains, open woodlands and shrublands, cliffs, and coastal environments around the site.

Mediterranean islands too have been colonized and transformed by people for a very long time. One of the most fascinating aspects of the Mediterranean history is the adaptation of shipping to exploit resources over the whole basin. Recent evidence from archaeological sites in Cyprus shows that human colonization of that island began as early as 10 000 years BP, or very soon after the end of the last glacial period (Simmons 1988).

10.1.2 A succession of civilizations

Following the last bursts of the last glacial period, the climatic improvement of the planet was certainly one of the major upheavals humankind has had to experience. The disappearance of the huge herds of large herbivorous mammals, the progressive rising of sea level, and, more importantly, the replacement of grasslands and steppes by forested areas, which progressively changed from open savannas to lush forests, the expansion of marshlands, all these events led to fundamental changes in life styles, food and foraging habits, as well as cultural and social practices. In short, humankind had to enter a completely new stage in our history.

Since the beginning of the Holocene, when the climate started to improve, the basin has been also a crossroads and meeting point for humans just as it is for plants and animals. Since the Neolithic, many historical waves of human migration, most of them from the east and the south, have been absorbed and have added and superposed their biological and cultural characteristics to the previous ones to produce mosaic-like assemblages of peoples. This made the Mediterranean Basin the theatre of the birth and blossoming, and then the collapse, of some of the most illustrious and powerful civilizations on record. Building on a thousand years of Sumerian, Akkadian, and other Semitic and non-Semitic civilizations in the lands between the Persian Gulf and the Mediterranean, the Phoenicians and Greeks extended their spheres of influence from the eastern Mediterranean to Iberia in the first millennium BC (see Box 10.2, below). Subsequently, Greek, Roman, and Ottoman Empires vied for power, influence, and domination (Fig. 10.1; see Fig. 1.1).

Islands were home to some of the most ancient and brilliant civilizations in the Old World. It was on Crete where, at the beginning of the second millennium BC, the first sustained civilization arose, the so-called Minoan culture, taking the name of the Cretan king Minos. The Minoans developed a huge trading network stretching from Egypt and all the islands of the Aegean Sea as far west as Lipari and Sicily (see Fig. 1.1). The Minoan frescos of Crete illustrate the sophistication of this civilization, which was interrupted abruptly by the eruption of the Santorini volcano. Then the Minoans were replaced by the Mycenians, who came to the island from Greece. They took over the devastated sites of Crete and developed their own trade to Sardinia, Ischia, and Sicily before collapsing about 1200 BC, probably as a result of the dislocation of their trading routes (Arnold 2008). Then, a revival of the Mediterranean Basin's economy started in the ninth to eighth centuries BC, thanks to the Phoenicians who expanded westwards from the Levant, travelling as far as Malta, Sicily, Sardinia, and the Balearic Islands.

Many islands endured repeated invasions and incessant wars because of their much-coveted strategic position for defence (e.g. Malta, Cyprus, Sicily, and Corsica), trade, or natural resources (see Box 10.1). Analysis of charcoal remains found in Crete suggests that during the middle and late Minoan culture (c.4000–3000 years BP), the landscapes around the city of Kommos consisted of intricate mosaics of cultivated fields and orchards alternating with semi-natural woodlands exploited for wood and other products (Shay et al. 1992).

All these civilizations and cultures profoundly shaped the landscapes everywhere in the basin. Apart from sheer, high cliffs, and some very remote mountainous areas, there is probably no square metre of the Mediterranean Basin which has not been directly and repeatedly manipulated and, one might say, 're-designed' by humans.

In the Early Bronze Age, the production of wheat, oil, or wine, in combination with metallurgy, localized metals and alloys, and the timber required for faster ships opened up the Mediterranean Sea for maritime commerce that even extended to India.

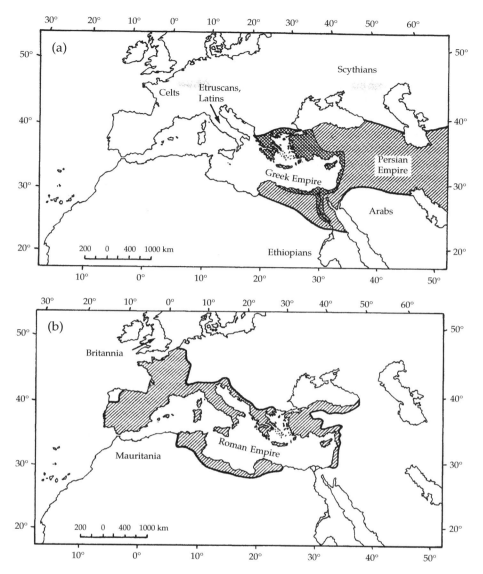

Figure 10.1 Greatest extension of (a) the Greek and Persian empires, about 500 years BC, and (b) the Roman empire in AD 44, at the death of Julius Caesar. The most powerful neighbouring kingdoms or tribes are indicated on each map.

Mediterranean polyculture is more than an agronomic category, but an integral part of a network of long-distance commerce, representing energy and information exchanges, predicated on heterogeneous natural and human resources (see Box 10.2). This early articulated system collapsed not as a result of 'abrupt climate change', but in response to warfare, destruction, and unrest during the third millennium (Butzer 1997). It was revived in a more integrated form during the Late Bronze and was again terminated for similar reasons after 1200 BC. During the eight and seventh centuries BC, it was resuscitated by 'Archaic' Greek and Phoenician colonization or commercial hegemony. Then after several ups and downs, the establishment of the *Pax Romana*, under Augustus, in 27 BC, ushered in a

> **Box 10.1. Corsica, a many-fold invaded island**
>
> According to the historian Michel Vergé-Franceschi (2002), Corsica has been invaded no less than 20 times over the last 2500 years and each time by different peoples. The first invaders were the Phoenicians (565 BC), soon to be followed by the Etruscans (540 BC). Since 270 BC, the Carthaginians (270 BC), Romans (259 BC), Vandals (AD 455), Byzantines (534), Goths (549), Sarrasins (704), Lombards (725), Pisanos (1015), Genoans (1195), Aragons (1297), the Genoans again (1358), 'Milanians' (1468), Franco-Ottomans (1553), French (1768), British (1794), and finally the German–Italian Axis during the Second World War have all invaded the island. This does not mean, however, that all these peoples successfully established a foothold. Only the Romans and Genoans stayed for long, seven centuries for the former and four centuries for the latter. Instead, most invaders established posts along the coast but did not venture far inland. Thus, it was in inland mountain regions that native peoples traditionally sought refuge and when forced to, fought back against invaders. This explains the memorable phrase of Sir Gilbert Elliot, when he was the English viceroy of Corsica, from 1794 to 1796: 'Corsica is an ungovernable rock'! More generally, in the troubled periods that have repeatedly marked the basin's history, risks always came from the sea and the nearby plains. At such times, the inland mountains served as a refuge, as we are reminded by the many *oppida* located at the top of hilltops throughout southern Europe. Traditional villages on islands and other fought-over ground were always situated on high, impregnable sites if at all possible. People retired inland where it was easier to resist invaders and wandering bands of robbers. This results in the strange paradox that, while a remarkable mixture of peoples inhabits coastal areas throughout the basin, in the interior mountains of each country, human populations are still extraordinarily distinct linguistically and behaviourally (McNeil 1992).

period of renewed growth and a new pattern of settlement, lasting until the death of Marcus Aurelius in AD 180. Prosperity probably peaked early in the reign of Justinian (527–565 AD), before a recessionary spiral fuelled by plague, warfare with Persia, the Arab conquests, immigration of Slavic pastoralists, and breakdown of the monetary economy, producing a second 'dark age', c.650–1000 AD. Destructive warfare, economic disintegration, or natural disasters such as earthquakes were probably important in decline and abandonment through feedbacks that would have undermined political authority and security to favour rural exodus or to open the gates to pastoral immigration and anarchy.

During the first century BC and for the unique episode of its entire history, the Mediterranean became one single political entity and benefited for several centuries from the *Pax Romana* (Fig. 10.1).

Then, in the fourth century AD, the Roman Empire split into several entities, one of which, the Roman Empire of the Orient, lasted until 1453, when it was replaced by the Ottoman Empire, which in turn collapsed in 1923. Yet well until the seventeenth century, as a legacy of the *Pax Romana*, the Mediterranean remained the centre of gravity of world economic activity (Braudel 1949). The Romans brought the art of land management to its highest known degree with a series of techniques particularly suited to conditions in the Mediterranean area.

The collapse of the Roman Empire in the western Mediterranean in AD 476 presumably had both positive and negative effects on biodiversity. While the Vandals and other invaders, spreading out in search of new lands, tended to sweep away the works-in-progress and the prosperity gained under the *Pax Romana*, the Visigoths, Sarrasins, and Francs, by contrast, appear to have sought to build upon Greco-Roman cultural foundations. They retained the land-use practices and resource-management schemes developed and disseminated by them. Moreover, a general return of ancestral cultures and regional specificities, mainly Slav, Greek, and

> **Box 10.2. A cyclic model of land use: history and archaeology in the Peloponnese**
>
> In a series of archaeological sites, the historian Karl W. Butzer (2005) found evidence for cyclical alternations of agricultural intensification and 'disintensification', or 'extensification', of land use. In other words, dramatic fluctuations in the levels of human activities are reflected in pollen sequences and macrofossil remains at sites such as Tyrens, in the Argolid plain of the Peloponnese. In the oldest portion of the soil cores, dating from 6200 to 5200 BC, broad-leaved oak pollen have been found along with many other tree species that prospered in partially open landscapes, such as those characterized by strong human presence and frequent disturbances. The typically Mediterranean matorral components of the pollen profile included evergreen oak (8–15%), mainly kermes oak, together with characteristic woody shrubs which tolerate frequent disturbance. This horizon suggests an open deciduous oak woodland interspersed with matorral, as well as fields and pastures, with their corresponding cohorts of segetal and ruderal weeds. Compared with the subsequent horizon (5200–3600 BC), this first horizon suggests a moderately disturbed environment, impacted by pastoralism and farming. The subsequent horizon is dominated by 60–80% deciduous oak pollen, with only 5–8% little evergreen oak and few weedy plants, suggesting dense deciduous oak forest, little disturbance, and no more than an incidental Late Neolithic presence.
>
> Based on the next youngest layers, the pollen sequence suggests that about 3500 BC, human pressure increased again, as indicated by a precipitous decline in deciduous oak pollen levels, whereas the portion of evergreen oaks, false olive, and rockrose pollen increased to 10–20%. During the Early and Middle Bronze period (2600–1600 BC), temperate species declined abruptly, while matorrals expanded, presumably at the expense of formerly cultivated but now abandoned land. During the Late Bronze period (1600–1050 BC), the steady decline of deciduous oak pollen continued, reaching values of 25% with a concomitant increase of olive, false olive, heathers, and pistachio.
>
> Bone fragments at the site indicate cattle and sheep were the most important livestock, which constituted fully 95% of the animal bones found for the entire soil profile. Wild animals constituted only 5% of the animal bones found, including the remains of three genera of deer, as well as boar, bear, lion, and lynx. This suggests that woodland and forest were still common in the landscape mosaic.

Turkish in the east, Arab in the south, and Latin and Celtic in the north-west, resulted in many changes in the design of landscapes and habitats, as well as in human pressures. Thus, it would be misleading to think that the history of the varying landscapes and regional biota we perceive today in the Mediterranean has been linear, or that ill-considered resource depletion has been its only theme.

10.2 Plant and animal domestication

Plant and animal domestication has been the human invention with the deepest consequences on the evolution of human societies, allowing the transition from hunter–gatherer to farmer societies and the establishment of permanent settlements (Diamond 2002; Diamond and Bellwood 2003). In turn, this revolution has had extremely important consequences for biological diversity in the Mediterranean region.

We will make a brief review of the major steps of plant and animal domestication without entering in the details of these processes, which have been subject of many recent studies. In particular, a combination of genetics and archaeology revealed the complexity of the relationships between domesticated plants and animals and their ancestors (Diamond 2002; Diamond and Bellwood 2003; Brown *et al.* 2008). New genetic data show the multiregional nature of cereal domestication, challenging the classical view that each crop species

Table 10.1 The principal cultivated plants originating in the Mediterranean area

Grain and pulse crops
Wheat, 5 species
Oat (*Avena*), 3 species
Barley (*Hordeum sativum*)
Canary grass (*Phalaris canariensis*)
Lentils, vetch, faba, erse (*Lens*, *Vicia*, *Lathyrus*, *Ervum*)
Peas, chickpeas (*Pisum*, *Cicer*)
Lupines (*Lupinus*), 4 species

Forage plants
Cock's head (*Hedysarum coronarium*)
Clover (*Trifolium*), 3 species
Gorse (*Ulex europaeus*)
Fodder peas (*Lathyrus*), 3 species
Serradela (*Ornithopus sativus*)
Corn spurrey (*Spergula arvensis*)

Oil-producing plants
Flax (*Linum*), 2 species
Safflower (*Carthamus tinctorius*)
White mustard (*Sinapis alba*)
Rape seed, colza (*Brassica*), 3 species
Garden rocket (*Eruca sativa*)

Fruit crops
Olive tree
Carob tree
Almond
Fig tree
Pomegranate
Jujube

Vegetables
Beets (*Beta*), 2 species
Cabbage (*Brassica*), 4 species
Parsley (*Petroselinum crispum*)
Artichoke, cardoon (*Cynara*), 2 species
Turnip, swedes (*Brassica*), 2 species

Purslane (*Portulaca oleracea*)
Onion, garlic, leek (*Allium*), 4 species
Asparagus (*Asparagus officinalis*)
Sea-kale (*Crambe maritima*)
Celery (*Apium graveolens*)
Endive, chicory (*Cichorium*), 2 species
Garden chervil (*Anthriscus cereifolium*)
Cress (*Lepidium sativum*)
Parsnip (*Pastinaca sativa*)
Oyster plant (*Tragopogon porrifolius*)
Salsify (*Scorzonera*), 2 species
Spanish oyster plant (*Scolymus hispanicus*)
Horse parsley (*Smyrnium olusatrum*)
Dill (*Anethum graveolens*)
Common rue (*Ruta graveolus*)
Sorrel (*Rumex acetosa*)
Blites (*Blitum*), 3 species

Condiments, dyes, and tanning agents
Black cumin (*Nigella sativa*)
Cumin (*Cuminum cyminum*)
Anise (*Pimpinella anisum*)
Fennel
Thyme
Hyssop (*Hyssopus officinalis*)
Lavender (*Lavandula vera*)
Peppermint (*Mentha piperita*)
Rosemary
Sage (*Salvia officinalis*)
Iris (*Iris pallida*)
Damascene rose (*Rosa damascena*)
Hops (*Humulus lupulus*)
Madder (*Rubia tinctorum*)
Sumac (*Rhus coriaria*)

Source: From Vavilov (1935); reviewed by Hawkes (1995).

was domesticated rather rapidly in a unique and geographically localized area (Brown *et al.* 2008).

10.2.1 Plant domestication

The Mediterranean Basin, especially its eastern part, is among the world's most important centres of origin for crop plants (Hawkes 1995; Zeder 2008; Table 10.1). In the 1920s, the Russian plant explorer Nikolai I. Vavilov noticed that many ancient cultures and many cultivated plants and their progenitors come from the Fertile Crescent, as well as some adjacent areas that stretch from eastern Mediterranean shores, the Jordan River/Dead Sea Rift valley, north and east to Syria, Turkey, and Mesopotamia; that is, modern-day Iraq and Iran.

Though an exact figure for crop diversity is not possible, there are easily more than 500 cultivated crop species in the Mediterranean Basin, including both indigenous and exotic species first cultivated here. The precise origins of agriculture in the region occurred between 9000 and 11 000 years BP, with barley (*Hordeum*) and wheat (*Triticum*) probably being the first plants to be domesticated in the Fertile Crescent. However, hunter–gatherers in the area may have been exploiting wild cereals from at least 23 000–20 500 BP (Kislev *et al.* 1992, 2004; Weiss *et al.* 2004, 2006).

Plant domestication was clearly associated with the domestication of animals. Fodder and pasture plants for livestock—such as lucerne or alfalfa (*Medicago sativa*)—were given an evolutionary 'push' by humans at about the same time—or earlier (Kislev 1985)—as the first cultivated cereals. Mortars and grinding tools, as well as sickles, have been found in the Near East that date much further back still, but they do not necessarily correspond to the sowing of crops around villages. Whatever its precise age, the invention of agriculture divided the plant world linked to agricultural practices into two parts, the segetal and the non-segetal (Zohary 1973). It also separated modern history into a segetal era, which began with the Neolithic domestication of plants and the settling process of human groups in farming villages, and the earlier, pre-segetal era.

10.2.1.1 CEREALS

Many fascinating details have recently emerged concerning the origins of the four Old World cereals first found and domesticated in the Near East; that is, wheat, barley, oats, and rye (*Secale*) (Pourkheirandish and Komatsuda 2007; Fuller 2007). Cultivated wheat, which was first domesticated some 9500 (Tanno and Willcox 2006) to 11 000 years ago (Salamini *et al.* 2002), derives from a number of wild progenitors (Harlan and Zohary 1966) with a complex history (Peng *et al.* 2003; Kilian *et al.* 2007). Diploid einkorn wheat (*Triticum monococcum*) and the tetraploid *Triticum timopheevii*, which are still cultivated on a small scale in the Balkans and Anatolia, both apparently derive from the species *Triticum boeticum*, which is widespread throughout south-western Asia and the eastern Mediterranean. Tetraploid emmer wheat (*Triticum dicoccon*) and hard wheat (*Triticum turgidum*) both derive from the eastern Mediterranean wild emmer (*Triticum dicoccoides*), first discovered by T. Kotschy in the anti-Lebanon in 1855.

The bread wheat we eat today (*Triticum aestivum*) is a hexaploid and derives from hybridization between hard wheat and its diploid relative, *Aegilops tauschii* (Zohary 1969, 1983). The great advantage of hard wheat is the fact that the 'ears' of grain are not brittle like those of emmer wheat and thus remain intact until harvest time. In addition, once harvested at full maturity, the spikelets of durum wheat separate readily from their hulls and thus can be easily threshed.

Cultivated barley (*Hordeum vulgare*) was probably domesticated at around the same time as wheat and served as the 'poor human's wheat' in areas of limited rainfall and poor soils. It ripens a full month before wheat, but the quality of the grain is inferior and, since the Middle Ages at least, has primarily been grown for forage and fodder rather than for bread. The wild ancestor of barley is now clearly established as being *Hordeum spontaneum*, which is native to the eastern Mediterranean and adjacent Irano-Turanian regions (Badr *et al.* 2000; Morrell and Clegg 2007). It appears that only one gene separates the two species, but it is critical for farmers since it controls the brittleness of the rachis, or stalk. Wild barley occurs in both primary and weedy habitats and shows a remarkable amount of biochemical variation (Nevo *et al.* 1979). It is an aggressive colonizer of disturbed matorral and a common roadside and field weed, which hybridizes freely with the cultivated species. The same situation is found in cultivated rye (*Secale cereale*), an important food grain throughout northern and eastern Europe, and also its wild eastern Mediterranean ancestor, *Secale montanum*.

The genetics of oat (*Avena sativa*, including the cultivars commonly called *Avena byzantina* and *Avena nuda*) are no less complex than that of wheat, barley, and rye (Weiss *et al.* 2006). The cultivated oat is closely related to an aggregate of wild hexaploid oats (*Avena sterilis*), which is widely distributed in the Mediterranean Basin. Here again, spontaneous hybridization between the two species is common and much wild **germplasm** remains to be collected and exploited by plant breeders. Two distinct modes of seed dispersal appear in the wild and weedy forms. One is the so-called *sterilis* form, which is synaptospermous, i.e. the whole spike drops to the ground upon maturity and, thanks to its special morphology, 'drills' itself into the soil where the seeds can germinate. In the so-called *fatua* form, by contrast, each floret shatters individually and is much harder to reap.

As in the case of the wheats and barleys, domestication of oats required a modification of the wild mode of seed dispersal in order to obtain a non-shattering crop plant. Oats were brought into cultivation later than the above-mentioned cereals, sometime between 3000 and 2000 years BP, after which it also spread rapidly in Europe and North Africa. This seems to confirm the notion of Vavilov (1951) that oats should be considered as a 'secondary' or derived crop. That is, modern oat cultivars probably started out as weedy races infesting Neolithic wheat and barley fields, and only much later were they domesticated as a new grain crop sown separately (Zohary 1983).

10.2.1.2 PULSES
Numerous protein-rich pulses were domesticated in Neolithic farming villages of the Near East and the eastern Mediterranean and accompanied cereals in their rapid spread throughout the Old World (Kislev 1985). This is supported by archaeological evidence, as well as the shared geographical distribution and ecological patterns of the two groups (Abbo *et al.* 2008). Priority was placed on the production of easy-to-store and highly nutritious seeds with a dual utilization of grass kernels rich in starch and leguminous seeds rich in protein. However, chickpea (*Cicer arietinum*), broad bean (*Vicia faba*), lupines (*Lupinus*), and other pulses, including of course alfalfa, were also grown as fodder, coffee substitutes, and green manure from very early times.

Among wild and domesticated fodder legumes of the Mediterranean area, alfalfa and the related annual medics (over 50 species) are the most widespread, but 50 annual and 25 perennial species of vetch (*Vicia*) should be mentioned, as well as sang-foin (*Ornithopus*), lentils (*Lens culinaris*), *Lathyrus* (35 species), *Lotus*, melilot (*Melilotus*), peas (*Pisum sativum*), *Trigonella*, and no less than 110 economically important species of trefoil (*Trifolium*). Just as the case in the cereal grasses, most Mediterranean pulses form part of a bewildering array of **feral** types of cultigens, weedy races, natural hybrids, and cultivated derivatives of all of the above. Three examples that have been studied in detail are the common vetch (*Vicia sativa*) group (Zohary and Plitmann 1979; Bouby and Léa 2006), the pea (Weeden 2007), and the 12 species of interfertile lupines occurring in the eastern Mediterranean and North Africa (Plitmann 1981). Muller *et al.* (2003) have recently elucidated the domestication history of alfalfa, providing a long-awaited and highly useful tool for alfalfa breeders (see Blondel and Aronson 1999).

Palatable and nutritious grasses in the region that also underwent early domestication and improvement by farmers and pastoralists include orchard grass, fescue, and ryegrass (*Lolium*).

10.2.1.3 MANY MORE ARABLE CROPS
A great many vegetables were domesticated and cultivated in the eastern Mediterranean from very early times. These included beets (*Beta vulgaris*), leeks (*Allium porrum*), lettuce (*Lactuca sativa*), cabbage (*Brassica oleracea*), carrot (*Daucus carota*), celery, radish (*Raphanus sativus*), globe artichoke (*Cynara cardunculus*), and aubergine (eggplant; *Solanum melongena*), as well as the subtropical melon (*Cucumis melo*) and watermelon (*Citrullus lanatus*) (Zohary 1983). Once again, genetic studies are revealing that the situation is more complex than previously thought. For example, Sonnante *et al.* (2007) conclude that 'historical, linguistic, and artistic records are consistent with genetic and biosystematic data and indicate that the domestication of artichoke (*Cynara cardunculus* var. *sativa* Moris, var. *scolymus* (L.) Fiori, and ssp. *scolymus* (L.) Hegi) and cardoon (var. *altilis* DC) diverged at different times and in different places. Apparently, artichoke was domesticated in Roman times, possibly in Sicily, and spread by the Arabs during early Middle Ages. The cardoon was probably domesticated in the western Mediterranean in a later period.'

Culinary herbs and spices of Near-Eastern and various regional Mediterranean origins constitute a list that is much longer still than that of legumes and includes black cumin, borage (*Borago officinalis*), coriander, dill, fennel, laurel, mustard (*Sinapis*), oregano, rosemary, rue, tarragon (*Artemisia dracunculus*), and thyme. Many of those elaborate plant chemical defence adaptations, as we discussed in Chapter 8, and those chemical compounds have been put to good use by humans, for cooking, medicine, and other uses (see Table 10.1). The list of Mediterranean plants used in traditional healing runs to more than 200, and dozens of these are still included in pharmacopoeias.

10.2.1.4 PERFUMES AND OILS

Sources of perfumes, balms, and religious-symbolic substances were also numerous among early plant domesticates. Plant resins, gums, and volatile oils of all kinds were sought to mask body odours and to sweeten the ambiance of tents, bathhouses, boudoirs, and palaces. Those which came into domestication include lavenders and myrtles (several species or chemotypes of each), lentisk, jasmine (*Jasminum*), henna (*Lawsonia inermis*), olive, as well as various 'balsams', including frankincense and myrrh. Wild populations of lentisk, rockrose, storax, and other aromatics were managed as well in wild stands or else exploited out of existence. Some of the earliest vegetable oils in history were obtained from safflower (*Carthamnus tinctorius*) and flax (*Linum usitatissimum*). Domesticated somewhere in the Near East, at least 4000 years BP, the thistle-like annual safflower was formerly an important cultivated and wild-harvested natural dye plant, and cultivated flax was the fibre of choice for the manufacture of fine textiles. Flax was also one of the earliest foodstuffs (Zohary and Hopf 2000), while safflower was an important oilseed crop. Some 16 species of *Carthamnus* occur in the Middle East, but the origins of the cultivation and domestication of safflower are only beginning to come to light. Although long neglected by breeders, safflower is a self-compatible, annual crop that thrives in hot, dry climates. As a result, it may well have increased importance in the future, and funding for its study is stepping up (Burke *et al.* 2005). In particular, using single gene phylogenetic analyses, Chapman and Burke (2007) found a close relationship between safflower and *Carthamnus palaestinus*, whereas *Carthamnus oxyacanthus* and *Carthamnus persicus* had also previously been thought equally likely to be progenitors. This discovery has major significance for breeding programs.

Far outshadowing safflower and all other edible oils, olive oil was so highly esteemed in ancient times that, in addition to its uses as food and lamp oil, it was also used in medicine and for ritual offerings and the anointment of priests and kings (Musselman 2007). It was unquestionably one of the most important and valuable trees of the ancient Hebrews. No other tree is mentioned so many times in the Bible, and olive orchards were clearly widespread throughout the Fertile Crescent and adjacent areas, in Biblical times, wherever climate and soils allowed it. In Greek and Roman mythology, the olive was the symbol of Athena and Minerva, goddess of medicine and health. The olive tree (see Box 10.3), which is the most emblematic plant species across Mediterranean cultures, currently constitutes a complex of many wild forms (*Olea oleaster*), as well as weedy types classified as *Olea europaea* var. *sylvestris* and many cultivars classified as *Olea europaea* var. *europaea* (Terral *et al.* 2004). Today, more than 600 local cultivars occur in the basin. The olive is not only a key economic crop, but its hardiness, longevity, and ability to survive in harsh and dry environments are taken to symbolize the strength and immortality of Mediterranean culture (Breton *et al.* 2003).

10.2.1.5 FRUITS

Archaeological evidence is very strong that fruit crops from trees were domesticated in very early times in the eastern part of Mediterranean and the Middle East. The four classical fruit trees of the region are the olive (see Box 10.3), grape—which is actually a vine—date palm, and fig. The early domestication of all four crops appear to go back as far as the Bronze (6500–7000 years BP) or late Neolithic ages (8000 years BP), if not further. Zohary (1973) argued that the origin of the date palm was in the oases, wild canyons, and salt marshes of the Near and Middle East. The exact origin of wild grape is obscure, but early signs of cultivation have been found from the early Bronze Age in Israel, Palestine, Syria, Egypt, and the Aegean area. The world's most famous vine was cultivated in the Near East at least as early as the period covered in the Old Testament (*c*.3800–5000 years BP). In Genesis (8:20), we read: 'Noah was the first tiller of the soil. He planted a vineyard.' The spies sent by Moses to explore the land of Canaan returned with 'a cluster of grapes and they carried it on

> **Box 10.3. The history of olive**
>
> For millennia, the olive has played a pivotal role for Mediterranean peoples both as a crop and as a cultural symbol of hardiness, longevity, and frugality. The Mediterranean olive is one of the six subspecies of *Olea europaea*, a species belonging to a widespread genus that is mostly tropical or sub-equatorial in distribution. Thousands of cultivars have been described based on fruit size, leaf shape, and colours, but the wild form, with its ability to survive in harsh environments, is a key component of landscapes throughout the Mediterranean region (Breton *et al.* 2006). In Greek mythology, olive was brought to Greece by the Goddess Athena. Botanists once thought that domesticated olive had been introduced in the Mediterranean and that self-sustaining forms, called oleasters, were feral varieties escaped from the cultivars. Historians suggested that Phoenicians, Greeks, and Romans not only introduced domesticated olive from the eastern to the western Mediterranean, but also that travellers later brought back locally selected western cultivars to the east.
>
> There has been much debate and controversy about the origin and history of the olive (Lumaret *et al.* 2004; Breton *et al.* 2006; Besnard *et al.* 2007). Some authors argue that the oleaster was present in the western Mediterranean before the arrival of cultivated olive; others challenge this.
>
> Molecular tools using a combination of approaches including DNA from the nucleus, chloroplast, and mitochondria shed new light on the early distribution of oleaster and the origins and spread of domesticated olive. Using 235 trees representing 27 oleaster populations from many parts of the basin, Breton *et al.* (2006) obtained three main groupings, corresponding respectively to the eastern Mediterranean, the western Mediterranean, and the complex Sicily/Corsica. The genetic diversity of oleaster is highly correlated with geography, with a clear differentiation between eastern and western populations of the basin. This disjunction suggests that oleasters survived in separate refugia during the Quaternary glaciations, one of these being Corsica/Sicily. A higher genetic diversity in the western part than in the eastern part of the basin suggests that oleaster differentiated in the west, presumably in southern Morocco. Then, cultivars were exchanged throughout the whole Mediterranean with an intense selection of local cultivars in the eastern part from where they were diffused. Therefore an eastwards route has been followed by oleaster **mitotypes** from the western to the eastern Mediterranean, and then a westwards route corresponded to the diffusion of cultivars from the primary centre of olive domestication in the east (see Breton *et al.* 2006 and references therein for more details).

a pole between them'. A conspicuous feature of the domestication of grape was the change in sexual reproduction achieved through selection. All known wild forms are dioecious, and fruit set thus depends on the transfer of pollen from male plants to female ones. In nearly all cultivated grape varieties, flowers are hermaphroditic and thus separate pollen donors are not needed, provided the variety is self-compatible (Zohary and Spiegel-Roy 1975).

Many other woody plant species were also domesticated for fruit crops from early times in the same general region. These include carob, mulberry (*Morus*), hazelnut, stone pine, pistachio, pomegranate (*Punica granatum*), and walnut, as well as dozens of stone fruits of Irano-Turanian origin. This latter group includes wild apple (*Malus sylvestris*), pear, and hawthorn, as well as medlar (*Mespilus*), mountain ash, quince (*Cydonia oblonga*), and serviceberry, almond, and apricot (*Prunus armeniaca*), three edible dogwoods (*Cornus*), and one species of oleaster (*Elaeagnus angustifolia*), known as Russian olive or Trebizond date. However, the record for antiquity for a cultivated tree crop may actually go to the fig. In a spectacular discovery at archaeological sites in the Lower Jordan Valley, Kislev *et al.* (2006) found evidence that fig trees may actually have been domesticated by 11 500 years BP that is as earlier than the bread cereals and alfalfa. As mentioned in Chapter 8, however, edible figs have been appreciated by humans for as much as 100 000 years, and the biogeography and genetics, not to mention domestication history of this archetypical Mediterranean tree are still only partially understood (Khadari *et al.* 2005b).

10.2.2 Animal domestication

The Mediterranean is apparently the region of the Earth where the first successful experiments in the domestication of mammal species took place. This area is thus exceptionally rich in livestock genotypes selected by people over generations in each microregion of their respective centres of origin, most of which are in the Near East and southwestern Asia (Georgoudis 1995). The diversity of livestock breeds in the Mediterranean reflects the diversity of environments where humans selected their animals. Once achieved there, the practice of domestication and the various domesticated species spread throughout the world in a remarkably short period of time. Domesticated mammals provided meat, milk, wool, and skins for the manufacture of tools, clothing, and tents, as well as an additional work force that multiplied the possibilities for working the land and the woodlands, for transport, and for trade.

10.2.2.1 THE FIRST STEPS OF ANIMAL DOMESTICATION: DOGS AND WILD BOARS

The first indication of animal domestication in the region dates back more than 10 000 years BP, just like for plants; that is, at the time when climate began to improve following the last Ice Age. The first domesticated animals were either predators, such as dogs, or generalist animals, such as the wild boar, which followed human groups to take advantage of their edible refuse.

According to archaeological data, dog has been domesticated as far back as 15 000 years ago in Asia (Sablin and Khlopachev 2002) from the wolf, as shown by mitochondrial DNA (mtDNA) analyses (Vilà *et al.* 1997). Independently or not, domestication of dogs also took place very early in the eastern Mediterranean, as testified by archaeological remains dating back to *c.*9000 years BP. Although bones of these early dogs resemble those of the dingo (*Canis lupus* ssp. *dingo*), a wild Australian dog, there is evidence that post-glacial human tribes established a hunting relationship with the wolf, and gradually succeeded in taming this species by capturing and raising young pups. Dogs were important because they allowed the development of new hunting techniques and the protection of all other domesticated animals and people, from wolves and other predators.

Pigs were also among the first mammals humans succeeded to domesticate. It is very easy to tame a young wild boar and then breed it, which explains that domesticated pigs appeared as early as 10 000 years BP (Giuffra *et al.* 2000). Pigs were most probably domesticated in the Fertile Crescent from wild boars; genetic analyses make it possible to reconstruct a rather complicated history of domestication with a mixture of lineages of Asiatic and European origin (Larson *et al.* 2005).

10.2.2.2 HORSES AND DONKEYS

Horses were domesticated in the steppes of Turkestan some 6000 years BP from the Eurasian wild horse, the tarpan (*Equus ferus*), which is now extinct. Studies of mtDNA reveal a very high genetic diversity, which is split into at least 17 main maternal lineages (Jansen *et al.* 2002). Horses allowed the development of rapid transportation systems across dry steppes, where cattle would have been too slow and food demanding. The horse has been a decisive companion of humans for territorial conquests and waging wars, as well as for many daily tasks in rural life.

Perhaps even more important than the horse has been the donkey, first domesticated some 6000 years ago from eastern African lineages of *Equus africanus* (Beja-Pereira *et al.* 2004). Wild donkeys from Asia were apparently not domesticated, at least in ancient times. The donkey quickly spread and reached the Near East by 3400 years BP and then rapidly extended to all the Mediterranean countries. The exceptional qualities of the donkey, including its physical endurance and its proverbial frugality, made this animal an invaluable daily companion in the harsh, dry, and mountainous Mediterranean environments. The donkey's services to humans include transportation of heavy loads on large pack saddles, pulling of ploughs, turning of mill stones, raising of well water, threshing of wheat, and innumerable others. One of the most familiar pictures in all Mediterranean countries until recently was that of people moving along narrow streets of small villages on the back of an overloaded donkey with their legs hanging down almost to the dusty ground. Additionally, the virtues of donkey's milk, very similar to that of human milk, are legendary. The wife of the Emperor Nero is said to have used the milk of 500 she-asses each and every day for her bath! When it reaches the end of its life, a dead donkey provides excellent meat, as well as tough hides useful for the manufacture of parchment, clothing, and drum skins.

10.2.2.3 CATTLE

Domesticated races of wild cattle, deriving from the auroch, appear in the fossil record more than 6000 years ago, from forms that were domesticated in Mesopotamia (Pfeffer 1973; Edwards *et al.* 2007). Cattle are quite often illustrated on ceramics from early Mediterranean civilizations and bull worship was long celebrated in many regions. This is best exemplified by the Egyptian Bull-God, Apis, and the famous Minotaur of Crete. Another Bull-God was celebrated in the Bronze Age in the western Mediterranean, at Mt. Bégo near Nice, where many carvings from this period are still visible in the rocks of the mountain. In fact, the co-existence of two distinct mitochondrial lineages suggests that domestication of cattle occurred in two centres of domestication, one in the Fertile Crescent for *Bos taurus* and the second one in India, giving birth to *Bos indicus* (Loftus *et al.* 1999).

Two types of water buffalo (*Bubalis bubalis*), first domesticated in India 5000 years BP, still occur in the Mediterranean: the 'riverine' and, somewhat less frequent, so-called 'Mediterranean' type. They are mainly bred for milk production and, only secondarily, for meat and to serve as draft animals for the cultivation of rice. These very large animals, which can forage in much deeper waters and on softer bottoms than traditional cattle, are still bred in Albania, Bulgaria, Egypt, Greece, Italy, Romania, Syria, Turkey, Tunisia, Slovenia, Croatia, Bosnia-Herzegovina, and Montenegro. In all countries, however, except Egypt, their numbers are decreasing, as are the areas of suitable rangeland.

10.2.2.4 GOATS AND SHEEP

The domesticated animals of most importance for all Mediterranean peoples and which have had the most widespread impact on Mediterranean ecosystems through grazing and browsing are undoubtedly goats and sheep. Goats were among the most ancient domesticated animals, some 10 500 years ago (Zeder and Hesse 2000; Fernández *et al.* 2006), strongly contributing to the Neolithic Revolution. Goats were first domesticated in south-eastern Anatolia and in the upper parts of the Tigris and Euphrates river valleys (Peters *et al.* 2005), as well as in the Zagros Mountains of Iran (Zeder and Hesse 2000). Data from mtDNA, as well as from the Y chromosome, clearly demonstrate that the only ancestor of the domestic goat is the wild goat (*Capra aegagrus*). All domesticated goats derive from the wild goat, which is extinct, even on Crete, where the distinctive goat population of *Capra agrimi* descends from prehistoric domestic goats. The 'goat complex' (genus *Capra*) clearly includes a large number of forms and hybrids, the phylogenetic status of which is still unclear. Much hybridization was carried out by early goat breeders with access to the many Palaearctic forms of ibex in the mountain ranges scattered from the Mediterranean Basin to eastern Asia (Fig. 10.2). A study of mtDNA polymorphism of the species demonstrated that no less than six maternal lineages occur, suggesting several distinct domestication events (Naderi *et al.* 2007). Genetic studies have shown that extant domestic goats (*Capra hircus*) have a much weaker geographical structure than other species of domesticated livestock, which suggests that they have been extensively transported worldwide far beyond the limits of the Mediterranean. From mtDNA sequences of ancient (7300–6900 years BP) goat bones, phylogenetic analyses revealed that all the extant lineages stemmed from two highly divergent goat lineages gradually dispersed from their centre of origin in the Near East and differentiated independently, in the course of their westward movement into Europe (see Chapter 12). Molecular studies show that high mtDNA diversity already existed more than 7000 years ago in European goats, suggesting a dual domestication scenario with two independent but contemporary areas of genetic differentiation (Fernández *et al.* 2006).

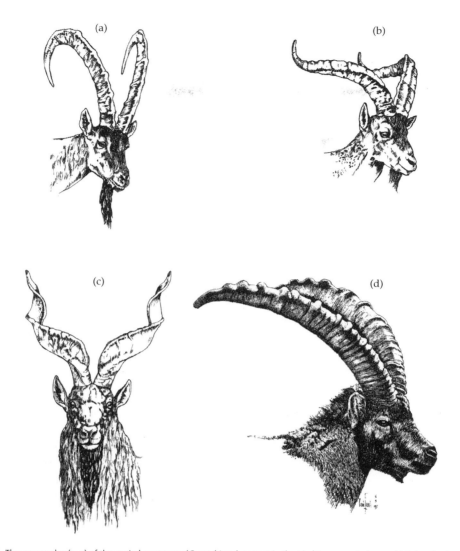

Figure 10.2 Three examples (a–c) of the myriad goat races (*Capra hircus*) present in the Mediterranean today, and (d) the closely related ibex (*Capra ibex*). Note the variety of horn patterns (R. Ferris).

Sheep domestication is very close in age to that of goat, dating back at least 10 000 years in southeastern Anatolia (Peters *et al.* 2005). An extensive comparative analysis of mtDNA of three species of wild Asian sheep (*Ovis orientalis*, *Ovis vignei*, and *Ovis ammon*) with domestic sheep (*Ovis aries*) shows that only *O. orientalis* of Anatolia and western Iran was used for domestication (Rezaei 2007). Today there are dozens of domesticated sheep races in the Mediterranean Basin, which differ in size, shape, coat, and horn patterns, all derived from this single ancestor.

One shared feature of sheep and goats that is of great interest for people is their fast rate of herd build-up. This is one of the main reasons sheep and goat production has always been an important part of rural economies throughout the Mediterranean Basin. Localized traditional systems of dairy sheep and goat production and the region's varied topography favoured the selection of genetically isolated populations of both animals, which progressively evolved fixed characters. Today, local varieties of sheep and goat occur in almost all the large Mediterranean islands, as well as in the oases on the borders of Morocco, Algeria, and Tunisia, and the oasis of El Fayum in Egypt (Georgoudis 1995). Several varieties of the western race of sheep are still widely used in Turkey and Iran as well.

10.2.2.5 SMALLER COMPANIONS OF HUMANS
Domestication of even smaller animals also began very early. The carnivorous species that was apparently used to control mice populations in human settlements, before the domestication of cats, was the genet, raised in captivity from young cubs captured in the wild. The cat itself, however, has long been thought to have been domesticated only much later than sheep, goats, and the other above-mentioned species; that is, by the Egyptians, some 5000 years ago. Indeed, as is well known, the ancient Egyptians practised a religious cult in honour of the cat. Millions of cats were carefully mummified and preserved in tombs in the Nile Valley. Alas, literally tens of tons of cat mummies were exhumed and transformed into fertilizer at the beginning of this century (Pfeffer 1973), depriving scientists of invaluable raw material for genetic studies. However, remains of a cat were discovered in 2004, in a human burial site found in Cyprus that dates from 9000 years BP (Driscoll *et al.* 2007). Using molecular phylogenetics, it has been shown that all known domestic cats cluster into various subspecies of the polytypic European wildcat (*Felis sylvestris sylvestris*), which includes the Near Eastern wildcat (*Felis sylvestris lybica*), the central Asian wildcat (*Felis sylvestris ornata*), and the Chinese desert cat (*Felis sylvestris bieti*) (Driscoll *et al.* 2007). These authors show that cats were domesticated in the Fertile Crescent and derive from five different progenitors from across this region. Their descendants were transported worldwide by humans.

Before closing this section, brief mention should also be made of the elephant, which was first used for purposes of war by Alexander the Great, the Egyptian pharaohs, the Carthaginians, and the Romans, but was never really fully domesticated. At best, they have been tamed into compliance, especially in East Asia. One subspecies, however, played an important role in the human history of the Mediterranean, as described below.

10.3 Forest destruction, transformation, and multiple uses

It is hard to imagine that the elephants, which were captured by Hannibal for his armies (218–220 BC), belonged to a forest-dwelling subspecies that lived in the huge forested areas of southern Tunisia. Mesopotamian cuneiform texts reveal that 3000 years ago, the Assyrian king Assurbanipal passed through large forests to reach Damascus in his conquest of Syria. In Roman times, more than half of northern Africa was still covered with dense forests or woodlands. Even as recently as the sixteenth century, the armies of Charles Quint travelled across Spain and France through lush forests, or woodlands, mostly dominated by deciduous trees. In this section, we sketch some of the huge oscillations in forest cover that have occurred in recent millennia, largely as a result of human activities.

10.3.1 Ups and downs

As explained in Chapters 2 and 6, a great many palaeobotanical, archaeological, and historical records demonstrate that the Mediterranean area was more forested in the past than today. Beginning as early as the Neolithic, the history of Mediterranean forests is one of terrible cycles of destruction and regeneration (see Table 10.2). The first significant deforestation began as early as 8000 years BP (Thirgood 1981) and dramatically increased by the end of the Neolithic (Yasuda *et al.* 2000).

Particularly well marked in archaeological and fossil pollen records are the periods of expansion of the Egyptian, Persian, and Greek civilizations (9000–2100 years BP), which were all eras of intensive land clearing and concomitant increase in the number and size of flocks and herds. This led to an expansion of cultivated fields at the expense of rangelands and an expansion of rangelands at the expense of forests (Trabaud *et al.* 1993).

In the north-western quadrant of the basin, pollen diagrams show that large-scale Neolithic deforestation in the Alps and the Pyrenees

Table 10.2 Forested areas in Mediterranean part of the 24 countries of the Mediterranean Basin

Country	Area with a mediterranean-type climate (ha × 1000)	Extension of forests and matorrals (ha × 1000)	Percentage
Spain	40 000	9200	23
Portugal	7000	n.a.	
France	5000	2186	43
Monaco	0.2	0	0
Italy	10 000	1570	16
Malta	32	0	0
Slovenia[1]	40	≈20	≈50
Croatia[1]	1725	905	52
Bosnia-Herzegovina	300	n.a.	
Montenegro[1]	100	≈50	≈50
Albania	2000	248	12
Greece	10 000	1568	16
Turkey	48 000	6051	13
Cyprus	925	171	18
Syria	5000	440	9
Lebanon	1040	95	9
Israel	1000	116	12
Jordan	1000	n.a.	
Palestine[2]	600	n.a.	
Egypt	5000	2	<1
Libya	10 000	501	5
Tunisia	10 000	840	8
Algeria	30 000	2424	8
Morocco	30 000	5190	17
Total	218 762.2	>31 577	>14

n.a., not available.
[1] Toni Nikolić, personal communication
[2] Palestinian territories (Gaza and West Bank)
Sources: After Le Houérou (1990), Quézel and Médail (2003), and T. Nikolif, personal communication.

coincided with a warmer, moister climate and the steady expansion of cereal culture as a result of human demographic expansion, first at low altitudes in southern France, throughout the Atlantic period (7500–4500 years BP), and at middle altitudes towards 5000 years BP (Triat-Laval 1979). Heavy human pressures occurred again as a result of large population increases during the Calcolithic period, especially the Sub-Boreal period (c.4500–2500 years BP), and continued right up to the decline of the western Roman Empire in the fifth century AD. Based on historical records, fossils, pollen, and charcoal remains at archaeological sites, it is clear that there have been many historical 'ups and downs' corresponding to the waxing and waning of human activities in the various regions. The decline of each major civilization was almost always followed by a wide-scale recovery of forested areas (see Box 10.2).

A spontaneous recovery of western Mediterranean forests occurred after the decline of the western Roman Empire. It lasted until a resurgence of human activities in the Middle Ages. In Corsica, for example, palaeobotanical data show a huge expansion of agriculture during the early Middle Ages accompanied by rapid deforestation after the colonization of this island by the Republic of Genoa (Pons and Quézel 1985).

Enormous amounts of wood were used in the course of history for such varied purposes as domestic firewood, furniture, charcoal, shipbuilding, other forms of construction, and clearing of land for agriculture and livestock husbandry. As early as the seventh century AD, the entire Mediterranean Basin was transformed to timber-based industries (Thirgood 1981). With the development of powerful empires, all easily accessible forests were heavily damaged and sometimes completely destroyed. (Some exceptions were made, however, for sacred trees or sacred groves; see below.) For example, when Spain and Portugal were major naval powers, in the fifteenth–seventeenth centuries, much of the Iberian forests were cleared to allow ship-building, especially near the coasts and along the major rivers. When industrial activities begun, enormous quantities of wood became necessary for glass manufacturing, mine shafts, and fuel for metallurgy and other modern industries.

Following the catastrophic second half of the fourteenth century, when the Black Plague struck southern Europe, a demographic renaissance took place, accompanied by renewed clearing and cultivation of lands, which had been abandoned during an entire century of famine and plague. Finally, the nineteenth and, especially, the twentieth centuries brought increasingly severe destruction of vegetation in many parts of the basin. In most Euro-Mediterranean countries, however, the massive substitution of fossil fuels in the place of wood after the First and, especially, the Second World War, resulted in a generalized recovery of forests, whereas forest destruction is continuing at alarming rates in all the African-Mediterranean countries to the south (see Chapter 11).

10.3.2 Regional histories

Each individual region, however, experienced a different history of forest destruction, as noted already for Crete and Corsica. In southern Europe, a spectacular demographic rise has taken place over the last 2000 years. In the eastern Mediterranean, a similar process took place but two or three millennia earlier. The Lebanese and Palestinian mountain forests were heavily exploited by Egyptian pharaohs, King Solomon, and other ancient Near Eastern rulers, starting at least 3000 years ago. In North Africa, the Roman legions also took a heavy toll on forests. When Julius Caesar's fleet was destroyed by a storm during the war against Pompeius (46 BC), legionnaires were sent to rebuild it from trees harvested in the Sousse region of Tunisia, which was therefore still heavily forested at that time. No trees at all can be found in this region today, except cultivated ones. A similar near-total removal of trees occurred in the Sharon plain between Tel Aviv and Haifa—the Hebrew word *Sharon* means 'forest' (Reifenberg 1955). Today, no vestiges whatsoever remain of these historic forests.

The beginning of the nineteenth century, in particular, was a very prosperous period for the peoples of south-western Europe, with high population density and widespread clearing of woodlands to increase agricultural production. That same period was a desolate, difficult time for peoples in the eastern Mediterranean and North Africa, which led to a certain amount of forest and woodland recovery. Naveh and Dan (1973) relate that, 200 years ago, Palestine—Israel and Palestine, today—had only 200 000 to 300 000 inhabitants, as compared to five million from the second to the fifth centuries AD and more than three times as many people today. Similarly, North Africa had only one half as many inhabitants in the eighteenth century—that is, 6 million people—than it did during the seventh century, at the beginning of the Arab conquest (Le Houérou 1981).

As noted in Chapter 6, Merlo and Pareiro (2005) estimated that at least 8.5% of the Mediterranean area is forested today, the rest being transformed into various types of farmland and in more of less advanced stages of deforestation and soil degradation. However, Table 10.2 provides higher estimates of the area occupied today by forests, open woodlands, and high matorrals in the various countries of the Mediterranean Basin. Yet, as mentioned earlier, a high proportion of forested areas are actually plantations of one or several species of pine, sometimes mixed with cypress, and cedars, among other species.

10.3.3 Successional dynamics

In areas of the Mediterranean where soil depth, rainfall, and ecosystem resilience allow, something resembling the 'primeval forest'—provided that this expression makes sense—is the expected end result of secondary succession. Quite often, in our region, this process is blocked at some stage of succession, as in the case of remarkable kermes oak shrublands (see Box 10.4), and can also occur in highly degraded steppes or grassland.

Indeed, most or all Mediterranean ecosystems can follow many different trajectories and occupy many alternative **steady states**, in the course of use, degradation, transformation, and secondary recovery. To give just one example, Fig. 10.3 shows some of the alternative states and trajectories that can be observed for an Aleppo pine woodland in Mediterranean France, under the influence of humans. This model is simplistic, of course, but does serve to indicate how human actions, when repeated over a long period, can have a lasting impact on ecosystem trajectories, and how positive **feedback systems** tending to block secondary succession can be set off by these actions.

10.3.4 Consequences of deforestation

Large-scale deforestation in the Mediterranean area has had two main consequences. The first was the progressive replacement of deciduous broad-leaved forests by open woodlands and matorrals of different physiognomy and composition, dominated for the most part by evergreen sclerophyllous shrubs and small trees. In the course of the last two to three millennia in particular, holm oak

> **Box 10.4. Dye-producing kermes oak**
>
> The kermes oak, whose scientific name, *Quercus coccifera*, means 'beetle-bearing oak', forms a dense shrubland in many parts of the Mediterranean (see Chapter 5). This knee-high formation was, in many cases, intentionally maintained by people to serve as pasture for the insects that produce a red dye to obtain the means for dying wool and other fabrics. This substance was produced by females of the cottony cochineal insect (*Kermococcus vermilio*), whose sac-like bodies contain carminic acid in high concentrations. As the female attaches herself permanently to a kermes oak and remains immobile throughout her life, this bright red and bitter substance no doubt serves as a defensive warning to ward off potential predators. When mature, the insect lays a number of eggs and dies, after which the young emerge and immediately begin feeding on the host plant. The use of these scale insects was widespread by the time of the Old Testament. The colour-fast, deep-red dye obtained was so highly prized by the Romans that it formed part of the tribute exacted from conquered nations. Feudal lords and monasteries in the thirteenth and fourteenth centuries also accepted it as partial payment of taxes and rent.
>
> Unfortunately, we know nothing about how the kermes shrublands were managed to optimize scale insect proliferation, but fire was certainly applied regularly. Although the cochineal-raising practice was progressively abandoned after the invention of aniline dyes in the late 1800s, in southern France and parts of Spain the kermes oak shrublands have remained remarkably stable in composition and dwarfed in size. They represented an example of arrested succession and an outmoded, now undesirable ecosystem state.

and kermes oak have progressively replaced the deciduous downy oak over wide areas in southern France, Spain, Morocco, Corsica, and elsewhere (Reille *et al.* 1980). Similar trends occurred in the eastern Mediterranean during a somewhat longer period. The substitution of deciduous trees by evergreen sclerophylls, as well as the increase of habitat patchiness over time, have had many consequences on the distribution of plant and animal species, as well as on their genetic diversity and population structure.

As we saw in Chapter 7, changes in the genetic make-up of populations also arise from changes in selection regimes across habitat mosaics, as a result of shifts in the composition of tree assemblages. For example, in the blue tit, shifts in life-history traits of the Corsican population have presumably occurred as an indirect result of human action 2000–3000 years ago. Before human impact on the island's vegetation became strong, most of the habitats suitable for blue tits on this island were deciduous, not evergreen as they are at the present time.

In some instances, widespread and colonizing pines accompany evergreen oaks in the process of secondary succession, such as the black, Aleppo, and Calabrian pines, which are all prolific producers of long-distance wind-dispersed seeds that readily colonize recently perturbed areas. Fire-sensitive species were also reduced or eliminated, and those not adapted to grazing and browsing were also increasingly confined to the few inaccessible habitats, where large ruminants could not reach them. In the eastern Mediterranean, a large number of Sudanian (Afro-tropical) species were probably eliminated or much reduced in distribution as a result of human action during the Holocene (Zohary 1983). Throughout southern Europe, an equally large number of Holarctic arboreal taxa were also lost during the same period (see Chapter 2).

The second consequence of forest destruction was a generalized desiccation of the Mediterranean as a whole, because of the rupture in water balance in many areas where forest cover was destroyed. Water-flow changes are among the most obvious consequences of deforestation. As plant cover decreases, surface runoff and stream flow increase. A dramatic increase in soil erosion occurred in

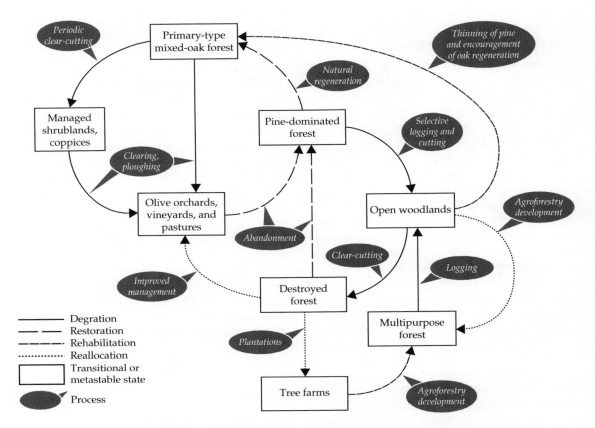

Figure 10.3 State-and-transition model illustrating the numerous alternative states that can be observed for an Aleppo pine woodland in Mediterranean France, with various degradation and regeneration trajectories depending on human activities. After Gondard et al. (2003).

many parts of the basin in conjunction with deforestation, especially in North Africa. Exposure of the soil surface and breakdown of soil structure increase the load of soil material that is carried away through runoff. Soil loss through gully erosion is several times higher when vegetation has been destroyed than on soils with a forest cover. For example, soil loss of a small area in North Africa, which was ploughed after its woody plant cover had been removed amounted to 50 t ha^{-1} year^{-1}, as compared to 0.4 t ha^{-1} year^{-1} on average in nearby forested areas (Dufaure 1984). In order to justify the huge reafforestation programmes that started in southern France at the end of the nineteenth century (see Chapter 13), foresters claimed that 600 000 m^3 of soil were washed away from the Mt. Aigoual each year as a result of forest clearing. These enormous amounts of soil were said to be partly responsible for the obstruction of the port of Bordeaux (Nègre 1931). In a region of southern France where vegetation spontaneously recovered from 7% of the surface area in 1946 to 49% in 1979 as a result of rural depopulation, Rambal (1987) calculated that the stream flow decreased by 11%. Deforestation and its consequences explain the large vulnerability of many Mediterranean hilly landscapes where steep slopes make the effects of erosion even more pronounced and often irreversible.

Despite its long-term destructive consequences, forest clearing was a prerequisite for the creation of a convivial living space for people. Deforestation allowed them to expand areas for their experiments in domesticating and 'taming' selected plants and animals.

10.3.5 Multiple uses of forest

Whatever the turbulent history of Mediterranean forests and, perhaps, as an explanation of this history, forests have always played an essential role for humans, providing many **ecosystem goods and services**. Indeed, rural peoples in all regions of the basin have managed them for a wide range of purposes: livestock grazing, bee-keeping, wood cutting, hunting, and gathering of useful plants, fruits, and fungi.

10.3.5.1 WILD FRUITS

From at least the second millennium BP, Greek landowners cultivated a wide range of fruit trees, including pear, pomegranate, apple, and fig, all found in the gardens of Alcinoos in Homer's *Odyssey*. They also cultivated quince, walnut, almond, and stone pine, as well as a number of Mediterranean trees that have more or less disappeared from modern usage in farms and gardens, such as the Christ's thorn (*Ziziphus spina-christi*) which was then widely used as a 'living fence' (Amigues 1980; see Saied *et al.* 2008).

Cropping in 'wild fruit forests' implies a prehistoric practise of managing natural forests for the optimal production of semi-domestic fruits and nuts. While no solid data or cave paintings support this hypothesis, it makes sense in light of many landscape formations common in the northeastern quadrant of the basin and adjacent regions. For example, extensive groves of pistachio grafted on wild *Pistacia atlantica* trees (the betoum tree, mentioned in Chapter 6) still occur in central Iran and, to a lesser extent, grafted pistachio on *Pistacia palaestina* in some parts of the eastern Mediterranean. Very early in history, western and eastern Mediterranean peoples grafted scions of productive almond trees onto non-productive ones or those producing bitter-tasting fruit. Wild olive and carob trees are also grafted in Turkey, Crete, and Cyprus, and in inner Anatolia, the most common wild trees over large areas are a hawthorn, *Crataegus laciniata*, and a pear, *Pyrus elaeagrifolia*. Along with apples and almonds, these trees are generally left when forests are cleared for farming or grazing purposes and frequently serve as stocks on which other cultivated varieties are grafted (Zohary 1983). How ancient those semi-cultural forests is not known, but there is no reason to doubt that the fundamental horticultural skill of grafting was discovered as early as the domestication of tree crops itself. Cultivated olives, after all, require grafting because ungrafted, they quickly revert and produce a small worthless fruit just like wild types.

Semi-wild 'forests' or woodlands were perhaps maintained for a variety of products and purposes other than fruits, oil, and nuts. The Vallonea oak (*Quercus macrolepis*) was for centuries widely planted in Anatolia, where it is native. In some areas of south-western Turkey, wild stands were still taken care of and treated as if they were planted until the middle of the twentieth century. The main revenue of this tree comes not from the wood, but rather from the galls and acorn burs, from which a highly sought-after tanning agent is derived. The acorns were eaten by livestock and some wood products were provided from regular pruning. Similarly, carob trees were once planted very extensively in warmer parts of both the southern and eastern shores of the Mediterranean, primarily as a forage and fodder tree in association with silvopastoral systems.

10.3.5.2 SACRED TREES

One important factor determining how early humans reshaped and 're-designed' Mediterranean forests and landscapes in ancient times was the role of certain 'sacred' trees and groves. Both oaks and various pistachios (see Chapter 6) were sacred trees in the belief systems and ritual practices of the ancient Near East, since both were associated with Abraham, or even with godliness, as Zohary (1962:211) points out: '*Alon*, the Hebrew word for oak, *Ela*, the word for terebinth, and *Ilan*, a collective name for trees in general, are all derived from El, one of the names of God.' Sacrifices, tribunes, and the burial of loved ones took place under these trees that, accordingly, were not to be cut or burned. Both olive and storax trees were also considered as sacred in the eastern Mediterranean, but some individual trees were considered more 'sacred' than others.

It is difficult nowadays to imagine how important sacred plants were in the past. Throughout southern Europe and the Near East, both pomegranate

> **Box 10.5. Sacred pomegranate**
>
> The pomegranate was long thought to constitute not only a genus but a separate family. Quite recently, a second species, *Punica protopunica*, was discovered on the island of Socotra, off the coast of northeastern Africa (Cronquist 1981). According to Zohary (1973), pomegranate 'grows fairly abundantly in the jungles [*sic*] of the Caspian coastline', as well as along rivers of this same area. Where the sacred pomegranate of the ancients was first domesticated is anybody's guess.
>
> The thick fleshy fruit of pomegranate, with its unusually deep tones of reddish purple, was highly prized for the edible pulp around its seeds and the delightful fermented beverage made from it. The skin was also used for tanning the finest leathers in ancient Egypt. This plant was so highly thought of that a detail of its fruit shape was reproduced by the designers of King Solomon's crown and, later, imitated in the crowns of many European monarchs. That hard fleshy fruit made it possible to transport the fruits long distances without damage, a trait of economic value. It was carefully tended, cultivated, and selectively improved.

(Box 10.5) and hackberry (*Celtis australis*) were respected and rigorously protected as 'safe' trees to sleep under, since no evil spirits could abide there. It is not unlikely that the 12 surviving cedar groves in Lebanon (Sattout *et al.* 2007; see Chapter 5) were protected as being sacred or at least as marking important burial spots.

Religious uses of plant parts, particularly those that give off good smells when burnt, were legion in ancient times. While 'sacred' trees were spared the woodsman's axe, many other species were systematically exploited for the 'services' they could provide. The word 'perfume' derives from the Latin *per fumum* (in the smoke), and rituals involving burnt fragrances and volatile oils of aromatic plants are still found in mountain villages of Anatolia, North Africa, Albania, and elsewhere in the basin.

In ancient and medieval times, several species of resin-bearing rockroses were so highly valued that they were managed as wild or semi-wild sources of the raw material of incense and perfumes. As the leaves of these species exude their resin in daytime, especially in summer, shepherds in Crete, Corsica, and elsewhere, are reported to have combed the beards of goats each evening in order to harvest the resin of rockroses as a supplement to their meagre daily income. When collecting firewood, these people harvested other shrubs and left the rockroses to grow.

Such usages and customs may have been transported as well as the plants. There is a very unusual disjunct occurrence of the above-mentioned *Styrax* in the Var, southern France, alone in all of the western Mediterranean. These small French populations of this unusual tree—the only species found in Europe of the tropical Styracaceae—are highly perplexing from a biogeographical point of view. The uses of the plant may explain the situation. In ancient and medieval times, the resin obtained from the trunk was used as an incense in rituals and also as a 'balsamic' healer of wounds. The ethnobotanist Pierre Lieutaghi (2005) suggests it could have been introduced by monks or other crusaders, returning to France from the Holy Land during the high Middle Ages. According to this hypothesis, the tree was planted around monasteries and became locally naturalized, but did not spread more widely.

10.3.5.3 OTHER SERVICES AND PRODUCTS

All the services and products provided by forests and woodlands offer to people are much more varied in the Mediterranean area than further north in Europe. A fine balance was often achieved among woodlots, pastoral grasslands, the myriad types of shrublands, and the open spaces reserved for cultivation. The resulting mosaic greatly contributed to the biological diversity of Mediterranean landscapes. The guidelines sought for, but not always achieved, were diversity, adaptability, and sustainability.

Forested areas in all Mediterranean cultures were used for wood and charcoal. With an average wood production usually not exceeding 0.5–1.5 m^3 ha^{-1} year^{-1}, annual productivity in most Mediterranean forests is low compared to that of lowland forests elsewhere in the temperate zones. Only in some particularly favourable locations does timber production reach 8–10 m^3 ha^{-1} year^{-1}. To give a comparison, in forests where an oak species is dominant, production values world-wide vary between 1.5 and 4.6 m^3 ha^{-1} year^{-1} and over 10 m^3 ha^{-1} year^{-1} in coniferous and oak wood temperate forests (Llédo et al. 1992).

Collecting firewood was a time-consuming priority of villagers until the use of fossil fuel became generalized after the Second World War. Wood collecting from orchard pruning and wood-lot coppices was accompanied by a number of other activities. For example, in the north-western quadrant, the bark of young branches of holm oak was widely exploited for the fabrication of tannin. A 20-year cycle of clear-cutting was found optimum for this purpose, as well as for fuelwood production. Thus, in many areas, matorrals were managed for a double production of tannin and wood. Additionally, small branches were collected for kindling, and leaves may have been used for a variety of purposes as well. While woodcutting was underway, mushrooms, berries, nuts, medicinal plants, resins, gums, oils, stimulants, dyes, and other forest 'by-products' were also assiduously sought and collected by villagers, hunters, and woodcutters. Most of these are now being lost, as evoked in Fig. 10.4.

In southern Europe, until the end of the eighteenth century, the downy and holm oaks were preferentially cut to make charcoal (see Box 10.6) and for industry (glasswork, metallurgy) and domestic heating and cooking. Because the downy oak resprouts less well from stumps than evergreen oaks, huge areas formerly dominated by this deciduous tree were gradually transformed into more or less depauperate forests dominated by evergreen oaks and other stump-sprouting sclerophylls that tolerate coppicing or pollarding. Since no other woody matorral species can compete with oaks for calorific value, these 'lesser' species were generally neglected or eliminated by woodcutters, except when local needs or market conditions provided an incentive. Examples include cade, boxwood, and mountain ash.

More than anywhere else in Europe, Mediterranean woodlands have traditionally been used for many services and non-timber products. These

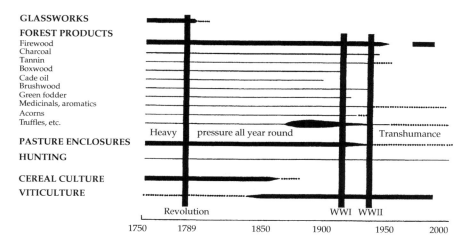

Figure 10.4 Evolution of the utilization of Mediterranean forests and matorrals in southern France over the past 250 years. Dotted lines indicate a gradual increase or decrease of utilization in recent years. WWI and WWII mean First and Second World Wars. Based partly on De Bonneval (1990) and Soulier (1993).

> **Box 10.6. Charcoal production**
>
> Since the Iron Age, charcoal has been the main source of energy for humans in the Mediterranean area, as testified by the myriad ancient charcoal production sites, up to 40 sites per hectare in some cases, that are still visible in many holm oak woodlands. These sites are small circular plots, approximately 5.5 m in diameter, sometimes sustained by a wall or embankment in hilly terrain. The soil is dark black to this day and covered with tiny pieces of residual charcoal. Old iron ovens and the ruins of charcoal makers' huts are still visible in many areas. Charcoal production greatly increased in the Middle Ages when proto-industrial activities, such as glassworks and iron metallurgy, required charcoal, which has much higher calorific properties than wood. Charcoal production sites are thus a vivid testimony of the past and are often used as archaeological sites to reconstruct the history of forest and its inhabitants (Bonhôte and Vernet 1988). Vegetation throughout the area has been profoundly marked by charcoal manufacture practices both in its floristic composition and structure, as shown by anthracological studies, which compare extant vegetation with that which prevailed in earlier periods, on the basis of dated charcoal remains combined with other techniques. Generally speaking, charcoal production over millennia resulted in the homogenisation of woodlands over large areas by favouring evergreen oaks at the expense of other less resilient species. Cyclical clear-cutting of these holm and *calliprinos* oaks over small areas of a few acres in size every 20 years resulted in a perpetual rejuvenation of coppices. In fact, the young stems, which do not exceed 5–6 m in height at the time of cutting, resprout vigorously, even from centuries-old underground root crowns. Although human action was severe and considered as leading to a progressive degradation of the former mixed evergreen/deciduous forests, charcoal production was nevertheless a sustainable forest use practice that resulted in a stable ecosystem (Vernet 1973). Charcoal production in southern Europe sharply decreased in the nineteenth century and completely disappeared in most European countries in the twentieth century, except in Portugal. In parts of North Africa and the eastern Mediterranean, however, the practice still exists.

include food (game, fruits, mushrooms, honey), shelter, medicinal plants, cork, tanning agents, and resins. Some fruits are of commercial importance, such as the large edible seeds (pine nuts) of the stone pine, which, to this day, provides a larger return than the timber itself in some countries, such as Tunisia and Portugal. Similarly, the maritime or cluster pine has long been cultivated in the western Mediterranean for its copious production of resin, which is tapped after the fashion of rubber trees. For centuries, this was the main source of turpentine for western Europe, as well as providing the tar used for caulking the hulls of ships during Roman and subsequent eras. Macedonian pine takes its name from its presumed identity with the *peukê*, as the ancient Greeks called the tree from which they made resinous torches and signal 'lamps' to light at night.

As we will discuss in the last part of this book, great efforts are now under way to re-evaluate the 'total economic value' of Mediterranean forests, taking into account the various supporting, regulating, provisioning, and cultural ecosystem services they provide, not all of which recognized as having a monetary or market value (Fabbio *et al.* 2003; Merlo and Croitoru 2005; Campos and Caparrós 2006; Berrahmouni *et al.* 2009).

10.4 In search of a long-lasting and convivial living space

Having reviewed some of the 'ingredients' used by people to create a convivial living space, in this section, we consider various historical approaches to land use and resource management, which provided a framework for the blossoming of

Mediterranean civilizations thanks to a variety of realized human niches, or 'oecumenes', a word which means living space, derived from an ancient word in Greek.

The first two of these ingredients—fire and grazing—can all produce as much harm as good for biota and ecosystems when ill-managed or abandoned. Even water-management techniques—like dams, canals, or terraces—can wreak havoc if ill-placed and especially when they are abandoned. The crucial points in all three cases is knowing the terrain before starting and finding that intermediate level of disturbance, which is called the 'golden rule' for Mediterranean agro-pastoral or agro-silvopastoral ecosystems (Gomez-Campo 1985; Pons and Quézel 1985; Seligman and Perevolotsky 1994). First we discuss fire, grazing, and water management separately, and then consider landscape-scale designs.

10.4.1 Fire

Fire was the very first tool used by humans in the region, as testified by the remains of ash and charcoal found in archaeological sites in Spain, Greece, and Israel, dating back some 20–25 millennia ago. In fact, the use of fire by Palaeolithic hunter–gatherers appears to go back at least 500 000 years in parts of the Mediterranean (Naveh 1974) and even more since Goren-Inbar *et al.* (2004) found evidence of the use of fire by early humans as early as 790 000 years ago in Israel. Fire was used for a variety of purposes, the first of which being to use fire for the development of grazing and browsing areas for domestic ruminants. Intentional burning was used to maintain and renew open spaces and mosaic landscapes essential for extensive or nomadic livestock grazing.

By the Iron Age, 2600 years BP, shepherds and farmers alike were using fire to attain more and better pasture and to increase unforested land available for cropping. Many passages in the Bible show the importance of fire in the life of the Hebrews. In the first century BC, Virgil also wrote of 'fires lit by shepherds in woodlands, when the wind is favourable' (Aenaid, X). Indeed, until quite recently (Kuhnholz-Lordat 1958), most human-set fires in modern times in the Mediterranean area were ignited for agricultural or pastoral purposes.

In many developing countries of the basin, this practice is still in use. Fire, as a threat to biodiversity and a driving force in ecosystem dynamics in mediterranean-type climates, will be discussed in detail in Chapter 11.

10.4.2 Grazing and range management

As noted already, the combination of forest cutting and the acquisition of control over fire opened the way for the development by early humans of grazing and browsing areas to be used by domestic ruminants. In the Mediterranean area, livestock husbandry and breeding have been of paramount importance to humans at least since the Neolithic revolution, so that fire represents one of the most important forces shaping Mediterranean landscapes. Some pastures in the eastern Mediterranean have been continuously grazed by domestic ruminants for more than 5000 years, whereas others are used much more sporadically, with consequences, in particular, on the spatial heterogeneity and, thus, on the biodiversity of entire landscapes (Henkin *et al.* 2007).

Domestic grazing regimes and pastoral systems under mediterranean-climate conditions can take three different forms, depending on resource availability, social structure, as well as local and historical factors. The first is sedentary livestock raising, involving both stall feeding and free grazing, which is by far the most common form today. Second, there is true nomad pastoralism, where the whole household moves with the herd or flock and which has now more or less disappeared from the basin and adjacent areas. Third, between these two extremes is 'transhumance', where only the herder moves with the livestock. This allowed cultivation of cereals or other seasonal crops as a supplementary source of food and revenues. The strategy of keeping mixed herds of small and larger ruminants was another adaptive element of the system that worked as 'drought-insurance strategy' (Le Houérou 1985), both for those practicing transhumance and true nomads.

In regions with bimodal weather patterns, like the Mediterranean, transhumance was a remarkably well-adapted form, which involved biannual movements of flocks between high summer pastures and winter grazing grounds at lower

> **Box 10.7. Transhumance**
>
> The word transhumance derives from the latin *trans* (beyond) and *humus* (land), and thus means 'beyond the land of origin'. Employed as a verb in Spanish (*transhumar*), the term entered French in the early ninteenth century and came to replace the former term *aestiver*, meaning 'to spend the summer'. In the most widespread form, ascending transhumance, flocks spend the autumn, winter, and spring in the lowlands and move to the cooler mountain areas in summer, where pastures stay green and where drinking water is available. Descending transhumance describes the system whereby flocks leave their usual high mountain homes in winter to pass the coldest season of the year lower down, where the climate is milder. With time, it came to incorporate burning of the vegetation in lowland areas in the late spring in order to 'renew' pastures prior to the return of the flocks and herds later in the year. This practice makes sense from the shepherd's point of view and can be repeated more or less indefinitely if all resources are properly managed.

altitudes, or vice versa (see Box 10.7). Indeed, the high plateaux and mountains of the Mediterranean have traditionally served as an 'escape zone' for animals, a space where herds and flocks could find refuge, food, and water during the dry, scorching Mediterranean summer. In such areas, continental and oceanic influences confer more reliable and higher rainfall than in the littoral, all throughout the year. This is well reflected in pasturelands, which are more productive, diverse, and actively growing throughout summer. Thus there evolved the system of transhumance, which seems to date back at least to the Bronze Age. Some authors suggest that the routes followed by shepherds and their herds were often those established by migrating wild animals, especially the large herds of deer, European bison (*Bison bonasus*), aurochs, mouflon (*Ovis orientalis*), wild goats, saiga antelope (*Saiga tatarica*), and the Eurasian wild horse (tarpan). Numerous petroglyphs and sculptures of large mammals in the mountains of North Africa and the central Sahara also suggest a very early origin for transhumance in those regions.

The distances covered in the biannual movement of herds in search of summer pastures were typically no more than 100–300 km in the moister areas of southern France and northern Italy. By contrast, in the semi-arid regions of southern Spain, southern Italy, the Balkans, and North Africa, each seasonal 'drive' could be as much as 700–1000 km and traverse a huge number of villages and property lines. From very early times, highly elaborate and ritualized networks of trails and footpaths (called *tratturi* in Italian, *cañada* in Spanish, and *drailles* in French) were employed for moving large herds between the Apulian plains and the Apennines, the plains of southern France and all the nearby mountain ranges (Pyrenees, Massif Central, Alps). Similarly, in interior Spain, with its huge ancestral emphasis on livestock, the transhumance routes were far-reaching and well maintained over a great many centuries. Seasonal movements of livestock in central Spain involved 4–5 million sheep, in the heyday of the early sixteenth century (Fig. 10.5).

In southern Europe generally, all forms of transhumance are on the decline today. For example, in the Department of Hérault, France, an area of 622 700 ha, there were approximately 400 000 head of sheep in 1894 and only 88 000 in 1964. This figure has remained stable ever since (Gintzburger *et al.* 1990). In Morocco and parts of the Near East, however, transhumance and pastoralism continue to play an important role in land use, but have undergone massive changes thanks to the growing use of trucks to transport livestock and to the availability of government-subsidized feed for the animal: barley, bread, cottonseed, etc.

These large herds of mammals attracted several other kinds of species, especially scavengers and their associated insect faunas, as well as plant species transported in the faeces of transhumant animals. For example, the griffon vulture used to belong to some kind of trophic community closely linked to migrating herds of wild mega-herbivores

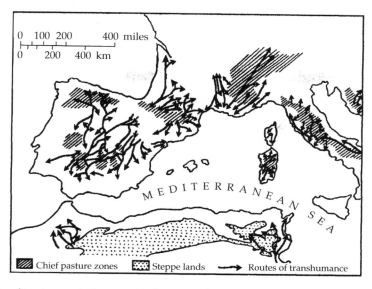

Figure 10.5 Major routes of transhumance in the western Mediterranean. After Houston (1964) and Brisebarre (1978).

and then domestic livestock. The carcasses and dung of these animals, which were scattered along migration routes, provided plenty of food for vultures, but also many other animals, such as corvids, mammals, insects (e.g. dung beetles), and others were associated with these resources. Large-scale transhumance persisted up to the middle of the twentieth century, providing plenty of food for vultures and associated animals in many parts of the basin. For example, at least nine large colonies of the griffon vulture have been reported to occur along the gorges of the Drina River in the Balkan Peninsula (Marinkovic and Karadzic 1999), as shown by many sub-fossil remains spreading from the upper Pleistocene and Holocene to modern times. Shorter migration routes and settling processes of cattle herds resulted in a decline and abandonment of transhumance. In the 1960s, laws forbidding transhumance in the former communistic federal state of Yugoslavia further dramatically reduced the abundance of nomadic herds, which subsequently completely disappeared. The breaking of the long migratory routes of cattle on the Balkan Peninsula started at the end of the nineteenth century, during the collapse of the Ottoman Empire and the forming of new impassable borders between countries. This, combined with a widespread use of poison against large carnivorous animals (e.g. strychnine), resulted in a complete collapse of the animal communities associated with this long-distance movement of animals.

10.4.3 Struggle for water

The struggle for water has always been a vital thread in the history of all Mediterranean peoples. When rainfall is scant or even entirely lacking for several months at a time, sedentary people must ensure a reliable supply of drinking water for themselves and their livestock. The Persians and various other Near-Eastern peoples built and maintained elaborate and nearly labour-free systems to collect and store rainwater in large quantities. Waterproof mortar was first invented around 3300 years BP, and this permitted the construction of permanent water cisterns. A particularly ingenious system for collecting water is the so-called draining gallery, which was devised for capturing underground water from inside cliffs and limestone rocks in tectonic fractures of karstic terrain, which often act as impressive water towers. This technique of interconnected galleries and wells was developed as early as the second millennium BC on the Iranian plateau and in Armenia. It is in the inner slopes and ground

> **Box 10.8. Coping with silt**
>
> An essential aspect of water engineering in the drier parts of the Mediterranean and the Near East is coping with silt. During the winter storms which characterize the eastern Mediterranean, desert wadis, which are bone dry in summer, suddenly fill with muddy torrents, which deposit boulders and stones but carry silt far away into the plains, where it can harm crops and agricultural lands. At Kurnub (Mamshit), in the central Negev, can be seen in sharp detail one of the most astounding examples of ancient engineering, carefully mapped and interpreted by archaeologists. In a region with 75 mm mean annual rainfall, a large caravan station flourished here in Nabatean and Byzantine times, thanks to dams and canals which impounded floodwaters in underground reservoirs. Based on experimental research carried out at Kurnub and two other Nabatean sites in the Negev, scientists were able to confirm the hypothesis that Nabateans engineers not only knew how to capture sufficient amounts of rain and flood water to support large populations, but also how to trap silt in order to prevent the clogging and choking up of fields (Evenari *et al.* 1982).

depressions that the oldest *kirez* or *quanats*, as they were called, are the most numerous and the most ancient. From this Iranian centre of origin, this technique spread westwards and has been somewhat modified or improved as *foggara* in Saharan oases, *khettara* in Morocco, and *minas* or *cimbras* in Spain (Reifenberg 1955).

Ancient peoples dug wells and canals, reinforced river banks, and, most impressively of all, built dams across wadi beds in the desert in order to impound floodwaters that might only occur once every year or less. Using rocks, gravel, and mud, they also designed and built extensive but simple systems to hold the water diverter from the flood to fill city and farm reservoirs, and for directly irrigating cultivated fields downslope from the impoundment areas. These works had to be sufficiently resistant so as not to be destroyed by exceptionally heavy floods that might occur only once in a decade. Additionally, they had to be sensitive enough to take benefit from the smallest possible rain events that trigger surface run-off.

In the Near East, sophisticated water-management and engineering practices date back at least to the Chalcolithic period, 4500–5000 years BP (see Box 10.8). The Nabateans, starting around 2500 years BP, were particularly organized and determined in the practice of rainfall harvesting, as testified by the extraordinarily large cities they built in some of the most arid and desolate regions of the Near East. These included the ancient Biblical city of Sela, in southern Jordan, which was renamed Petra by the Nabatean Arabs when they made it their capital for seven or eight centuries. A series of Nabatean water works of varying dimensions and intentions were also laboriously excavated in Israel, for example south of the city of Be'er Sheva in the northern part of the Negev desert. They have been restored to working order by a small team of scientists over a period of 30 years (Evenari *et al.* 1982). The results of this work help bring to life that far greater number of unexplored archaeological sites are known in the 15 steep wadi beds, which plunge into the eastern shore of the Dead Sea on the Jordanian side. Innumerable remains of runoff agriculture and waterworks also exist in the desert fringes of North Africa. Both the Phoenicians and the Carthaginians were well versed in sophisticated techniques long before the arrival of the Romans. The Phoenician colonists who brought the eastern art of writing also may have brought to southern Europe and northern Africa the art of irrigation (Reifenberg 1955; Hillel 1994). A few centuries later, the conquering followers of Mohammed also brought many innovations and new crops to Iberia during their four-century-long occupation of the Spanish Levant.

It was the Romans who made run-off agriculture and water-diversion systems into a high art. With their vast, centralized government and large pools of forced labour, the Roman rulers were especially well placed to undertake large-scale construction, water transport, and irrigation systems under a variety of conditions all around the basin. The *Pax Romana* provided favourable conditions for intensive cultural exchanges, favouring the flow of know-how throughout the entire Mediterranean world. In coastal areas, where monocultural production of cereals and other rainfed crops dominated, the mastery of hydrology and building by the Roman engineers is illustrated in the preserved remains of their monumental and remarkably astute hydraulic works. Some of the most spectacular examples are found in southern France, including the thousand-year old aqueducts at the Pont du Gard, the industrial mills at Barbegal, and the public baths of the Emperor Constantine in Arles. In half a dozen other countries around the basin, and many of the islands as well, similar marvels exist, particularly in proximity to cities. In Italy, the spectacular waterworks remains also reveal the intricacies of Roman farming, and land-use planning, in all its fascinating detail (White 1970).

10.4.4 Terraces

From very early times, the history of Mediterranean peoples, both in the mountains and at the border areas between hills and plains, has been heavily marked by cultivated terraces. Hand-built stone terraces permitted cultivation on slopes ranging from 20 to 75% and this sometimes required carrying soil up from the valleys on peoples' or animals' backs (Lepart and Debussche 1992). The construction of tailor-made water distribution systems added to the quality of terrace cultivation. Mainly fruit trees, vines, cereals, flax, and vegetables were grown on terraces. The construction and maintenance of terraces, which are labour-intensive, strongly reduce potential damage from soil erosion in the wake of de-vegetation and is an efficient water device preventing run-off. It represents one of the earliest forms of effective conservation. Terraces are a popular and domestic counterpoint to the monumental works of the Romans and their vast imperial cities and farms. Until the early twentieth century, terrace cultivation remained a hallmark speciality of Mediterranean landscapes, from the mountains right down to the coast.

On a small scale, wherever steep hillside terraces are abandoned, the processes of erosion and gully formation soon set in and strip the slopes of their topsoil. All around the Mediterranean, vast hilly regions occur, where formerly terraced and intensively farmed hillsides near human settlements are overgrown, with neighbouring remnant seed-bearers, such as oaks, rockroses, maples, and junipers. In other sites, formerly terraced hills are now entirely denuded of woody vegetation despite total abandonment since many decades. On a hillside near the former settlement of La Lauze (Hérault, southern France), for example, a Visigoth community built an impressive series of terraces in the fifth and sixth centuries AD. This site was definitively abandoned sometime in the twelfth or thirteenth centuries, and farming was never resumed. By the fifteenth century already, these hills were known in land tenure documents as *Monts-Chauves* or the 'Bald-hills'. Since that time, woody vegetation has still not recovered. This is partly testimony to the ill-managed use of fire and grazing since the 1600s, but ultimately the desertification and ecosystem degradation was caused by the construction and, above all, the subsequent abandonment of the terraces.

10.5 Traditional landscape designs

For centuries or millennia, traditional Mediterranean lifestyles were based on self sufficiency at both family and village levels. This was largely imposed by the isolation of the many small river basins and other habitable sectors offered by a mountainous terrain, in epochs when transportation systems were limited. Over time, a large number of local land use systems, more or less clearly delineated in space, arose in the various Mediterranean hinterlands. They varied greatly, according to regions and ethnic groups and to the relative importance of grazing, water management, and the use of fire, among other things. Among these traditional designs, the two best known are the '*sylva–saltus–ager*' and the '*dehesa–montado*' systems. They differ in the spatial organization of the three main activities: cultivating, grazing, and forest products harvesting. Traditionally, these activities occurred in separate areas in the first system, whereas they were—and still are—all combined in a single area in the second. Human activities and interventions integral to both these land use systems have existed for so long that it is often difficult to distinguish the non-human from the human components, at population, community, and, indeed, ecosystem levels (Foster 2002). For example, Dupouey et al. (2002) have shown that a 'memory' of Roman activity still persists in forests in the form of soil characters such as concentrations of nitrogen and phosphorus in combination with the persistence of certain plants (e.g. *Vinca minor*, *Ribes uva-crispa*) which are associated with ancient human settlements. This memory of Roman activity has been discovered in forests of central France but they most certainly exist in Mediterranean forests as well (J.L. Dupouey, personal communication).

10.5.1 *Sylva–saltus–ager*

Perhaps the most influential of all ancient land use systems in the Mediterranean area was the triad called *sylva–saltus–ager* (woodland-pasture-field), sometimes also called *ager–saltus–sylva*. This system appears to be the first scientific; that is, deliberately methodical system of land use in western civilization (Houston 1964:107). The triad of land uses was first explicitly formulated, so far as we know, by Theophrastus, the father of botany, in 313 BC (see Box 8.1). It sought to optimize microclimatic and edaphic variations at the scale of a single farm holding. As the possibilities of overland transportation were limited, most farmers and villagers were more or less self-sufficient; that is, their needs were all primarily fulfilled by their own growing, hunting, and animal breeding practices. The plot (or field) was the smallest section of rural space used by farmers, and optimum use had to be found for each. Over the centuries, the individual field also became the basis for cadastral registration and taxation (Lepart and Debussche 1992).

For each of the different crops in use, specific landscape units were identified as being the best suited. Therefore, plot limits often coincided with geomorphologic limits. Rocky areas were often covered with set-aside woodlands used for wood, charcoal, edible mushrooms, etc., while stony plateaux were used primarily for extensive grazing of livestock. On thinner soils, at the foot of the hills, olive groves were recommended with vineyards to be planted either below on stony ground, where olives fare less well, or else at higher elevations, where grapes resist better than other crops to frequent frost in winter. Cereal crops, by contrast, were cultivated on the deepest, most fertile soils, in the plains or in inter-montane valleys, as well as a variety of fruit trees, textile plants (hemp, flax), dye plants (madder and indigo), and vegetables. The most energy-demanding crops, the ones requiring irrigation, fertilizers, and a large amount of manpower were grown closest to the villages. Over the years, land management led to the installation of intricate networks of linear elements, such as fence walls, sustaining walls, hedges, ditches, embankments, irrigation and drainage canals, and paths. Many Mediterranean landscapes, especially in Italy, are so indelibly marked and shaped by these artificial constructs that can remain for so many centuries that the Roman cadastre is still visible today from aerial photographs (Chouquer et al. 1987). Clearly, one goal of this management approach was to avoid the 'ups and downs' with *ager* or *saltus* expanding out of control at the expense of *sylva*. It also helped avoid or reduce the inevitable conflicts between nomadic or semi-nomadic shepherds and farmers or foresters. To wit, traditional common grazing lands permitted extensive livestock raising and the biannual transhumance of large herds of sheep and goats, in conjunction with the ancient rotation of cereal cropping and fallows, which dates back at least to the days of Homer. Domestic animals were of course regularly brought into fallow fields to clean off crop residues and fertilize the land. And alterations in patterns of the triad of land uses occurred in response to local needs and wider markets, just as they do today. In some regions, however, huge areas of forest or grazing lands belonged to a king, a lord, or the monks of an abbey, who allowed villagers to use them for grazing or collecting timber or firewood under carefully controlled conditions (Lepart and Debussche 1992). Traces of those large holdings can still be seen today in some areas, in rural architecture and even in the size and shape of the trees.

10.5.2 *Dehesas* and *montados*

In some parts of the basin, the *sylva–saltus–ager* triad was and still is replaced by a different landscape design, known today as agroforestry or agro-silvopastoralism. Here the three main rural activities of wood-gathering, livestock husbandry, and agriculture are practised altogether in a single space. Under this scheme, livestock grazes acorns or chestnuts and grass under open forest or woodland cover, while annual or perennial crops are sown between planted or protected fruit or forage trees, where they take advantage of the shade provided in summer. Many other possibilities for mutual benefits among soil, crops, and animals are also generated by agro-silvopastoral systems that, in effect, mimic natural ecosystems such as Mediterranean woodlands far better than the *sylva–saltus–ager* approach or design, which segregates each economic pursuit into a separate landscape unit. Mixed, agro-silvopastoral systems appear to do a better job at maintaining ecological equilibria or, at least, resilience, thanks to a better balance and diversity of herbaceous and woody plant layers and a greater diversity and integration of resource uses in time and space at both ecosystem and landscape levels.

Some examples of ancient agroforestry in the Mediterranean Basin are found in Sicily and the Calabrian Mountains of southern Italy, central Corsica, and the Cévennes Mountains of southern France, among others. In all these regions, chestnut and mulberry tree-based systems were maintained over for several centuries with a range of small livestock, subsistence crops, and that remarkable cash crop, the oriental silkworm (*Bombyx mori*), all providing income and sustenance for local populations. Either coppiced for wood production or left to grow tall and produce annual crops of carbohydrate-rich food and fodder, the chestnut tree formed the 'backbone' of these mountain systems until a rust disease decimated most populations in southern Europe. More spectacular still are the *dehesa* and *montado* systems, which cover more than 6 million ha in the plains of the southern Iberian Peninsula (Joffre *et al.* 1988; Bugalho *et al.* 2009). These and related systems in Sardinia, Greece, Algeria, Tunisia, Morocco, Turkey, and elsewhere combine extensive grazing of natural pastures and intermittent cereal cultivation (oats, barley, and wheat) in park-like woodlands consisting of cork oak, holm oak, and smaller numbers of deciduous oaks (*Quercus faginea* and *Quercus pyrenaica*). These savanna-like formations result both from selection and protection of superior, well-shaped trees occurring among natural stands and, in some regions, from the intentional planting of acorns chosen from selected trees. Tree density is maintained between 20 and 40 per hectare, and mature trees are pruned regularly to remove infested branches, broaden their canopy cover, and increase acorn production. These cultural, anthropogenic woodlands provide goods and services that include pasture and browse for livestock, cereal cropping, firewood, charcoal, fruits, oils, berries, mushrooms, and, especially in the last two centuries, cork. Wild animals of many kinds provided game for food and recreational hunting.

According to several sources (Bugalho *et al.* 2009 and references therein), the word *dehesa* originates in the late Latin word *deffensa*, which, in the context of the medieval transhumance systems of Spain, especially Castilla, referred to an enclosed pasture protected from grazing by migratory sheep flocks and also maintained for feeding and resting of labouring cattle. Some authors suggest that *dehesa* may come from the Arabic word *dehsa*, designating a landscape that is dominated by neither the dark green colour of a dense forest nor the brownish colour of a desert. However, the most accepted origin of the term *dehesa* seems to be that of 'private land', independent of vegetation type. Similarly, the equivalent Portuguese term, *montado*, derives from the name of a medieval tax that was paid per head of livestock for the use of a particular area for grazing in different regions of the Iberian Peninsula. In North Africa, the Berber word *azaghar* refers to a range of similar systems.

A well-managed cork oak *dehesa* with 40–50 mature trees ha^{-1} produces on average 600–1000 kg of cork every 9 years (Joffre *et al.* 1988). In addition, cereal crops are produced, and a variety of grazing animals are raised, including pigs, sheep, goats, cattle, and bulls (see Plate 9b). A special breed of semi-feral 'Iberic' pigs has been in use since the end of the Middle Ages, which lives primarily on acorns between October and February and achieves the remarkably high weight increase of 1 kg for 9 kg of acorns eaten.

Joffre *et al.* (1988) report that herbaceous plant assemblages in Andalusian *dehesas* differ depending on whether they are under tree canopy or not. Under tree cover, there is greater moisture and more organic elements in the soil from leaf shedding and animal excretion. As a result, there is twice as much potassium, phosphorus, nitrogen, and carbon as in soils in the same field not under tree canopy. Consequently, a mosaic-like structure of herbaceous plant assemblages is promoted, which is characterized by increased botanical diversity, longer growing seasons, and greater pasture productivity. This enhances animal productivity, which in turn probably favours long-term growth and vitality for the trees. Hence, a mosaic-like structure of plant assemblages.

Dehesas and *montados* are diverse, heterogeneous, and well adapted to the unpredictability of Mediterranean climate. Never highly productive—except for the cash crop related to cork—they provide land owners with greater flexibility and options than a system with a fixed land use for each landscape unit. However, today, they are increasingly threatened by many difficult-to-control driving factors, both regional and global (Ovando *et al.* 2009; Aronson *et al.* 2009).

Thus these iconic systems and landscapes are being increasingly altered or abandoned today, however, because of many factors. A decrease in agro-silvopastoral activities due to rural depopulation is the main problem in some areas. But increasing deforestation and semi-industrial clearing in order to extend mechanized cropping lands are also creating conflicts, largely driven by the impetus of global and European economic pressures. In many regions, *dehesas* and *montados* are being replaced by large stands of eucalyptus (*Eucalyptus*).

Yet *dehesa*-like systems, artificially maintained in an open woodland state of ecosystem development, are increasingly being recognized as well-adapted and economically viable multiple-use agro-ecosystems for promoting sustainable development while also maintaining biodiversity in many rural areas of the Mediterranean Basin (Berrahmouni *et al.* 2009).

Recently, however, Urbieta *et al.* (2008) showed that efforts of forest management in the Iberian Peninsula favoured the cork oak during the twentieth century in several parts of southern Spain, mostly at the expense of the introduced Canary pine. As a result of forest management, the realized niche of this oak has been considerably enlarged, providing further evidence of humans as major drivers of forest composition across the Mediterranean and shedding some light on the restoration of *dehesa*-like systems. The extant monospecific stands of cork oak have been shown to be a secondary forest structure because palynological data suggest that, in the absence of human influence, cork oak would mostly develop into mixed woodlands forests, sharing the arboreal stratum with other sclerophyllous and deciduous tree species (Carrión *et al.* 2000).

A final example of an ancient agroforestry system well adapted to the hottest Mediterranean life zone is the argan 'forests' of the Sous region in south-western Morocco (Benchekroun and Buttoud 1989). Until the mid-twentieth century, over 1 million ha in this dry corner of the basin were occupied by a centuries or millennia-old agroforestry system, based on open parklands featuring the endemic argan tree, as well as local varieties of olives, almond trees, and an endemic acacia tree whose large pods were used as animal fodder. What allowed the system to survive is the argan tree, whose hard seeds yield a culinary oil prized locally even more highly than olive oil. In this semi-arid region where no permanent water is available for livestock or irrigation, this *dehesa*-like system made life possible for local populations and their herds. In favourable years, when rainfall was adequate, the ground between trees was lightly ploughed and short-season cereal crops were sown. In wet and dry years alike, the argan and olive trees produced edible oils of high commercial value. Foliage of the argan trees and the acacias provided supplementary fodder to goats and other small livestock, which were present nearly year-round. Once again, a system 'mimicking' nature was created, incorporating a multi-tiered vegetation structure and firm integration of plants and animals, albeit in lower densities than in the Iberian systems. From over 1 million ha, only about 600 000 ha are left today. Like the *dehesas* and *montados* of Andalusia and Portugal, the argan agro-forests are now in danger of collapse. This unique and ecologically well-adapted polycropping system built on endemic plants is gradually being replaced by intensive, irrigated farming methods.

10.5.3 Cultural landscape mosaics

Traditional land-use systems, such as the *sylva–saltus–ager* and *dehesa–montado* which represent the most typical cultural landscapes of the Mediterranean, have been in use for many centuries without resulting in significant depletion in the production of resources. They are certainly a demonstration that sustainability may be achieved, provided that management techniques do not result in dramatic changes in major ecosystem functions. Mediterranean habitat mosaics are increasingly recognized as well-adapted and economically viable multiple-use agro-ecosystems for promoting sustainable modern development. In contrast to the 'Ruined Landscape Paradigm' sometimes advocated for describing the consequences of human action (see Chapter 11), the long-term management of Mediterranean landscapes up to the beginning of the twentieth century did not result in calamitous decreases in biodiversity. Overall, it appears that human activities have in fact been beneficial for many components of biological diversity in the basin. For example, Pons and Quézel (1985) and Seligman and Perevolotsky (1994) have argued that the highest species diversities in the Mediterranean Basin are found in areas that have experienced frequent but *moderate* disturbances. We will come back to this issue in the last chapter of this book. However, the deep socio-economical changes that characterize modern times have already had and will have profound consequences on Mediterranean cultural landscapes. The current trends of habitats and landscape changes vary markedly across regions of the basin (Mazzoleni *et al.* 2004). On the northern shores of the basin, the near-collapse of traditional land-use systems and rural depopulation destructured traditional cultural landscapes, as they were replaced by modern intensive agriculture. In contrast, in North Africa and most parts of the eastern Mediterranean, pressures on natural habitats by human populations and livestock are still strongly on the increase, degrading soils and ecosystems, especially through soil erosion.

Summary

This chapter is an attempt to give a brief overview of the human contribution in the shaping of Mediterranean ecosystems and landscapes. From the first appearance of hominids in the basin, 790 000 years ago, to the present day, an impressive succession of proto-human and then human societies contributed to an endless design and re-design of communities, ecosystems, and landscapes at all spatial scales. Much waxing and waning of civilizations and many wars and invasions also contributed to the fact that most habitats and landscapes were highly dynamic with many ups and downs. Some examples are given with regional-specific histories of human impact on living systems.

The history of Mediterranean peoples has been a long epic to domesticate plants and animals, a prerequisite for the establishment of permanent sedentary societies. In addition to landscape management, domestication of animals and plants, which began more than 10 000 years ago, significantly added to diversity resulting in the context of a so-called segetal era. The Mediterranean is one of the most important regions in the world as a centre of domestication of both plants and animals. The remarkable combination of plant domestication and selection of cultivars along with the domestication of animals have facilitated the rapid spread of herding and farming economies westwards throughout the rest of the Mediterranean Basin. Once the breeding of plants and animals was achieved, the next step was to establish a permanent and comfortable living space.

Human activities repeatedly resulted in the decline and sometimes the recovery of the most important types of habitats such as forests, providing a near-infinity of successional dynamics. Mediterranean forests were among the first living spaces or 'oecumenes', of humans in the region, providing them with a wide range of products that they used for their primary needs, such as food, shelter, or firewood.

Humans had to control fire, water supply, gully erosion in the fragile and heterogeneous areas that characterize the Mediterranean, as well as living spaces for livestock. This progressively resulted in the establishment and management of sophisticated and highly resistant land-use systems, such as the *sylva–saltus–ager* and *dehesa–montado* systems.

The long-lasting management of Mediterranean living spaces did not necessarily result in calamitous decrease in biodiversity, the exceptions being a few animal and plant species that directly compete with humans. Human activities have in fact often been beneficial for many components of biological diversity.

CHAPTER 11

Biodiversity Downs and Ups

In view of the mass extinction of plants and animals underway everywhere in the world, one would expect large declines in biodiversity in a region like the Mediterranean that has for so long been managed, modified, and, in places, degraded by humans. Surprisingly, the current rate of decline is not as high as in many other regions, perhaps due to the fact that human-induced decline of biological diversity started many thousands of years ago, as shown by many palaeontological and archaeological records. Although there is no doubt that large-scale destruction has taken place and that much of the shrublands in the Mediterranean are modified or derived—not to say degraded—forms of former forests and woodlands, it is indisputable that the exceptional diversity and dynamic structure of Mediterranean ecosystems and communities result in part from human influence (see Chapter 10). Intertwined processes of human and non-human factors have affected Mediterranean ecosystems and their biodiversity, yielding systems not only endowed with stunning biodiversity, but also with exceptional resilience and resistance to disturbance.

Two contrasting theories or paradigms exist concerning the relationships between humans and ecosystems in the Mediterranean Basin (Blondel 2006). The Ruined Landscape or Lost Eden theory, first advocated by historians in the sixteenth and seventeenth centuries and later by a large number of ecologists, foresters, and land managers, assumes than human-caused deforestation and overgrazing resulted in the cumulative degradation and desertification of many or most Mediterranean landscapes, including a multitude of formerly magnificent forests. For example, Marsh (1874), Naveh and Dan (1973), Thirgood (1981), Quézel (1985), Attenborough (1987), and McNeil (1992) all assert that unsustainable use and resource depletion best describe the interactions between humans and Mediterranean ecosystems over the millennia and, especially, in recent centuries. A second school of thought challenges this view of wholesale detrimental effects of humans. For example, Grove and Rackham (2001) argue that it is not so simple; an imaginary past, they say, has been idealized by nostalgic artists and scientists, who do not fully appreciate or acknowledge human contributions to the maintenance, diversity, and even embellishment of Mediterranean landscapes since the last glacial period. Grove and Rackham stress that savanna-like or woodland landscapes, as well as largely treeless steppes, occur naturally and are fairly characteristic of the Mediterranean Basin. They also note that a mosaic of woodlands, shrublands, and grasslands can frequently be found at the landscape scale, resulting from non-human, as well as human determinants. They emphasize that many Mediterranean shrubs and trees grow right back when cut (see Chapter 8), and that in many parts of the region, especially in southern Europe, woodlands and forests are now growing back spontaneously, thanks to inherent resilience, in a context of agricultural abandonment and rural exodus; this is a subject we will return to in Chapter 13.

In this chapter we will briefly review the history of human impacts on biodiversity and ecosystem functioning in the Mediterranean region, starting with the bad side of the coin. Then, we will show that, indeed, human impact is not always detrimental, especially when far-sighted resource and land management practices are employed. In a last section, we discuss the beneficial and detrimental effects of fire, which is one of the most important abiotic ecological factors in this region. Thus, in a sense, this chapter will give support to both of the above-cited theories on the role of humans on biodiversity. As always, it is important to specify the specific scales of space and time under consideration in order to tease apart and evaluate the drivers and factors contributing to current status of biodiversity and ecosystem dynamics.

11.1 Losses

We begin with a brief overview of the most salient aspects of recent biodiversity loss at the levels of habitats, ecosystems, and landscapes and, then, discuss the mass extinctions that occurred in earlier, prehistoric times. Thereafter we will review the losses for some of the major groups of organisms.

11.1.1 Loss and degradation of forests, wetlands, and coastal areas

In his remarkable and influential book, *The Earth as Modified by Human Action*, George Perkins Marsh (1874) decried human mismanagement and failure to replenish resources through restoration. Taking the Mediterranean region as one of his primary examples, he deplored the destruction of primeval forests, which had, he said, contributed to violent floods, malaria, and a drier climate. Marsh attributed the perceived sterility and destruction of Mediterranean landscapes to 'tyranny, ecclesiastical misrule, and slovenly land use by papists', among other things (Butzer 2005). However, other well-travelled and knowledgeable authors, like Huntington (1911), attributed forest regression and abandonment of farmland in the region to a progressive decline of rainfall, an argument which was later inverted to claim that Mediterranean soils were destroyed by heavy rains related to climatic anomalies! In fact, an archaeological/historical overview supports a model of non-linear change, namely a cyclic alternation of agricultural intensification and extensification (Butzer 2005).

As seen in the previous chapter, Mediterranean forests have always been a major resource for people. They have been so long and heavily exploited for many different uses (see Fig. 10.4) that what remains today, from a forester's viewpoint, can only be seen as a 'heritage of forest depletion and degradation' (Thirgood 1981). A recent study by the World Wildlife Fund (WWF 2001) reports that no more than 5–10% of native, post-glacial Mediterranean forest remains. Besides outright deforestation of huge areas since the Neolithic, additional causes of forest degradation and transformation have been the steady or punctual overexploitation of selected species or areas, for both timber and non-timber products, including not only wood fibre and charcoal, but also tannin, tree saps, and grazing resources for domestic livestock. A particularly dramatic situation is that of the forests and woodlands of North Africa, formerly dominated by a mixture of many species of conifers and oaks and that are now reduced to much less than half of their former extent, complexity, and biodiversity. The once extensive and beautiful stands of cedar, fir, red juniper, Barbary thuja, and deciduous oaks of the Atlas and Rif Mountains, North Africa, are now almost completely gone (M'Hirit and Blerot 1999; Quézel and Médail 2003). In other regions, especially in Morocco, scattered stands of trees do remain, since they are protected by law from cutting, but there is little or no undergrowth and no tree regeneration at all, as a result of over-grazing and harvesting of acorns, both of which are allowed. In parts of south-western Europe, by contrast, various kinds of holm oak and cork oak forests and woodlands are regenerating (e.g. Urbieta *et al.* 2008), but elsewhere formerly prosperous and diverse forests of cork oak are now much reduced and suffering from pests, climate change, and lack of natural regeneration (Aronson *et al.* 2009).

In south-eastern European countries and most countries of the southern banks of the sea, there is still a continuing trend of forest decline. We will see in the next section, however, that in some circumstances, in recent decades, there has been a trend for forest recovery, especially in countries of the northern shores, such as France and Italy, where the annual rate of forest recovery is about 1.7–2% (Quézel and Médail 2003). To summarize, a common feature of Mediterranean forests is their fragility, instability, and degradation (M'Hirit 1999; Merlo and Croitoru 2005), but with a pronounced capacity for regeneration over time in many or most cases.

The second group of ecosystems at high risk is wetlands. Because wetlands are very productive ecosystems, they are also much coveted and often transformed by humans for cropping, grazing, fishing, hunting recreation, and other activities. In addition, wetlands have for long been considered as insalubrious areas generating various diseases including malaria. The draining of wetlands goes back several centuries when many freshwater lakes and coastal wetlands progressively disappeared. Marshes around Rome have been drained as early as the fifth century AD (Pearce and Crivelli 1994). In Turkey, most of the 250 bodies of water of central Anatolia have disappeared in the 1950s and1960s, in order to eradicate malaria and to expand agricultural lands. Between 1984 and 2002, 90% of the Anatolian marshes of the Konya plain were drained, including the 10 000 ha of the Yama marshes which were completely dried up within 10 years (Gramond 2002). Many now-drained Mediterranean wetlands, which served as refugia during glacial times for Euro-Siberian and boreal species, have been dried up around the basin, dooming many local populations of relict flora and fauna to extinction (e.g. Blondel and Médail 2009). In Macedonia alone, 1151 km^2 of wetlands out of a total of 1572 km^2 have been drained since 1930 (Catsadorakis 2003). Due to the expansion of the city of Thessaloniki, the Axios River delta in Greece is progressively shrinking, despite the fact that its mussel beds and ricefields provide 90 and 70% of total Greek production, respectively. Even in well-preserved areas such as the Camargue, southern France, similar trends are underway since the nineteenth century. During the second half of the twentieth century, a net loss of approximately 40 000 ha has taken place (Tamisier and Grillas 1994). The driving forces for wetland reduction now are high consumption of water for both agriculture and tourism, which induces chronic over-exploitation of groundwater in wetlands all around the Mediterranean. Water consumption peaks during the summer, precisely when resources are at their lowest, with each Mediterranean tourist using much more water per day than permanent residents (De Stefano 2004).

The third category of ecosystems of great concern is those found along the coasts. Rapid changes in land-use practices in the twentieth century have had disastrous consequences for coastal ecosystems, where more than 60% of people in the Mediterranean region currently concentrate and where 75% are expected to live by 2050 (see Chapter 12). This is particularly the case in islands where tourist encroachment is a real threat for many habitats and their plant and animal communities.

11.1.2 Mass extinction: from the late Pleistocene to modern times

There is much evidence that biodiversity decline at the hands of humans in the Mediterranean actually started several millennia ago. The first significant impact of humans, well before the Neolithic revolution and the establishment of permanent settlements, was probably their role in the extinction of a large number of large mammals at the end of glacial times. In the Northern Hemisphere, including the Mediterranean Basin, this epoch was characterized by an extraordinarily number of large mammals, including herbivores, as well as predatory species. Most of them are now extinct, but a testimony of this magnificent late Pleistocene mammal fauna is provided by large quantities of fossils in various sites in southern Europe, as well as by the superb wall paintings in ornate caves of southern France and Spain. These include bison, horses, bears, deer, mammoths, hyenas, panthers, lions, rhinos, reindeer, aurochs, and ibex (see Chapter 2). A debate still exists over whether this mass extinction was caused by humans, as suggested by the so-called Blitzkrieg overkill hypothesis (Martin 1984), or rather resulted from environmental and climatic changes, which led to the transformation of grasslands to forests. Recent studies emphasize the importance of human impact (e.g. Brook and Bowman 2004), but it is probable that rapid climate changes in the late Quaternary also contributed to the huge faunal collapse, in combination with human impacts, including indirect factors, such as disease, biological invasions, and habitat alteration (Burney and Flannery 2005).

Box 11.1. The story of dwarf elephants

The human-induced extinction of the so-called mega-nano-mammals of the Mediterranean islands, especially the dwarf hippos, elephants, and deer of Cyprus, Malta, and Sicily, following the colonization of these islands by humans, is a sad story indeed (Simmons 1988, 1991; Diamond 2000). No fewer than 12 species of dwarf descendants of the ancestral, full-sized elephant *Palaeoloxodon antiquus*, which was common on the European mainland in the Pleistocene, inhabited Mediterranean islands until they were killed off by early hunters (Lister 1996). These species were variable in size and all evolved independently from *P. antiquus*. Even the tiny Aegean island of Tilos (64 km^2) had its own species. The smallest of them all, the Sicilian *Palaeoloxodon falconeri*, was less than 1 m high, and may have given rise to the myth of the Cyclops, immortalized in Homer's *Odyssey* (Lister 1996), because of the large frontal hollow of the nostrils which must have looked like an enormous frontal single eye.

For several centuries, the bones of the extinct dwarf mammals of Cyprus were revered as the remains of saints and early Christian martyrs, with pilgrimages being organized to the caves where they rested. It was not until the early 1900s that they were correctly identified as pygmy mammals, including a pig-sized hippopotamus and a pony-sized elephant. Geological observations suggest that the colonization of Cyprus and other islands occurred sometimes in the late Pleistocene, between 100 000 and 250 000 years BP. Many adaptations, especially the diminutive stature of these mammals and the loss of foot pads, gave them greater mobility in the mountainous environment of their island homes and allowed them to reach much larger population sizes that were less prone to natural extinction.

The decline of large mammals continued without interruption during the Holocene. The last European wild horses (*Equus caballus*) were slaughtered in the middle of the nineteenth century in the Ukraine. Aurochs somehow survived in the vast forests of eastern Europe until the beginning of the seventeenth century, but the last survivor died in 1627 in Poland. The Mediterranean Basin was unique in harbouring populations of two species of modern elephants, the Asian (*Elephas maximus*) and the African (*Loxodonta africana*). The last herd of Asian elephants that survived in Syria was killed by an Assyrian king about 2800 years ago, and the African species, which had been semi-domesticated by the Romans, disappeared much more recently, having survived in southern Morocco until the eleventh century AD (Blondel and Aronson 1999).

In North Africa, moreover, many species of the rich mammal fauna which occurred in this region survived in the Holocene well into the Neolithic period, but became progressively extinct, mostly as a result of human influence (Kowalski and Rzebik-Kowalska 1991). For example, from the six species of gazelle that occurred during the middle–late Pleistocene in North Africa, only two (*Gazella dorcas* and *Gazella cuvieri*) survive today. We will come back later in this chapter to the decline of mammals in recent times.

During the Quaternary, most of the larger Mediterranean islands were populated with what now seem odd assemblages of animals, including tortoises, giant rodents, flightless owls, dwarf deer, hippos, and elephants (Box 11.1). Many arose from archaic Tertiary lineages and evolved striking adaptations, such as gigantism or dwarfism, which are well-known adaptations for living on islands (see Chapter 7). These animals most probably reached the islands by swimming during periods when average sea level was much lower than today so that the distance of open water to be managed was much shorter. Palaeontological and archaeozoological data from the Neolithic (7000 years BP) show that these endemic and disharmonic upper Pleistocene faunas of Mediterranean islands were quickly decimated once humans arrived, as in many other parts in the world. A large series of archaeological sites, such as the famous Eagle Cliff in Cyprus, shows that human settlements occurred well before 10 000 years BP (Simmons 1988); that is, soon after the end of the last glaciation, when human societies were based entirely on hunting and gathering. In this site, which is one of the oldest on any Mediterranean island and one of the earliest providing evidence for human seafaring in the Mediterranean Basin (Reese 1990; Simmons 1991), the remains of over 200 different animals have been discovered, including dwarf hippos and elephants (Box 11.1). Many bones were broken and most bore marks of cooking, which indicates that these animals were hunted for food. This deposit also yielded a large number of manufactured tools and necklaces, as well as more than 20 000 shells of edible sea molluscs, suggesting that they were deliberately transported by people and discarded after eating. The site also included many bones of large birds, such as great bustards, geese, doves, and ducks. Dating of this material by radiocarbon method gives an approximate age of 10 465 years BP, which provides the oldest archaeological evidence that these animals did not die off before humans arrived, but rather were hunted to extinction by the first humans to arrive, just as happened in Madagascar, New Zealand, and other isolated islands and archipelagos in more recent times. All the highly endemic and disharmonic upper Pleistocene mammal faunas of Mediterranean islands were decimated by humans and replaced by the extant species assemblages. After the mass extinction of this archaic endemic mammal fauna that started in the late Pleistocene, only a few endemic mammal species were left. They include three species of shrews, one in Sicily, *Crocidura sicula*, one in Crete, *Crocidura zimmermanni*, and presumably one in Cyprus, *Crocidura cypria*, as well as some endemic rodents, such as the recently discovered *Mus cypriacus* on Cyprus (Bonhomme *et al.* 2004). These small mammals presumably differentiated recently and have been discovered thanks to genetic studies. In any case these are not 'old endemics' like most reptiles and amphibians. These endemic species apart, a nearly complete turnover has occurred in the mammalian island fauna since the Pleistocene.

The story of mammals on Mediterranean islands is a telling demonstration that biodiversity decline linked to human activities started many millennia ago. Corsica and Sardinia taken together provide an excellent case in point (Vigne 1990; Blondel and Vigne 1993; Blondel and Aronson 1999). These two islands share much the same history because they were connected during most of the Pleistocene, and even today they are only 11 km apart and the channel separating them is no more than 60 m deep. The Cyrno-Sardinian complex, as it is called, is a continental microplate that parted ways from the European plate in the late Oligocene (35–30 mya) (see Chapter 1 and Thompson 2005). It is the largest island complex of the Mediterranean area, extending over more than 32 000 km^2, and is also the nearest to the mainland, which perhaps explains some of the faunal specificities, such as the absence of the unusual 'elephant-deer' assemblages (Sondaar 1977) that characterized other large islands, such as Cyprus. By the early Holocene, just before humans arrived, the insular mammal fauna included only six endemic species of mammals. As soon as humans reached Corsica and Sardinia, however, between 7000 and 9000 years ago, hunting pressure rapidly pushed several species, such as the *Megaloceros* deer and probably a carnivorous dog, *Cynotherium*, to extinction. Some mammals succeeded in resisting hunting and competition with new invaders, introduced by humans over several millennia, but they finally went extinct as well. Others, especially small rodents, persisted until Roman times and then disappeared. The rabbit-like *Prolagus* managed to survive much later still in the predator-free islet of Tavolara off the coast of Sardinia, but also ultimately disappeared.

Late Pleistocene/Holocene extinction events have been less severe in bird faunas although in the Pleistocene (350 000–150 000 years BP), the bird fauna of Corsica included several species that are now extinct, such as a giant barn owl (*Tyto* cf. *alba*) and a dwarf great owl (*Bubo insularis*), both of which persisted into the Holocene. Almost every large Mediterranean island had its own endemic species of little owl: *Athene angelis* in Corsica, *Athene cretensis* in Crete, *Athene* cf. *noctua* in the Balearic Islands, and one still undetermined species in Sicily. They persisted till the upper Pleistocene but finally went extinct as well.

11.1.3 The decline of diversity in recent times

At the global scale, the best sources of information on the conservation status of plants and animals are the IUCN Red List of Threatened Species (IUCN 2008). Except for the large mammals that have paid a heavy price for human hunting in the last few centuries, few extinction events have occurred recently for most groups of Mediterranean plants and animals, but this does not mean that decline in biological diversity did not occur. Definitive loss of a species is one thing; impending declines in distribution, population sizes, genetic diversity, and threat of extinction is another for which we have far less reliable information. Fragmentation of ranges of distribution and declines in population numbers are real threats for many Mediterranean species today. The conditions are very different in marine environments, as we will see below.

11.1.3.1 EXTINCTIONS AMONG PLANTS

Despite a large number of endemics that are narrowly distributed in a single or few localities, no more than 40 Mediterranean plants are considered to have gone extinct in recent times, which is a mere 0.15% of the recognized taxa in the region (Greuter 1994; Blondel and Médail 2009). Such a level of extinction is perhaps not significantly higher than during geological times. The most numerous reported extinctions are in Turkey (10), Greece (six), and Italy, where six taxa are also reported to be extinct (F. Médail, personal communication). Threats faced by Mediterranean plants are manifold. They include destruction of habitat, notable as a result of urbanization, pollution, climate change, biological invasions, selective harvesting, and overgrazing (Médail 2008a; see Chapter 12 for detailed discussion). Despite these diverse threats and given the huge number of narrowly distributed endemic plant species, it is surprising that so few of them have gone extinct in the Mediterranean region. Perhaps the strong local persistence and resilience of Mediterranean plant species is mainly due to their life strategies, with a high tolerance to stress and disturbance, including human-induced disturbance. However, the number of known extinction events among plants in some countries of North Africa and the Middle East must be disproportionately low, and many species may actually be in the verge of extinction because of a dramatic decline in population sizes due to habitat destruction and degradation. Leon *et al.* (1985) estimated that nearly 25% of the Mediterranean flora may be threatened in the decades to come. Mathez *et al.* (1985) reported that an astonishing—and probably overestimated—50% of plant species known to occur in Algeria have not been seen for at least 20 years. The conservation of certain groups of species, notably bulbous plants, is another concern in many locations around the Mediterranean. Intensive harvesting of geophytes is a serious threat for many endemic localized species (see also Box 11.2 and Chapter 13). For example, about 57 million pseudobulbs of 38 terrestrial orchid species are dug up each year in Turkey alone (Sezik 1989), where they are used to prepare a milk-based drink called 'salep', popular throughout the eastern Mediterranean.

Fortunately, some integrated conservation programmes including *in situ* bulb propagation have reduced by half these huge removals and constitute an alternative source of bulbs for international trade (Entwistle *et al.* 2002). From two surveys carried out in 1886 and 2001 in southern France, Lavergne *et al.* (2005) tried to decipher the role of ecological and anthropogenic factors in the spatial distribution and the dynamics of rare plant species since the late nineteenth century. They found that, on average, rare species tended to occur at higher altitude in areas with semi-natural open habitats where human impact decreased over the last 30 years. During the twentieth century, rare species were the most prone to extinction in zones where human population density, cultivated areas, and livestock density had increased. In addition, rare species of Euro-Siberian distribution were more prone to extinction than rare Mediterranean species. Restricted endemic species mostly occurred on steep slopes or areas with very low human population density. Urbanization and modern industrial-style agriculture caused most local extinctions of rare species during the last century. Surprisingly, a detailed comparison of life form spectra for the floras of south-eastern France and Corsica indicates that endemic and very localized plant species are

> **Box 11.2. Cork oak woodlands: a sanctuary for Mediterranean plants and animals**
>
> Land-use management is a crucial issue for both plants and animals, and cork oak woodlands of the western Mediterranean are a prime example (Berrahmouni *et al.* 2009; Plate 9b). Fully 30% of the herbaceous plant species of the Iberian Peninsula are found primarily or exclusively in managed cork oak woodlands (Pineda and Montalvo 1995). Throughout the Iberian Peninsula, for example, multi-purpose, multi-user cork oak landscapes are thought to provide habitat for the few remaining populations of the critically endangered Iberian lynx (*Lynx pardinus*). Unfortunately, some thinks that it is extinct in open oak woodlands (J. S. Carrion, personal communication). Many migratory birds, such as black stork (*Ciconia nigra*) and 60 000–70 000 common cranes (*Grus grus*), use these woodlands as a stop-over site or overwinter there. In addition, there are many large avian residents as well, including Iberian imperial eagle (*Aquila adalberti*) (see Box 3.3). In North Africa, the highly endangered Barbary deer (*Cervus elaphus barbarus*), which is the last surviving deer in Africa, is currently found only in the cork oak woodlands on either side of the border between Tunisia and Algeria. Cork oak woodlands also harbour nearly a hundred other animal species listed in the annexes of the EU Habitats and Birds Directives, including species that are only rarely found elsewhere. In the Maamora cork oak woodland of Morocco, for example, there are at least 160 bird species (Thevenot *et al.* 2003). Cork oak woodlands also support more diverse communities of butterflies and passerines than nearby dense woodlands, grasslands, or arable areas (Díaz *et al.* 1997).
>
> A menace hovers over these sanctuaries, however, as the economic outlook for private maintenance of most cork oak woodlands today is not very bright. For cork oak woodlands managed as *dehesas* or *montados* (see Chapter 10), public subsidies are likely to be required in exchange for ecosystem services provided to society as a whole (Vallejo *et al.* 2009).

not more prone to extinction than other species (Verlaque *et al.* 2001).

In fact, it is in the 12 000 islands and islets where the situation of endangered Mediterranean plants is of greatest concern (Montmollin and Strahm 2005; Médail 2008b), largely because of transformation and destruction of coastal habitats as a result of the ongoing housing and tourism boom. Problems of waste treatment and excessive demand on freshwater supplies affect inland areas, as well as the coasts (see Chapter 12). In coastal areas particularly, plant species are often threatened by aggressive invasive species, as we will discuss in Chapter 12.

On large islands, the percentage of taxa that are threatened ranges from 2% on Corsica to 11% in Crete (Delanoë *et al.* 1996). Ten of the known extinct plant species are insular endemics. Two of them, which are extinct in the wild, are *Lysimachia minoricensis* from Minorca and *Diplotaxis siettiana* from islets in the Alboran Sea, but as many as four endemic species disappeared from Sicily (*Allium permixtum*, *Anthemis abrotanifolia*, *Carduus rugulosus*, and *Limonium catanense*). Other species considered as extinct in islands of the eastern Mediterranean are *Geocaryum bornmuelleri* and *Paronychia bornmuelleri* from the island of Thasos, and *Dianthus multinervis* from the small islet of Jabuka, off the coast of Croatia (Médail 2008b). An alarming species decline is also underway in the Canary Islands and in Madeira, where 24 and 20%, respectively, of native plant species are considered to be threatened with extinction (Whittaker and Fernandez-Palacios 2007).

Considering Mediterranean trees, 61 taxa including 42 endemics subspecies and cultivars are considered to be threatened (Quézel and Médail 2003), some of which are progenitors of cultivated trees in the genera *Malus*, *Olea*, *Phoenix*, *Prunus*, and *Pyrus*. One of the most emblematic of threatened tree species in the region is the Cretan date palm (*Phoenix theophrasti*), a distinct species recognized long ago by Theophrastus (see Box 8.1). The distribution of this palm was much more extensive in the past but it persists today only in small, disjunct populations at low altitudes, in riparian habitats in Crete, Peloponnesus, and south-western Turkey (Boydak 1987). The famous Lebanon cedar has been exploited for several millennia by the Egyptians, Phoenicians, Assyrians, Romans, and Turks, especially for ship-building. Only a handful of isolated and fragmented stands of this species still remain in Lebanon, covering at most a surface of 2700 ha (Quézel and Médail 2003; Sattout *et al.* 2007; see Chapter 5). Similarly, the emblematic Atlas Mountain cedar of North Africa has also suffered dramatic range reduction and fragmentation.

In fact, however, many presumed extinctions among Mediterranean plants are uncertain because supposedly extinct species may actually survive in unnoticed localities or else as part of dormant seed banks. This is especially true in the Mediterranean area, where few species are monitored adequately in the field. In Greece, for example, no less than three species reported as being extinct have been rediscovered in the last few decades (Greuter 1994), including the Peloponnesian pheasant's-eye (*Adonis cyllenea*), which was recently rediscovered 130 years after its first discovery on Mt. Killini in 1854. Another example is the small genus *Plocama* in the tropical family Rubiaceae, whose curious story was told in Chapter 3.

In addition, baseline data allowing a predictive overview of plant extinctions in relation to patterns of rarity and threats in the Mediterranean are still lacking. Future work should be inspired by the detailed study on e.g. *Centaurea corymbosa*, a narrow endemic and cliff-dwelling Asteraceae of southern France (Colas *et al.* 2001). As metapopulation dynamics of this species are very low and depend on only six natural populations located within a very small area, the survival of this species is certainly at risk.

Another interesting example of a rare and localized Tertiary relict is the Sicilian zelkova (*Zelkova sicula*), a small deciduous tree or large shrub of the elm family (Ulmaceae), of which 200–250 individuals were recently discovered in a population occurring in a cork oak woodland (see Box 11.2) near the city of Syracuse, in south-eastern Sicily (Garfi 1997; Fineschi *et al.* 2004). Like the other Mediterranean endemic species in the genus, *Zelkova abelicea*, which is only found in Crete, *Z. sicula* is included in the IUCN Red List of Endangered Species.

11.1.3.2 INVERTEBRATES

We know very little about the distribution and abundance status of most invertebrate species in the Mediterranean Basin, because almost no long-term monitoring schemes exist and many groups are poorly investigated. However, there is a growing body of evidence of an ongoing decline in several groups (Collins and Thomas 1991; Pullin 1995). A dramatic decline during the past few decades is obvious for large conspicuous insects, such as butterflies, large bees, dragonflies, and many groups of scarabeids, cerambycids, and large moths, including the flamboyant peacock-of-the-night (*Saturnia pyri*). Along the French coastline, several large scarabeid species were mentioned since the nineteenth century, but have recently disappeared (e.g. the emblematic and sacred scarab beetle), because of human pressure and abandonment of traditional land-use practices, including the abandonment of livestock raising. Many other species, such as carabids (see Chapter 3), e.g. *Scarites*, *Eurynebria*, *Carabus clathratus arelatensis*, and *Macrothorax rugosus*, and dynastids, including *Callicnemis latreillei*, are in serious decline. Several species of large beetles, including the beautiful *Polyphylla fullo*, have also become extremely scarce in recent years. On the other hand, there does not seem to be any decline in the cicadas, of which seven species occur in southern France alone (e.g. *Tibicen plebejus*, *Cicada orni*, and *Tettigetta pygmaea*; see Box 5.1). Reduction of species diversity and population abundance of most groups of large insects translates in a parallel decline of large insectivorous birds, such as the scops owl (*Otus scops*), little owl, and all species of shrikes. However, other birds, which prey upon similar large insects, such as the roller, have maintained pretty wealthy populations. Long-term monitoring programmes are badly needed to evaluate the extent of the decline of invertebrates (see Chapter 13). One well-documented and worrisome example of declining invertebrates is that of earthworms, as described in Box 11.3.

> **Box 11.3. The decline of earthworms**
>
> Over recent decades, a steady impoverishment of earthworm communities has occurred in many parts of Europe, due to leaching and nutrient export by logging, fire, acid rains, heavy metal pollution, and conifer plantations, leading to litter and top-soil removal (Granval and Muys 1992). The combination of soil acidification and decline in the number of earthworms has resulted in a steady decrease of biological activity and compaction of soils, reduced formation of humus and natural regeneration of trees, weakening of trees due to superficial rooting, and accelerated erosion processes (Hildebrand 1987). Current levels of heavy metal soil contamination may also have important consequences for major physical and biogeochemical ecosystem processes, since they adversely affect earthworms. Given their vital ecological functions in ecosystems, especially in respect to physical properties of soils (see Chapter 8), the decline of earthworm communities contributes to increased soil erosion and the frequency of severe flooding that periodically plagues various Mediterranean regions. This is especially true in agricultural areas, such as those with large numbers of vineyards, where copper is widely used as a fungicide, or mining areas, where various heavy metals can be found in water courses and groundwater reservoirs. In this context, the regional eradication of the earthworm genus *Sclerotheca* from many areas of southern France is especially alarming (Abdul Riga and Bouché 1995). These impressively large creatures, over 25 cm long and 1–2 cm thick, are active ground feeders and play an important role in water filtration and decomposition of organic matter. They are much more sensitive to heavy metal pollution (Cu, Zn, Pb, Mn, Cd, and Fe) than any other co-occurring group of earthworms.
>
> However, successful experiments in earthworm reintroduction have lead to increased primary productivity in depauperate soils and an increase in forest productivity by more than 50% after enrichment of soils with fertilizers in the presence of anecic earthworms (Toutain *et al.* 1988). Moreover, liming and/or fertilizing degraded forest soils proved to be more efficient in the presence of **endogeic** and anecic earthworms (Granval and Muys 1992). In most degraded forest soils, earthworm reintroduction appears to be the best way to promote bioturbation and the regeneration of soil activity.

As noted already for plants, for many invertebrates and micro-organisms, most presumed extinctions are shrouded in uncertainty. For example, from an intensive, 5-year sampling of ants in a small area of 400 m², Espadaler and Lopez-Soria (1991) found 40 different species, which is an exceptionally high figure for a Mediterranean habitat. Many of them are considered by myrmecologists as 'interesting' and rare species, which means they are unusual in published faunistic accounts. Actually they are probably present in many habitats but remain unnoticed, because a much higher sampling effort than that which is achieved in most faunistic surveys would be required to detect them.

11.1.3.3 FRESHWATER FISH

Although only seven species of freshwater fish are considered as extinct in the Mediterranean area in recent times, more than 56% of the endemic species are classified as threatened by the International Union for Conservation of Nature, of which 18% are considered critically endangered, 18% endangered, and 20% vulnerable. Only 52 species (21%) are assessed as being of little concern with regards their conservation status (Crivelli 1996; Smith and Darwall 2006; see Table 3.3 for more details). The main threats to freshwater fish are: (1) eutrophication, resulting from urban, sewage, and agricultural runoff; (2) increasing water pumping for irrigation and domestic use; (3) dam construction, which limits sediment and nutrient flow downstream and affect the dispersal and migration of species, as well as fisheries productivity; (4) overfishing; (5) industrial and domestic pollution; and (6) introduction of invasive species, especially predators (see Chapter 12). For example, in the Mikri Prespa and the Megali Prespa lakes, at the border between Greece and Albania, eight of the 20 fish species present have been introduced: two from North America (*Lepomis gibbosus* and *Oncorhynchus mykiss*), five from Asia, and one, the sheatfish (*Silurus glanis*), from central Europe (see Chapter 12). All of these represent a threat for native species, which include two endemics, a barbel, and a variety of common trout. Based on experience elsewhere, these introduced species will compete with native species for food and spawning grounds, or else will prey directly upon them. The greatest concentration of threatened Mediterranean fish species is in the Rio Guadiana basin in southern Spain and Portugal. (7) Finally, freshwater fishes will certainly be one of the groups of animals most threatened by global change in the forthcoming decades in the Mediterranean region which is expected to be particularly affected by climate warming and a strong decrease of rainfall.

11.1.3.4 MARINE FISH AND MEDITERRANEAN FISHERIES

Compared to freshwater fishes, the situation is quite different in the sea. As soon as humans had access to the Mediterranean coast, they began to collect seafood, including fish. Many archaeological sites provide evidence for fishing activity and sea fauna consumption since several millennia on both sides of the basin. However, even if the pressure on living marine resources can be heavy—as indeed it is— it is more difficult to eradicate marine food stocks than freshwater fishes. This sustainability of sea species is due to the spatial and ecological continuity of the sea environment. In addition, compared to aerial and terrestrial life forms and food webs, aquatic life is very different (see Chapter 9). The combination of sea-specific patterns of gametes and larvae dispersal, more stable environments, and fewer barriers to movements and migrations makes the functioning of sea communities quite different from those in terrestrial environments. Particularities of the marine environment reduce specific diversity but make unlikely the complete eradication of a species. In their comprehensive review, Dulvy *et al.* (2003) found no sign of a global extinction in recent times of any marine bony fish species in the world. Over the last 300 years, there is evidence for extinction of only three marine mammals and four gastropods. Another 18 low-taxonomic level taxa could in the future be declared extinct if their status as valid species is confirmed (Carlton *et al.* 1999).

However, quite a few more marine species have disappeared from specific areas, and two Mediterranean arthropods (in the Mysidae family) and 12 cartilagenous fish species in the Chondrichthyes (Dulvy *et al.* 2003) have disappeared in recent times. However, the effects of global warming, already well documented in the Mediterranean region (see Chapter 12), will presumably threaten many more species, especially those that live in localized areas with a stable temperature. In marine caves, the influence of global warming will undoubtedly threaten the Mysidae fauna (Chevaldonné and Lejeusne 2003; Lejeusne *et al.* 2006).

Today, Mediterranean and Black Sea fisheries together yield landings of about 1.5 million t (Mt) of fish per year (1.62 Mt in 2006; FAO 2008). The countries most involved in fisheries are Turkey, Italy, Algeria, Tunisia, Spain, and Greece, but Libya and Croatia have recently increased their captures quite significantly as well. The total catches were quite stable during the 1950s and 1960s at 0.8–1.0 Mt year^{-1}, and then increased during the 1970s and 1980s to a range of 1.17–1.65 Mt, with an unusual peak of 2.08 Mt in 1988. Since then, total catches have decreased and, for the last 15 years, the figure has not changed substantially, ranging from 1.31–1.69 Mt year^{-1} (Fig. 11.1).

The total number of fishermen and fishing boats is decreasing in the Mediterranean region, but more and more efficient fishing tools and devices are developed, including both stock detection and capture systems. Recently, formidable fishing boats have been built and launched, but they did not result in an increase in the marine harvest. Thus, the average figure of 1.5 Mt probably represents the maximal potential score for a sustainable resource, but a smaller yield would undoubtedly be preferable. This corresponds to a value of 1 billion € but it is questionable whether such a heavy harvest is sustainable in the long term. The main species targeted and captured in Mediterranean fisheries are presented in Table 11.1.

It is well known that these 'official figures' are not always reliable. Depending on the species, the large differences between the 1985 and 2006 data

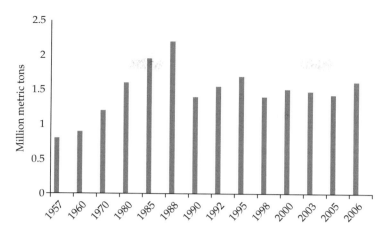

Figure 11.1 Fisheries landings in Mediterranean and Black Sea. Data from FAO (2008).

can be due to a new interest for a given resource (often due to the transfer of fishing activity from one species to another), overfishing, or else setting up of protection regulations. Interannual stability of a figure should indicate optimization of a fishery or, very often, respect for a legal quota. However, an apparent stability as reported does not necessarily correspond to the truth and that, for species like the bluefin tuna, captures are clearly underestimated (Box 11.4).

Another very important Mediterranean food resource, the European hake (*Merluccius merluccius* var. *mediterraneus*), is also threatened. Looking at the size records registered since the beginning of modern fisheries reveals that all the biggest species ever caught have been recorded before the 1950s. Since invertebrates and fish grow during their entire lives, this means that extant individuals of many species will never live long enough to reach such sizes! Today, the FAO estimates that 76 % of the living stocks of marine resources are fully or overexploited, and Cury and Morand (2004) mention that 50–90% of large pelagic fish have disappeared in the last 15 years. The large-scale exploitation of benthic communities by trawling and dredging has made ecosystems highly vulnerable to biological invasions.

Overexploitation has also been a major factor in the decline of coastal ecosystems in association with other negative factors, such as eutrophication, biological invasions, and climate disruption, among others (see Chapter 12). If we do not agree with the catastrophic predictions of Worm *et al.* (2006) on the disappearance of the major fishery stocks over the next 30 years, the situation is nevertheless worrisome and requires urgent action (see Chapter 13). Fish discards reach almost 30 Mt worldwide and are never taken into account in official statistics. While capture methods clearly target large individuals, landings include smaller and smaller fish, which results in changes in population structure. Fishes reproduce at younger ages and smaller sizes, a fact which is very clear for example for the Mediterranean hake. Faced with such pressures, marine species will reach and cross very low population thresholds and may eventually become extinct. In the Mediterranean, recent efforts to target deep water species, such as shrimps or Norway lobster are also quite ominous (see also Chapter 9 for deep-sea resources). Fishermen argue that 'economic quenching' always precedes ecological extinction. The idea is that when a resource becomes too rare for its exploitation to become profitable, then economic actors will look for substitutes. In fact, in the fishing industry, the opposite result is often observed; that is, that very rare and highly prized species bring ever higher prices and accessory captures continue until the species is driven to extinction, either locally or globally (Cury and Morand 2004). Box 11.4 provides an illustration of this problem.

Table 11.1 Major marine species landed in the Mediterranean Basin

Species	Amount landed (t)			
	1985	1993	1999	2005
Whiting	23 203	21 279	14 643	10 723
Blue whiting	10 119	18 842	23 530	10 969
Mediterranean hake	47 985	51 539	24 144	27 430
Mullet	17 811	21 083	45 745	35 098
Red mullet	13 705	12 643	5618	15 289
Surmullet	23 403	25 250	9985	14 390
Bogue	27 601	38 541	24 800	30 583
Common sole	9396	8871	4179	5397
Gobie	6877	2530	2682	14 426
Sardinella	21 310	35 397	78 518	64 001
European sprat	51 793	14 878	39 396	53 936
European anchovy	512 380	310 571	442 346	253 651
European pilchard	246 138	257 965	212 244	203 541
Atlantic bonito	18 456	26 312	25 655	77 460
Bluefin tuna	19 128	24 345	22 856	23 886
Swordfish	13 220	11 120	13 686	14 582
Bluefish	8524	18 973	4367	20 273
Horse mackerel	192 119	65 157	46 777	75 907
Mackerel	41 385	22 600	31 801	41 091
Rose shrimp	2733	13 669	9261	16 326
Decapod	28 861	11 512	14 974	15 605
Blue mussel	90 491	39 050	44 453	22 898
Clam	n.a.	3010	2245	1496
Cuttlefish	19 702	17 361	10 500	9744
Octopus	16 119	19 266	13 393	12 856
Total	1 978 564	1 534 968	1 533 360	1 438 301

n.a., not available.
Source: FAO (2007).

11.1.3.5 REPTILES AND AMPHIBIANS

Of the 355 species of reptiles and 106 species of amphibians which currently occur in the Mediterranean Basin (see Chapter 3), 46 species of reptiles (13%) are of conservation concern. Lizards are more threatened than snakes, mostly as a result of habitat loss. For example, five of the seven species of the genus *Iberolacerta* that occur in the Iberian Peninsula are globally endangered. Many endemic insular species of the genus *Podarcis* are of conservation concern and three of them are threatened (Cox *et al.* 2006). In the Canary Islands, where the giant lizards of the genus *Gallotia* experienced an explosive adaptive radiation with as many as eight species occurring today, one species—*Gallotia auaritae* from La Palma—is already extinct, and three others are critically endangered (Cox *et al.* 2006). Three of the 12 Mediterranean terrestrial turtles (*Testudo kleinmanni*, *Testudo werneri*, and *Rafetus euphraticus*) are seriously threatened. In addition, many species of reptiles that occur only marginally in the Mediterranean region are not globally threatened, but their Mediterranean populations are often endangered. Cox *et al.* (2006) mention 21 such species, including *Crocodylus niloticus*, eight snakes, several geckos, and one tortoise.

As expected from the poor ecological state of marshland and freshwater lakes and rivers in the region, the conservation status of amphibians is even worse than that of reptiles, with 25.5% of the species being threatened and one species—*Discoglossus nigriventer* from Israel/Palestine—now declared extinct (Cox *et al.* 2006). However, the different groups of amphibians are not equally threatened. Salamanders and newts are more threatened than frogs and toads, with 42.9 and 14.1% of the species, respectively being at risk. However, none of the 11 species of newts (genus *Triturus*) is globally threatened, whereas most species of salamanders are in poor condition. Among frogs and toads, six species of frogs (*Rana*) and two species of Discoglossidae (*Alytes*) are of concern. Again, the main threat for reptiles and amphibians is habitat loss, which has negative effects on at least 200 species of reptiles and all species of amphibians in the region (Cox *et al.* 2006). But there are many additional causes of decline, including invasive species, harvesting, pollution, and other kinds of human disturbance (see Chapter 12). A severe new disease due to the chytrid fungus *Batrachochytrium dendrobatidis* could become a serious threat for many species of amphibians in the Mediterranean (Daszak *et al.* 2003). This disease was discovered for the first time in Spain in 1997 and is responsible for a severe decline of *Alytes obstetricans* (Bosch *et al.* 2001) and *Salamandra salamandra*. Pollution is a particularly serious threat for amphibians and invasive species may be extremely aggressive. For example, the invasion of many Mediterranean wetlands by the Louisiana crayfish (*Procambarus clarkii*; see Box 12.3 and Plate 5b) resulted in a significant decline of the European pond terrapin and most species of amphibians, not to mention

> **Box 11.4. A threatened species in the Mediterranean Sea, the bluefin tuna**
>
> Scientists and conservationists are concerned about the future of the bluefin tuna, an emblematic Mediterranean fish species of great ecological and economic importance. This large predator can reach 3 m in length with a body mass of 800 kg or more and its flesh is firm but tender, and delicious. Around the middle of the twentieth century, several females weighing more than 1 ton each were landed, but none since the 1950s. There are 14 species of tuna, including eight species of *Thunnus*, the skipjack (*Katsuwonus pelamis*), three bonitos, and two frigates; only six species of tuna occur in the Mediterranean, of which the bluefin is the largest. Today, the bluefin tuna is not common in the Mediterranean Sea and several indications of over-fishing are already obvious (Fromentin and Powers 2005). The International Commission for the Conservation of Atlantic Tunas (ICCAT), established in 1969, determines the amount of each species of tuna which is allowed to be caught, including in the Mediterranean Sea. Since 2007, the yearly quota for the bluefin tuna is 29 500 t for the eastern Atlantic Ocean and the Mediterranean Sea combined. But this quota is far from being respected, i.e. in 2007, the landing figure was 61 000 t. A recent study of the Tuna Research and Conservation Centre (Block *et al.* 2005) used tagged tunas for collecting information on their biology, dispersal, and migrations between the Atlantic Ocean and the Mediterranean Sea. They demonstrated that Mediterranean fisheries can have serious impact on the western Atlantic resources. In fact scientists estimate that 15 000 t year^{-1} is the maximum catch that may be considered as sustainable for the species. In addition, they recommend a complete protection of this species during the breeding period that extends between May and July. Unfortunately, fishing pressure has never been as high as during the past few years so that the future of the species is at risk. In 2006, 6.5 Mt of tuna—all species combined—were caught worldwide (FAO 2008), one of the highest harvests in history. Of this total, 88 725 t were declared from the Mediterranean for the year 2006. Of this, nearly 23 000 t were declared for the bluefin tuna in the Mediterranean, probably less than half of the true landings. So, respectful measures have to be taken to protect the resource. Many attempts are being made to feed and breed tuna in floating offshore fish farms, especially in Spain and Croatia. But this is not a satisfactory or sustainable solution for many reasons. Among the many problems is the impossibility to check the young captured fish placed in the floating farms. As a result, these captures are not taken into account in calculating capture against the fixed quota. Secondly, natural food (fish and squids), fresh or frozen, must be used for farmed tuna, unlike farmed salmon, which eat pellets. This fresh food is transformed into tuna biomass at extremely low rates: 12–20 kg for a gain of 1 kg; that is, a conversion rate four to five times less efficient than for salmon. Thirdly, fish farms contribute to sea pollution, and finally, tuna do not reproduce in captivity. For all these reasons mariculture of tuna cannot be considered sustainable at the present time (see Chapter 13).

insects whose larvae are aquatic, such as dragonflies and several families of Coleoptera (e.g. Ditiscidae, Hydrophilidae), which are highly vulnerable to predation by crayfish.

11.1.3.6 POPULATION DECLINE IN BIRDS

Among birds, very few species have disappeared from the basin in historical times, apart from three large species in North Africa: the lappet-faced vulture (*Torgos tracheliotus*), the helmet guinea folw (*Numida meleagris*), and the Arabian bustard (*Choriotis arabs*). The demoiselle crane (*Anthropoides virgo*), a bird of steppe regions and high valleys of central Asia, used to breed in some parts of Turkey and the Middle Atlas Mountains of North Africa until the 1930s. Today, only a few pairs still breed there. Another threatened species is the bald ibis (*Geronticus eremita*), which used to breed in many parts of the basin but is restricted today to a few, scattered colonies. Even if only a few species of birds have become extinct in recent times, many formerly widespread species are now threatened and confined to small, scattered populations. A major concern is a general trend of sharp decline in population sizes among nearly one-quarter of European bird species (Tucker and Heath 1994; BirdLife International 2004). Species that suffer from the highest decline are farmland species and long-distance migrants (BirdLife International 2004). For birds, as well as for many insects and reptiles, the main consequences of human-induced changes in Mediterranean habitats, especially large-scale deforestation, have been a tremendous advantage for species adapted to dryland and shrubland, as opposed to forest-dwelling species. As birds are known to be good indicators of environmental conditions, the picture they provide of the state of the environment for wildlife probably reflects what is happening for many other life forms as well.

The impact of the general use of pesticides and fertilizers is a well-known cause of population decline that has been documented in a large number of studies (e.g. Balmford *et al.* 2003; BirdLife International 2004). In spite of the most harmful pesticides such as DDT having been banned in Europe, pesticides are still a threat for many species, especially migratory species that overwinter in Africa, because of their widespread use in many Sahelian countries. This may be the direct cause of a sharp population decline of the lesser kestrel throughout the Mediterranean. This species declined in Spain from approximately 100 000 breeding pairs in the 1970s to less than 5000 in 1988, coincident with a period of heavy pesticide use both in Spain and Africa (González *et al.* 1990). For most migrant species, pesticide use in North and Central Africa to fight against locusts might affect birds through reduction in prey availability rather than through acute or sublethal poisoning. Other causes of declining bird populations in the Mediterranean include climatic changes (see Chapter 12), for example through reduction of bodies of surface water as a result of drainage and irrigation. Such causes have been suggested to explain population decline in several species, for example the white stork (*Ciconia ciconia*) (Kanyamibwa *et al.* 1990) and the purple heron (*Ardea purpurea*) (Bibby 1992). As reported by the European Union for Bird Ringing (EURING), changes in the survival rates of several species are strongly linked with the severity of drought in the Sahel region of western Africa. In years following severe African droughts, fewer birds are counted in European breeding areas, and survival rates measured by ringing are much lower than in years of normal rainfall.

The Mediterranean area is visited twice a year by billions of migrant birds that shift seasonally between the Palaearctic region and their Mediterranean and tropical winter quarters (see Chapter 7). This traditionally represents 'manna' for Mediterranean peoples, who developed a panoply of trapping systems to catch them. All species, irrespective of their size and protection status, are caught, cooked, and eaten including the tiny kinglets! Woldhek (1979) pointed out that indiscriminate shooting and trapping of protected species occurs on a wide scale in all Mediterranean countries. Recent estimates suggest that up to one billion birds are killed annually in the basin (Magnin 1991; see Box 11.5).

It has often been argued and even demonstrated for some game species that mortality due to hunting pressures has no significant effects on population densities because increased mortality due to hunting is compensated by a decrease in natural mortality rates, as a result of density-dependent processes of population regulation. However, there are limits to density-dependent mechanisms, and critical parameters of population dynamics are unfortunately too poorly known for most species to allow an accurate estimation of the rate of annual harvest of populations that would be compatible with the long-term health of migratory birds. The current rate of harvest is obviously too high for most species.

One well-documented example of the decline of bird diversity has been provided by Brosset (1990), who worked in eastern Morocco in the 1950s and then returned to the region in 1989. The survey of the most important habitats revealed a tremendous decline of large raptors and the bald ibis, 35 years after the first survey was conducted. The two most badly degraded habitats, in terms of bird species diversity, were the alfa steppe and the large cliffs that characterize the huge mountain ranges of approximately 15 000 km^2 between Tlemcen and Debdou, near the Algerian/Moroccan border. The alfa steppe in this region was characterized by large isolated betoum trees that formerly served as nesting sites for many birds, including brown

> **Box 11.5. Bird decline in Europe and the Mediterranean region**
>
> The overall conservation status of birds is worsening throughout Europe, including Mediterranean countries. According to BirdLife International (2004), farmland birds, long-distance migrants, and waders are doing particularly badly. Forty-three per cent of Europe's 526 bird species are currently in trouble; that is, 5% more than in 1994. Globally, birds are suffering from intensive land-use practices, especially the Common Agricultural Policy which is very harmful for farmland species. Among waders, 50% of the species are now in poor conditions, and the decline of long-distance migrants affects 60% of the 161 species that winter south of the Sahara and cross the Mediterranean during biannual migrations. Unfortunately, 72% of the species in the Annex 1 of the EU Bird Directive are still in trouble.
>
> Based on the analysis of recoveries of ringed birds, a study of the British Trust for Ornithology shows that a decline in survival rates of many migratory birds throughout Europe may be directly related to excessive hunting and trapping pressures in Mediterranean countries (Crick and Jones 1992). Such pressure is particularly high in islands, such as Malta, Cyprus, and most of the Aegean Islands. Magnin (1987) estimated that, in Cyprus alone, approximately 2 million birds are killed each year by illegal mist-netting and liming. In Malta, the figure is estimated at 3 million birds, including well over 5000 birds of prey. Due to intensive bird shooting in recent decades, Malta lost all its breeding birds of prey: peregrine falcon, kestrel, and barn owl. The Balkans, most Mediterranean islands, and coastal countries of the Middle East (except Israel) and North Africa remain regions of unabated hunting of migratory birds. From a detailed analysis using recovery data from 19 European bird ringing schemes for 20 migratory species (five raptors and 15 passerines), McCulloch *et al.* (1992) showed that the majority of populations of these species are subject to considerable hunting pressure during migration and wintertime in the Mediterranean. Other threats to raptors include illegal traffic of birds captured for falconry and for exhibition centres, leading to a sharp increase in prices. The cost of a young peregrine falcon may reach 4500 € and that of an imperial eagle or a lammergeyer, 10 500 € (Kurtz and Luquet 1996), on the Middle East markets.

crow (*Corvus ruficollis*), lanner falcon, kestrel, hobby (*Falco subbuteo*), black kite (*Milvus migrans*), long-legged buzzard (*Buteo rufinus*), short-toed eagle (*Circaetus gallicus*), and, more rarely, African eagle owl (*Bubo ascalaphus*). However, most of the betoum trees have been cut down in recent decades, and all the above-mentioned birds have disappeared from the region. The cliffs in this region used to be one of the most famous habitats in the western Palaearctic for large raptors and the bald ibis (see Chapter 6). But, by 1989, most large bird species had disappeared or became rare in the case of the cliffs of eastern Morocco (Table 11.2). Surveys of the status of large raptors at broader spatial scales suggest that their decline is generalized throughout eastern Morocco and indeed in most parts of North Africa as well. In contrast, none of the small passerines that occur near cliffs and other rocky habitats, for example Moussier's redstart (*Phoenicurus moussieri*), black wheatear (*Oenanthe leucura*), trumpeter finch (*Rhodopechys githaginea*), crag martin (*Hirundo rupestris*), and rock bunting (*Emberiza cia*), show any trend of loss or decline. This is a clear indication that the main cause of decline among large raptors has not been climate or habitat change, but rather direct persecution by people. One species which dramatically declined in North Africa as a result of overshooting is the houbara bustard (*Chlamydotis undulata*), which is a traditional quarry of Arab falconers and continues to be hunted throughout its entire range. This species is also poached and affected by habitat loss and degradation (Le Cuziat *et al.* 2005) so that conservation plans are urgently needed for this species.

Finally, several sea birds are at risk mostly in their breeding grounds as a result of predation by cats and rats. This is particularly the case of shearwaters, for example the storm petrel (*Hydrobates pelagicus*)

Table 11.2 Decline of raptors, corvids, and the bald ibis in the Zekkaras cliffs and Djebel Mhasseur of eastern Morocco, between 1953 and 1989

Species	Number of pairs in			
	1953–9	1966	1975	1989
Golden eagle	1	0	0	0
Booted eagle	1	1	0	0
Bonelli's eagle	1	1	0	0
Black kite	3	2	1	0
Long-legged buzzard	1	1	0	0
Peregrine falcon	1	1	1	0
Lanner	1	1	1	1
Kestrel	4	2	2	0
Lesser kestrel	5	5	13	11
Egyptian vulture	2	0	0	0
African eagle owl	1	0	0	0
Raven	2	1	1	0
Jackdaw	80	0	0	0
Alpine chough	85	5	10	6
Bald Ibis	20	10	0	0
Total	208	30	29	18

Source: Brosset (1990).

and the endemic Mediterranean yelkouan shearwater (*Puffinus yelkouan*), whose population size probably does not exceed a few thousand breeding pairs (Bourgeois and Vidal 2008). This species has recently been added to Annex 1 of the European Union (EU) Birds Directive.

11.1.3.7 THE HEAVY PRICE FOR MAMMALS
The combination of habitat changes and direct persecution in recent times has doomed many large species of mammals to extinction. That fact compounds the effects of the mass extinction that occurred at the turning point between the late Pleistocene and the Holocene, with the result that the present-day mammal fauna of the basin is but a pale reflection of what it was in late Pleistocene times (see Chapters 2 and 3). Many Egyptian, Greek, and Roman artworks testify to the impressive animals that were captured in Spain, the Balkans, and North Africa for games and spectacles. The elephant was still abundant in North Africa in Roman times and Hannibal used it to help transport his armies through the Pyrenees in AD 218 and then the Alps, where most of the 37 individuals pressed into service died from starvation and cold (see Chapter 10). In North Africa, a large number of big terrestrial mammals have become extinct since Antiquity: the lion, the elephant, the African horse (*Equus africanus*), one antelope (*Alcephalus busephalus*), one gazelle (*Gazella rufina*), the African ass (*Equus asinus*), and the onager (*Equus onager*). Only five large-hoofed mammals still survive there today: the wild boar, two species of gazelle (*Gazella dorcas* and *Gazella cuvieri*), the aoudad (*Ammotragus lervia*) and the red deer. Among large carnivores, only a few individuals of serval (*Felis serval*) and panther still remain in remote forested areas of Algeria and Morocco, respectively (Cuzin 1996). Others such as the brown bear still survive—often just barely-scattered in isolated populations in Spain, the Pyrenees, Italy, Greece, and Bulgaria (Posillico *et al.* 2004; Preatoni *et al.* 2005; see Chapter 2). Among marine mammals the monk seal (*Monachus monachus*) is now on the verge of extinction (Box 11.6).

11.1.4 Subtle evolutionary and ecological changes

Hybridization between closely related species that come in contact with one another as a result of human-induced changes in the environment, or for other reasons, may cause subtle changes in biodiversity. For example, Mediterranean orchids have a wide range of isolation barriers that lower the risks of interspecific recombination under normal circumstances. But, in many regions of the basin, such barriers have been weakened or eliminated over recent millennia, partly because of human activities that have altered landscape structure and created innumerable disturbed sites conducive to hybridization (Kullenberg and Bergström 1976). Infra- and inter-generic hybrids are now common within the orchid genera *Ophrys*, *Orchis*, *Dactylorhiza*, and *Serapias* (see Plate 3). For example, at least 100 hybrids involving the 50 or more species of *Ophrys* have been documented and no less than 15 hybrids between *Serapias* and *Orchis* have been recorded as well (Baumann and Kuenkele 1982; Cozzolino *et al.* 2006; see Chapter 6). However, the behaviour of the dominant insect pollinators helps prevent hybridization. For example, isolation between the *Ophrys fusca/Ophrys lutea* complex from

> **Box 11.6. The decline of the monk seal**
>
> The monk seal provides a useful case study of a declining mammal because its status is well known, and much effort is currently devoted to saving it from extinction (Pastor *et al.* 2007). This seal is a large species of approximately 3 m long, weighing 240 kg (males) to 300 kg (females). Its original range extended throughout the whole Mediterranean area and the Black Sea, the Atlantic coast of Mauritania, and Macaronesia. This seal was intensively hunted in classical Greece when huge herds were described by Homer. It was even still fairly common in the mid-nineteenth century, when a burst of seal-hunting caused all populations to dwindle to mere smatters. Ongoing fragmentation of populations can mostly be attributed to persecution by fishermen, who consider the monk seal as a competitor for fish and damaging for their fishing gear. No more than 600 individuals still survive, in very small disjunct populations in Greece (80–400 individuals at most), along the Mediterranean coasts of Turkey (50–100 individuals), and North Africa (50–100 individuals), in a few sites in the Black Sea and Madeira, and in the north Atlantic coastal regions of Mauritania (Panou *et al.* 1993). In the summer of 1997, a severe disease, presumably caused by the combination of saxitoxins produced by a dinoflagellate bloom and/or a morbillivirus, resulted in mass mortality of seals in the colony of coastal Mauritania. It has been estimated that only around 90 individuals from the former 300 seals have survived. This means that nearly one-quarter of the extant world population of this species disappeared in just 2 months (Reijnders 1997). Chances for saving the Mediterranean monk seal seem very slim indeed. Simulation models show that, as is often the case in large, long-lived animals, the most critical factor for population dynamics is assuring adult survival. More than half of reported adult and juvenile mortality is caused by fishermen and hunters deliberately killing animals. An additional one quarter dies accidentally in fishing nets. Normal birth rates are insufficient to counterbalance such heavy losses. The only hope for conservation of this species lies in a drastic reduction of human-related mortality.

other species of *Ophrys* may result from the fact that pollinia are deposited on the abdomen of the pollinating insect in the former, whereas they are carried on the insect's head in the latter (Dafni and Bernhardt 1990; see Chapter 8).

11.1.5 Loss of cultivars and human-selected species

An additional aspect of biodiversity change to consider is that of the 'regression' of species of economic value that are now neglected after centuries of selection and attention. In Greco-Roman times, the attractive wood of the Barbary thuja was so highly prized that by the first century AD, Pliny reported that Cicero paid a sum equivalent to 275 000 € for a table made from thuja and that many wealthy Roman nobles competed to have their own thuja coffee table 'trophy', for its beautiful and unusual burl wood produced by the fire-resistant lignotubers and rootcrowns of this tree (Amigues 1991). As a result of this fad, large stands of thuja in coastal Libyan foothills (Cyrenaica) were wiped out. The surviving populations of thuja in Tunisia, Morocco, and the extreme south of Spain are but tiny, scattered fragments of formerly extensive mixed forests in those countries (see Chapter 3). Similarly, the yew tree, formerly common throughout the basin, was reduced almost to nil during the fifteenth and sixteenth century to supply the market in the UK for the fabrication of crossbows.

Not so long ago, a huge range of Mediterranean fruit, nut, herb, and also fodder plants varieties were propagated, cultivated, and continually 'improved', according to the needs and tastes of growers and consumers. Over the centuries, hundreds of varieties of olive, almond, pear, pomegranate, wheat, barley, alfalfa, grape, etc., were passed down and preserved *in hortus* (see Chapter 10). In all too many cases, the horticultural and agricultural selections and discoveries of past generations have been lost in very recent decades. One hundred years ago, 382 named cultivars of almond were cultivated on the island of Mallorca alone (Socias i Company 1990). How many are there today? Throughout the Mediterranean, a huge number of edible fruit, nut, vegetable, condiment, medicinal herb, and other plant varieties identified, selected, and domesticated in the past disappeared. Their loss and the loss of the know-how that went with them should be of great concern to one and all. Fortunately, much effort is devoted today to maintaining what is left—in seed banks, living collections, and working farms—and to protecting them from the risk of extinction (see Chapter 13).

11.2 Gains

In this section, we will provide examples showing that biodiversity loss as a result of human action is not inevitable. Even if the rationale of Grove and Rackham (2001) seems overly optimistic, it is true that many aspects of the design of landscapes by humans have at times been beneficial to biological diversity. In addition, the growing concern about environmental issues and the biodiversity crisis, especially since the 1992 Rio Conference and the Convention on Biological Diversity, resulted since some years in active and growing action plans that succeeded in improving the status of many ecosystems throughout the basin (see also Chapter 13).

Not all groups of animals and plants of the basin are in decline, even in ecological systems that have been highly modified by humans, so that we must be careful when formulating a global assessment on the status of Mediterranean biodiversity. Many examples of spectacular population recoveries and sudden range expansions of formerly rare and endangered species, as we will highlight in this section, show that, in many cases, biodiversity losses are not irreversible.

11.2.1 Long-term effects of land-use practices

Over the past 10 millennia or so, biological diversity in the Mediterranean Basin has depended in increasing measure on the many interactions between people and natural ecosystems. In other words, human activities have had positive, as well as negative effects on biodiversity—perhaps more so in this region than anywhere. Traditional land-use practices described in Chapter 10, especially the *sylva–saltus–ager* and *dehesa–montado* systems, are examples of such complementary effects on biodiversity. It is likely that these two types of land-use systems both had positive albeit different effects on various components of diversity. Clearing forest to plant pastures and crops allowed many species of shrubby and grassland habitats to colonize the area and thus increase biological diversity at the scale of landscapes—the so-called gamma diversity (see Chapter 5)—in a most dramatic way. One may imagine that the other two components of diversity, namely alpha and beta diversity, differed greatly between the two systems as a response to a different distribution of habitat patches: coarse-grained in the *sylva–saltus–ager* triad and fine-grained in the *dehesa–montado* system. The former was presumably characterized by moderate alpha diversity and high beta diversity, whereas *dehesa–montado* was presumably characterized by still higher alpha diversity but much lower beta diversity (Fig. 11.2). The collapse or near collapse of these systems undoubtedly had or will have consequences on biodiversity at the scale of landscapes and entire regions.

If forest recovery as a result of recent abandonment of traditional land-use practices in most Euro-Mediterranean countries is beneficial for forest-dwelling species, it results in habitat reduction for species of open habitats (Sirami et al. 2008). Within less than 30 years, we have witnessed no fewer than six species of birds that have re-colonized areas of the Mont-Ventoux region where forest is growing back, including the buzzard, honey buzzard (*Pernis apivorus*), black woodpecker, marsh tit (*Parus palustris*), and song thrush, which are all species of central European forests. However, population recovery by these species has been counterbalanced by a retreat of species of shrubby and open habitats, such as warblers, pipits, or buntings. One example is the spectacled warbler, which is a typical bird of low scrubland and shrubland (see Chapter 5). The species has steadily declined in number over recent decades and the most likely cause is forest and woodland recovery, which reduces the areas

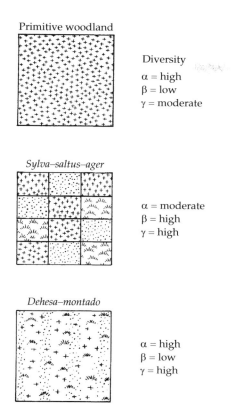

Figure 11.2 Hypothetical changes in values of the three components of diversity as a result of land management of former holm oak woodlands in the *sylva–saltus–ager* and *dehesa–montado* systems.

with suitable, open habitat for this bird (Gilot and Rousseau 2008).

Thus, habitat changes have not had the same effects on all groups. Although intensive use of herbicides and insecticides has caused a reduction in bee species diversity in many parts of the basin (Dafni and O'Toole 1994), forest clearing, agriculture, and even the establishment of sand pits, roads, and paths have had beneficial effects for many species that prefer or require open habitats. Bees find a new range of nest sites in these new habitats, as well as in dwarf shrub communities, such as low matorrals, *bath'a*, and *phrygana* (Table 11.3).

As a general rule, the periodic changes in vegetation cover that result, over the long term, from changes in land-use patterns by people have produced changes in the distribution patterns and populations sizes of species, depending on whether they are 'at home' in closed woodlands or in open habitats.

We have emphasized several times in this book that forest clearing—which has been for centuries the main means used by humans for organizing their living space (see Chapter 10)—has more or less stopped on the northern shores of the sea, because high income levels and low demographic growth have resulted in an increasing rate of land abandonment. Indeed, forest currently recovers at a rate of approximately 2% per year in countries,

Table 11.3 Effects of forest clearing and management of Mediterranean ecosystems on the diversity and abundance of solitary bees

Habitat type	Structure of the vegetation	Habitat characteristics	Diversity of bees
Oak woodland	Dense vegetation, many small low-reward flowers, few annuals and geophytes	Scarcity of nest sites, low diversity of plants as well as pollinators	Few solitary bees
Matorral	Sparse vegetation, high insulation and degree of patchiness	High availability of nest sites, high diversity of annuals and geophytes	High diversity and dominance of solitary bees
Traditional agricultural habitats	Farmland	Frequently disturbed habitats, high availability of nest sites, spread of ruderals and segetals	Proliferation of solitary bees
Industrial agriculture	Crop fields, orchards	Use of biocides, apiculture, reduced floral diversity	Few solitary bees, competitive exclusion of solitary bees as pollinators

Source: After Dafni and O'Toole (1994).

> **Box 11.7. The return of forest in Tunisia**
>
> Some 2000 years ago, at the beginning of Christian times, Tunisia, which was named Afriquia, included huge blocks of lush forests covering about 3 million ha. However, anarchic use of wood for ship construction, fire wood, clearing for livestock, and many other uses resulted in a reduction of forests areas to some 1.5 million ha at the end of the eighteenth century and less than 400 000 ha when Tunisia became independent in 1956. Efficient reforestation programmes were decided in the 1970s after a Forestry Code has been established in 1966. Forested areas steadily increased to 670 000 ha in 1987 and more than 1.25 million ha in 2007, with a rate of 12.5% of the land being forested instead of 4% in 1956. Several species of oaks (cork oak, holm oak, Spanish oak) have been replanted, as well as pines (Aleppo pine, maritime pine) and cedars (Kouki 2007). The annual planting rate in Tunisia is estimated at about 14 000 ha with the aim of enhancing agroforestry practices (with *Acacia*, *Atriplex*, and *Medicago* species) and fighting desertification (FAO 2003). Multi-purpose species, such as walnut, pistachio, pecan (*Carya illinoinensis*), hazel, and carob, have also been widely planted (Merlo and Croitoru 2005).

such as France and Italy (Quézel and Médail 2003). Although the process of forest clearing is still in progress in most countries of North Africa and the Near East, there is currently a consensus in several countries of North Africa and the Middle East for developing new forestry practices and reforestation programmes (see Chapter 13). This is the case, for example, in Israel and Cyprus, where mean annual plantation rates reach 500–1000 ha (Scarascia-Mugnozza *et al.* 2000). In Tunisia, active reforestation programmes are in progress (Box 11.7). The primary objectives of reforestation are soil protection and runoff control, but tree plantations of fast-growing species, such as eucalyptus and several species of pines (e.g. Monterey pine (*Pinus radiata*) and maritime pine) are expanding in several northern Mediterranean countries to meet wood and timber needs.

In several countries, increasing attention is also being paid to use native trees and to move beyond reforestation to ecological restoration, as will be discussed in Chapter 13. There are many reasons for this, starting with the fact the large-scale plantations of eucalyptus, especially *Eucalyptus camaldulensis* and *Eucalyptus gomphocephala* that now cover over 1.5 million ha in the Mediterranean Basin, especially in Spain (400 000 ha), Portugal (700 000 ha), and North Africa (280 000 ha), are biological 'deserts' with dire consequences for native communities and landscapes over the long term (Quézel *et al.* 1990). Although they may be economically very rewarding in the short term, with timber production that may reach 5–10 m^3 ha^{-1} year^{-1}, such plantations will incur ecological costs in the long run. For example, the utilization by birds and insects of eucalyptus leaves, flowers, and fruits is minimal outside Australia, probably because native fauna in the Mediterranean and elsewhere are incapable of using these resources. In Portugal, no more than 13 species of birds breed regularly in eucalyptus plantations, as compared to 30–35 species recorded in oak forests nearby. Additionally, eucalyptus plantations in the Mediterranean and elsewhere are infamous for extracting all available water and nutrient resources in a given site. Furthermore, although none have yet become truly harmful invasives, a few species of beetles in the Chrysomelidae and Cerambycidae families have been introduced with eucalyptus trees during the last 50 years, and some are becoming quite frequent in the area. Finally, monocultural plantations of eucalytus and of pines are highly susceptible to total devastation by fire. As an alternative, combined pine and oak plantations are increasingly proposed for degraded land restoration and multipurpose-multi-user production systems, on the basis of the complementary features of both groups of species (Pausas *et al.* 2004; Berrahmouni *et al.* 2009). Better still, full-scale ecosystem restoration is increasingly being pursued in a variety of Mediterranean countries, in forest, woodland, riparian, coastal, and wetland habitats. In

Chapter 13, we will briefly describe some of these activities.

11.2.2 Spectacular increases in biodiversity

Increases in biodiversity have sometimes been observed in habitats that are considered as highly threatened everywhere in the Mediterranean such as wetlands. For example, Table 11.4 gives an idea of population sizes of gulls, terns, and avocets in the Camargue and their variation since 1956. The overall numbers of breeding pairs for all species combined increased between the 1950s and the 1970s, mostly as a result of a sharp increase of some species. Examples of increasing species are the slender-billed gull, the Mediterranean gull—which did not occur at all in the Camargue as breeding birds in the 1950s—and the sandwich tern. There is hardly any significant trend of decline for any species, except perhaps the pied avocet (*Recurvirostra avosetta*), where a huge interannual variation of population sizes renders the observed variation of numbers ambiguous or insignificant.

The slender-billed gull, the Mediterranean gull, and the sandwich tern have also increased elsewhere in the Mediterranean, for example in the Po River delta, in Italy, the Ebro delta, in Spain, the Greek Evros delta, and the salt pans of Sfax, Tunisia. Other gull species that were considered threatened some decades ago have undergone a spectacular demographic increase and have recently colonized many new sites. This is the case of Audouin's gull, which forms large colonies in Corsica, the Ebro delta, the small Chaffarina islands off the Mediterranean coast of Morocco (more than 3000 breeding pairs), and some other areas. For example, this gull which was absent from the Ebro delta in the 1970s colonized the area in the 1980s, with approximately 200 breeding pairs in 1987 and 11 000 in 1997! The precise demographic mechanisms responsible for this spectacular increase are still obscure.

The rates of increase in the Camargue of the slender-billed gull (22% per year between 1967 and 1994) and the Mediterranean gull (20% per year between 1969 and 1994) are too high to result only from local demographic processes, because local recruitment of offspring is not enough to account for this large increase. Modelling population dynamics of these species, Sadoul (1996) showed that such spectacular population increases would imply an average yearly production of viable offspring of 1.42 and 1.60 young per breeding pair for the slender-billed gull and the Mediterranean gull, respectively. Yet, the observed offspring production of these species in the Camargue is usually less than 0.5 offspring per nest! One explanation for the discrepancy between reproduction levels and population increase of these two species is immigration from eastern populations, possibly those from the badly-polluted Black Sea area, where populations have sharply declined between 1985 and 1994 from approximately 30 000 to 6000 and from 300 000 to 50 000 breeding pairs for the slender-billed gull and the Mediterranean gulls, respectively (Sadoul 1996). It is presumed that some of these birds migrated to the Mediterranean area in search of better habitats. In fact, good news about increase of bird populations in the western part of the Mediterranean must be mitigated for two reasons. First, applying a **Living Planet Index** to water birds gives contrasting results. The Living Planet Index for water birds shows an increase of 38% between 1970 and 2002 in the western half of the basin (MedWet 2008), but monitoring efforts have

Table 11.4 Numbers of breeding pairs of the main bird species breeding in the lagoons of the Camargue between 1956 and 2005

Species	Year			
	1956	1976	1991	2001–5[1]
Black-headed gull	2900	8000	5200	1400
Yellow-legged gull	300	2700	4000	4000
Slender-billed gull	0	12	200	600
Mediterranean gull	0	7	120	1700
Common tern	3000	1500	1100	600
Sandwich tern	15	1000	1430	700
Little tern	400	450	370	100
Gull-billed tern	250	200	340	300
Avocet	750	850	500	100
Total	7615	14 719	13 260	9500

[1] For technical reasons, censuses for the period 2001–5 are strongly underestimated, especially for terns and avocet.

Sources: Sadoul (1996), Kayser *et al.* (2008).

> **Box 11.8. The greater flamingo (*Phoenicopterus roseus*)**
>
> The greater flamingo, a strange bird which descends from a very ancient group of birds whose ancestors can be traced back to the Cretacean era, is one of the most emblematic and eye-catching birds of the Mediterranean. The pink colour results from both the abundance of carotenoids in their diet, which includes many small crustaceans, and a particular efficiency in the metabolic process of these compounds (see Box 6.1). Many aspects of the biology of the bird are really fascinating; for example, their manner of feeding by filtering water with bill and tongue, which are a unique and complex mechanism designed for filtering minute organisms from mud and water. The bill is curved so that the upper mandible becomes the lower during feeding in the lagoons, which constitute their habitat. Other oddities include their habit of raising their chicks in a nursery or crèche and even more surprising for a long-lived species, is their mating system since the two partners of a pair almost never re-mate in spite of close proximity between the birds all year round and especially before breeding when they colonise the breeding site. Flamingos divorce each year and sometimes within a breeding season, a behaviour which remains largely unexplained.
>
> Flamingos are gregarious and live in flocks numbering hundreds or thousands of individuals. They breed in large colonies preferably established on small islands or ancient dykes surrounded by shallow water, which affords protection against predators. Breeding success is often dependent on abundant rainfall in the catchment zone before breeding. As a consequence of unpredictability of rainfall in the Mediterranean, breeding failure is frequent, but the life history of the bird has evolved in such a way that poor breeding success is compensated by a high life expectancy of adults since annual survival rates reach 97% with some individuals surviving over 30 years and even more.
>
> The range of the greater flamingo extends from the Mediterranean Basin and south-western Asia to eastern and southern Africa. In the Mediterranean Basin, flamingos can be observed in almost any wetland and as many as twenty breeding sites have been reported in various countries of the two sides of the sea. Population estimates amount to 290 000 birds in the eastern part of the basin and 80 000 birds in the western part (after Johnson and Cézilly 2007).

been applied to gregarious water birds, which are easier to count than other groups, such as amphibians, reptiles, fish, and invertebrates, which apparently declined during the same period, especially freshwater fish. Second, calculating Living Planet Indexes for three Mediterranean subregions that differ in biogeography and socio-economic history shows that biodiversity has evolved very differently in the western and the eastern parts of the basin. Whilst the species tied to wetlands have increased by 150% in the west in the early twenty-first century, they have declined by 30–40% in the east. These opposite trends between east and west did not occur at the same time in the different regions: the decrease in the index is obvious from the early 1980s in the Black Sea, while it occurs only from the middle of the 1990s in the eastern part of the basin (MedWet 2008). It is highly probable that these opposite trends result from different conservation practices and habitat use in the different parts of the basin, as suggested by the demography of the slender-billed gull and the Mediterranean gull discussed above. If this is the case, this is an interesting demonstration that whenever properly managed biodiversity can recover quite quickly.

The flamingo also showed population increases and colonized new sites in Spain and Sardinia during this same period, probably from the core colony of the Camargue, which has been protected and highly productive for the last 30 years (Johnson and Cézilly 2007; see Box 11.8).

Another example of spectacular increase in local diversity is the Moulouya River delta, in northern Morocco. Some decades ago, this was a small marshland of 50 ha at most with no more than four breeding species of birds. The size of this marshland increased at least tenfold over the past few decades, thanks to the building of large reservoirs in the upper course of the river and the expansion of irrigated farmland, which raised the water table. Today, this area is one of the best breeding sites in North Africa for many rare bird species, such as the purple gallinule (*Porphyrio porphyrio*), the marbled teal, the pratincole (*Glareola pratincola*), several species of herons, and many marsh warblers, such as the rare Savi's warbler (*Locustella luscinioides*). The growing concentration of birds has of course attracted raptors, for instance the black-winged kite (*Elanus caeruleus*). Moreover, this marshland is also an excellent stopover area for thousands of migratory waders and ducks (Brosset 1990). The large improvement of this marshland is an indirect, unintentional effect of human action, but it demonstrates that biodiversity can sometimes recover quickly once favourable conditions reappear.

For example, in most member states of the EU, many bird populations showed a significant overall increase since the establishment of the EU Birds Directive in the 1990s. This was especially the case for protected species listed on Annex 1 of this Directive, species of inland wetlands, Mediterranean forest, and montane grassland. Of the 14 species whose status has improved from unfavourable to favourable in the last decade, 10 are on Annex 1 of the Bird Directive. As a result of more stringent regulation policies and active action plans launched by BirdLife International for the protection of raptors, many species are now in a phase of population recovery, especially in Euro-Mediterranean countries, where there is a growing concern for nature conservation. A spectacular example of recover is that of the peregrine falcon. The population of this species decreased so dramatically everywhere in Europe in the 1970s and 1980s, largely as a result of generalized contamination by DDT, that it was considered endangered. But during the 1990s, this species increased so that it now has a favourable conservation status (BirdLife International 2004).

Other emblematic bird species benefiting from conservation and action plans are the large vultures, especially the griffon vulture, the black vulture, and the lammergeyer. Thanks to a long-term action plan of population reinforcement, the lammergeyer can now be considered as successfully reintroduced in the Alps, where several pairs breed in the wild. Other species that are increasing in numbers include the golden eagle, the Iberian imperial eagle (see Box 3.3), Eleonora's falcon, and several other species of raptors (Muntaner and Mayol 1996).

Sometimes, biodiversity recovery may be a by-product of undesirable human-induced changes in the composition of communities. This is the case of certain invading species which, just like the two-faced Roman god Janus, may have both positive and negative effects on biodiversity. For example the Louisiana crayfish (see Box 12.3 and Plate 5b) became a favourite prey for many bird species, especially herons whose populations are increasing in many wetlands, such as the Guadalquivir, Spain, and the Camargue. Thanks to this new 'manna', the Camargue is the only location in Europe where the nine European species of herons are currently breeding. However, the bad face of Janus is that this crayfish has negative, albeit poorly known, effects on many species of freshwater animals, such as amphibians, as explained earlier, aquatic beetles, dragonflies, and presumably many other groups of invertebrates. We will come back in Chapter 12 to the problem of invasive species.

Several large mammals, notably ungulates (deer, wild boar, red deer, chamois (*Rupicapra rupicapra*), ibex), are also increasing in population sizes and range in many parts of the basin, notably as a result of increasingly stringent hunting regulations and conservation policies. For example, the population of the mountain gazelle (*Gazella gazella*) in Israel, which had sharply decreased by the 1950s as a result of severe poaching, built up again from a few hundred to about 10 000 individuals in the 1980s. In some districts, the density may now reach 35 animals km^{-2}. Incidentally, the grazing habits of this gazelle have resulted in a recovery of plant species diversity, including species that are heavily grazed (Kaplan and Gutman 1989; see Chapter 13). Other mammal species have spontaneously returned, thanks to generalized forest recovery and agricultural land abandonment. One spectacular example in France is that of the wolf, which colonized the National Park of Mercantour, in the southern Alps, from adjacent Italian mountains, where populations of this species are large and in good health (Valière et al. 2003). The wolf, which had been eradicated from the French Mediterranean region in the nineteenth century, was first observed in this park in 1992. Five years later, there were at least 19 individuals in four groups. Traces of wandering individuals have been observed at least 300 km further north. The population in France is estimated in 2007 to have reached more than 100 individuals, mostly in the Alps, but spreading west in landscapes that become more and more suitable for this species, as rural exodus and agricultural abandonment progress. They have recently reached the Pyrenees and a few individuals originated from the Alps are present in the Carlit Massif. This is definitely a happy event, even if predation pressures on sheep raise problems and spur hot debates among livestock managers, legislators, and conservationists.

11.3 Fire: a threat and a driving force

Climate and human activities are the two main drivers of landscape dynamics since the beginning of the Holocene, but little attention has been paid so far to the ecological and evolutionary role of fire (Carrión 2002). Fire is an ecological factor to which species have had to develop specific life-history traits in a region where fire events have been relatively frequent since the beginning of the Pliocene or before (Moreno and Oechel 1994; Dubar et al. 1995; Carcaillet et al. 2002). Recurrent and periodic wildfires have been documented since the beginning of the Holocene (Wick et al. 2003; Vannière et al. 2008). They are definitely a key ecological factor in all Mediterranean ecosystems, but their effects are mitigated because of a complex combination of positive and negative effects. Fire statistics are often difficult to interpret (see Grove and Rackham 2001), and it is far from easy to determine the magnitude, frequency, and effects of fire events in space and time. As a natural disturbance, fire contributes to the dynamics of ecosystems, making the moving mosaic of habitats running in space and time, as illustrated in Chapter 7, with the succession of bird communities in relation to fire. Fire has been used by humans since many millennia, with the most ancient mention of human control of fire dating back to 790 000 years in the Acheulean site of Gesher Benot Ya'aqov, Israel (Goren-Inbar et al. 2004). Fire has been used for millennia for clearing forests and establishing pastures and croplands so that ecological systems had to adapt to this factor (see Chapter 10). But the detrimental effects of wildfires may result in a severe degradation of ecosystems whenever fire return is too high and wildfires occur on large areas. Therefore, fire is a complicated issue with both positive and negative effects on living systems.

11.3.1 Adaptive responses of vegetation to fire

Fires, which have been shown by Dubar et al. (1995) and Carcaillet et al. (2002) to naturally occur at least since the late Miocene, with the progressive establishment of a seasonal climate with summer high temperature and strong winds, have been an important factor in plant evolution in the Mediterranean (Thompson 2005). As a result of evolutionary responses to fire, post-fire resilience of Mediterranean vegetation is a well-established adaptive characteristic (e.g. Trabaud and Prodon 2002). Indeed, the consequences of wildfires in the reproductive biology of plant species are a major factor in Mediterranean ecosystems, because of their importance in determining life-history traits, vegetation dynamics, and dispersal patterns. Many Mediterranean plant species were preadapted—'abadapted' according to Harper (1982)—as a result of their having evolved life-history traits under the pressure of a high regime of fire. These so-called pyrophytes (Kuhnholz-Lordat 1938) include two main types. The first are 'resprouters' that have the ability to resprout quickly after fire thanks to the existence of lignotuber and a strong and diversified root system, which often constitutes most of the living biomass of the plant. Examples are the holm oak, kermes oak, and other evergreen oaks, as well as lentisk, many heath family species, and the various strawberry trees. In some resprouter species, fire defence mechanisms also include a thick and corky bark, as in cork oak, low inflammability due to high mineral content of wood (e.g. *Tamarix* and *Atriplex*) (Le Houérou 1981), or underground regenerative organs, which resprout after fire. These enlarged root crowns or 'burls' are found in Barbary thuja and false olive, for example, and show a highly unusual and ornamental structure in cross-section. Cork oak shows the very unusual feature of being able to resprout from the tops of burnt branches following a crown fire (Moreira *et al.* 2009; Pausas *et al.* 2009).

The second types of active pyrophytes include the 'seeders', like rockroses (*Cistus* and related taxa), whose seeds require or at least benefit from a thermal shock to germinate (Trabaud and Oustric 1989; Roy and Sonié 1992). For example, some *Cistus* which normally produce approximately 30 seedlings m^{-2}, which is already quite a lot, can produce as many as 5000 seedlings m^{-2} shortly after a fire (M. Etienne, personal communication). Other examples are several species of legumes (Fabaceae), for example *Genista linifolia*, *Vicia altissima*, and *Vicia melanops*, whose seed germination is enhanced by high temperatures, which is also the case for most geophytes (e.g. *Acis*, *Iris*, *Orchis*, *Serapias*, and *Tulipa*). In an interesting study from Switzerland, Moretti *et al.* (2006) examined the impact of fire as a natural factor or a management tool influencing the local distribution of *Cistus salviifolius*, a Mediterranean pyrophyte that currently reaches its northern range limit in the southern Swiss Alps. In Chapter 12, we will return to the question of the complex interactions of land use, other human activities, and climate change.

In contrast to resprouters, seeders must recolonize the recently burnt area and seedling recruitment is restricted to a narrow 'window of time', during the first growing season after fire. Thus, there is a clear advantage for the so-called 'temporal dispersers' among 'fire recruiters' (Keeley 1991), which spread the risk of seedling mortality in time rather than space. For such temporal dispersers, the bulk of the seed pool is deposited near the parent plant. Seedling recruitment after fire is often superabundant, but since germination events are rare, an equilibrium of sorts is established. Temporal fire dispersers also tend to be highly drought-tolerant and physiologically able to withstand extreme water stress. As a consequence, they have rapid growth rates on open sites but are shade-intolerant. Granivorous insects and vertebrates may affect this narrow 'window' of reproductive opportunity, through direct consumption that reduces the seed bank or by redistribution of seeds through dispersal.

With one or two exceptions such as the Canary Island pine, which, like the cork oak, has evolved a thick, partly fire-resistant bark which protects the tree from mortality during fires, most pines do not resprout after fire and are readily killed. The relatively fast-growing but short-lived pines rely on their small wind-dispersed seeds to colonize new sites. The most common strategy of pines is to occupy recently perturbed areas, such as those that are provided just after fire. With the many seeders that colonize recently burnt areas, they are often the pioneering species that will constitute the first stages of post-fire succession.

Summarizing the evolutionary response of plants to fires, five main strategies have evolved, as discussed by Thompson (2005). These include: (1) evolution of traits, such as thick insulating and protective bark (e.g. cork oak); (2) underground reserves allowing rapid resprouting while most of the aboveground parts are burnt (e.g. kermes oak); (3) seed adaptation with traits, such as hard coat favouring resistance to fire or long-term persistence in the soil bank (e.g. rockroses); (4) long-distance dispersal of small wind-dispersed seeds and recolonization (e.g. Aleppo pine); and (5) serotiny, which is the capacity to keep old closed cones for several years in a seed bank, which remains in the canopy of the tree (e.g. many conifers, such as wild cypress, Barbary thuja, and Calabrian pine).

11.3.2 Fire as a natural disturbance event

Fire is a key disturbance event in Mediterranean ecosystems and much progress on fire ecology has been achieved during the past three decades or so (Trabaud 1981; Trabaud *et al.* 1993; Carcaillet *et al.* 2001). Fires contribute to maintain the 'moving mosaic' of communities and ecosystems at the scale of landscapes, so that they have been a driving force in adapting and shaping species and communities, keeping ecosystems functioning (see Chapter 7).

As natural disturbance events, fires contribute to maintain habitat heterogeneity and biological diversity at the scale of landscapes. Some studies have shown that forest productivity in Mediterranean regions, for example in Spain, is one of the main factors determining fire regimes (Vazquez et al. 2006). These authors reported fire occurrences in 4–23% of the years in most of the sites they surveyed on a 10-km grid, a pattern that depends on the fuel available for fire. In arid conditions, the scarcity of fuel makes fire recurrence lower, but human presence and activities tend to increase it. Therefore, they must be integrated in management policies, and people must learn to live with occasional fires. The problem is to devise techniques for controlling their frequency and intensity. One strategy, for some areas, is to manage forests and woodlands with proscribed fires, so as to reduce the risk of catastrophic fires *before* the next fire breaks out. In other areas, fire is not the right management tool, and repeated fires combined with soil erosion can cause drastic changes in post-glacial soils. There seems little doubt they have contributed to a strong decline in Mediterranean old-growth forests formed in part by laurifolious (e.g. holly, yew, laurel) and deciduous trees (e.g. oaks, European hackberry (*Celtis australis*), and the Judas tree) (Quézel and Médail 2003). Still, proscribed fire should not be ruled out *a priori*. As shown in other parts of the world, if properly managed, fire has just as important and positive a role to play as grazing in the rational maintenance of open spaces and the management of mosaic landscapes that are biologically diverse (see Chapter 13).

11.3.3 Wildfires as devastating anthropogenic disturbance events

In most Euro-Mediterranean regions today, rural depopulation combined with encroachment of woody vegetation result in the accumulation of inflammable material over large areas. Hence the risk of very large fires with catastrophic consequences over millions of hectares. Whenever the magnitude and frequency of fire events surpass the capacities of resistance and resilience of ecological systems—that is, the natural regime of regional disturbance (Rundel et al. 1998)—they have negative effects on species and communities. Unfortunately, large destructive wildfires are more and more common, presumably because of the combination of increasing human population pressures and the effects of global warming. Forest fires number over 60 000 each year, 95% of which are started by people (Moreno and Oechel 1994). In the 1960s and 1970s, an average of 200 000 ha of forest and matorral were burnt each year in the Mediterranean Basin (Le Houérou 1981), with a high inter-annual and inter-regional variability. In September 1994, over 150 000 ha burned in 1 week in northern Spain and during the summer of 2003, which was exceptionally dry throughout the area, nearly 2 million ha of forests and shrublands burned down. In eastern Spain, the shrublands and Aleppo pine forests around Valencia are among the most seriously affected by wildfires within the Mediterranean Basin, and this is one of the regions where the danger of fire-induced desertification is greatest. Over the past 30 years, the yearly average has been about 2% of remaining forests and shrublands, with the result that the portion of the province considered as 'badlands' has increased by 14% (Vallejo and Alloza 1998).

Large wildfires tend to occur more frequently since some years, with more than 400 000 ha burnt in Portugal in 2003. On average, about 100 000 ha of forest areas have been burnt each year in this country in the last 10 years. Since matorrals and woodlands combined include about 150 m^3 of wood ha^{-1}, a 10 000 ha fire destroys on average 1.5 million m^3, which is roughly the same amount of wood as is exploited commercially each year in the entire French Mediterranean region. Although only some 0.3% of the total forested area in southern Europe is affected annually by fire, in some areas the figure may reach 10% and more. In the so-called red belt of southern France, for example, fires rage out of control every 25 years on average in the same place (Le Houérou 1990). When the return rate of fires is too high, the degradation of vegetation leads to dwarf formations, asphodel 'deserts', and the so-called leopard's skin with an alternation of bare ground and stones with low bushes of dwarf oaks (*Quercus coccifera*) or spiny xerophytes (e.g. *Astragalus, Sarcopoterium, Ulex*). As far as we know, however, no plant species has become extinct as a result of fire.

In fact, some rare and threatened plants are actually favoured by fire events (Diadema *et al.* 2007; Bondel and Médail 2009).

Summary

Various threats and conflicting trends characterize the status of most landscapes, ecosystems, and species in the Mediterranean Basin. Contrasting schools of thought traditionally considered the impact of humans on landscapes and ecosystems. One explanation for the difficulty in properly assessing the role of humans on biodiversity is that their impact shows both positive and negative aspects. Downs and ups in biological diversity have been quite dynamic though time, partly as a result of variation in the socio-economic wealth and density of human populations. Generally speaking, extant biodiversity suffers from various threats in most landscapes and habitat types, but three of them are of special concern—forests, wetlands, and coastal areas—especially on islands. Several components of marine biodiversity are also seriously threatened today, as a result of the cumulative and synergistic effects of global changes, including the severe overfishing referred to above, which is directly linked to human demography and increase in per-capita consumption of resources. Over past millennia, the large number of products used by human societies in forests and woodlands and the huge fluctuations of human needs have led to many changes, with alternation of positive and negative effects on biodiversity. Most groups of plants and animals are threatened in various ways, but, in many circumstances, biodiversity recovery is observed, resulting from active restoration and protection or from spontaneous regeneration which may be spectacular. Finally, fire is shown to be a complex ecological factor, with both beneficial effects as a natural disturbance event and driving force in ecosystem functioning and detrimental effects whenever it frequency and magnitude becomes too high. However, it is also clear that too-frequent fires constitute a real danger for both flora and fauna, and the increasing occurrence of wildfires threatens houses and humans as well.

CHAPTER 12

Biodiversity and Global Change

As a transitional or frontier zone between the humid and cold, extra-tropical zone to the north and the hot desert and semi-desert subtropical zone to the south, the Mediterranean region's terrestrial and marine fauna and flora are highly sensitive to environmental and climate changes (Ortolani and Pagliuca 2006). Furthermore, as a cultural and socio-economic crossroads area, the region is strongly influenced by other kinds of global change. Indeed, due to their long-standing exposure to multiple and intense human activities (Blondel 2006) and their inherent sensitivity to fluctuating climatic conditions (Peñuelas *et al.* 2002; Thuillier 2007), Mediterranean biota and ecosystems may be especially susceptible to global change of various sorts that, among other things, modify long-established biodiversity patterns and disturbance regimes (Lavorel *et al.* 1998; Sala *et al.* 2000).

However, as we have seen in previous chapters, Mediterranean biota and ecosystems are highly dynamic and resilient. They have, for a very long time, been adapting to environmental changes and new species arriving from far and near. There is evidence of rapid collapses of populations (Chapter 11), both at local and regional scales, often in islands, which are mainly due to a combination of changes in climate and human impacts. Yet the large number of Tertiary relicts and palaeoendemic species in the Mediterranean biota is proof of the capacity of the region's ecosystems to adapt to strong environmental changes (Rodrigues-Sanchez and Arroyo 2008; see Chapters 2 and 7). The question is how populations, species, and communities will be able to respond to changes which are much more rapid than those they experienced in the course of their history (Møller *et al.* 2004; GIEC 2007; EEA 2007).

In view of the complexity of Mediterranean biota and landscapes, it would be out of the scope of this book and beyond our expertise to discuss in detail all of the issues raised by these various kinds of global change. According to Vitousek *et al.* (1997), the six main components of global change are as follows: (1) the global demography and increase of human populations, (2) over-exploitation of natural resources, (3) change and fragmentation of habitats, (4) pollution and the spread of toxic chemicals, (5) biological invasions (including genetically modified organisms), and (6) climate change. A seventh component, which could become crucial in the near future, is the emergence of new pathologies that emerge or re-emerge as a result of several socio-economic factors which interact in combination with new dispersal potentialities of pathogens and diseases. Here we will only mention some of the best known and expected consequences of these components, focusing on the first and, particularly, on biological invasions and climate change. In particular, invasive species of terrestrial and marine plants, algae, and animals are on the rise in the Mediterranean lands and the sea. Some of them, especially some predators and microbes, represent threats to biodiversity and food webs, as well as to the fishing and mariculture industries, to sea-based tourism, and human health. Finally, the effects on, and responses of, resident biota to global warming will round out our discussion in this chapter.

12.1 Human demography

Population dynamics of human societies has long been a major factor in the economic, social, and environmental evolution of the Mediterranean Basin (see Chapter 10). The total population of the basin is now rising quickly and unevenly. For one thing, population density also varies widely within the basin, from a high of 1080 inhabitants km^{-2} in Malta—which is the highest national density anywhere, except for Monaco and Singapore—to a low of 28 inhabitants km^{-2} in Corsica. Demographic trends are, moreover, diametrically opposed in the north-western quadrant of the basin and the other three. In 1950, the 140 million inhabitants on the northern banks of the Mediterranean represented 66.7% of the basin's population. By 1990, with some 190 million, those same countries accounted for only 50% (di Castri 1998). By 2025, there will be between 520 and 570 million residents (Charpentier 1998; Benoît and Comeau 2005), concentrated in large part in the southern and eastern countries of the basin, from Morocco to Turkey, where the human population has grown five-fold compared to the 1950 census.

This demographic discrepancy is driven by the fact that human populations of the non-European Mediterranean countries are also much younger than those of the richer, northern populations. To wit, the ratio of 15–24-year-olds to 55–64-year-olds is about 2.5 times higher in the southern Mediterranean countries than in the Euro-Mediterranean region (di Castri 1998). Conservative analysts suggest that, by the year 2025, the population of southern Europe will stabilize at around 170 million, while that of the Near Eastern and North African countries will have increased by about 70% (see Table 12.1), thanks to annual population growth rates between 2 and 3.5% since the 1960s, when most of these countries achieved independence. These growth rates are among the highest in the world, and, as a result, by 2025, the non-European countries of the basin should be home to two-thirds of the region's human population. However, recent demographic trends in North Africa suggest that population size will stabilize sooner than previously expected.

Table 12.1 Human population sizes, rate of increase, and gross domestic product (GDP) per inhabitant in the 24 countries and territories of the Mediterranean region

Country	Population size (in million inhabitants)		Current rate of increase (%)	GDP/inhabitant (€ in 2005)
	2007	2025		
Spain	45.3	46.2	0.3	20 190
Portugal	10.7	10.4	0.0	15 430
France	61.7	66.1	0.4	23 880
Monaco	0.03	0.04	0.9	n.a.
Italy	59.3	58.7	−0.1	22 550
Malta	0.41	0.40	−0.2	14 830
Slovenia	2.0	2.0	0.0	17 330
Croatia	4.4	4.3	−0.2	9970
Bosnia-Herzegovina	3.8	3.7	0.0	6090
Montenegro	0.63	0.64	0.3	n.a.
Albania	3.2	3.5	0.8	4240
Greece	11.2	11.3	0.0	18 470
Turkey	74.0	87.8	1.3	6590
Cyprus	1.0	1.1	0.6	17 380
Syria	19.9	27.5	2.4	2920
Lebanon	3.9	4.6	1.4	4490
Israel	7.3	9.3	1.5	19 770
Jordan	5.7	7.8	2.4	4130
Palestine[1]	4.0	6.2	2.9	n.a
Egypt	73.4	95.9	2.1	3470
Libya	6.2	8.1	2.0	n.a
Tunisia	10.2	12.1	1.1	6180
Algeria	34.1	43.2	1.6	5290
Morocco	31.7	38.9	1.5	3410

n.a., not available.
[1] Palestinian territories (Gaza and the West Bank).
Source: Pison (2007).

As shown in Table 12.1, economic conditions as reflected by gross domestic product also vary enormously within the region. But the general impact on ecosystems and biota corresponds to the second form of global change cited above; that is, steadily increasing overexploitation of natural resources. Note, however, that some figures in the table are skewed. For example, in the French Riviera, population density can reach 2500 inhabitants km^{-2}, as compared to 108 inhabitants km^{-2} for France as a whole, which is coincidentally close to the average for the Mediterranean region as a whole: 111

inhabitants km^{-2} (Blondel and Médail 2009). Percapita income is also much higher there than the French national average.

Another important global trend concerns urbanization, mostly concentrated in coastal areas. Human population growth in the coastal Mediterranean areas, from Iberia to the Near East, is increasing more rapidly than anywhere else, even in Italy where 70% of the coastline is already urbanized (Arnold 2008). In Greece, 60% of the population, 40% of agriculture, 70% of industrial activity, and 90% of tourism are concentrated in coastal areas (Catsadorakis 2003). By 2025, the level of urbanization of coastlines around the Mediterranean as a whole, including North Africa, is likely to reach 70 or even 80%. Very simply, the demand for coastal landscapes for their beauty and tourist attraction makes them particularly vulnerable to unsustainable population increase, especially on the islands (Delanoë et al. 1996). Each summer, the population swells even further, thanks to tourist influxes from around the world. For example, on the French Roussillon coast it is not uncommon to observe small towns of 10 000 year-round inhabitants swell to 200 000–300 000 during the summer months of July and August, with deleterious impacts on the local surroundings, in terms of contamination, pollution, habitat destruction for plants and animals, and overexploitation of limited natural resources.

12.2 Habitat degradation and pollution

For brevity, we will treat these two important forms of global change together, as in Chapters 11 and 10 we have already stressed the many negative aspects of human impact on ecosystems and landscapes. Also, we will not consider the impacts of mining or industrial activities here. The interested reader is referred to, for example, Martínez-Sánchez et al. (2008) and Carrer and Leardi (2006).

12.2.1 Contamination from agricultural and horticultural chemicals

It would be out of the scope of this book to discuss this problem at length, but some specific points should be mentioned, such as the widespread use of veterinary products for protecting livestock against parasites. The systematic treatment of animals with vermifuge products dramatically reduces the diversity of dung animals, especially dung beetles, which, as we saw in Chapter 8, are of paramount importance for recycling dung and incorporating it in the soil (Lumaret and Kirk 1991). Some molecules of these products are extremely toxic. Rapidly defecated by cows, sheep, and goats, veterinary products slow down the process of dung recycling from some weeks to several years (Martinez and Lumaret 2006) and make pastures covered by dry dung, reducing sometimes significantly the areas available for grass to develop.

Massive use of various chemicals, such as pesticides, herbicides, and inorganic fertilizers, in the Mediterranean region has also expanded enormously over the past 20–40 years. In Mediterranean countries, many aquifers are located in areas of intensive agriculture and ground water is increasingly contaminated with nitrates and pesticide residues. This is especially pronounced in areas like southern Spain, parts of Italy, France, Israel, Algeria, and Morocco, where irrigated and fertilized field and orchard crops are produced for export to markets in Europe and elsewhere. Following a detailed study of aquifers in the Balearic Islands, Candela et al. (2008) report that, in addition to nitrate pollution, contamination from pesticide residues is also on the rise. This is not surprising since pesticides are increasingly used in traditional non-irrigated crops like grapes as well, such that significant amounts of residues of many different kinds of toxic pesticides can be found in most wines produced in southern Europe (and Germany, Austria, Chile, etc. as well). It is no wonder that a growing market for so-called 'organic' or biowines is developing in many countries (www.paneurope.info/Media/PR/080326.html).

Without going into detail, we note that, once again, birds are very useful sentinels for monitoring and, hopefully, addressing these issues. Generally speaking, farmland birds exhibit a sharp decline in all European countries, including Mediterranean countries, whereas forest birds remain more or less stable and sometimes slightly increase (BirdLife International 2004; Seoane and Carrascal 2008; see Chapter 11). Life-history attributes of the species may be critical for explaining the population dynamics of species in response to environmental changes (see Chapter 8). Generalist species with large ecological amplitude tend to thrive better than specialist species (Julliard et al. 2004). For example, there is a sharp decline of many insectivorous birds that prey upon large insects, including the little owl (*Athene noctua*), the scops owl, and all species of shrikes (*Lanius*).

12.2.2 Marine pollution

There are serious problems of high quantities of nutrients from excess fertilizers in many coastal waters, the most productive areas being at the outflows of great rivers like the Ebro, Rhône, Po, and Nile. The decaying of human-produced massive biomasses can remove oxygen with many problems, such as death of fauna, production of toxic gas, or impairing of coastal activities. Some blooms can also be of toxic algae, causing fish mortality and serious poisoning, particularly when filter-feeding shellfishes are eaten by humans. Death of sea mammals and turtles are often linked to the discharge of untreated sewages. The introduction of pathogenic bacteria, the presence of industrial effluents, heavy metals, pesticides, or detergents increase the damage against marine biotopes.

The most eutrophic waters are more numerous along the northern coastline, particularly in the Adriatic Sea, and this problem will increase with the arrival of new industrial and housing structures, causing irreversible damages to the coasts. In the south, the Nile delta is already much polluted, but the expanding populations along the African coast could worsen the situation.

Moreover, the Mediterranean has a very important shipping industry, and it was estimated in 1977 that about 1 million t year^{-1}, 20% of the global oceanic oil pollution, was discharged in its waters (Turley 1999).

Lastly, the regulation of great rivers (including those of the Black Sea) induces an important reduction of silicon supplies and the shift observed in the phytoplankton population is dangerous as well. Most of the surface primary production was by siliceous diatoms and there is now an occurrence of change in the phytoplankton communities to the benefit of non-siliceous species of flagellates and dinoflagellates. The anchovies and sardines represent a great part of the Mediterranean fisheries and their food consists of crustaceans grazing on the diatoms. The stagnation of these fisheries, which is evident over the last decade, can be due to human pressure (overfishing), but perhaps also to the decrease of diatom populations (Béthoux et al. 2002). Red tides triggered by dinoflagellates are more and more common and may locally pose serious problems in shellfish mariculture (see Chapter 13). Several major fish resources today, such as swordfish and even tuna, are invariably contaminated by heavy metals or dioxin in small quantities, and it is recommended for those species not to eat them more than once a week.

In short, there is much evidence that the health of the Mediterranean Sea is a subject of great concern. This situation is known since a long time. The first purpose of the Blue Plan, launched several decades ago by the United Nations for Environment Programme, was to improve the quality of seawaters (see Box 13.5). In the next section we will discuss biological invasions, which are another important change that represents an increasingly challenge for the Mediterranean Basin as a whole.

12.3 Biological invasions

The maintenance and welfare of biological diversity is threatened, among other things, by invasive alien species, a component of global change that is widely considered as one of the six major causes of diversity decline in those studies that have made a close analysis (e.g. Wilcove et al. 2000). While perhaps less dramatic than pollution and habitat destruction and less pervasive than urbanization or anthropogenic global warming (see below), the accidental or intentional introduction of species that become pests, weeds, or transformers is a very serious problem. Many examples around the world have shown that they can cause irreversible disruption and damage to native communities and ecosystems, as well as considerable damage to human health and economies (Simberloff 1981; Drake *et al.* 1989). This is particularly true for the marine environment (see Chapter 11) and islands (Sax and Gaines 2008).

The geographical location of the basin and its long tradition of marine and terrestrial trade with all parts of the world make it potentially vulnerable to biological invasions. The constant growth in international trade and transport activities and the increasing movement of people means that the number of introduced and invasive species, including pathogenic species, is rising. Yet literally thousands of plant species have already been introduced over millennia as medicinal herbs, crop plants, or horticultural subjects, while dozens of animal species have been brought in as livestock, game species, or pets. Large numbers of invading plants and animals have also been introduced unintentionally in ship ballast and as hitch-hikers in bags of grain and fertilizer, or bales of wool, cotton, straw, and hay. Humans themselves and their introduced plants and animals carried even more invaders in the form of bacteria, fungi, viruses, nematodes, and parasites. By introducing alien species, humans have strongly influenced the resident biota, often with unfortunate consequences for native communities and ecosystems in many cases. Therefore caution is required when discussing these issues, starting with the choice of words, especially as the subject tends to provoke strong emotional reactions among many people, often presenting an obstacle to rational debate (Simberloff 2003).

A biological invader is any alien species of plant, animal, fungus, or micro-organism that colonizes new territories as a result of intentional or unintentional introduction by humans. Terms like pests and noxious weeds are good labels for invaders that have harmful effects. Moreover, for those invaders that change the character, condition, form, or nature of ecosystems over substantial areas, an even stronger term—transformers—can be used (Richardson *et al.* 2000). Neither the terms invasive nor transformer should be used, however, when discussing natural processes of colonization of areas that occur or have occurred over large scales of time as a result, for example, of geographical or climatic upheavals, such as those that repeatedly occurred since the Miocene/Pliocene and during the whole Quaternary (Peng *et al.* 1998). Nor do they apply to long-term processes of range shifts of native species even if these shifts are more or less influenced by human activities, nor to the case of resident plant species arriving in a new habitat as a starting point for a new vegetation succession (Richardson *et al.* 2000).

Two additional words that are relevant for our discussion are naturalized and feralized. The former is applied to intentionally or accidentally introduced plants or animals that overcome former abiotic and biotic barriers to survival and regular reproduction. Frequently, but not always, it concerns intentionally introduced animals like the carp and domesticated plants, such as forestry trees or ornamentals, that escape cultivation and then naturalize and may invade unmanaged areas and habitats. Similarly, domestic animals like sheep or goats or aquarium fish that 'go wild' are called feral and said to have become feralized.

In recent years, there has also been a growing awareness of, and concern about, how the combination of species invasions, partly driven by naturalization or feralization, and local extinctions sometimes related to invasions can change the distinctiveness and heterogeneity of long-term resident faunas and floras, through a process called biotic homogenization (McKinney and Lockwood 1999; Leprieur *et al.* 2008). Homogenizing effects increase similarity between habitats, which may result in a simplification of ecosystem structure, composition, and function and what we may call a 'banalization' of the biota across biomes. McKinney and La Sorte (2007) contend that invasive alien species have a particularly strong homogenization effect because they are more likely to occur across multiple regions and habitats than long-term natives. This is clearly the case of two invasive species we will discuss later in this chapter, the Louisiana crayfish and the South African ice plants (*Carpobrotus*), and several marine invaders as well.

One additional pair of terms to clarify is exotic and translocated. Generally speaking, ecologists distinguish exotic species—those introduced from outside the study area, for example the Douglas fir or the rainbow trout (*Oncorhynchus mykiss*) from North America into the Mediterranean area (see below)—and translocated species, those species originating within the Mediterranean basin, for example the Atlas cedar from Morocco frequently planted in southern Europe. Yet exotic and translocated species may generate distinct geographical patterns of biotic homogenization because of their contrasting effects on changes in community similarity among regions. For example, Leprieur et al. (2008) showed that non-native fish species introduced in various parts of the Mediterranean region have caused diversification among neighbouring basins whose faunas were formerly quite similar, while contributing to overall homogenization among distant ones which had previously been dissimilar. Thus the relative impact of invasions and translocations may depend on scale effects with the result that patterns of homogenization versus differentiation may be explained by the scale of the study. Decrease in beta diversity resulting from homogenization of the faunas and floras is expected to have important evolutionary and ecological consequences (Olden and Poff 2004), so that it is an important research agenda for ecologists because the combination of species invasions and extirpations are two key components of the modern biodiversity crisis (Olden 2006).

One last issue to raise, before we begin to catalogue some of the more notable invaders and transformers, is the question just how important are invading species in Mediterranean habitats in today's changing world, and given the region's long history of past introductions, migrations, and invasions. In other words, to what extent do intentionally introduced organisms or accidental invaders today become harmful for long-term native species assemblages and ecosystem functioning in the Mediterranean? In the final chapter of this book, we will come back to this question when we consider possible recommendations for policy-makers and governmental agencies charged with conservation. For that purpose and to answer the questions just raised, it will be helpful to briefly review the biological features that are likely to help a species to become invasive. Conversely, we also address the question of just how easy to invade the Mediterranean region is, as compared for example to the four other mediterranean-climate regions.

12.3.1 What makes an invader successful?

To be successful, a biological invader must have ecological, physiological, genetic, and morphological characteristics, which promote long-distance dispersal of offspring and propagules, rapid colonization rates, and a high competitive ability (Lambdon et al. 2008). In plants, attributes of obvious importance are morphological characteristics of seeds that facilitate their transportation by wind or by animals (di Castri et al. 1990), but many other traits contribute as well, including phenology (Godoy et al. 2009) and relative ecological amplitude or adaptability of introduced species (Traveset et al. 2008). Long ago, Baker and Stebbins (1965) identified a series of biological attributes likely to facilitate the invasiveness of a species. Reviewing the physiological, demographic, and genetic attributes of invaders, Roy (1990) showed that the invasive flora of a country is composed of a large array of plant types with a wide range of traits. Recently, Gassó et al. (2009) examined sets of species attributes of 106 species of plant invaders in Spain and found that wind dispersal and minimum residence time favour invasion success. They also found that human-induced disturbance, low altitude, dry and warm environments, and short distance to the coast favour invasiveness. Global warming should facilitate the spread of invasive plant species, especially wind-dispersed species. It is also very well documented that high biodiversity of long-term residents in a community provides a 'line of defence' against the spread of biological invaders by inhibiting the establishment of newly arriving, non-native species (Kennedy et al. 2002).

In fact, despite a growing concern about the negative effects of invasions, relatively little is known of the determinants of distribution, abundance, and population dynamics of invading organisms. The fate of introduced species seems to depend mostly on context-specific processes within the target community, such as interspecific competition, herbivory, resource availability in space and time, and human choice about introduced species. If such is the case, invading species-community interactions determine invasion success. Therefore, it is likely that small biological differences among species interact with habitat characteristics to produce distinct patterns of distribution and abundance. One example is provided by two closely related species of annuals that have invaded a huge variety of biogeographic regions throughout the world: the horseweed (Conyza canadensis) and the fleabane (Conyza sumatrensis) (see Box 12.1).

Box 12.1. The invading horseweed and fleabane

In the Mediterranean region, the horseweed is restricted to recently perturbed areas, while the fleabane typically invades old fields in early stages of secondary succession. Thébaud *et al.* (1996) experimentally demonstrated from cross-habitat transplant experiments that a variety of factors, including interspecific competition, availability of soil nutrients, and water resource, all influenced the demographic performance of the two species. They differed sharply in their survivorship and reproductive success depending on these factors. The relative performance of the species during the experiment nicely matched their regional distribution, with the fleabane doing better in most habitats. The higher reproductive effort but lower competitive ability of the horseweed explains why this species invades only the most recently perturbed areas, where their seedlings are not confronted with resource limitations. A greater competitive ability allows the fleabane to invade and persist in more 'advanced' communities as well. This illustrates how subtle biological differences between closely related species can determine the patterns of distribution and abundance of invaders in a recently invaded area. This elegant experiment shows that invasion failure or success may depend on many traits interacting with many features of the target community.

In some circumstances, invasive species may benefit from resources that are unexploited by other species and themselves constitute valuable resources for native or migratory predators, which use them as prey. This is the case of the zebra mussel (*Dreissena polymorpha*), a mollusc originating from the Caspian Sea. This species has invaded most bodies of water of Europe where it may cause much damage to hydraulic installations. On the other hand, it is a valuable prey species for diving ducks, several of which overwinter more and more frequently in central Europe thanks to the new 'manna' provided that the bodies of water do not freeze.

Having set the scene for this discussion of biological invasions, we now begin a survey, group by group.

12.3.2 The few invading plants and where they come from

Surprisingly, invading plant species do not seem to be a major problem in most Mediterranean ecosystems, except islands and wetlands (Blondel and Médail 2009). Of the 25 000–30 000 taxa of vascular plants found in the Mediterranean region today, approximately 1% are considered non-native. However, very few of these taxa are considered as harmful for natural communities, despite the fact that many have been present for a very long time. Indeed, a century ago, at a time when there was much international trade in wool and other sheep products, the local flora in the region contained hundreds of recently adventitious species, as testified by the famous *Flora Juvenalis* of Montpellier (Thellung 1908–10). Not a single one of them succeeded in massively invading the region. In succeeding decades, none has penetrated far from roadways, ditches, and cultivated fields. Indeed, most plant invaders in the Mediterranean Basin survive only in unstable and human-made habitats, such as cultivated fields, old fields, orchards, urban and peri-urban wastelands, and roadsides. Even the two aggressive species of *Conyza* described in Box 12.1, both of which invaded the Old World some 150 years ago, are only seen in heavily disturbed sites, which they share in general with other ruderals. However, some alien plant species are currently spreading rapidly and tend to threaten Mediterranean communities through the homogenizing effects referred to above (see Lambdon and Hulme 2006).

Most invasive plant species of the Mediterranean area are from temperate climatic zones rather than from the other mediterranean-type regions. Apart from ornamental plants, such as *Pelargonium*, *Arum*,

Nigella, and *Impatiens*, which occasionally escape cultivation in some regions, several species in the Mediterranean come from the Cape Province of South Africa: *Senecio inaequidens*, *Senecio angulatus*, *Polygala myrtifolia*, as well as the succulent creeper, called ice plant (*Carpobrotus edulis* and *Carpobrotus acinaciformis*), which may extend over large areas on dry coastal rocky substrates and sands, and the Bermuda buttercup (*Oxalis pes-caprae*), which frequently becomes a very weedy pest, mostly in disturbed areas as well as along roads and in gardens. One accidental introduction from Chile is the spiny-fruited spiny cocklebur (*Xanthium spinosum*), which has become somewhat naturalized in sandy soils in the Mediterranean Basin, and, because of its spiny fruits, it can be quite noxious. From Australia, many species of *Atriplex* and nitrogen-fixing *Acacia* have been voluntarily introduced, but, here again, none has succeeded in invading forest or matorral in the basin. The only exception is the Australian wattle, *Acacia dealbata*, called mimosa in France, which has been cultivated for 150 years in the French Riviera for its beautiful yellow flowers, which are harvested and sold throughout Europe in early spring. This tree is now escaping in some areas of southern France and invading native woodlands and matorrals. Efforts are being made by the National Botanical Conservatory of Porquerolles (see Box 13.3) to limit the use of this exotic invasive in gardens and promote instead the use of substitute flowering species. This, however, meets resistance from many people who consider the plant delightful, ignoring the fact that it is an exotic and that it may well have deleterious effects on native biota.

Several plant invaders come from temperate North America, but few of them have had any major impact. Examples are the pigweeds (*Amaranthus albus* and *Amaranthus retroflexus*), false indigo (*Amorpha fruticosa*) in wetlands, black locust (*Robinia pseudoacacia*), evening primrose (*Oenothera biennis*), and the two *Conyza* species mentioned above. Some succulent species introduced from the Americas for ornamental purposes occasionally escape cultivation along the coasts, including a few agaves (e.g. *Agave americana*) and several prickly pear cacti (*Opuntia*). One specific habitat type that is particularly vulnerable to invasion is riparian gallery forests, which are often invaded by alien trees and shrubs, such as black locust, paper mulberry (*Broussonetia papyrifera*), European ash, maple ash (*Acer negundo*), Japanese honeysuckle (*Lonicera japonica*), butterfly bush (*Buddleja davidii*), tree-of-heaven (*Ailanthus altissima*), and various herbs, such as *Amorpha fruticosa*, *Reynoutria japonica*, and two or three species of *Impatiens*. This is no doubt because seeds arrive in quantity from upstream and because they are much wetter than the surrounding matorrals. Other exotic plant invaders are specifically a threat to coastal ecosystems, with Mediterranean islands being more vulnerable than most mainland areas. For example, recently arrived, exotic plant species represent 17% (473 taxa) of the Corsican flora, of which 1.4% (38 taxa) are considered to have become naturalized (Natali and Jeanmonod 1996).

In a study carried out on eight Mediterranean islands, Vilà *et al.* (2006) demonstrated that significant impact on native plant species and soils may result from individual invasive plants. Using as models the two invasive species of ice plant present in the islands, which are both herbaceous perennial subshrubs, the cosmopolitan tree-of-heaven, and the Bermuda buttercup, an annual geophyte, they showed that, at the scale of a few square metres, species loss resulting from these aggressive invaders averaged a remarkable 23%. However, caution must certainly be taken when interpreting this result, as the spatial scale of this study is too small to allow broad generalizations. Still, it is important to note that the few species of introduced plants capable of invading and dominating Mediterranean coastal habitats, such as dunes and cliffs, may be particularly harmful because of the high number of endemic, rare, and vulnerable species that occur in these habitats and which may be displaced by them. However, in an *in situ* sowing experiment, devised to examine establishment after germination and hence risk of invasion of these same four species, and carried out in a series of Mediterranean islands differing in climatic conditions and local species richness, Vilà *et al.* (2008) came to the conclusion that Mediterranean island ecosystems were generally quite resistant to invader establishment, except in the case of the Bermuda buttercup!

Lambdon *et al.* (2008) reported that alien floras are not intrinsically simpler than natives, mostly because they include strong competitors, such as, for example, ice plants that are likely to dominate communities and doom local natives to extinction. These authors came to the conclusion that initial concerns over the 'simplifying power' of aliens are largely unfounded. In their study, they found that taxonomic diversity between habitats and functional diversity within habitats were remarkably equal among aliens and natives. The observed differences in the response of native communities to invasive species clearly deserve further studies. As Vilà *et al.* (2008) put it, divergent responses to invaders may result from idiosyncratic events associated to the variability of climate, water availability conditions, and the structure and species-specific composition of the invaded community.

Except for some particular context-specific situations, such as the riparian and coastal habitats just discussed and ruderal communities, which may include unusual or even endemic segetal plant species (Heywood 1995), the relatively low invasibility of Mediterranean ecosystems contrasts with the situation in many other regions in the world where invading species have displaced whole cohorts of native species and disrupted ecosystem dynamics. Between 57 and 82% of plant species introduced to the four other mediterranean-climate regions of the world originated in southern Europe or the Near East, with the percentage of strictly Mediterranean species being highest in Chile. In California as well, as much as 20% of the extant flora consists of invading species, mostly of Eurasian origin (Raven and Axelrod 1978). In the South African Cape Province, an equally massive invasion has taken place, with trees and shrubs coming from Australia, and hoards of annuals and biennials arriving from Eurasia. As well as sharing a mediterranean-type climate, California and Chile were first settled by people from the Mediterranean itself. Thus, in both these cases, most early introductions probably came from the Iberian Peninsula. Groves and Kilby (1991) considered that the success of introduced plants from the Mediterranean to other mediterranean-type climate regions of the world is due as much to their 'preadaptation' to a set of factors, such as fire (see Chapter 11), low soil nutrient levels, and grazing pressures operating in their new environments (see Chapter 8), as to their response to a particular climatic regime.

The reasons why the Mediterranean is relatively resistant to invasion are still obscure and largely conjectural. Several hypotheses have been proposed to account for the susceptibility of ecosystems to be invaded by alien species (di Castri 1990; Quézel *et al.* 1990). One possible explanation relevant to Mediterranean ecosystems is that they have been subjected to continuous disturbance of fluctuating regimes and intensity over many millennia, accompanied by thousands of spontaneous colonization events. According to this view, these ecosystems have become progressively more resistant since 'old invaders' prevent access to potential 'new invaders' (Drake *et al.* 1989). Fox and Fox (1986) also argued that mature communities characterized by complex interspecific interactions among a full set of native species should be less susceptible to invasion because all available resources will be optimally used, and, as mentioned above, this may play a role in preventing or hampering the establishment of alien invasive species at the local scale (Kennedy *et al.* 2002).

In an attempt to explore whether habitat type may be a predictor of the level of plant invasion, Chytrý *et al.* (2009) demonstrated from a survey of a large number of study plots across Europe that the highest levels of invasion were predicted for highly human-modified habitats. Interestingly, they found that the lowest levels of invasion were predicted for sclerophyllous vegetation, heathland, and peatlands. They concluded that low levels of invasion were likely in the Mediterranean region, except its coastline, riparian forests, and in irrigated or recently abandoned agricultural land, which fairly well fits with what is actually observed.

Similarly, in a landscape-scale study in the Canary Islands, Arévalo *et al.* (2005) found evidence that ecosystems occurring at mid-level altitudes were more susceptible to invasions than either coastal areas or high-altitude zones, precisely because there is greater overall disturbance and prior transformation of ecosystems in that zone. A simulation study by Gritti *et al.* (2006) on the vulnerability of five Mediterranean islands to climate change and invasive plants also tends to support this general conclusion that only heavily disturbed habitats in the Mediterranean region are highly susceptible to invasive plants.

Thus, we conclude by saying that invasive plants are not the most important immediate threat to resident biodiversity in the Mediterranean, as compared to some of the other groups of invaders to be discussed below. As Davis (2003b) put it, invasive plants for the most part represent at most a new source of competition for long-term native plants and primarily in disturbed habitats, whereas introduced or invasive animals are often predators that wreak havoc on indigenous fauna at a broad landscape scale. Introduced pathogens that arrive with plants or animals, or in soil, seeds, or animal feed also represent a much more serious threat for the most part than exotic plants. However, given the rate of transformation of habitats and ecosystems being carried out by people, it seems very likely that the importance of invasive plants in the Mediterranean will grow considerably in coming decades. There is a clear need for reflection and consultation on this issue.

12.3.3 Invertebrates

There are few cases of well documented invasion processes by insects in the Mediterranean but the number of invading species is rapidly growing. One important case worth mentioning is that of the domesticated bumblebee (*Bombus terrestris*), which represents a very real threat for native bees for two reasons. In Israel, for example, increasing numbers of this bee—first introduced in the 1930s—have been associated with a reduction in native honey bees and several species of solitary bees (Dafni and Shmida 1996). The bumblebee is a generalist species able to harvest floral resources from a large spectrum of unrelated plant species in various habitats. It is active early in the morning and reduces nectar availability, which results in food shortage to other bee species. This is especially true in late spring when the pollen market switches reversed from a surplus of flowers to a surplus of bees (Shmida and Dafni 1989; see Chapter 8). Bee censuses conducted on several dominant plant species of matorral on Mt. Carmel, Israel, have shown a reduction in all other species of bees as numbers of bumblebees increased. Although honey bees have shown a constant retreat due to another invader—the parasitic mite *Varroa*—observations on bee activity revealed a regular exclusion of solitary bees from several plant species in the presence of *Bombus*, suggesting competitive exclusion (Dafni and O'Toole 1994). Harnessing bumblebees for commercial pollination in agribusiness thus represents a real threat for native bees. In addition, a second threat is through interbreeding between domestic bee and native bees, as documented throughout the entire Mediterranean Basin (Rasmont *et al.* 2008). Several species of beetles (Cerambycidae, Chrysomelidae, and some others) have also been introduced with eucalyptus trees from Australia, but without having caused any known problems to date (see Chapter 11).

12.3.4 Freshwater fish: a more severe threat

A group of vertebrates with many highly 'successful' and frequently aggressive invaders in the Mediterranean is freshwater fish, dozens of which have been introduced accidentally, or intentionally, and now occur in most of the lakes and rivers in the region, including those on the major islands, of which Corsica is a well studied example (Box 12.2).

Originally from the western coast of North America, the rainbow trout has been introduced in most European and Mediterranean countries for aquaculture. Italy and France in particular were for many years world leaders in freshwater culture of this fish. It very frequently escapes from hatcheries in large numbers and has also been intentionally released in many lakes and rivers. However, surprisingly, it very rarely reproduces 'in the wild' and has therefore never become naturalized in the Mediterranean region, apart a few areas such as a few lakes in the eastern Pyrenees where it remains quite localized (G. Boeuf, personal observation). In Israel, no less than 41 species of freshwater fish have been introduced, of which 27 have become established (Roll *et al.* 2007). Among them, an exotic trout, salmon, and sturgeon escaped from aquaculture ponds into the

> **Box 12.2. Fish invaders in Corsica**
>
> There are 32 freshwater fish species in Corsica but only 12 of them are native. Of the 20 introduced species, some arrived as early as the beginning of the nineteenth century (Roché and Mattei 1997). Native species belong to nine families (Salmonidae, Blenniidae, Gasterosteidae, Anguillidae, Clupeidae, Atherinidae, Mugilidae, Moronidae, and Cyprinodontidae). Species have been introduced either as predators of mosquitoes for fighting against malaria or for sport fishing. Most of them never leave freshwater, but some species are euryhaline and amphibiotic, which means they may live also in brackish and even marine waters. This is of course the case of the common eel (*Anguilla anguilla*), aloses (*Alosa fallax*), and four species of mullet (*Chelon labrosus*, *Mugil cephalus*, *Liza aurata*, and *Liza ramada*). Non-native species belong to eight families—Acipenseridae (one species), Poeciliidae (one), Cyprinidae (nine), Siluridae (one), Ictaluridae (one), Esocidae (one), Salmonidae (three), and Percidae (three). Unfortunately several species, for example sheatfish, pike-perch, bass (*Micropterus salmoides*), and perch (*Perca fluviatilis*), are ferocious predators threatening the native fish fauna. In some small streams, introduced *Salvelinus* are predators of the endemic newt (*Euproctus*). Introduced salmonids (*Salmo trutta*, *Oncorhynchus mykiss*, and *Salvelinus fontinalis*) may hybridize with the local endemic variety of trout (*Salmo trutta macrostigma*), seriously threatening its genetic structure and integrity. In addition, introduced fish may carry pathogens and parasites. This is the case of *Anguillicola crassus*, a parasite of Japanese origin, which infests many populations of eels in Corsica.

Jordan River system, leading to overpredation of several local endemic fish species (Goren and Ortal 1999). Exotic species led to homogenization of freshwater fish communities, especially in southwestern Europe, notably in the Iberian Peninsula which differs from the rest of Europe by its low number of native species and high level of endemism (see Chapter 3; Elvira 1995). Many freshwater fish species have been intentionally introduced in Mediterranean streams and marshes more or less recently. Several of them, such as carp and pike-perch, became naturalized and are now well-integrated components of extant communities everywhere in the basin. As mentioned above, these introduced fish have been shown to result in the decline or extinction of several native species, including some endemics. For example, in Spain, no less than 19 introduced fish species are considered to threaten native endemics. The most harmful are the large predators.

The mosquito-fish (*Gambusia affinis*), from southeastern North America, has been introduced in almost every Mediterranean lake, lagoon, and lowland river, where it was hoped that it would serve as an anti-malaria agent by controlling mosquitoes. However, there is no evidence that this fish may cause any problems to native communities. The common European carp (*Cyprinus carpio*) was originally only localized in eastern Europe. From this area it was introduced to western Europe, including several Mediterranean countries, during the Roman Empire, and then quite massively during the Middle Ages. Several other species originating in Asia were also introduced into Europe many centuries ago, like the silver carp (*Hypophthalmichthys molitrix*) from China, in the Danube delta in Romania, sometimes causing problems of competition with local species (Cowx 1997). Even when introduced in closed fish ponds, however, some alien species may well have detrimental effects when individuals escape human control. For example, when, in the 1970s, Israeli farmers also introduced the silver carp for commercial production in aquaculture systems, they were running a great risk. This is an extremely aggressive species with a high potential to escape and become invasive. Fortunately, the species does not appear, as yet, to reproduce in the wild (Roll *et al.* 2007).

The introduction in 1983 of herbivorous carps in Lake Oubeira, Algeria, resulted in the rapid destruction of half the native reed beds (Pearce and Crivelli 1994), with dire consequences on the native biota. Fortunately, Mediterranean wetlands have not experienced the catastrophic consequences of invasion by exotic fish species such as those that have so badly decimated cichlid fish communities in the great Lakes of East Africa. However, ongoing fish invasions (Clavero and García-Berthou 2006) and escapes from aquaculture farms (see Chapter 13), combined with the highly seasonal Mediterranean climate in southern Europe, may increase the risk of extinction for endemic fish that are already threatened (Griffiths 2006). Indeed, throughout the Mediterranean region, special attention and legislation should be given to control the spread of invasive fish species in the wetlands and rivers, all of which are recognized as hotspots of fish diversity and providers of multiple ecosystem services to people. The same is true for other aquatic organisms such as the Louisiana crayfish (see Plate 5b), even though, paradoxically, some of them may be beneficial to one or more native species, as described in Box 12.3.

12.3.5 Reptiles and amphibians

Two alien species of amphibians and reptiles are or will presumably be serious threats to native biodiversity. The first is the bullfrog (*Rana catesbeiana*), from the USA, and the second is the red-eared slider (*Trachemys scripta*), a freshwater

---

Box 12.3. The paradox of invasions: the Louisiana crayfish

The Louisiana crayfish, first introduced in fish ponds in Spain in 1973 (Habsurgo-Lorena 1983), rapidly invaded Europe and tends to be superabundant whenever present (Arrignon 2004). This species is particularly well adapted to Mediterranean wetlands because it can tolerate long dry spells thanks to its burrowing habits and dispersal behaviour. In addition it is much less sensitive to salt than other crayfish species. The Camargue has been invaded by this species since at least 1995. Exotic crayfish species are an example of the paradox of invasions because they may benefit populations of emblematic and popular predators such as herons that feed upon them. For example, herons in the Camargue have flourished as it is now. For example, among the six common prey categories of the Eurasian bittern (*Botaurus stellaris*), the Louisiana crayfish contribute to up to 80% of the diet. This invading species contributes to the increase of populations of this rare species, as shown by a correlation between bittern density and crayfish abundance (Poulin *et al.* 2007). The Camargue is probably the only location in the basin where the nine European species of herons (Ardeidae) currently breed—bittern, little bittern (*Ixobrychus minutus*), night heron, cattle egret (*Bubulcus ibis*), squacco heron (*Ardeola ralloides*), little egret (*Egretta garzetta*), great white egret (*Ardea alba*), grey heron (*Ardea cinerea*), and purple heron. In fact several species of mammals, reptiles, fish, and birds have switched from a fish diet to a crayfish diet (Barbraud *et al.* 2001). Unfortunately the 'negative side of the coin' is undetected and irreversible damage to ecosystem functioning, causing severe damage to many communities of invertebrates, especially larvae of dragonflies and several families of aquatic beetles, such as Ditiscidae and Hydrophilidae. This crayfish is also responsible for a strong decline of amphibians, both frogs and newts. In addition, because the crayfish also grazes on higher plants, it has transformed many marshes from macrophyte-dominated clear-water bodies to turbid phytoplankton-dominated waters (Rodriguez *et al.* 2005).

Coming across an angler some years ago near a canal in the Camargue, one of us saw him pulling a bass out of a thick carpet of *Ludwigia*, a semi-aquatic relative of the evening primrose. It appeared that the belly of the fish was full of Louisiania crayfish! All three players—the bass, the *Ludwigia*, and the crayfish—were invasive exotics, and all of them had probably flourished at the expense of native species!

turtle also native of North America. The bullfrog, which was introduced in Italy in the 1930s and has since expanded northwards, is a formidable predator that can reach 20 cm in size. It attacks a wide range of prey, including invertebrates, other species of frogs, small fish, small snakes (e.g. *Natrix maura*), and young water birds. This animal was originally introduced as a pet but is now found in most rivers and lakes of western Europe, as a result of feralization. In addition, this invasive frog is a frequent carrier of the fungus *Batrachochytrium dendrobatidis* which decimates native populations of amphibians.

Regarding the red-eared slider an experimental study on competition between this turtle and the pond terrapin (see Box 3.1) revealed that both body mass and survival of the native pond terrapin were lower in experimental fish ponds where the invasive slider was introduced than in ponds including only the native species (Cadi and Joly 2004). This alien turtle can become a real threat because it survives harsh winters in the Mediterranean and successfully reproduces in the same habitats as the native species.

In addition to invasive species which, according to our definition, are not native to the Mediterranean, we will nevertheless mention two translocated turtles—*Testudo graeca* and *Testudo marginata*—which are Mediterranean species but were intentionally introduced in Sardinia, outside of their natural distributional range. Sometimes, especially on islands, such translocated species may displace local endemics. This is the case of the mainland lizard, *Podarcis sicula*, introduced probably inadvertently in several islands. Its introduction in the Balearic Islands, for example, resulted in the extinction of the endemic *Podarcis lilfordi* of Minorca. In Corsica, *P. sicula* displaces the local endemic *Podarcis tiliguerta*, and, in several islets of the Dalmatian coast, it out-competes the local *Podarcis melissellensis* (Cheylan and Poitevin 1994). It is also quite likely that the extinction of certain species such as the Minorcan midwife toad (*Alytes talayoticus*) and the quasi-extinction of the Majorcan midwife toads (*Alytes muletensis*) are due to the recently introduced green frog (Cheylan and Poitevin 1994).

12.3.6 Invasive birds

There are very few bird species from other continents that occur regularly as established breeders in the Mediterranean region, despite the fact that hunting clubs and agencies repeatedly introduce exotic species in hopes of increasing and diversifying their game scores. Examples in the western Mediterranean are California quail (*Lophortyx californicus*), Reeve's pheasant (*Syrmaticus reevesii*), and black francolin (*Francolinus francolinus*), from the Middle East. Even the pheasant, which was introduced by the Romans, has not succeeded in establishing self-sustaining populations in the wild. It occurs only in localized lowland areas where humans continuously breed it and release it for hunting. Attempts to translocate native species from one region into another are also frequent, for example, the partridges *Alectoris barbara* and *Alectoris chukar* which were repeatedly introduced from nearby mainland areas into Mediterranean islands, but usually unsuccessfully.

Exceptions to this general rule are some cases of intentionally introduced birds which escaped human control. The first is that of small free-living populations of the monk parakeet (*Myopsitta monachus*), in Spain, and the ring-necked parakeet (*Psittacula krameri*), which breeds in some large Mediterranean cities, such as Nice, Madrid, Lisbon, and others. Other examples are the tiny red munia (*Amandava amandava*), from South Asia, and the common waxbill (*Estrilda astrild*), from Africa, both of which have developed populations of more than 1000 individuals in the Spanish province of Extramadura (De Lope *et al.* 1984; Real *et al.* 2008).

Several other exotic bird species have escaped from captivity but are not considered a threat for native species. These include Indian silverbill (*Lonchura malabarica*), Reeves's pheasant, Fischer's lovebird (*Agapornis fischeri*), masked lovebird (*Agapornis personatus*), red-billed leiothrix (*Leiothrix lutea*), the Chilean flamingo (*Phoenicopterus chilensis*), and the lesser flamingo (*Phoeniconaias minor*). The mute swan (*Cygnus olor*), from central Asia, is currently increasing in numbers in many parts of Europe and parts of the Mediterranean region. This large species probably contributes to the eutrophication of marshes by its abundant droppings.

Kark and Sol (2005) challenged the view that few bird species have invaded the Mediterranean, arguing that 'the proportion of bird species successfully established was high. The probability that a species will become established was highest in the Mediterranean Basin and lowest in Mediterranean Australia and the South African Cape'. However, of the 40 species they list for the Mediterranean, more than 30 are not firmly established and have only very small localized unstable populations. Other species listed as native have actually been translocated quite recently, such as the partridges mentioned above. These cannot be considered as invaders *sensu stricto*.

One alien species which is indeed problematic is the ruddy duck (*Oxyura jamaicensis*), first introduced in the UK from the Caribbean in 1949. It has since escaped from captivity and reached Mediterranean shores, especially Spain, where cases of hybridization with the native white-headed duck have been reported since the 1990s (Pascal *et al*. 2006). Unfortunately, progress in the recovery of the white-headed duck is limited due to inadequate eradication of the ruddy duck.

An interesting case study is that of the blue magpie (*Cyanopica cyanopica*). This beautiful bird lives in small flocks in the open forests of southern Spain and Portugal, in populations that are totally disjunct from the majority of populations of this species, which occur in China and Manchuria. Explaining such a pattern has long been a brainteaser for biogeographers. It was long thought that the species was brought from Asia to Spain sometime during the seventeenth century by European navigators or missionaries. But fossil remains of this bird have recently been found in southern Spain (Cooper 2000), which proves that the bird is in fact native to Iberia. Two last species to mention are the sacred ibis (*Threskiornis aethiopica*), which has recently developed small breeding feral populations in the Camargue and a few other places of the basin where it can very possibly out-compete colonial species of herons, and the African collared dove (*Streptopelia roseogrisea*), which now breeds in small numbers in Spain and should be closely monitored.

12.3.7 Invasive mammals on Mediterranean islands

Starting around 7000 years BP, people progressively introduced all the extant mammal species on the various Mediterranean islands. Some were intentionally introduced as game species (e.g. deer and fox) or as domestic animals (sheep, goats, pigs, horses, donkeys, cattle, and dogs). Many smaller mammals (e.g. *Crocidura*, *Suncus*, *Apodemus*, *Mus*, *Rattus*, *Eliomys*, and *Glis*) were introduced accidentally, often as stowaways on ships, a process mentioned already.

Although the diversity of carnivores on Mediterranean islands is fairly high (Cheylan 1984), large species, such as the wolf, the wildcat (*Felis silvestris*), and the lynx, were never introduced, probably because they were perceived as potential competitors for people. Introduced predators were all small species, such as the fox (sixth millennium BC), the dog (fifth millennium BC), the weasel (*Mustela nivalis*), and the domestic cat (*Felis libyca*) (see Chapter 10). Most of the 24 species of mammals introduced in Corsica are ecological generalists strongly associated with humans, both ecologically and culturally (Vigne 1990). Overall, extinction-immigration processes have resulted in a three- to five-fold increase of mammal species diversity but a dramatic loss of indigenous species and genetic diversity.

Some large mammals became an important part of the present fauna through feralization. This process is exemplified by the Corsican mouflon (*Ovis ammon musimon*; Fig. 12.1), which was absent from western Europe during the Pleistocene. As described in Box 12.4, the contemporary Corsico-Sardinian mouflon is nothing other than a relictual Neolithic domestic sheep that escaped human control before the late Neolithic (Vigne 1988). A similar process of feralization probably occurred in the Corsican boar and the wildcat *Felis silvestris libyca* var. *reyi*. In addition, competition from introduced ungulates, such as sheep, goats, and cattle, indirectly led to habitat loss, as a result of forest and matorral clearing undertaken by people for

Figure 12.1 The Corsican mouflon (R. Ferris).

Fossilized bones of humans have been found together with remains of this antelope, which suggests the species was protected, managed, or else thoroughly domesticated. The replacement of an impoverished, endemic, and disharmonic mammal fauna by a **supersaturated** and monotonous fauna during the Holocene occurred on all the larger Mediterranean islands.

The only relictual Pleistocene endemic species on islands in the whole Mediterranean are three shrews, a rodent, and a porcupine (see Chapter 3). Even the Cretan goat (*Capra aegagrus cretica*), long thought to be endemic, is a feral population of the domestic goat.

Recent invasive species of mammals include several rodents: the coypu and the muskrat from North America, which had been introduced for their fur (see Chapter 2). The American mink (*Mustela vison*), which has also been introduced for its fur, may threaten the rare European mink (*Mustela lutreola*). Three species of mongooses have been locally introduced: *Herpestes ichneumon* in the Iberian Peninsula and Croatia, *Herpestes auropunctatus* in several Adriatic islands, and *Herpestes edwardsi* in Italy, but they do not seem to expand their range from the first locality of introduction (MacDonald and Barrett 1995). Other mammal species that could become pests in the future, if they reach the Mediterranean,

the benefit of their domesticates (Vigne 1983; see Chapter 10).

The story of mammals on Corsica may be extrapolated to other Mediterranean islands, except in the Balearic Islands where human colonization was delayed until the Mesolithic, some 8000 years BP. Domestication of an endemic antelope, *Myotragus balearicus*, apparently saved it from extinction until the Bronze Age or Iron Age (Alcover *et al.* 1981).

Box 12.4. The Corsican mouflon

The Corsican mouflon has long been considered native in Corsica and Sardinia. In fact, archaeozoological studies show that it derives from a sheep that was domesticated in the Middle East about 10 000 years BP. This sheep, *Ovis aries*, spread rapidly throughout the Mediterranean Basin, and, in some places, especially on large islands such as Corsica, a few animals escaped human control. These feral animals sometimes gave rise to genetically altered populations through a process whereby ancestral characters progressively revert to type in the absence of on-going human selection. Much later, in the twentieth century, the Corsican mouflon was introduced as a game animal in many parts of Europe, North America, and even such remote places as the Hawaiian and Kerguelen islands.

This mouflon is a rather small animal, weighing about 30 kg for females and 40–45 kg for males (see Fig. 12.1). Adult males have spiral horns that vary greatly in shape and size. Surprisingly, some females have horns as well. This species lives in small herds varying in size. During the autumn rutting period, a rigid social hierarchy exists among males, resulting in fights, which may be fierce; the dominant male monopolizes most of the females. Around 400–600 individuals presently live in Corsica, moving according to seasons, and grazing in a variety of habitats from the mountaintops in summer to lower altitudes in winter.

include the raccoon dog (*Nyctereutes procyonoides*), which was massively introduced in the former USSR for its fur between the 1930s and the 1950s, and the raccoon (*Procyon lotor*), from North America. A very serious problem could emerge as well with regard to the grey squirrel (*Sciurus carolinensis*), which was introduced in Italy in 1948 and has become thoroughly naturalized. It seems likely that it could outcompete and replace the native Eurasian red squirrel (*Sciurus vulgaris*). The grey squirrel is currently expanding its range in Europe, but any project to eradicate it is fiercely rejected by the public which considers it attractive! This is similar to the problem conservationists face with regards to the cultivated and invasive mimosa in southern France, as mentioned above.

Several species of mammals have been more or less successfully introduced as game species. They include the aoudad from desert areas of Africa in several regions of Spain (Real *et al.* 2008), the fallow deer from the Middle East in several countries of the western part of the basin, and the mouflon, *Ovis gmelini*. In the 1950s, after the collapse of most local populations of the native rabbit *Oryctolagus cuniculus*, resulting from an attack of the myxomatosis disease, hunting agencies tried to introduce the American cottontail rabbit (see Chapter 2). Fortunately, most attempts to establish permanent populations failed, except in Italy where some populations have survived for more than 30 years (Pascal *et al.* 2006).

12.3.8 Marine aliens

Having made a brief tour of terrestrial invaders, let us now turn to the Mediterranean Sea itself, where approximately 12 000 species of marine organisms have been recorded thus far (see Chapter 4). How many of these are recent arrivals, of the last century or so? The answer is at least 500 and perhaps as many as 1000. As Briand (2002:7) put it: 'Although constricted..., the Straits of Gibraltar and the Suez Canal represent open gates to marine invaders, and only stop deep-sea migrants. Moreover, hundreds of cargo ships ply the route every day, bringing along cohorts of uninvited passengers arriving from distant harbours, either attached to the hull or in ballast waters. Transports by anchor fouling, recreational vessels, plus inadvertent releases from mariculture, aquariums, or scientific research constitute other vectors of introduction and further mechanisms for dispersal'. It is clear that we cannot put Suez and Gibraltar at the same level, Gibraltar being a route into the Mediterranean and remaining the major one to this day. The Suez Canal is a totally new route, opened by humans in the recent past (see Chapter 2), but, since then, it has had a very strong impact in terms of alien invasive species from the Red Sea. Altogether with its very high rate of endemism and heavy use by people, the Mediterranean Sea appears highly vulnerable to biological invasions, which are now widely recognized as one of the most significant components of global change in marine environments as they are in terrestrial ecosystems (Valéry *et al.* 2008), where they can have far reaching and often harmful effects on native biodiversity and the functioning of natural ecosystems.

Over 700 alien marine species, including seaweeds, invertebrates, and fish, have been recorded and are prominent in most coastal areas in the Mediterranean region (Galil 2007). With few exceptions, the ecological impact of these invasive alien species on native Mediterranean marine biota is poorly known, but it appears they may cause major shifts in community composition and ecosystem dynamics. Although no extinction of marine species has recently been recorded (Boudouresque 2004), there is much cause for concern.

Of course, a main avenue of invasion, mainly for the eastern part, is the Suez Canal, that was completed in 1869. For that, marine biologists speak of Erythrean or better still Lessepsian alien species, in reference to Ferdinand de Lesseps, the promoter, engineer, and overseer of the 166 km-long Suez Canal, which re-established direct communication between the Mediterranean Sea and the Indo-Pacific region for the first time in millions of years. The *raison d'être* of the canal was, of course, to facilitate the movement of economic goods. But the biological and ecological consequences were massive as well.

Despite impediments, such as the canal's length, shallowness, current regimes, temperature, and salinity extremes, hundreds of Lessepsian species have passed through the canal and settled in the Mediterranean in the past century and a half, many of which have established thriving populations (Galil et al. 2002). Overall, 5% of Mediterranean marine species are of Lessepsian origin, and, in the south-eastern part of the basin it is more than 10%, including fishes, invertebrates, and plants. Often, to ensure the validity of the records in the Mediterranean Sea, authors choose to include species of Indo-Pacific origin that were recorded after 1920, a convenient starting date because the Cambridge expedition to the Suez Canal in that year marked a turning point, not to say the dawn of Mediterranean marine science (Galil et al. 2002). But several invasive species had in fact been recorded before 1920, as some of them had proliferated extremely rapidly in the first decade after the opening of the canal.

Today, shipping-mediated bioinvasions are considered as the largest single vector for non-indigenous species movements. But, in the eastern Mediterranean, ballast and hull transported exotics, plus intentional and unintentional mariculture transfers (see Chapter 13), lag far behind the Lessepsian invasions in the number of introduced species. In recent decades, the arrival of exotic species from the tropical Atlantic through Gibraltar has been also well documented. Whether this reflects a warming trend of Mediterranean waters, an expansion of maritime traffic, or a simple artefact, remains to be seen (Briand 2002). In any case, recent invasions in the Mediterranean are today well-documented (review in Galil 2007).

12.3.8.1 PLANTS AND ALGAE

The Mediterranean Sea harbours the largest number of introduced macrophytes in the world: marine botanists at work in the CIESM Exotic Task Force have drawn up a list of 110 taxa (22 Chromobionta, 71 Rhodobionta, 16 Chlorobionta, and one Magnoliophyta), which they consider as recent arrivals to the Mediterranean Sea (see the CIESM website, www.ciesm.org).

Two invasive green algae have been especially well studied (Boudouresque et al. 1995) and both have been found to be particularly harmful to Mediterranean marine ecosystems (see Chapter 4). *Caulerpa taxifolia*, sometimes known as 'killer seaweed', was first reported in 1984, off the shores of Monaco, having somehow escaped from the national aquarium. Within 20 years, it had spread to coastal areas of Spain, France, Italy, Croatia, and Tunisia, where it grows rapidly and forms dense meadows—with up to 14 000 leaves m^{-2}—on various infralittoral bottoms. This has lead to the formation of homogenous microhabitats and the displacement of native species.

This process of homogenization can reduce species richness of native hard substrate algae by 25–55% and the killer seaweed may also, very specifically, out-compete native species, such as *Cymodocea nodosa* and *Posidonia oceanica* (Boudouresque et al. 1995). Another problem is that fish assemblages are less rich in *Caulerpa* beds than in native *Posidonia* seaweed beds. The repellent endotoxins produced by *Caulerpa* also widely affect associated invertebrate faunas (Longpierre et al. 2005).

The second species of invasive seaweed is *Caulerpa racemosa* var. *cylindracea*, which was introduced into the Mediterranean by ships coming from Australia (Carriglio et al. 2003). Today it is found along coastlines from Cyprus to Spain and in the Canary Islands. It is known to attain total coverage in certain zones, within a mere 6 months after introduction. In Cyprus, where it was first noticed in 1991, it replaced the dominant *Posidonia oceanica* community within 6 years. Invertebrate communities have also been profoundly affected through proliferation of polychaetes, bivalves, and echinoderms, and reduction of gastropods and crustaceans (Carriglio et al. 2003; Piazzi and Balata 2008).

Another Lessepsian invasive plant is *Halophila stipulacea*, widely distributed along the western coasts of the Indian Ocean and the Red Sea, which also competes with native *Posidonia*. Previous studies have revealed both high phenotypic and genetic variability in *Halophila* populations from the western Mediterranean basin (Valeria Ruggiero and Procaccini 2004).

In the Thau lagoon of southern France, more than 60 large seaweed species from the North Pacific have been unintentionally introduced and have become naturalized (Mineur et al. 2007). To date, no detailed information is available about introduced macro-seaweeds, such as *Sargassum muticum*,

Undaria pinnatifida, *Chondrus giganteus*, *Litophyllum yessoense*, and *Porphyra yezoensis*, all of which are originally from Japan and are now developing rapidly in the lagoon. Not only is the Thau lagoon already very heavily modified and transformed by human activities—divers there may think they are really in Japanese waters!—but also many oyster producers there exchange oyster spat with other growers in other parts of France and indeed neighbouring European countries as well. As a result, the Thau lagoon is a strategically critical area for the entire western Mediterranean.

12.3.8.2 FISH, MUSSELS, OYSTERS, AND LIMPETS

Of the 650 fish species presently described from the Mediterranean, at least 90 from 56 different families are migrants or invasive (Golani *et al.* 2002). Of the known 350 decapod species, 60 are newcomers (Galil *et al.* 2002), and most have arrived from the east, via the Suez Canal. For molluscs, the region is home to approximately 1800 species (Zenetos *et al.* 2003), many of which are more or less recent immigrants. For example, the exotic mussel *Brachidontes pharaonis* and the pearl oyster *Pinctada radiata* were reported in the region as early as 1877–8 (Zenetos *et al.* 2003).

A typical Lessepsian fish species, the goldband goatfish (*Upeneus moluccensis*) was first observed along the coasts of Israel and Lebanon in the 1930s. Since then, it has established populations along the coasts from Rhodes to Libya, where it actively competes with red mullet (*Mullus barbatus*), a native species adapted to cooler, deep waters and which is thus highly threatened by global warming (Golani *et al.* 2002; Galil 2007).

Yet another recent invasive of note is the lizardfish (*Saurida undosquamis*), from the Indo-Pacific Ocean. First recorded from the Mediterranean coast of Israel in 1952, today it occurs as far east as Albania. Later, this species, and also the Indo-Pacific goldband, proliferated enormously after the unusually warm winter of 1955 (Galil 2007). By the mid-1960s, it formed the main catch of trawlers, and, in the late 1980s, it represented two-thirds of the fish-landing biomass. It competes directly with the European hake (see Chapter 11), which is also demersal, which means that it lives close to sea bottoms consisting of flat sand and muddy substrates at depths of 30–70 m, rarely more than 100 m.

This is an example of what Pérez (2008) calls the 'southernization' of the Mediterranean marine biota as a result of global warming, a process that appears to be particularly pronounced, at least in the well-studied north-western quadrant of the basin. He cites as indicator species of this process the ornate wrasse (*Thalassoma pavo*), the grey trigger fish (*Balistes carolinensis*), the sea urchin *Centrostephanus longispinus*, and the dusky grouper.

Another example of disruption is related to hybridization, by which an endemic fish genotype has been lost within the last 25 years. To wit, the killifish *Aphanius dispar*, which originated in the Red Sea but is unusually euryhaline—capable of migrating back and forth between bodies of fresh water and seawater—and its various hybrids have totally replaced *Aphanius fasciatus*, the native Mediterranean species, along the Mediterranean coast of Israel (Goren and Galil 2005). Similarly, the native meager (*Argyrosomus regius*) was replaced by the narrow-barred mackerel (*Scomberomorus commerson*), and two species of dragonet, *Callionymus pusillus* and *Callionymus risso*, have been supplanted by *Callionymus filamentosus* along the Levantine upper shelf.

There are also known cases of transformers among fish invaders. For example, two Lessepsian species of the rabbitfish group (Siganidae), present in the Mediterranean since the 1940s, nowadays proliferate as far west as the southern Adriatic Sea, Sicily, and Tunisia (Galil 2007). One of these, the marbled spinefoot (*Siganus rivulatus*), has replaced the native herbivorous cow bream *Sarpa salpa* in various areas of Lebanon, Israel, Syria, and Libya, profoundly affecting native algal communities. The second species, *Siganus luridus*, is now common along the eastern coasts of the Mediterranean, from Libya north to Turkey, and has also reached Tunisia. The grazing pressure of these two rabbitfish on the intertidal rocky algae may be responsible for the proliferation of the above-mentioned mussel, *Brachydontes pharaonis*, which is also of Lessepsian origin, by providing suitable substrate for its establishment. Over the past century, this miniature mussel has replaced the native *Mytilaster minimus*, formerly abundant throughout the eastern Mediterranean (Galil 2007). This has had dramatic effects on the biota of the hard substrates. For example, populations of the mussel predator, the whelk *Stramonita haemastoma* have grown enormously.

Another example is the Lessepsian limpet, *Cellana rota*, first collected in the Mediterranean in 1931. Today, it out-competes the native Mediterranean limpet, *Patella caerulea*, dominating the upper rocky littoral. Along the central coast of Israel, for example, it now occupies 40–50% of the available space in that habitat (Mienis 2003). Yet another telling case is that of *Pinctada radiata*, a pearl oyster originally from the Indo-Pacific, including the Red Sea. First sighted in Alexandria in 1874, it gradually invaded coastal areas in Greece and Turkey and now occurs as far west as Tunisia (Zenetos *et al.* 2003). Recently, it has been reported in Toulon, France, and Trieste, north-eastern Italy, as well. This invasive oyster is remarkably resistant and is sometimes used as a bioindicator of pollution. It is also overcollected in search of its pearls and is cultivated in Greek waters as well. The related species, *Pinctada margaritifera*, which is much more interesting for the quality of its pearls, has recently been intentionally introduced for mariculture along the Calabrian coast in southern Italy and in Egypt. Thus, this is a complex and rapidly evolving group of organisms in the Mediterranean, some of which may well prove to be transformers.

12.3.8.3 SHRIMPS

Eight species of Lessepsian shrimps in the Penaeid family have been recorded in the Mediterranean to date, two reaching as far west as Tunisia. *Marsupenaeus japonicus*, *Metapenaeus monoceros*, and *Penaeus semisulcatus* today are of great interest for local fisheries, new trawlers being specifically designed and launched to catch these introduced species (Galil *et al.* 2002; Galil 2007). However, the economic boom associated with these new species comes at the expense of the native shrimp *Melicertus kerathurus*. In particular, *Melicertus japonicus* has almost evicted the native shrimp from the eastern part of the Mediterranean, and the rapid advent of *Metapenaeus monoceros* into the Gulf of Gabès in Tunisia has raised concern over the fate of *Melicertus kerathurus* fisheries there (Galil 2007). Similarly, the Lessepsian snapping shrimps *Alpheus inopinatus* and *Alpheus audouini* are more common in the eastern Mediterranean rocky littoral regions today than the native *Alpheus dentipes*. On muddy bottoms as well, the invasive *Alpheus rapacida* is more abundant than the native *Alpheus glaber*.

12.3.8.4 OTHER MARINE CREATURES

In other marine groups as well, a similar pattern emerges. Over recent decades, the Indo-West Pacific starfish, *Asterina burtoni*, has replaced the native Mediterranean *Asterina gibbosa* (Galil 2007), the Erythrean spiny oyster (*Spondylus spinosus*) out-competes the native congener *Spondylus gaederopus*, and the Lessepsian jewel box oyster, *Chama pacifica*, outnumbers *Chama gryphoides*. Meanwhile, the native Mediterranean cerithiid gastropods, *Cerithium vulgatum* and *Cerithium lividulum*, have been supplanted by the Lessepsian Cerithiids *Cerithium scabridum* and *Rhinoclavis kochi* (Mienis 2003). We could continue this list almost indefinitely, thanks to three recent books (www.ciesm.org/atlas/) describing almost 300 species of alien invaders in the Mediterranean Sea.

In sum, there is much evidence that invasive alien organisms of various kinds have led to significant reduction or loss of native species in the Mediterranean Sea (Galil 2007). Additionally, significant interference with the functioning, dynamics, and stability of marine communities is underway, including the modification of marine food chains and the simplification or enhancement of local specific diversity (Briand 2002). Some experts think that the establishment of alien biota and the consequent changes in marine communities and food webs prefigure catastrophic and clearly anthropogenic ecosystem shifts in the sea.

The contracting parties to the Barcelona Convention (1995) specified that 'we have to take all appropriate measures to regulate the intentional and non-intentional introduction of non-indigenous into the wild and prohibit those that may have harmful impacts on the ecosystems, habitats or species'. A few attempts have been made to eradicate a few migrant species, but it is extremely difficult, or impossible, in the sea. On the other hand, projects to introduce non-solid barriers—such as bubble fields, sound or electric fields, various kinds of particles, etc.—have been advanced to limit migratory fluxes through the Suez Canal, from the Red Sea into the Mediterranean. Time will tell what this approach brings.

12.4 Climate change

According to the Intergovernmental Panel on Climate Change (IPCC 2007) and the European Environment Agency (EEA 2007), global climate change is unequivocally demonstrated by (1) a significant increase of temperatures of both air and seawater on a worldwide scale, (2) a generalized melting of snow and ice, and (3) a significant rise of sea level. Moreover, global warming in particular probably represents one of the most significant threats to biodiversity, worldwide, given its proven potential to affect areas far from any human habitation (Malcolm et al. 2006). Thomas et al. (2004) tentatively forecast the loss of 1 million living species by 2040 due to global warming alone. However, this conjecture has been much challenged in many studies.

12.4.1 Global warming

During the twentieth century alone, the Earth warmed up by around $0.74°C$, with regional variations of course, especially in summer. In the Mediterranean Basin, air temperature was observed to have risen by $1.5–4°C$ depending on the sub-region. Over the same period and with clear acceleration since 1970, temperatures in south-western Europe—Iberian Peninsula and southern France—rose by almost $2°C$. The same warming effect can also be seen in North Africa, albeit more difficult to quantify given the more patchy nature of the observation system (EEA 2007). Even if the European Union's objective of not exceeding a global average temperature increase of $2°C$ is met, temperature increases in the Mediterranean are likely to be above $2°C$, because of the ecological and socio-economic characteristics of the area. For the sea, surface temperature have increased in the last 30 years, between 0.5 and $3.5°C$, depending on the site, especially in the eastern part, around Cyprus particularly (EEA 2007).

According to the different scenarios of greenhouse gas emissions, the IPCC predicts a global temperature increase between 1.8 and $4°C$ before the end of the twenty-first century and perhaps as much as $6.4°C$ (IPCC 2007). Concurrently, we can expect increasing occurrences of violent climatic events, such as floods, storms, droughts, and increases in diseases, pests, and invasion by non-native species that compete with native species populations. The projected climatic changes for the twenty-first century will be faster and more important than any experienced in the previous 40 000 years (Bush et al. 2004) and probably in the previous 100 000 years.

For the Mediterranean Basin specifically, the various models proposed by the IPCC (2007) all suggest that (1) there will be a strong increase of average annual temperature, (2) especially in summer, when there will be higher peaks of maximal temperatures; (3) mean annual rainfall will also decrease, (4) as will summer rainfall, especially in the second half of the twenty-first century. In addition, (5) there will be a decrease in the number of rainy days and therefore more prolonged droughts, as well as (6) a decrease of river flows; (7) a strengthening of the Azores anticyclone will reinforce the positive phase of the North Atlantic Oscillation (NAO), and (8) there will be a decrease in the number of depressions and associated violent winds. The increase in temperatures and a strong decline of rainfall will make ecosystems and organisms more vulnerable to water shortage, large wildfires, changes in the distribution envelopes of species, and loss of agricultural potential. In sum, the Mediterranean region is expected to be more severely hit by the various components of climate changes than temperate regions further north (IPCC 2007), making it a 'hotspot' for climate change (Giorgi 2006). For the sea, climate change will very likely cause large-scale alterations in sea temperature, sea level, sea-ice cover, the dynamics of currents and the chemical properties of seawater (CIESM 2008).

12.4.2 Impact of climate change on community composition and distribution ranges

Observed biological impacts include altered growing seasons and shifts in species composition and distribution. Further impacts could include the loss of marine organisms, especially those with carbonate shells as a result of acidification. Adaptive policies should include measures to reduce non-climatic impacts in order to increase the resilience of marine ecosystems and the coastal zones to climate change.

As shown in Chapter 11, effects of global warming are perhaps already apparent in making forests and matorrals increasingly vulnerable to large wildfires. The dreadful summer of 2003 could be the first signal of what will happen more and more frequently. In that year, temperatures were the highest ever recorded in southern Europe, with a maximum of 45.4° C in the Algarve, southern Portugal (Clement 2008). An exceptional episode of dryness and heat also occurred in Greece in the 2007 summer, with catastrophic consequences for people and forests.

The increase in intensity and duration of summer drought such as that of 2003 will almost certainly have serious long-term consequences, including a decline of key tree species, such as cork oak, holm oak, beech, and some species of pines, especially maritime pine. Climate change may also be an important driver of biological invasions by affecting the global distribution of invasive species of plants, animals, and micro-organisms (Thuillier *et al.* 2005). The question thus arises, how will Mediterranean biota and ecosystems respond?

Species can respond to climate warming in a number of ways, including range shifts, phenotypic adaptation, or local specialization through evolutionary responses to new selection regimes (Médail and Quézel 2003).

Range shifts lead to a turnover of species, i.e. community reassembly, whereas local adaptation allows species to persist in the same community composition. Climate warming has already many effects on populations and communities, and the three mechanisms mentioned above have already been described at a broad scale. Changes in the distribution of species with distributional shifts in latitude and longitude, as well as changes in many fitness-related life-history traits, have been reported in animals and plants (Parmesan 2006). Bioclimatic envelope and ecological niche modelling have proved to be a successful tool for inferring plant distribution as a response to climate change or other environmental drivers (Pearson and Dawson 2003). Many modelling studies have attempted to depict the putative consequences of future climate change (e.g. Chuine and Thuiller 2005; Thuillier *et al.* 2004, 2005). Although there has been some controversy about the use of bioclimatic envelopes to forecast the response of plant species to global warming, the study of Rodrigues-Sanchez and Arroyo (2008) on the dynamics of *Laurus* species in response to the Plio-Pleistocene climatic changes supports the scenarios that predict the northward expansion of biotas as a response to global warming. Few studies to date have documented distribution shifts of Mediterranean species, but the fingerprint of global change is already apparent in all species and communities that have been studied so far, including in the Mediterranean region (Parmesan *et al.* 1999; Parmesan 2006). A progressive replacement of cold-temperate ecosystems by Mediterranean ecosystems has been reported in the Montseny Mountains (Catalonia, north-eastern Spain), where beech forests have shifted upwards by approximately 70 m. Both beech forests and heather heathlands are being replaced by holm oak forest (Peñuelas and Boada 2003). Using forest gap dynamics simulators for anticipating vegetation shifts in north-eastern Mediterranean mountain forests as a response to climate warming, Fyllas and Troumbis (2009) showed that fire events should increase in numbers and that they will be associated with elevational shifts of the dominant tree species. Current scenarios also predict a global shift of species and communities northwards, resulting in an increase in global biodiversity in northern Europe, with the Mediterranean holm oak reaching Scandinavia, and a significant decrease in the Mediterranean Basin (Chuine and Thuiller 2005). The overall picture of several global change scenarios (Thuillier *et al.* 2004) is that significant biological diversity loss is likely in the Mediterranean mountains, due to habitat tracking problems and interspecific competition.

12.4.3 Responses of communities and population to climate change

To date, very few studies have addressed community-level responses to climate change across elevational gradients of mountain ranges (e.g. Kazakis et al. 2007).

One exception is the demonstration of an uphill shift in butterfly species richness and composition in the Sierra de Guadarrama (central Spain), between the early 1970s and 2005 (Wilson et al. 2007). Butterfly communities shifted uphill approximately 300 m. Changes in species richness and composition included a loss of species of lower elevation and an increased domination in communities of widespread species, resulting in a homogenization of the communities along the gradient. Indeed, like birds, butterflies appear to be a very sensitive indicator group to study responses to climate change. Stefanescu et al. (2003) discuss the consequences that an increase in aridity in the Mediterranean Basin, caused by current climatic warming, may have on butterfly phenology and, as a result, on various species' population abundances, migratory patterns, and overall distribution.

In a study on the effects of multi-year droughts on fish assemblages of seasonally drying Mediterranean streams, Magalhães et al. (2007) reported that while present-day droughts cause relatively small and transient changes, 'longer and more severe droughts, expected under altered future climates, may result in declines or local extinctions of the most sensitive species and their potential replacement by more resistant species'. Furthermore, by simulating changes in the distributions and species richness of 120 native terrestrial non-volant European mammals, under two of IPCC's future climatic scenarios, Levinsky et al. (2007) conclude that as many as 5–9% of European mammals may be at risk of extinction, while 32–46% or 70–78% may be severely threatened (i.e. lose more than 30% of their current distribution). However, these authors suggest that if mammalian species richness becomes dramatically reduced in the Mediterranean region, it may also increase towards the north-east and at higher elevations. Very cogently, however, they point out that 'Bioclimatic envelope models do not account for non-climatic factors such as land-use, biotic interactions, human interference, dispersal or history, and our results should therefore be seen as first approximations of the potential magnitude of future climatic changes' (Levinsky et al. 2007: 3803). An excellent illustration of this caveat was also provided by Kéfi et al. (2007), who showed how the impact of grazing provides an equally important external stress on the spatial organization of vegetation in arid or semi-arid Mediterranean systems in southern Spain, Greece, and Morocco, and how overgrazing can accelerate the effects of global warming to produce desertification.

Establishing how past climates have shaped distributions and brought extinctions is of great relevance for providing insights on expected future changes as a response to global changes. Phylogeographic studies on the nature of re-colonizing populations following glacial episodes could be useful for predicting the responses of populations to global climate change. However, the long-lasting idea that almost all biotas had to find refugia in the three main peninsulas of the Mediterranean Basin during glacial episodes (e.g. Taberlet et al. 1998; Hewitt 1999) has been mitigated by the discovery of higher-latitude refugia (see Chapter 2) and the existence, besides the three large peninsulas, of many smaller refugia within the Mediterranean Basin (Médail and Diadema 2009), which leads to a re-evaluation of the migration rates of species based on the apparent re-colonization from low-latitude refugia (Provan and Bennett 2008). The existence of such higher-latitude refugia could influence predictions on the responses of various species and species groups to climate change (Svenning and Skov 2007; Valdiosera et al. 2007). In fact, responses of organisms to climate change assume that species do 'track' their habitats; that is, migrate in search of refugia when the local climate changes. However, as pointed out by Provan and Bennett (2008), even if phylogeographic approaches have clearly identified patterns of re-colonization during expansion phases following glacial maxima, they cannot give information on population shifts during contraction phases. Therefore, it is unclear whether organisms do carry out habitat tracking, or whether populations outside refugia simply go extinct. If that is the case, population extinction, rather than migration, might be the primary driver of range shifts. This is an important issue in current climate change scenarios, especially for populations that live at the limits of their range.

In birds, one case study has been provided by Lovaty (2008), who documented a northward shift of Mediterranean warblers which recently colonized new areas north of the conventional Mediterranean region. Subtle differences may occur, depending on the migratory behaviour of the species. For example, increasing winter temperatures are expected to lead to declines in the proportion of migratory bird species, whereas increases in spring temperatures and decreases in precipitation should lead to increases in this proportion (Schaefer *et al.* 2008). These authors also forecast that changes in the proportion of migratory species will be modest and that responses will more probably involve adaptations of the migratory behaviour rather than a turnover of species. A number of studies document changes in breeding patterns of birds, including advance in the onset of breeding and changes in associated life-history traits, such as clutch size and breeding success. The response often depends on whether populations are resident or migratory. One common explanation for the between-species variation in the response to climate change is that resident populations are more able than long-distance migrants to track changes in the availability of resources. Resident birds can adjust their breeding time to the spring advancement of resources, whereas migrant species cannot detect this on their wintering grounds, which may be thousands of kilometres away from their breeding sites (Crick 2004; Coppack and Both 2002; Rubolini *et al.* 2007). This is especially true in the Mediterranean region where climate warming in spring and summer is higher than further north (Sanz *et al.* 2003).

A consequence of global warming is already apparent for long-distance migratory birds, which usually spend the winter in sub-Saharan Africa. For many of these species, populations stay in the Mediterranean and therefore do not cross the Sahara. On average they have increased by a factor of 14 since 1970 (MedWet 2008). However, as discussed in Chapter 11, many of these species are not in good shape, ecologically speaking, as the index of their population sizes as nesting of migratory stopover did not increase in the same period. For example, some partial migrants such as the little egret became sedentary, whereas some populations of central or northern Europe, for example, the little ringed plover (*Charadrius dubius*) or the squacco heron, now winter in the Mediterranean, which up to now was a stopover area on their migratory journey. The importance of Mediterranean wetlands could increase in the future, especially if Sahelian wetlands continue to degrade.

12.4.4 Climate change and marine life

Finally, returning to the sea (Boeuf and Bodiou 2008), the effects of rapid climate change and water warming are more and more pronounced and worrisome (Bianchi and Morri 2000; Morri and Bianchi 2001; Zenetos *et al.* 2003; Garrabou *et al.* 2003). In addition to rising sea level, these trends have also triggered mortalities or ecological shifts in communities of marine invertebrates during recent years (Pérez *et al.* 2000, Pérez 2008). Of course, the impacts are more acute for species from the deep sea or those acclimated to marine caves, where temperatures are normally very stable throughout the year. In the Mediterranean, a warming trend with a higher frequency of exceptional meteorological-hydrological events has been well documented (Béthoux and Gentili 1999; Salat and Pascual 2002; Pérez 2008).

A species shift between two cave-dwelling mysids (Crustacea, Mysidacea) could be related to the more thermophilous characteristics of *Hemimysis margalefi*, which successfully out-competes its congener *Hemimysis speluncola* (Chevaldonné and Lejeusne 2003; Lejeusne and Chevaldonné 2005). Since the severe 1999 thermal anomaly, *H. margalefi* has become dominant in most dark marine caves of the north-western Mediterranean. This species was chosen as a biological model for further studies since heat-shock proteins (HSPs), which are produced by these organisms when facing a stressful condition, often increase with temperature, among other responses. For example, Lejeusne *et al.* (2006) studied the capacity of various species to adapt to changing temperatures and the great influence of temperature on **stenotherms**, leading to species turnover or clear risks of extinction. They may prove useful in developing a diagnostic approach of the fine-scale mechanisms, leading from a thermal anomaly to a disease outbreak or a shift in distribution (Pérez *et al.* 2000; Chevaldonné and Lejeusne 2003). Increasingly frequent cases of mass disease and mortality are occurring, most often as a result of a period of abnormal warming of water masses, such as occurred in 1999 and 2003 (Boury-Esnault *et al.* 2006). Today, it is clear that the mean temperature of the Mediterranean Sea is rapidly increasing, leading to multiple ecological impacts. Of course, deep sea-dwelling species, not accustomed to temperature fluctuations, are the most directly threatened. Information about very deep-sea thermal changes is lacking, in fact, and it is urgent to begin investigations in several areas of the sea. Along the coasts, many new species are appearing, as a direct result of global warming, and ecosystems are undergoing southernization, to use that odd word again. Another well-documented and intriguing problem to ponder concerns mobile species, actively migrating to the north in search of colder waters. What will they do in their tracking of still colder water once they reach the southern coasts of the northern Mediterranean countries, such as France, Italy, or Greece? Eventually, they will need to migrate even further to the north—that is, to the Atlantic—but there are only two exits available from the Mediterranean and, for a fish, they are not easy to find!

Summary

Rapid environmental changes are under way and will have profound consequences on species and communities in both terrestrial and marine ecosystems. Population dynamics of human societies is a major driver of changes in the Mediterranean Basin, a situation which is especially challenging because of the accelerated rate of urbanization and habitat fragmentation, mostly in coastal areas. Due to their high exposure to human activities and sensitivity to climatic conditions, Mediterranean ecosystems appear to be especially susceptible to the impacts of global change.

Biological diversity is threatened by invasive species, with homogenizing effects on communities. Invading plants are not a major threat in most habitats, but there is currently a trend of non-native plant species spreading rapidly in various ecosystems, especially along the coast. Apart from the emblematic example of biological invasion by mammals on Mediterranean islands, dating back several millennia, the only group of vertebrates with successful and sometimes harmful invaders in the Mediterranean area is fish. Many non-native fish species have been introduced in most of the major river basins in the Mediterranean, including those on islands. Few alien species of amphibians and reptiles occur as yet, but two species, the bullfrog and the red-eared slider, are or presumably will be serious threats to native biodiversity.

In the sea, well-known invading processes are those of Lessepsian invasions, which occurred after the Suez Canal connected the Mediterranean Sea to the Red Sea, allowing many species of the latter to invade the former. This process is called southernization, and its consequences will no doubt be profound. Recently, many species from the northwestern coasts of Africa have also invaded the Mediterranean Sea.

Predictions on climate change in the Mediterranean Basin lead to a series of expectations, which could make the basin more severely hit by climate warming than other parts of the Palaearctic. Climate warming has already many measurable effects on populations and communities, including shifts in the distribution of species and changes in life-history traits of many species.

CHAPTER 13

Challenges for the Future

Current trends are worrisome when we consider the prospects for the future of biodiversity and ecosystem functioning in the Mediterranean Basin, not to mention the future of cultural landscapes and indigenous cultures and languages, and ecological knowledge relevant to biodiversity. All known components of global change (Vitousek 1994; Vitousek *et al.* 1997) threaten biodiversity and ecosystems in this region to varying degrees, but the component of most concern is human population growth, as noted in Chapter 12. The deep-rooted Mediterranean mindset that 'culture' is mainly a humanistic affair and that 'nature' is somehow different and, of course, second in importance to culture has resulted in far too little attention being paid in the past to environmental quality, ecosystem 'health', biodiversity maintenance, and sustainable supply of natural ecosystem services.

After 10 000 years of co-habitation, not to say co-evolution, most Mediterranean ecosystems are so inextricably linked to human activities that the future of biological diversity and ecosystems cannot be disconnected from the realm of human affairs. In other words, we had best adopt the notion of socio-ecological systems when considering the fate of Mediterranean biota, landscapes, and peoples. In addition, biodiversity and ecosystems of all kinds should be grouped together and called 'renewable **natural capital**', as per the Millennium Ecosystem Assessment reports (MA 2005), in order to build bridges between ecology and economy in our increasingly human-dominated biosphere (Palumbi 2001).

Biological diversity is but one of the many pieces of the 'puzzle' that global society faces. Perhaps the most urgent of all is to alleviate the unacceptably large economic discrepancies between people on the northern and the southern shores of the basin. In this final chapter, therefore, we briefly discuss the problems, constraints, and prospects at a regional scale in the Mediterranean region, which is clearly a microcosm of world problems and conflicts. We address socio-economic, demographic, and natural capital issues, and end on a cautiously optimistic note. In the near future, however, the situation is worrisome indeed and demands serious attention, especially on islands and in coastal areas, as a result of greatly increased impacts from urbanization, habitat destruction, and biological invasions, among other factors (Hulme *et al.* 2008). Notably, human populations in the southern and eastern parts of the basin are becoming younger and poorer, compared with richer and older people in the north.

13.1 A microcosm of world problems

At the crossroads of three continents with their respective biotas, the Mediterranean region has eight main cultural and linguistic groups, three major religions, and many distinct political areas, ranging in size from tiny Malta and Monaco to the massive nations like Algeria and Turkey. It is also a meeting ground for the geopolitical north and south, with half a dozen affluent, industrialized countries, and a dozen and a half much poorer and much less economically developed countries to the south and the east. Geographically, it combines fertile lands, mountains, forests, islands, and desert fringes. It is all in all a vast frontier zone of divisions and conflicts and of convergence and cooperation. In sum, it is a socio-ecological mosaic like no other in the world.

The Mediterranean region also presents the full range of major socio-economic, political, and environmental problems that are faced by the whole planet. It may also have some lessons or models to offer to other parts of the world, thanks to the long human presence and the deeply remodelled landscapes and ecosystems that have resulted.

Today, 24 states totalling 474 million people constitute two sharply contrasting worlds within the basin, each with its separate histories (see Fig. 13.1). To the north-west, six countries, totalling 190 million inhabitants, adhere to the European Community and enjoy an average yearly income of about 17 000€ per capita. In contrast, the average income of some 277 million people along the eastern and southern shores (with the exception of Israel) is four times less and barely reaches 3700€ per capita per year (see Table 12.1). Emerging slowly from recent wars, the four redefined Balkan states and Albania all resemble the North African and the poorer Near Eastern countries, economically speaking, much more than Spain, France, and Italy.

In addition to the many efforts that should be made to maintain native biodiversity and restore or rehabilitate degraded ecosystems, future challenges include a thorough investigation of the consequences of the components of global change, because all of them threaten to various extents the biodiversity and ecosystems in this region (see Chapter 12). Certainly the components of most concern are human population growth, runaway urbanization, and increasing resource extraction, and their consequences for species and ecosystems. In our rapidly changing world, global change will affect the distribution and abundance of species (e.g. Walther *et al.* 2002; Balmford *et al.* 2003; Parmesan and Yohe 2003; Thomas *et al.* 2004). Besides foreseeable trends and events, consequences of global change could include unpleasant surprises resulting from unpredictable threshold effects on the dynamics of living systems (Schneider and Root 1996; Biggs *et al.* 2007). This is especially true in the Mediterranean region, which is highly vulnerable to global warming (IPCC 2007; EEA 2007; see Chapter 12). Let us now consider the problems of the north and south in turn, with special emphasis on the impacts of tourism and rural exodus, and the prospects of resource depletion, desertification, and overuse of water.

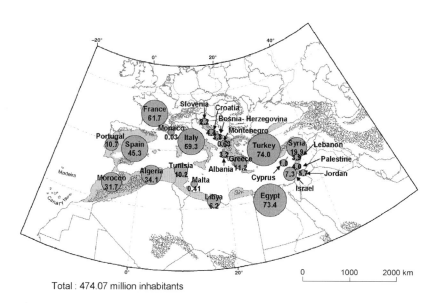

Total : 474.07 million inhabitants

Figure 13.1 Twenty-four nations around the Mediterranean Sea, with human populations given in millions.

13.1.1 The northern shores: urbanization, tourism, and rural exodus

Along the northern banks of the Mediterranean, human pressure on many ecosystems has steadily decreased because of agricultural abandonment and rural depopulation dating back to the end of the nineteenth century, but accelerating greatly since the Second World War. Across the entire range of life zones and habitat types we described in Chapters 5 and 6, a progressive recovery of forest and matorrals is proceeding at a rate of 1–2% per year. Abandonment of fields and pastures favours the re-colonization and spread of plant species that were formerly scattered in the landscape. For example the surface areas covered by the **anemochoric** Aleppo pine increased threefold between 1878 and 1904 and 2.6 times again between 1904 and 1978 (Achérar *et al.* 1984). In a study area of approximately 1000 km^2 in southern France, the amount of area occupied by vegetation over 2 m high and more increased from 7% in 1946 to 49% in 1979 (Lepart and Debussche 1992). In the same period, cultivated and grazed areas declined from 22 to 11%. Other expansive plant species that colonize many old fields are bird-dispersed species, such as *Pistacia*, *Rhamnus*, *Phillyrea*, and junipers. These processes of plant recovering lead to a decrease in habitat patchiness and the typical 'moving mosaic' landscapes that are so characteristic of the Mediterranean area and beneficial for biological diversity (see Chapter 8). Certain species are expanding their ranges whereas many other are disappearing locally from many regions. Since the potential productivity in most Mediterranean lands is insufficient to justify a reallocation of abandoned land to forestry production, more and more inland areas are increasingly becoming abandoned.

On the other hand, there is a growing demand for tertiary activities, especially near the coast, from promoters, speculators, and entrepreneurs of all sorts. From one end of the basin to the other, whole regions and landscapes are losing their age-old configuration and contours. Hence emerges a gloomy dichotomy: far from the coast, there are deserted fields, orchards, and pastures progressively encroached upon by shrublands and increasingly dense, unproductive, and ill-managed woodlands. Along the coast, urban sprawl and industrialization are increasing steadily in many areas, with densely urbanized, human-dominated, and homogenized zones, erasing all traces of the ecological and cultural richness of the Mediterranean history.

Well over 8800 km of Mediterranean sea coasts (19%) are now occupied by tourist installations, concrete structures, and diverse roadways (Henry 1977; see Box 13.1). In Italy, France, and Spain, including the Balearic and Canary Islands, large parts of Mediterranean coastal areas have already disappeared under concrete and macadam. According to some estimates, by 2025, fully half of the coastline of the entire region will be under concrete. On the island of Mallorca (Balearic Islands), for example, 48% of the coastline has been irreversibly 'artificialized' in this fashion, and the result is catastrophic. The same thing is true for several places in continental Spain, for example Benidorm. One result, of course, is that such sites become less and less attractive for the very activity for which they were artificialized! In the northeastern quadrant of the basin, in Greece, Turkey, Cyprus, Lebanon, and Israel, the same process is under way.

In 1990, the total coastal population in the basin was 140 million, with 95 million living in cities. In 2025, it may reach 200 million, with 170 million in the sprawling cities (Charpentier 1998). The ever-growing human agglomerations, combined with ever-denser industrial zones, result in high levels of pollution on land, and in rivers, lakes, lagoons, estuaries, and the sea. The Cousteau Foundation and many others warn of increasing maritime pollution as the Mediterranean Sea has become the dustbin for ever-expanding megacities along the coasts and on the islands. Only a few of them, unfortunately, are equipped with adequate waste and sewage treatment systems. This pollution is bacteriological, organic, metallic, radioactive, and chemical. When flying over the Mediterranean coasts in summer, one sees a brown ribbon several hundreds of metres wide all along the coast, accompanied by dense crowds of people basking in the sun or occasionally 'enjoying' the polluted waters nearest the beaches (Box 13.1).

> **Box 13.1. Tourism in the Mediterranean**
>
> The Mediterranean Basin is the world's leading tourist destination, with approximately one-third of international tourism revenues and 40% of total international arrivals (Apostolopoulos and Sönmez 2000). More than 265 million people visited the region in 2005 alone. The Mediterranean region appeals to tourists for four main reasons: a benign climate and attractive coastal fringe; exceptional variety and richness of natural, cultural, historic, and artistic attractions; proximity to the main European pool of potential tourists; and its image as a desirable tourist destination. Between 1999 and 2005, average annual growth of international tourism in the region was 3.2%, with a veritable boom for relatively new destinations, such as Turkey (+19.7%), Croatia (+14.3%), and Egypt (+10.2%). These three countries, among many others, made considerable investments in their tourist industries during those recent years, both with public and private funds.
>
> Mediterranean islands alone are visited by more than 37 million visitors annually (Arnold 2008). The top 10 islands by tourist capacity are Mallorca, Crete, Sardinia, Sicily, Cyprus, Ibiza, Rhodes, Djerba, Corsica, and Menorca. These 10 islands include 85% of the total Mediterranean island population and receive 76% of the visitors (Arnold 2008). Needless to say, such high seasonal populations result in excessive demands on services regarding transport, healthcare, freshwater supply, refuse collection, etc., which all weigh heavily on the resident population, while at the same time dramatically increasing their economic revenues.
>
> Tourism represents the mainstay of the economy in almost all Mediterranean and Macaronesian islands, which is a liberating force in some ways, because in the past almost all these islands, except Sicily, were at an economic disadvantage and strongly dependent on the mainland. Of course, such heavy influxes of people during summer months have many highly negative consequences on habitats and biodiversity, especially in coastal areas, not to mention freshwater supplies and other ecosystem services. This is all the more crucial because most tourists stay almost exclusively along the coasts. For example, 90% of the tourists who visit Greece stay on the coast where 75% of the local population also lives there. In addition, while providing a short-term bonanza, the tourist industry usually leads to a loss of traditional lifestyles and the corresponding cultural landscapes and biota. New, less harmful forms of tourism are developing, however, in combination with educational programmes on environmental issues, including ecotourism, agricultural tourism, and cultural tourism (Blangy and Mehta 2006). Furthermore, although each Mediterranean state traditionally has viewed its tourist product as competing with that of neighbouring states, Mediterranean countries are beginning to view their traditional competitors as 'partners' in regional collaboration, in order to secure a stronger position in the global tourist market in the coming years (Apostolopoulos and Sönmez 2000).

One of the biggest issues related to tourism and urbanization and changing land use in coastal areas is the search for sustainability and social justice in the distribution and use of limited supplies of fresh water. Bellot *et al.* (2007) provide an outstanding socio-ecological study of four municipalities of the Almería region of south-eastern Spain, one of the driest regions of Europe and one which receives a huge annual influx of tourists. In particular, they studied the changing proportions and web of interactions related to water use among five major categories of land use: urban, dryland crops, irrigated crops, shrubland, and woodland. Average water flow ($m^{-3} ha^{-1} year^{-1}$) for each of the main land uses was calculated in terms of three inputs—precipitation, irrigation, and urban supply—and the three major outputs—evapotranspiration, infiltration into the aquifer, and surface runoff. By taking such a holistic approach and considering both socio-economic and purely ecological factors and fluxes, a much improved planning process can be developed at municipal and micro-regional levels.

At a broader regional level, overexploitation of aquifers was already substantial by the early 1990s in many Mediterranean countries (Blue Plan 1999) and has gotten much worse since then. Indeed, coastal ground water has been reduced to below sea level by excessive pumping in Cyprus, Greece, Israel, Italy, Libya, and Spain (Blue Plan 1999), and also Turkey (EEA 2003). Today the estimates of overdraft are 13% in Cyprus, 24% in Malta (in 1990), 29% in Gaza, 32% in Israel (in 1994) and 20–25% in Spain (Blue Plan 2004). Aquifer overexploitation was also documented in Egypt, Libya, Morocco, and Tunisia. There is also severe salt-water intrusion into many coastal freshwater aquifers due to overexploitation of groundwater reserves (De Stefano 2004). None of this, of course, is sustainable or desirable; new planning, new morals, and new laws are needed and must be applied.

13.1.2 The southern shores: desertification and degradation of natural capital

In the overpopulated, drought-prone southern and eastern banks of the Mediterranean, the situation is just the reverse from that described above, from a natural capital perspective. Disturbance regimes in farm, pasture, and forest lands are moving towards still greater intensity of land and resource exploitation for short-term survival of local people. In the absence of sustainable land-use systems updated to meet the needs of young and growing local populations, ongoing forest clearing and land degradation through excessive, non-sustainable land use, are proceeding at a rate of about 2% per year (Marchand 1990), which is just about the same rate at which agricultural lands are being abandoned in southern Europe! Degradation has proceeded to the stage of desertification in many areas, and food security is a very real issue for large segments of rural populations (e.g. López-Bermúdez and García-Gómez 2006).

A visit to most mountainous areas of North Africa today reveals ongoing demographic growth combined with a highly conservative rural economy, often disconnected from outside markets. Crop yields and overall farm productivity are generally low, which can lead to the all-too-familiar cycle of increasing ploughing and grazing areas, followed by soil erosion and then new clearing elsewhere. In the Ouarsenis region, in Algeria, for instance, and in the Rif Mountains of Morocco, slopes of up to 50° inclination are increasingly ploughed and cultivated without any special measures being taken to retain soils. This results in catastrophic gully erosion and soil loss. Negative trends on soils are accelerating with a loss or productive sediments of 30 million t year^{-1} and more. In Algeria and Morocco, forested areas now cover less than 30% of their potential territory (Rubio and Calvo 1996; Brandt and Thornes 1996). The rest has been permanently and, in all probability, irreversibly removed. Of the remaining 30%, the quantity of wood and grazing material harvested annually far exceeds primary production. In Table 13.1 are provided revealing and frightening statistics from a fairly representative region of Morocco. The problem of water is a most challenging issue especially in the southern and eastern parts of the basin, where water shortage is going to increase, potentially resulting in serious conflicts. Irrigation currently represents up to 90% of the water withdrawals in the south, and a 50% increase in urban consumption is expected in the north and 300% in the south, where the amount of exploitation is already much higher than acceptable in several countries (Charpentier 1998).

In 1971, wood consumption for cooking and heating was estimated at 55 million m^{-3} year^{-1} in North

Table 13.1 Statistics relevant to forest destruction and over-grazing in the Azizal Province, Morocco

Size of the area	1 million ha
Annual human population growth	1.5%
Remaining forested area	340 000 ha
Primary annual wood production	230 000 m^3 (0.7 m^{-3} ha^{-1} year^{-1})
Annual rate of wood harvesting	≈490 000 m^3
Yearly loss of woody 'capital'	≈260 000 m^3 (i.e. 1.7% of remaining)
Percentage of territory grazed	50%
Sustainable carrying capacity	0.8 head ha^{-1}
Current grazing load	1.6 head ha^{-1}

Source: Marchand (1990).

Africa; that is, half a cubic metre of wood per capita per year. This represented 41% of the total energetic consumption of the region. Wood requirements for domestic use (cooking and heating) in these countries are still roughly 0.5–1 m^{-3} per capita per year, even if butane gas and solar heaters are gradually becoming more common in the rural areas.

The situation is similar in the steppe areas of the south where bush-dotted grasslands represent the primary resource. Elimination of alfa grass by overgrazing from the steppes, which cover more than three million hectares in Algeria alone and similar areas in both Tunisia and Morocco, undoubtedly has profound implications for ecosystem stability, resilience, and potential productivity (see Chapter 6). The sheep population in Algeria increased from 3.8 million animals in 1963 to 16.1 million in 1987 (Aidoud and Nedjraoui 1991). As a result, most of the steppe regions have been so badly degraded that livestock breeders have reverted to stockyard breeding and the motorized transport of animal feed and water. Table 13.2 summarizes the dramatic changes that have resulted from overgrazing and systematic overcutting of woody plants in an alfa steppe of central Algeria. Since 1991, the situation has gotten much worse, and field monitoring of degradation and loss of natural capital has been stopped as a result of political instability (A. Aidoud, personal communication).

Summarizing the overall situation, a Tunisian government minister reportedly said 'the northern [Mediterranean] countries can build upon rich and fertile plains whereas we have only the desert to build upon. Where is sustainable regional development to come from when you have huge natural resources and overindustrialization on one side, and sparse resources and overpopulation on the other?'

In summary, the Mediterranean Basin is a striking microcosm of current world problems. The opening of the European Union to southern and eastern countries, the growing use of various kinds of energy and water resources, the development of tourism, progress in research, and new technologies are all leading to rapid changes in land, water, and other resource use. These changes are also provoking problems, such as overpopulation in coastal areas, degradation of landscapes and ecosystems, spread of invasive species, shortage of fresh water, overfishing, and increase in the pollution and the volume of toxic wastes that are discharged into landfills, rivers, and the sea. All these threats and challenges make the Mediterranean a highly suitable region for testing and implementing sustainable development strategies. Drinking water is a tremendously challenging problem in this effort, especially as the impacts of climate change will lead to still less annual rainfall in the Mediterranean region (IPCC 2007). Human beings need around 75 m^3 of drinking water per capita in an average lifetime and this cannot change! Providing this quantity of water to every person in the forthcoming decades will definitely be a crucial challenge.

Table 13.2 Evolution of the alfa steppe in central Algeria, based on long-term studies in a series of permanent study plots (Aidoud and Nedjraoui 1991)

	1976	1989
Bare ground	16%	83%
Vegetation cover		
Ephemerals	11%	9%
Perennials	39%	6%
Alfa	34%	2%
Total	84%	17%
Phytomass (kg dry matter ha)	2,100	750
Production (FU ha year)[1]	130	60

[1] FU, food units (1 FU = 1, 650 kcal for ruminants)

13.2 Conservation sciences

To achieve goals related to biological conservation and sustainable development such as those that are proposed by the Blue Plan (see Box 13.5, below), a useful first step is to analyse the feedback mechanisms that keep ecosystems 'running'. Studies of positive- or negative-feedback cycles at local or regional levels may be an appropriate strategy in the long term for planning new management policies for ecosystems and biological resources. This idea is illustrated in Fig. 13.2. Note

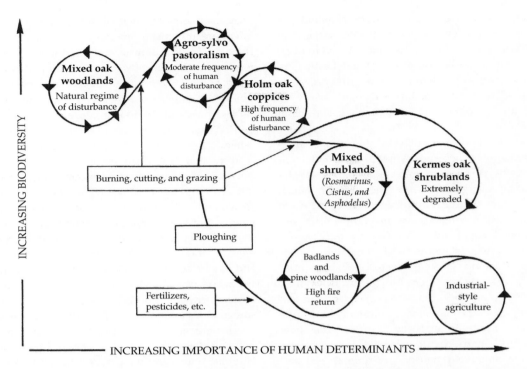

Figure 13.2 Schematic representation of some human-induced changes in mixed oak woodlands in the Mediterranean Basin. Under pressure from prolonged disturbances of different kinds, these ecosystems may 'shift' across ecological thresholds from one trajectory to another. This results in a change in the dynamics of energetic, hydraulic, and elemental fluxes, as well as in the composition of species assemblages. Numbers of arrows on the circles indicate the relative intensity of ecosystem dynamics (inspired from Woodward 1993), as indicated by the number of functional groups, interspecific interactions, etc. (modified from Blondel and Aronson 1999).

that, as discussed in Chapter 10, the highest biological diversity in Mediterranean ecosystems probably does not occur in forests, not even the pristine ones depicted by the first circle in Fig. 13.2, but rather in the landscape mosaics where various agro-silvopastoral systems, such as the traditional *sylva–saltus–ager* and *dehesa–montado* systems were once combined with transhumance (Fig. 13.2, second circle upper left).

The ultimate goal would be to identify, maintain, and, where needed and possible, restore an optimally 'running' system, involving as many as possible of the native constituents of living systems from populations to whole communities. However, compromises must be made between fully preserving biological diversity and native ecosystem functioning, on the one hand, and satisfying the immediate and long-term socio-economic needs of people on the other.

Therefore, relying on the potential natural vegetation (PNV) for devising conservation strategies would probably not be the best strategy in the Mediterranean, where ecosystems and landscapes have been transformed and re-designed by humans over several millennia. The PNV system was defined by Küchler (1964) as 'the vegetation that would exist today if man were removed from the scene and if plant succession after his removal were telescoped into a single moment'. A strategy based on PNV is a useful approach in areas or countries that have been only recently transformed by humans, but certainly not in the Mediterranean area where ecosystem trajectories leading to PNV are difficult or impossible to determine.

It is thus within a very general, realistic, and flexible framework that action plans should be devised to simultaneously promote the conservation of biological diversity and the sustainable management of ecosystems for the generation of ecosystem goods and services to people. In addition to a range of basic and applied sciences pertaining to this goal, Geographic Information Systems (GIS) are a good tool for achieving this ambitious goal, since they help integrate and manage the wealth of climatic and environmental data with databases on species distribution and ecology (Scott *et al.* 1991). By storing and manipulating varying types of mapped data on soils, vegetation types, species distribution, etc., this tool can highlight correlations among elements of a given landscape, and help in planning the management of ecosystems, landscapes, and biogeographical regions.

Coming back to classical issues about conservation such as they have been put in practice since many decades, biological conservation relies on two strategies. The first consists in assisting threatened individual species to survive in at least some localities. The second aims at the conservation of habitats, ecosystems, or larger areas. The best strategy for the long-term protection of biological diversity is the preservation of natural communities in the wild that is *in situ* or on-site preservation. A relatively important amount of scientific work and money has already been devoted to biological conservation of selected species in the Mediterranean and many new projects are getting underway.

13.2.1 Conservation of species

Except for some groups of plants and large animals, such as birds and mammals, we lack information on the distribution and abundance for most species in the Mediterranean (see Chapters 3 and 4). Species to be protected are often chosen more on the basis of aesthetic or cultural criteria, such as size, colour, or notoriety, rather than on scientific grounds. Many protected areas have been created to protect 'charismatic vertebrates' that capture public attention, have symbolic value, and are crucial to ecotourism (Primack 1993). This is the case of many bird reserves. Even for certain insect species, much effort has been made. For example, in Italy and France, attempts to conserve the beautiful carab beetle *Carabus olympiae* and *Carabus solieri* are underway, but nothing at all has been done for their much rarer and more threatened relatives, *Carabus clathratus arelatensis* and *Carabus alysidotus*, both of which are rather drab and inconspicuous (Baletto and Casale 1991). Nothing at all has been done so far for preserving legions of minuscule and microscopic organisms like these, which nonetheless play vital roles in making ecosystems 'turn'.

One problem is that many rare endemic species occur in localized areas that are not likely to be included in large reserves; conservation of such species requires specific action plans. For example, in a detailed study on insular plants both in the western and eastern Mediterranean islets, Lanza and Poggesi (1986) showed that several species occur only on small satellite islets off larger islands. They have either gone extinct on the large island, as a result of changes in their environment, or else they never occurred there because the species in question live only in the relatively inhospitable islets where few competitors occur. In this latter group are a wild onion, *Allium commutatum*, and a perennial grass, *Parapholis marginata*, of a genus endemic to the Mediterranean. Strikingly, there were strong similarities in the ecological distribution of these species on islets off Corsica and those found on islets near several of the larger islands of the Aegean Sea. The showy hairy pink (*Silene velutina*) disappeared from Corsica in recent times but still persists in a few small nearby islets. This gives an idea of the complexity of species conservation in the Mediterranean in spite of much effort to adopt stringent conservation measures in the framework of international conventions (e.g. CITES and the Bern Convention). There is also a two-faced aspect of conservation, as mentioned before and discussed in Box 13.2, below.

13.2.2 Species reinforcement and reintroduction

The survival of very rare and threatened species often requires active intervention, including reinforcement of populations, translocation, reintroduction, and captive breeding. Indeed, the smaller

> **Box 13.2. Preserving species: the bad side of the coin**
>
> Sometimes the road to Hell is paved with good intentions. The inclusion of rare or endangered species in 'red books' can have detrimental effects by attracting clandestine collectors and promoting illegal trade of specimens whose price on the black market is directly related to the species' growing rarity. For several species of rare carabid beetles and butterflies, for example, recent price hikes have prompted local collectors to search ever more avidly for this unexpected source of income under the pressure of unscrupulous dealers. This is particularly true for the wonderful *Carabus solieri* in the southern Alps, and for *Carabus olympiae* (see Chapter 3). Some local populations of the rare butterfly called Spanish festoon (*Zerinthia rumina*), which is closely linked to its patchily distributed host plant, an *Aristolochia*, have been decimated right down to the last individual by collectors who are not always aware of the gravity of their actions. A similar threat exists for several other showy species, such as Corsican swallowtail, southern swallowtail (*Papilio alexanor*), and the superb moth *Graellsia isabellae*. On the other hand, preventing a young child from catching a few stag beetles (*Lucanus*) or horn beetles (such as *Cerambyx*) may also short-circuit some promising entomologists and naturalists from fulfilling their vocations!

a population becomes, the more vulnerable it is to the so-called vortex effect (Gilpin and Soulé 1986): demographic variation, environmental stochasticity, and genetic factors that tend to drive the population to extinction. Saving very small populations is a difficult task, because the success of reintroduction depends on population viability, which involves complex genetic, behavioural, and demographic processes. The success of reintroductions relies on careful studies of such issues as the genetic consequences of inbreeding, behavioural consequences of captivity for those species that are released after having been bred in captivity, and the population consequences of infectious diseases (Sarrazin and Barbault 1996). In some cases, the consequences of outbreeding depression must also be considered because hybrid offspring born from divergent genotypes or populations may no longer have the precise mixture of genes that allow individuals to survive in a particular local environment. This often raises difficult problems. For example, if we seek to reinforce the fast-dwindling population of the brown bear in the Pyrenees, where this species is on the brink of extinction, should new animals be reintroduced from Italy, Greece, or Scandinavia? The Italian and Greek bears live in ecologically similar conditions, but the Scandinavian populations are genetically closer to the indigenous Pyrenean stock (see Fig. 2.8). Outbreeding of populations of different geographic origin may cause genotypes and co-adapted genes which evolved under certain local conditions to disappear altogether.

Hundreds of species could be reinforced or locally reintroduced in the Mediterranean Basin, but reintroduction policies are faced with two problems. First, if properly carried out on scientific grounds, reintroduction programmes are extremely expensive and cannot be applied to all species. They require a careful preparation of the awareness of the public. Second, economic or political criteria very often prevail over scientific (and moral) arguments for biological conservation. Whatever their motivations, conservationists prefer to choose 'flagship' species, such as large birds, mammals, or fish, rather than the cohorts of disappearing invertebrates, partly because it is easier to raise money for saving such species than small inconspicuous ones. In this context, several projects of reintroduction and reinforcement of raptor populations are in progress for 11 species in four Mediterranean countries (France, Italy, Spain, and Israel). Fifteen of them involve large vultures (griffon vulture, black vulture, lammergeyer) and may be very successful with indirect beneficial effects for other species. For example, a project of reintroduction of the griffon vulture started in 1968 in the Cévennes, France, where this species was extirpated at the beginning of this century. After a long process of captive breeding, the first individuals were released in the wild in 1981. Nowadays a self-sustaining population, including more than 250 free-living individuals, produces enough young each year to make the population sustainable in the long term (Terrasse 1996). Such successful projects should prompt ecologists to develop action plans for many other species. An additional example comes from the sea, where the grouper *Epinephele marginatus* was successfully reintroduced into the French marine reserve of Cerbère-Banyuls, from the Spanish reserve of Médes, following total eradication by French scuba divers in the 1980s.

For species that became extremely rare and endangered, *ex situ* facilities for their preservation include zoos, game farms, aquaria, and captive-breeding programmes. However, certain big fishes and marine mammals are so large that the facilities necessary for maintaining and handling them are extremely difficult and prohibitively expensive. This is why projects for reinforcing the Mediterranean population of monk seal have not yet yielded significant results (see Box 11.6).

A guild of species which is currently receiving a certain amount of conservation, restoration, and revised management efforts is that of large ungulates. This prominent group in the Mediterranean includes several species of deer (roe deer, red deer, fallow deer, and two gazelles in Israel, *Gazella dorcas* and *Gazella gazella*), and several subspecies of ibex, goats, and mouflon. In addition to their aesthetic and cultural values, they are of great hunting interest and clearly play important roles in many functional processes in forest, shrubland, and steppe ecosystems. Thus, a sustainable management scheme for ungulate populations should strive to conserve genetic diversity while also making rationale use of this resource. Most regions, especially in the northern part of the basin, clearly have a great potential for increasing population density and spreading the range of various ungulates.

13.2.3 Preserving ancient varieties of domesticated plants and animals

How are we to preserve the extraordinary variety of ecotypes and gene pools of cultivated plants and domesticated animals selected over millennia by traditional agriculturists and pastoralists? The 'option value' of a species or an ecotype is its potential to provide an economic benefit to human society at some point in the future. The growing biotechnology industry is finding new uses for many species and varieties, yielding a wide range of economic benefits within the context of traditional cropping and livestock raising practices, but also in medicine, pharmacology, as well as biological control. Often resistance to a particular disease or pest is found in only one or a few varieties of a crop that is grown in only small areas.

Preserving this genetic variability is especially critical in the Mediterranean Basin, where traditional farmers are widely abandoning their local varieties and land races. Thus, an important challenge is to identify and save wild ancestral gene pools in the Mediterranean before they are irretrievably lost along with their habitats. In the past 30 years, there has been a burgeoning of interest in the need and value of biological diversity of ancient races and cultivars by United Nations agencies (Food and Agriculture Organization (FAO), United Nations for Environment Programme (UNEP), United Nations Development Programme (UNDP), United Nations Educational Scientific and Cultural Organization (UNESCO)) and other governmental and private institutions. The Mediterranean Basin includes 45% of the bovid varieties and 55% of the goat varieties of Europe and the Middle East. Of 145 varieties of domesticated bovids and 49 varieties of sheep that occur in the basin, 115 and 33 species, respectively, are considered as in danger of extinction by the FAO World Watch List of Domestic Animal Diversity (Georgoudis 1995), mostly as a result of the diffusion of European dairy breeds, which dramatically reduced the number of local varieties.

Much effort is being made today by international, governmental, and private organizations to conserve the genetic resources of 'old native' breeds of several species, such as the Mediterranean buffalo, cattle, sheep, and goat. There are many examples of well-managed 'herd books' and programmes based specifically on the storage of semen. This is especially important in the framework of sustainable development, because the disappearance of local races may seriously limit the exploitation of marginal lands where only locally selected breeds can survive harsh environments and prolonged drought. In several Mediterranean countries, active conservation programmes employ pedigree, progeny testing, and data recording for keeping the local diversity of goats.

Many plants of economic interest occur in small fluctuating and poorly dispersed populations, such that they are at risk of extinction or severe genetic loss. Hundreds of programmes in progress aim at conserving genetic resources of fruit trees, grape, field crops, forage species, vegetables, and ornamental plants (Charrier 1995). For cultivars that are no longer used by rural farmers, but which nevertheless constitute highly valuable genetic resources, as well as for wild relatives of cultivated species that are heavily threatened in the wild, one strategy is to preserve them *ex situ* or off site. Facilities for preserving plants *ex situ* include botanical gardens, arboreta, botanical conservatories, and seed banks, which all have great potential for preserving genetic variability for the future (see Box 10.3; Avishai 1985). Integrating research on the current status of a species with conservation *in situ* and cultivation *ex situ* provides good tools of preserving biodiversity. Several botanical gardens have developed gene banks specifically designed for the storage of germplasm of wild plants. Some of the world's most ancient botanical gardens were created in the Mediterranean during the sixteenth century, mostly for the conservation of herbs and medicine plants. Today there are approximately 100 botanical gardens and arboreta around the basin (42 in Italy alone) (Du Puy and Jackson 1995). Most of them are established in Euro-Mediterranean countries with the notable exception of the Jardin d'Essais du Hamma in Algeria, which maintains a collection of 8000 taxa of native Algerian plants. A great many botanical gardens consider today conservation as a major theme in their activity and maintain living collections of economically useful plants, such as fruit tree cultivars and medicinal and aromatic plants. For example, the beautiful botanical garden of Gibraltar, established in 1991, already contains an impressive collection of plants, aiming to become an important Mediterranean centre for native plant conservation and environmental education. In France, the Domaine du Rayol is an important centre of horticultural education and demonstration, and the Botanical Conservatory of Porquerolles has a programme dedicated to native plant conservation (Box 13.3).

Box 13.3. The Botanical Conservatory of Porquerolles

Established in 1979 on the island of Porquerolles, southern France, the aim of this Botanical Conservatory is not only to inventory and monitor wild plant species that are rare or threatened in Mediterranean France, but also to intervene directly to assist in saving them from extinction in any way possible. Achieving this goal is pursued through cytotaxonomic and ecological research on target species. Conservation programmes include both *ex situ* conservation of genetic resources of wild species and cultivars, in the form of a seed bank that currently holds more than 2000 species, and *in situ* conservation of rare and endangered populations of plant species. In the case of populations on the verge of extinction, this can take the form of population reinforcement or, where local extinction has already taken place, attempts at reintroduction, with attention being paid to the dangers of 'genetic pollution' in a regional context. In both situations, it is the *ex situ* seed bank and living collections which makes the *in situ* efforts possible. This conservatory also maintains a project to document, record, and conserve the traditionally grown fruit trees of the Mediterranean Basin. A major discussion is currently underway as to the proper attitude to adopt *vis-à-vis* non-native species in France that are now increasing their territory as a result of global warming and other global changes, as well as increasing use by gardeners of well-adapted ornamental species that occasionally escape from gardens and become naturalized. These are not easy issues to resolve in such a rapidly changing world.

A promising international initiative is the International Centre for Advanced Mediterranean Agronomic Studies (CIHEAM; www.ciheam.org), which includes the Mediterranean Agronomic Institutes of Bari (Italy), Chania (Greece), Montpellier (France), and Zaragoza (Spain). This institute established a working group on 'underutilized fruit crops', such as figs, loquats (*Eriobotrya japonica*), Japanese persimmons (*Diospyros kaki*), pomegranates, and Barbary figs (*Opuntia*). These species have often been considered as marginal crops, but they are of great value in some situations, especially as a result of an increasing demand for their fruits in industrialized countries and in local markets.

In particular, the fig tree and the pomegranate are two of the seven Biblical plants—along with wheat, barley, olive, date, and grape—that represent ancient Israelite agriculture (see also Chapter 10). At a number of CIHEAM centres, especially the one in Zaragoza, Spain, much genetic and applied horticultural research is underway on the olive tree, pistachio, and almond tree, among other trees. This research involves both *in situ* and *ex situ* study of wild relatives and local land races of these traditional Mediterranean fruit and nut crops, whose economic future in the region depends entirely on maintaining a high level of research and conservation commitment at national and international levels.

13.2.4 Hotspots

As already pointed out in Chapter 3, Myers *et al.* (2000) and Mittermeier *et al.* (2004) rated the Mediterranean Basin as one of the world's 34 'hotspots'. However, the basin is far too extensive and heterogeneous to be treated as a single hotspot area. Identifying key sectors or regional hotspots that warrant special treatment (see for example Fig. 3.1 for vascular plants) would be a first step for developing conservation strategies at a regional scale. However, to be biologically meaningful as candidates for special attention, these regional hotspots should present two characteristics: they should be rich in species of many different groups of organisms (Gaston 1996) and there should be a positive relationship between species diversity and overall rarity within each of the groups. The first condition is rarely met, as shown by Williams and Gaston (1994) and Prendergast *et al.* (1993), who demonstrated that areas of maximal richness for different taxonomic groups rarely coincide. For example, in the Mediterranean area, the regions of greatest species richness are the eastern Mediterranean and Iberia for reptiles and the western Mediterranean for amphibians (Meliadou and Troumbis 1997). Nevertheless, examples of obvious candidates for regional hotspot areas in the Mediterranean are the Baetic Cordillera of south-eastern Spain and the mountains of southern Greece. In these regions, there are exceptionally high concentrations of endemic species among plants, fishes, reptiles, and insects. For example, in the 336 ha Nature Reserve of the Massane Forest, created in the Eastern Pyrenees in 1973 (Travé 2000), 10 km from the sea, over 6000 living species have already been identified, one-third of which are beetles, and many of which are unknown outside this forest reserve (J. Garrigues and G. Boeuf, unpublished data).

Another consideration, of course, in the designation, recognition, and priority assessment of hotspots is the degree of menace to biodiversity. On this account, it is highly noteworthy that Véla and Benhouhou (2007) propose an eleventh regional hotspot for Mediterranean flora, in addition to the 10 proposed by Médail and Quézel (1997) and Médail and Diadema (2006), as depicted in Fig. 3.1. This area in northern Algeria, which the authors describe as the 'Kabylias-Numidia-Kroumiria' hotspot, comprises an unrecognized regional hotspot with very high levels of endemism and high levels of threat from human activities. They call not only for intensified studies but also for reinforced national and international policies of conservation and protection, which is the subject of the next section.

13.2.5 Protected areas and ecotourism

Preserving habitats and their biological communities is the most effective way to preserve biological diversity. Except for some well-known threatened species, such as the monk seal, the European lynx, or the bald ibis, habitats should be the target of conservation efforts because conservation of habitats can preserve large numbers of communities, species, and local populations in self-maintaining units. Much effort has been made recently to raise the number of preserves in the Mediterranean Basin so that the surface of protected areas has increased by 26.7% between 1985 and 1996. The total coverage of protected areas was 4.3 million ha in 1996, including 3.4 million ha of terrestrial ecosystems and 0.9 million ha of coastal areas (Ramade 1997). The great majority of protected areas are located in the northern side of the basin, with 84% of them in European Union member states (Benoît and Comeau 2005). Protected areas range from minimal to intensive use of habitats and resources. Following the classification of the International Union for Conservation of Nature (IUCN 1985), they range from strict nature reserves to multiple-use management areas that are not primarily managed for conservation but still may contain most of their original species. Such multi-use management areas may be particularly significant, especially in the Mediterranean, since they are often much larger than strict nature reserves. In a region like the Mediterranean Basin, characterized by highly dissected landscapes with thousands of small private properties, it is difficult to establish reserves of appropriate size. Ideally, the size of any reserve should be big enough to really serve for targeted species' survival. It is also important that reserves not 'stand out' too emphatically with respect to their surrounding environment, like oases in a desert.

In this respect, a good strategy is to embed strict nature reserves in a matrix of managed areas as it is done in Biosphere Reserves and Regional Natural Parks, which are certainly the best tools for implementing sustainable development issues. The UNESCO's Man and Biosphere reserves, of which there are currently 36 in 10 countries of the Mediterranean Basin, mostly in the western part of the basin (12 in Spain alone) (Table 13.3), have been conceived precisely for this purpose. The biosphere reserve concept proposes to coordinate the functions of the conservation of biodiversity, the long-term monitoring and study of changes in ecosystems, and the contribution to sustainable development of local human populations. Thus, they are protected areas flexible enough to cope with the varied ecological

Table 13.3 Protected areas in some Mediterranean countries

Country[1]	Size of Mediterranean area ($\times 10^3$ ha)	Protected areas (inland and coastal, $\times 10^3$ ha)	Percentage
Spain (79, 3, 11)	40 000	1604	4.0
France (39, 2, 3)	5800	498	8.6
Italy (75, 2, 2)	20 000	415	2.1
Malta (1, 0, 0)	32	0.3	0.9
Croatia (16, 4, 1)	1720	199	11.6
Greece (28, 4, 2)	10 000	107	1.1
Turkey (35, 9, 0)	48 000	517	1.1
Cyprus (8, 1, 0)	925	102	11.0
Lebanon (3, 2, 0)	1040	4.8	0.4
Israel (8, 2, 0)	2200	20	0.9
Tunisia (9, 6, 4)	10 000	35	0.3
Algeria (13, 6, 1)	30 000	202	0.7
Morocco (6, 0, 1)	30 000	52	0.2

[1] The three figures in the first column correspond to total number of reserves, national parks, and Man and Biosphere reserves, respectively.
Source: After Ramade (1997) and Toni Nikolić, personal communication.

and socio-economic conditions that prevail in each particular region. Their aim is not so much to preserve large areas of the most common habitat types as it is to include representatives of all habitats on a regional scale. They are or should be laboratories for experimenting and supplementing sustainable development in such a way that this expression is no longer an oxymoron. Unfortunately, except for strict nature reserves, too many 'protected areas' exist only on the map and receive little protection from habitat degradation, overharvesting, and pollution.

Establishing large multi-use managed areas should be especially rewarding in the Mediterranean because of the huge potential for ecotourism based on activities, such as hiking or bird-, dolphin-, or whale-watching. Ecotourism may provide one of the most immediate justifications for protecting biological diversity in the framework of sustainable development, including investments in eco-restoration (Blangy and Mehta 2006). Although not widespread as yet in the basin, ecotourism can represent a considerable amenity value in most Mediterranean tourist areas with exceptional scenic beauty, cultural significance, and often important archaeological sites as well. There is a large potential for ecotourism in islands and coastal areas which often include extremely valuable habitats in terms of biological diversity, but which are the most vulnerable ecosystems in the Mediterranean.

The International Ecotourism Society states that 'sustainable' tourism could grow to 25% of the world's travel market within the next 6 years, taking the value of the sector to £250 billion (US$ 473.6 billion or 361 billion €) a year (www.ecotourism.org/). The coastal fringe is the area of most human pressure with urbanization, industry, agriculture, transportation installations, and tourism. Only 1% of Mediterranean coasts are currently protected and the World Wildlife Fund (WWF) aims to raise this to 10% over the forthcoming 10 years. In some countries, however, large efforts are made at a governmental level to preserve as much as possible of coastal ecosystems. For example, in 1975, the French government created a 'Conservatory for Littoral and Lacustrine Shores', which has to date purchased or received as legacy more than 400 coastal properties totalling 103 000 ha or nearly 20% of the total French coastline, for the express purpose of nature conservation. The goal is to set aside one-third of the Atlantic and Mediterranean littoral by the year 2050, despite the ever-growing demand and rising land value of these lands. Furthermore, existing marine reserves, located primarily in Spain, France, Italy, and Greece, are much rarer than terrestrial ones and should be greatly extended.

13.3 Steps towards sustainability

Sustainable economic development will only work in the Mediterranean area, as elsewhere, if it is coordinated with appropriate efforts to conserve and restore ecosystems and biodiversity at the landscape and regional scales. In addition to set-aside areas, such as natural parks and reserves, agriculture and other production systems need to evolve as well. If the Mediterranean region is a microcosm of world problems, it also has many models to offer for sustainable development and human well-being in a beautiful and biologically diverse environment. An approach that is attracting interest in many parts of the world is designing and operating agricultural production systems that work 'in nature's image'. The working premise is that natural ecosystems of any region are adapted to fluctuations in key resources, as well as to the constraints imposed by the environment. They therefore provide regionally specific models for sustainability if well mimicked by new land-use practices. Several experiments in progress in Australia, the USA, and various tropical countries as well are aimed at redesigning land-use systems in structural and functional terms so as to better mimic nature. Successful mimic systems should look for complementary species according to the M5 golden rule: Making Mimics Means Managing Mixtures (Dawson and Fry 1998). Mediterranean habitat mosaics are an ideal place to put into practice and test these ideas, given that an historic model exists in the form of the 800-year-old *dehesa* systems described in Chapter 10. *Dehesas* are examples of a production system whose sustainability derives from accepting moderate yields and rotating among a range of different products, including tree crops, annual cereals and pulses, and various forms of livestock (Joffre and Rambal 1993).

The problem is to identify plant and animal species that will provide a diversity of functional roles, accommodate environmental variation and grow well in mixtures. They must also offer economic returns that justify investment, and provide adaptability to changing socio-economic conditions over the long term.

For Mediterranean ecosystems that have been subjected to long periods of human perturbations and in a time of dramatic and worrisome climate and socio-economic changes and upheavals, such an approach seems essential to complement a growing public sense of the need for greater conservation of ecosystems and maintenance of native biodiversity. For the landowner, the challenge arises when short-term economic interests do not cover the costs of maintenance and labour. The concept of ecosystem services, referred to above, can help solve this problem, as it is recognized that sustainable production systems render service to society as well as to individual land owners. Increasingly, markets offering 'payments for ecosystem services', such as carbon sequestration and protection of aquifers, are being developed and this should help considerably.

One relatively new tool that economists are using for this purpose is the valuation of Total Economic Value (TEV) of a resource or service which combines market and non-market values or benefits to people, including the value of non-monetarized ecosystem or environmental services provided by ecosystems. In other words, the TEV includes both commercial and environmental benefits, as illustrated in three studies of cork oak woodlands in Portugal, Tunisia, and Spain, wherein both commercial and environmental benefits were calculated with the same TEV approach (Coelho and Campos 2009; Campos et al. 2009; Ovando et al. 2009).

In all three studies, the TEV process helps find ways to reconcile agricultural land use, biodiversity conservation, and, where needed, ecological restoration (see Rey Benayas et al. 2008). Indeed in conjunction with plans and measures to protect species and communities and to provide payments for ecosystem services to society, another important building block for achieving sustainability is greater investment in ecological restoration.

For this to take place, it is essential to first take into account that much degradation has taken place (see Chapters 10 and 11), and that ecological restoration may be required, not only to maintain biodiversity but also to improve the health and well-being of people. The fact of climate change and other global changes (Chapter 12) only reinforce this conclusion. Restoration can be costly, but ultimately the benefits often far outweigh the costs.

As defined by the Society for Ecological Restoration International (SER), ecological restoration is 'the process of assisting the recovery of an ecosystem that has been degraded, damaged, or destroyed' (SER 2002).

A somewhat larger concept relevant to our discussion is restoring natural capital, which refers to all investments in renewable and cultivated natural capital stocks and their maintenance in ways that will improve economic well-being of people. This can be done through (1) restoration of degraded ecosystems, (2) ecologically sound improvements to lands managed as production systems for useful purposes, (3) improvements in the utilization of biological resources, and (4) the establishment or enhancement of socio-economic systems that incorporate awareness of the value of natural capital into daily activities (Aronson et al. 2007; Clewell and Aronson 2007).

Although considerable innovative and promising work on the restoration of Mediterranean forests and woodlands is currently taking place (e.g. Maestre and Cortina 2004; Pausas et al. 2004; Bowen et al. 2007; Rey Benayas et al. 2008), here we will discuss only one of the longest running programmes, the so-called RTM—*Restauration des Terrains en Montagne*—programme, which began a century and a half ago. With such a lifespan, it is now possible to evaluate the successes and failures of these programmes, and consider fine tuning measures of both economic and ecological relevance. Afterwards, we will also discuss restoration of natural capital in Mediterranean wetlands.

13.3.1 Restoring forests

Mediterranean forests traditionally provided a wide range of products and services, as explained in Chapter 10. From the time of Roman Emperors to the present, many attempts have been made to restore forests following periods of heavy exploitation, especially in mountainous areas. The French programme RTM was an early and exemplary initiative to modernize forestry and to develop new techniques for restoring heavily degraded mountain slopes in the Mediterranean area. Sufficient time has now elapsed to allow a thorough analysis of the methods used and results achieved.

Nearly 150 years old, the effort at Mont-Ventoux, southern France, is worth summarizing as an example of the numerous RTM programmes undertaken around the same time in several mountainous areas. In the middle of the nineteenth century, nearly all the southern slopes of the Mont-Ventoux range, just like those of most other mountains in southern Europe, were completely deforested as a result of several centuries of woodcutting and overgrazing. In the mid-1850s, the French botanist Martins wrote that 'most of the massif is the sole realm of thyme and lavender, almost devoid of trees'. A stony desert, with only scattered subshrubs and large expanses of bare ground exposed to wind and surface erosion, was all that was left of the dense forest mantle which formerly included stands of pines, oaks, beech, maples, and many others according to the life zones occurring from about 350 to 1800 m (see Chapter 5). Devastation was the only word to describe the scene. The second Empire was a period of great prosperity in France and among monumental public works, the decision was taken in 1861 to reforest the Mont-Ventoux. The main reason given was to stop soil erosion and, if possible, to 'restore' some of the lost soil. An enormous amount of money was devoted to this project. Between 1861 and 1873, 2500 ha were planted with pine seeds and tree saplings on eroded and rocky slopes. Workers had to climb up each day from their base camps at the foot of the mountain, carrying with them an armload of plants and jerrycans of water. The sequence of tree species planted proceeded from Aleppo pines and holm oaks, at the foot of the mountain, to montane pines near timberline, at 1500 m altitude, with downy oaks and beech in between. Moreover, many non-native species were experimentally introduced, such as Atlas cedar (*Cedrus atlantica*), spruce, larch (*Larix*), cluster pine, and black pine. Remarkably, this huge and still ongoing project has succeeded in reconstituting dense forest cover of mixed ages and even on the most exposed southern slopes of this mountain, autogenic succession has clearly been reinstated. The original plantations of native trees, complemented by Atlantic cedars of Moroccan origin, have contributed to the gradual reconstitution of soils and the re-establishment of a mixed canopy under which native trees, shrubs, and herbs could re-colonize.

Since the main aim of the programme was to protect soils against erosion, management did not involve any kind of major interference with the development of vegetation. However, a large series of interdisciplinary studies conducted in the 1970s (du Merle 1978) have shown that plant and animal communities of the reconstituted forests do not differ greatly from those that occurred in the few places where the forest cover had not been destroyed. Bird communities of the planted areas in particular are now very similar to those of relatively undisturbed forest patches used as a reference, as shown by in-depth quantitative studies. The RTM reforestation programme has led in this case to the return of a European woodland avifauna and the reintroduction of many Mediterranean species, which had secondarily occupied these areas after destruction of the primeval forests.

As part of a detailed study of another RTM programme, Vallauri (1997, 1998) studied the structural and functional diversity and 'health' of an Austrian black pine plantation (*Pinus nigra* subsp. *nigra*), 120 years after planting began in the then-denuded Saignon Valley of the south-western Alps. He evaluated restoration success by measuring earthworm activity, as an indicator of soil biological activity, natural regeneration of native woody plants under the pines, and infestation by mistletoe (*Viscum album* subsp. *austriacum*), which may have considerable negative consequences for the economic value of the black pine forests. The results of this 4-year study (Vallauri *et al.* 2002) clearly indicate that the initial goals of halting soil erosion and re-establishing plant cover have largely been attained, thanks to initial plantings begun in 1876 and subsequent plantings made during the 1960s (see Table 13.4). However, the study showed that considerable improvement could now be made, in the light of current knowledge and know-how, in order to finetune the restoration efforts today and better achieve both ecological and economic goals. For example, by thinning the dense black pine stands, especially in those areas most affected by mistletoe, revenue can be generated from the sale of timber while the openings created will favour increased colonization by the dozens of native woody plants that have already begun to regenerate in these artificially created forests.

13.3.2 Restoring wetlands

Mediterranean wetlands are among the most threatened and the most productive ecosystems in the Mediterranean region. Historically, they were considered as virtual wastelands with their only perceived value being conversion for grazing or—after costly draining—a limited range of cropping. Yet fresh water in the Mediterranean Basin is of huge economic, environmental, and livelihood importance, as mentioned already. Formerly extending over very large areas, Mediterranean wetlands have been so widely drained that an estimated half of

Table 13.4 Vegetation dynamics in the Saignon Valley experimental watershed (Haute Provence, France) between 1836 and 1995 (percentage of total area)

Vegetation types	1836–1875	1948	1995
Degraded and rock-dominated lands	50	31.7	23.7
Bare marls	n.a.	19.1	5.0
Colonized marls: cover <50%	n.a.	5.4	8.8
Colonized marls: cover >50%	n.a.	3.1	6.7
Agriculture and grazing	42	38.0	19.9
Ploughed land	7	0	0
Vegetable gardens and vineyards	<0.1	0	0
Meadows	2.4	9.8	0.2
Meadows with shrubs and shrublands	32.5	28.2	19.7
Woodlands	8	30.3	55.4
Early woody stage	0	4.6	7.4
Austrian black pine forest (exotic)	0	19.2	32.1
Broad-leaved forest	8	6.5	14.1
Mixed forest (broad-leaved+native and exotic conifers)	0	0	1.8
Wetlands		0	1.0

Data were obtained from Napoleonic cadastral surveys (1836, 1987), aerial photographs (1948, 1995), and numerous land surveys carried out in 1995.
n.a., not available.
Source: Modified after Vallauri *et al.* (2002). Reproduced with permission.

all the region's wetlands has been lost. Major wetland drainage begun in Italy during the Etruscan period (fifth century BC) and accelerated during the time of the Roman Empire to expand agricultural areas at the expense of riverine wetlands in all the major valleys of Italy and in imperial provinces in France, Spain, and North Africa. In Roman times, there were 3 million ha of wetlands in Italy alone. At the beginning of the twentieth century, only 1 300 000 ha were left, a figure that by 1991 had dropped to 300 000 ha. It is now estimated that only 7% of Italy's natural wetlands remain.

After the collapse of the Roman Empire, wetlands recovered but drainage was renewed again under the leadership of large and powerful monasteries in the Middle Ages. Wetland drainage still increased after the Renaissance and a further acceleration took place in the 1850s with the introduction of steam-powered machinery. A new impetus for drainage occurred at the end of the nineteenth and the beginning of the twentieth century to eradicate malaria, which was present in all Mediterranean countries.

No more than 21 000 km^2 of wetlands are left in the basin today (Pearce and Crivelli 1994), including 4700 km^2 of coastal lagoons, 2800 km^2 of lakes and natural mashes, and 10 000 km^2 of artificial wetlands, mostly dammed lakes. Salinas are present in all Mediterranean countries, with a total coverage of 621 km^2, but the largest ones occur in the more industrialized countries of southern Europe (570 compared with 51 km^2 in North Africa). In spite of much warning against the destruction of wetlands over the last 50 years and despite several projects aiming at conserving and restoring surviving Mediterranean wetlands, dredging and draining have continued unabated until a decade or so ago to make new space for intensive agriculture, aquaculture, intensive grazing lands, salt pans, industrialization, and tourist installations, all resulting in major ecosystem degradation. In addition, nearly every important river in the Mediterranean Basin has been dammed. In the Camargue (see Plate 10b), which is one of the best protected wetlands of the whole basin, no less than 40% of natural habitats have been lost in the last 50 years (Tamisier and Grillas 1994).

However, there is now a decelerating trend in wetland destruction in most parts of the basin, and some positive steps to restore natural capital and habitats are underway (e.g. Mauchamp *et al.* 2002; Berberoglu *et al.* 2004; Papayannis 2008). The main problems are mismanagement, pollution, overexploitation, and the increasing impact of invasive plant and animal species. Additional threats are sedimentation, siltation, episodes of hypertrophic anoxia, and sea-level elevation, which proceeds at a rate of 1–2 mm year^{-1} and which is expected to strongly increase in the upcoming decades. The International Panel on Climate Change forecasts that this will increase to between 3 and 8 mm per year by 2030 (IPCC 2007). Many coastal wetlands will be at risk of overflooding, with the Mediterranean sea level increasing today three times more rapidly than during the 1990s.

Yet, Mediterranean marshlands perform a range of functions, which deserve conservation and better management, not to mention compensation for all four of the types of ecosystem services recognized by the Millennium Ecosystem Assessment team (MA 2005), namely provisioning, regulating, supporting, and cultural. In simpler terms, wetlands provide valuable seasonal grazing land, fisheries, agricultural land, reeds for thatching, and hunting grounds, as well as less direct benefits, such as flood control, storm protection, groundwater recharge and sediment, pollution alleviation, and nutrient retention. Coastal wetlands act as a sponge, rapidly soaking up huge quantities of water during heavy rainfalls, and then serve as a source of surface water during dry periods. Thus they can help buffer and alleviate the potentially devastating effects of storms, if properly managed. Wetlands are habitats for wildlife. Many of them are hotspots of diversity for many rare species of plants, insects, fishes, and birds. Nearly 50% of Europe's bird species and 30% of the plant species depend more or less exclusively on wetland habitats. Hundreds of rare and endemic species of insects in the basin are characterized by at least one aquatic stage in their development. For example, among ground beetles, no less than 500 species are exclusively wetland dwellers in Italy alone (Baletto and Casale 1991; see Chapter 3).

They also provide income at both an artisanal and commercial scale through ecotourism, fishing, and hunting. In most Euro-Mediterranean countries, hunting is in fact a 'life saver' for wetlands. Estimated annual harvest of hunters in France alone amounts to 1–3.5 million ducks, with an annual value of the carcasses equalling 10.5 million €. Many of the largest and most famous wetlands of southern Europe owe their existence to economic benefits derived from hunting and grazing by local landraces of horses and cattle. For example, experiments in progress in the Camargue aim at rehabilitating abandoned ricefields through management of standing water for improving habitat quality for wildlife and grazing potential for horses and cattle. By monitoring water levels and salt concentration, Mesléard *et al.* (1995) found the best compromise for the development of an herbaceous plant cover, which allows grazing by livestock and provides feeding and breeding habitats for birds. A long-term process of preserving and restoring Mediterranean wetlands began in the early 1960s with the IUCN's MAR project. Another important vehicle for wetland conservation is the Ramsar Convention (see below). A symposium held at Grado, Italy, in 1991 defined an Action Plan to stop and to reverse the loss and degradation of Mediterranean wetlands (MedWet; see Box 13.4). The Tour du Valat Biological Station is deeply involved in basic research and conservation biology in these habitats (see Box 13.4). Finally, a large part of the Mediterranean Basin is included within the European Union and now falls under the legislation of the Water Framework Directive, which is following the Integrative River Basin Management (IRBM) approach. The purpose of the directive is to establish a framework for the protection of inland surface waters, transitional waters (estuaries), coastal waters, and groundwater.

13.3.3 Managing and living with fire

Although most people consider fires as devastating scourges, they may have both negative and positive effects on living systems depending on their return rate in any given place, as explained in detail in Chapter 11. It is unrealistic and biologically unsound to strive to totally prevent their occurrence in Mediterranean lands. Fires are natural disturbance events that contribute to maintain the 'moving mosaic' of communities and ecosystems at the scale of landscapes (see Chapters 7 and 11). As such, they participate to a large extent in maintaining biological diversity and ecosystem 'health'. On the other hand, fires are permanent threats for landscapes and peoples. Therefore some kind of equilibrium must be found for keeping them at the same time useful for ecosystems and not too devastating for human affairs. They must be carefully considered and properly integrated in management policies, and people must learn to live with occasional fires. The problem is to devise techniques for controlling their frequency and intensity. The best strategy would be to manage forests and woodlands so as to anticipate and reduce the

> **Box 13.4. The Tour du Valat Foundation**
>
> The mission of the Tour du Valat Foundation (http://en.tourduvalat.org/) in the Camargue is to 'halt and reverse the destruction and degradation of Mediterranean wetlands and their natural resources, and promote their wise use'. For more than 50 years, this institution has conducted ecological studies for the conservation of Mediterranean wetlands with the aim to achieve 'better understanding for better management'. Convinced that it will only be possible to preserve wetlands if human activities and the protection of the natural heritage can be reconciled, the Tour du Valat has for many years been developing programmes of research and integrated management that promote interchanges between wetland users and scientists. Three main projects are conducted by a team of 60 or so employees and researchers to (1) develop an observatory of Mediterranean wetlands at the scale of the basin, (2) improve integrated management in the framework of detailed studies on ecosystem dynamics, and (3) address research projects on global change and species dynamics. The Tour du Valat has much expertise in the study of population biology of key animal and plant species and the ecosystem functioning of Mediterranean wetlands. Long-term studies have been conducted on birds, fishes, mammals, including semi-wild herds of horses, and plants in an attempt to integrate population dynamics in food webs at the level of ecosystems. These long-term studies provide invaluable databases for investigating the response of ecological systems to global change and to attempts at restoration.
>
> With the aim of stopping the loss of Mediterranean wetlands, restoring or rehabilitating degraded ones, and piloting rationale use of the remaining wetlands at the scale of the whole basin, a long-term Action Plan called MedWet (www.medwet.org) was launched under the auspices of the Ramsar Convention supported by the European Economic Community. This Action Plan, monitored by the Tour du Valat Biological Station, with partners, such as the World Wildlife Fund (WWF), Wetlands International, and the Greek Centre for Wetland Conservation, promotes international cooperation to preserve vital or key specimens of these pivotal habitats, while also updating resource management and environmental protection plans in the framework of sustainable development and payments for ecosystem services.

risk of catastrophic fires *before* the next fire breaks out. Although the idea of intentionally setting fires is anathema for most people in the basin, fire itself, if properly managed, has just as important a role to play as grazing in the rational maintenance of open spaces, and the management of mosaic landscapes that are biologically diverse.

Concurrently, given the large number of accidental or intentionally set fires which occur each year in the basin and the too-rapid fire-return cycle this entails, it is crucial to develop effective revegetation techniques for restoring burned lands. New approaches to reforestation are also being developed in light of ecosystem services provided by mixed oak and pine forests, both in terms of amenities, biodiversity, and protection against wildfire (see for example Siles *et al.* 2008; Moya *et al.* 2009).

13.4 Present threats and conservation efforts in the marine environment

The situation of the sea is of course quite specific, and human impacts and global changes are very different as compared to those in terrestrial environments. Lack of a clear common Mediterranean fisheries policy is therefore a cause for worry. More and more landings declarations are imprecise or false, quotas are not respected, contamination of marine trophic chains continues, and the effects of global warming are more and more visible. All these causes of degradation make it urgent to take this problem very seriously if we want to conserve our marine biodiversity heritage and continue landing those 1.5 million t of seafood that are taken each year by the 24 bordering countries and some others that fish in the Mediterranean Sea

Table 13.5 Impacts related to main pressures on the coastal and marine environment

Pressures	Main impacts
Climate change	Increased risk of floods and erosion, sea-level rise, increased sea-surface temperature, acidification, altered species composition and distribution, biodiversity loss
Agriculture and forestry	Eutrophication, pollution, biodiversity/habitat loss, subsidence and salinization of coastal land, altered sediment balance, increased water demand
Industrial and infrastructure development	Pressure on coastal lands, urbanization, eutrophication, pollution, habitat loss/fragmentation, subsidence, erosion, altered sediment balance, turbidity, altered hydrology, increased water demand and flood-risk, seabed disturbance, thermal pollution
Urbanization and tourism	Pressure on coastal lands, highly variable impacts by season and location, artificial beach regeneration and management, habitat disruption, biodiversity loss, eutrophication, pollution, increased water demand, altered sediment transport, litter, microbes
Fisheries	Overexploitation of fish stocks and other organisms, by-catch of non-targeted species, destruction of bottom habitats, large-scale changes in ecosystem composition
Aquaculture	Overfishing of wild species for fish feed, alien species invasions, genetic alterations, diseases and parasite spread to wild fish, pollution, eutrophication
Shipping	Operational oil discharges and accidental spills, alien species invasions, pollution, litter, noise
Energy and raw-material exploration, exploitation, and distribution	Habitat alteration, landscape changes, subsidence, contamination, risk of accidents, light disturbance, barriers to birds, noise, waste, altered sediment balance, seabed disturbance

Sources: Based on EEA (2007).

(see Chapter 11). In a recent Report (EEA 2007), the main pressures on the coastal and marine environments have been specified, as summarized in Table 13.5.

There is an urgent need to establish many more marine protected areas than those that exist today. These areas must be significant in size and located in strategic locations within the Mediterranean Sea. There are today some reserves, for example along the French coast, in the Banyuls area, along the Gulf of Lion, around the National Park of Port Cros off the Riviera coast, in Corsica around the Scandola Reserve, and some others. The Pelagos National Park in the Corsica-Sardinia region is an interesting initiative, as well as several other projects off the coasts of Italy and the new countries of the Balkan region. In Greece, a large marine reserve has been established for the protection of the few remaining monk seals (see also Mabile and Piante 2005). Several important marine reserves should definitely be established along the southern shores of the sea. In addition, connections among the various marine reserves are necessary to make them truly effective.

13.4.1 Aquaculture (mariculture) in the Mediterranean Sea

Aquaculture is an ancient activity in many countries; in both China and Egypt there is proof of tilapia and carp cultures dating back at least 4000 years. Mariculture is more recent, however, existing apparently for only 2000 years. The Mediterranean was one of the original venues of mariculture experiments, for example with oysters in Greece (Boeuf 2002). The endeavour is fast gaining momentum in today's crowded world. The 'new aquaculture' began during the 1970s with the domestication of marine fish, shrimps, and new molluscs, scallops, and clams (Boeuf 2000, 2003). In particular, marine fish species, mainly sparids (porgies), have been tested over the last three decades for aquaculture—or more precisely, mariculture—in the Mediterranean Sea. Today two species emerge as the best candidates for captive breeding, namely sea bass and sea bream. According to the FAO (FAO 2008), from 20 000 t in 1996, sea bass production reached 57 000 t in 2006 for the Mediterranean region, including additional production in

Portugal and on the Atlantic coast of France. For sea bream, it surged from 32 000 t in 1996 to 106 000 t in 2006. These two species are mainly produced in Greece, Turkey, Italy, France, and Spain. The total market value for the two species in 2006 was more than 800 million €, with sea bass bringing a higher price than sea bream. The first attempts began at the end of the 1970s, in France, Italy, and Spain, and significant production began in all three countries, and, after the early 1990s, in Greece. Today, for both species, mariculture yields largely surpass fisheries landings: five times more for sea bass and 20 times more for sea bream in 2006 (FAO 2008).

Several hatcheries produce alevins (juvenile fish, fries) of these two commercial species, in Greece, Spain, Italy, and France. The numbers produced are enormous: over 800 million a year for the two species combined. Big companies operate hatcheries and growing sites, as well as net pens or tanks, where the growing fish are maintained under very specific, highly controlled conditions, and regularly monitored and fed. Density in net pens ranges from 10 and 30 kg biomass water m^{-3}, which represents between 1000 individual fish at the beginning of the period to 50 fish at harvest time, as the fish grow bigger and are separated into more and more pens. Fish are generally fed with artificial pellets, but far too much natural or frozen diet (fish, squids, and shrimps) is used, as was discussed in Box 11.4 on bluefin tuna. The pellets contain fish meals and oils, more and more vegetal meals, and a vitamin premix. Fish are commonly harvested after 3 years, sea bream growing more rapidly than sea bass. Both species are appreciated by consumers throughout the Mediterranean region, but also in Northern European markets. The interest of aquaculture, compared with fisheries, is to produce a well-calibrated fish of a stable quality throughout the year.

Another advantage of mariculture consists in releasing juveniles of native species into wild waters in order to help restock specific areas where populations have become depleted. This is done with several species in the Mediterranean, but not presently at such a large scale as is practiced with salmon in the North Pacific. This activity probably will develop in the future, in conjunction with fisheries, in order to maintain the stocks.

The impacts on the marine environment depend on the bathymetric conditions and local currents. In a sea without tides, this has to be more carefully checked and controlled. Many data exist on the production of wastes by marine farms: for example, nitrogenous excretion in sea bass corresponds to 150–450 mg N kg living weight^{-1} day^{-1} and 13 mg of organic matter day^{-1} l of water^{-1} (Blancheton and Canaguier 1995). Today, the tendency is to minimize the production of nitrogen and phosphorus from marine fish culture and to decrease the quantity of proteins and increase the lipids contents in the pellets.

Fish escapes also may represent a danger for wild populations, as is well documented for salmonids in several countries, including Norway, Scotland, and British Columbia. The main problems are inbreeding with wild fish, spatial and trophic competition, and, above all, the transmission of parasitic diseases to wild fish (Krkosek et al. 2007; Rosenberg 2008; http://news.bbc.co.uk/2/hi/science/nature/4391711.stm). For Mediterranean aquaculture to date, the problem does not seem too serious, but vigilance has to be maintained. Mediterranean fisheries—like fisheries worldwide—are at the tipping point. They are now unable to produce more without great danger to the stocks and overfishing. Aquaculture and mariculture are inevitably going to be developed in the future to meet the huge and growing demand, worldwide, for edible, succulent fish and seafood (Boeuf 2003). This must be reconciled with respect for the environment and indigenous biodiversity through active, adaptive coastal management, including all the maritime activities.

Another recent activity having developed in the Mediterranean is production of the bluefin tuna, mainly in Croatia, Spain, France, and Italy (Boeuf 2003). This is not without problems. Juveniles (20–40 kg) are caught at sea by purse seine boats and transferred from capture nets into transport nests, then towed for several days to the growing site. Afterwards, they are reared in large net pens during several months, fed on natural or frozen diet. They are harvested before or during the Christmas holiday seasons (6000 t for 72 million € in 2006; FAO 2008). As mentioned in Box 11.4, however, there are many problems associated with tuna breeding, and its sustainability is very doubtful. In the Mediterranean, a few other fish species are produced but at smaller scales. They include shi drum (*Umbrina cirrosa*), meagre, reaching almost 1000 t in 2006 for 5.4 million €, other sparids, such as the common pandora (*Pagellus erythrinus*), the common dentex (*Dentex dentex*), and some others. Molluscs are also reared in the Mediterranean in several countries. For example, in 2005, 3000 t of flat oyster (*Ostrea edulis*) were produced, mainly in Spain, but today this activity is threatened by two parasites introduced in the early 1970s from North America. Additional harvests from mariculture include 30 000 t of Japanese oyster (*Crassostrea gigas*), mainly in France and Spain; 115 000 t of Mediterranean mussel *Mytilus galloprovincialis*, mainly in France, Spain, Greece, and Italy; 5000 t of grooved carpet shell (*Ruditapes semidecussatus*), in France, Italy, and Spain; and 67 000 t of Japanese carpet shell (*Ruditapes philippinarum*), in Italy and France (FAO 2008). The latter species and the Japanese oyster were introduced from Pacific areas, which resulted in the 'accidental' introduction of more than 60 species of Japanese seaweeds which have now thoroughly invaded the Thau Lagoon of southern France (Mineur *et al.* 2007; see Chapter 12).

Today, it is clear that Mediterranean aquaculture may grow much more, to meet the ever-growing market demand. We already saw that Mediterranean fisheries cannot extract more from the sea. And human population density is increasing steadily through the Mediterranean Basin, as well as the demand for seafood. To be sustainable, mariculture has to be more respectful of the environment and the limits to growth and pollution are patent. What is the 'carrying capacity' for mariculture in the Mediterranean? The major problems come from (1) the use of wild fish for feeding the captive fishes and fossil energy to produce carnivorous species (marine fish), (2) the impacts on the coastal marine environment, including both ecosystem destruction and contamination by organic matter produced by farms, and (3) the introduction of exotic species, including pathogens and parasites. Mariculture could support local fisheries by releasing high quantities of juveniles, then harvested by fisheries. This is a very attractive possibility in the future through diversification of present or new species in culture.

Destructive fishing practices continue although it is difficult to assess their extent. Bottom trawling is a permanent disturbance which keeps benthic ecosystems in a juvenile stage with low biodiversity. This also affects fish and the whole marine ecosystem negatively. Bycatch and the discard of non-target fish, including birds, marine mammals, and turtles, also contribute to the large-scale impacts of fisheries on marine ecosystems.

Mariculture may be a reasonable alternative, but care must be taken with regards the environmental impacts on coastal areas and the fact that rearing carnivorous species can cause severe environmental problems. Molluscs and herbivorous fish are more promising.

13.4.2 Whales

Whales, in particular, and cetaceans, in general, are classified by ecologists as keystone species; in other words, if they disappear, many other species will also. They are critical components of food chains and ecosystems of which they are a part. The whales are also highly emblematic mammals rightly considered by conservation biologists as 'umbrella species' useful in the struggle to raise awareness of the need for greater efforts and investments in conservation and habitat restoration. In particular, there is urgent need to improve and strengthen regulatory systems aimed at protection, sustainable use, and active replenishment of the 'stocks' or populations of the cetaceans themselves and indeed all of the astonishing and highly endangered Mediterranean marine biodiversity.

To protect and conserve whales and, mainly, to protect specific reproductive and feeding areas, different decisions have been specified in international measures (Raga and Pantoja 2004):

- prohibition of killing and commercializing them;
- prohibition of contaminating and polluting, in general in the entire sea, but more particularly in highly sensitive areas;
- regulation of maritime traffic, prohibition of fisheries and tourism traffic in specific sensitive areas, and prohibition of the extraction of marine granulates.

Unfortunately these recommendations of the Barcelona Convention are not respected. All whale species are strictly protected in the Mediterranean. Several areas are already protected and many others are proposed. For example, for the western Mediterranean, in Spanish, Italian, and French waters, specific zones enter in the Habitat Directive Law or in Natura 2000 areas. There are areas between Corsica and France, Italy and Corsica, north-eastern Spain, Alboran Sea. Such zones are specified according to their interest for feeding and juveniles breeding, migration routes, high population densities, high specific diversity, specific zones of residence, etc. Interactions between cetaceans and fisheries are closely studied and monitored, and the impact of nets estimated on the stocks (Hall and Donovan 2002; Raga and Pantoja 2004). To reduce cetaceans being accidentally caught by fishing boats is also essential, for example, through modification of acoustic detection and nets. Notably, sundown sets are prohibited in the tuna purse-seine fishery. Nevertheless, bycatches that is 'accidental' capture of non-targeted species in a fishery result from a complex combination of environmental, ecological, biological, and gear factors, as well as the motivation and the ability of the fishermen themselves (Hall and Donovan 2002).

It has been recently recognized that the modern underwater acoustic environment represents a tremendous problem for cetaceans, as they are highly sensitive to acoustic signals (Würsig and Evans 2001). Areas with high densities of cetacean should be kept off-limits to maritime traffic. Of course, collisions with big boats are also common, when the whale emerges after diving.

ACCOBAMS (the Agreement on the Conservation of Cetaceans of the Black Sea, Mediterranean Sea and Contiguous Atlantic Area) addresses data on fishery bycatch and habitat degradation. The IUCN SSC Cetacean Specialist Group (IUCN 2003) specifies the following urgent tasks for Mediterranean waters: (1) assess population sizes and threats to survival of harbour porpoises, (2) investigate the distribution, abundance, population structure, and factors threatening the conservation of shortbeaked common dolphins, (3) investigate the distribution and abundance of bottlenose dolphins, and (4) evaluate threats to their survival, to develop and test approaches to reduce conflicts between bottlenose dolphins and small-scale fisheries. Finally, they recommend conducting a basin-wide assessment of sperm whale abundance and distribution in the Mediterranean Sea.

Furthermore, global warming, so prominent in the Mediterranean region (see Chapter 12), may also create specific disturbances for cetaceans in terms of drastic changes in prey distribution and abundance (Würsig et al. 2001). Global climate change is but one threat among many, including overfishing, net entanglements, toxin pollution, noise, and habitat destruction. The cumulative effects of these may spell extirpation for cetacean populations and species already depleted or restricted to ever-smaller geographic areas (see Chapter 4).

Due to its geography at the entry of the Mediterranean, in 2007, Mediterranean whale and dolphin scientists agreed to recommend the area as the Alborán Sea Special Protected Area of Mediterranean Interest (SPAMI). This would establish the area on the high seas while national laws and the European Union Habitats Directive would take care of selected waters within 12 nautical miles. In March 2008, the countries of the Mediterranean and Black Seas agreed in principle to consider this area and other proposals in the Mediterranean for designation to help fulfil worldwide 2012 biodiversity and Marine Protected Area (MPA) targets. Now Spain, Morocco, Algeria, and Gibraltar (UK) need to take further action to designate this MPA and prepare a management plan (Whale and Dolphin Conservation Society 2008).

13.5 International cooperation

As a result of a growing concern about the ongoing degradation of the Mediterranean Sea, which is a natural link and a common property for all Mediterranean peoples, all countries bordering the Mediterranean met at Barcelona in 1976 under the auspices of the UNEP and launched an Action Plan, the Barcelona Convention, for stopping and reversing the degradation of the environment. Initially designed to struggle against pollution of the sea, this project was soon extended to terrestrial ecosystems because most pollution comes from the land. This Action Plan, known as the Blue Plan (Box 13.5), which is funded by the World Bank, is brought into play by all the Mediterranean countries and the European Community (Grenon and Batisse 1989). The Barcelona Convention appointed a Coordination Unit, which is located at Athens. Specialized institutions of the United Nations and international non-governmental organizations, such as WWF and IUCN, are involved in the Blue Plan. These organizations have an extremely active role relative to biodiversity in the Mediterranean: many hotspots of plant diversity, for example the Baetic Cordillera of southern Spain, the mountains of central and southern Greece, north-eastern and southwestern Anatolia, and several others, are among the 231 sites throughout the world identified as the most important centres of plant diversity which must be protected and properly managed (Médail and Quézel 1997). Implementation of the Barcelona Convention in 1995 aimed to institutionalize recommendations of the Rio Conference of 1992 at the scale of the Mediterranean Basin.

As mentioned above, another tool for the conservation of wetlands of international importance is the Ramsar Convention established in 1971 to halt the destruction of Mediterranean wetlands and to promote instead their ecological, scientific, economic, and cultural value (Kusler and Kentula 1990). Up to now, 89 wetland sites totalling 46 000 km^2 in 12 Mediterranean countries have been nominated for inclusion in this convention. In Turkey alone, 12 wetlands totaling 200 000 ha are Ramsar sites. Together with the UNEP Mediterranean Action Plan and the UNESCO Man and Biosphere programme, this convention is actively concerned in promoting sustainable development of wetlands that includes the main function of these sites of exceptional biological diversity. There are many other governmental and private initiatives to stop environmental degradation of both marine and terrestrial ecosystems in the Mediterranean. The Barcelona Convention mentioned above and the Agreement on the Conservation of African-Eurasian Migratory Water Birds (AEWA in 1999) have all been effective driving forces in identifying and protecting wetlands of major importance in the Mediterranean Basin. Another initiative is the Alghero Convention organized at Alghero, Sardinia, in 1995 by the association MEDVARAVIS. It is supported by 35 non-governmental organizations under the auspices of the Bern Convention, Council of Europe, IUCN, and UNEP (Mediterranean Action Plan). More species-oriented projects include action plans designed for the protection and restoration of populations of large mammals and birds. For example, in 1992 BirdLife International launched a project to prepare action plans for all the globally threatened birds which occur in the Basin. As a result of these efforts, many species of large raptors are in a process of population recovery in many countries of the northern bank, sometimes at spectacular rates (Muntaner and Mayol 1996).

Conservation efforts of the marine environment have long been neglected, but a growing concern about conservation issues is currently emerging. Sadly, no more than 200 km^2 of marine areas are currently under full protection. However, the first and still only example of marine protected areas out at sea is the Pelagos Sanctuary, established in 1999 on the basis of an agreement among France, Italy, and the Monaco Principality (Benoît and Comeau 2005).

Many other areas need greater cooperation as well, and some promising beginning steps are being made. The common problem of water shortage may also help bring about greater regional cooperation, for example in developing

Box 13.5. The Blue Plan

The Blue Plan has been launched by 20 bordering countries of the Mediterranean Sea and the European Community, which were the Contracting Parties of the Barcelona Convention on the Mediterranean Sea. This Convention, signed in 1976, 4 years after the United Nations Conference on Environment than met in Stockholm in 1972, aims at reducing pollution of the sea and protecting marine environments. After the Agenda 21 was adopted in Rio de Janeiro in 1992, the Mediterranean countries involved in the Blue Plan decided at Tunis in 1994 to formulate an Agenda MED 21 for their region. The Blue Plan is a centre of prospective studies for the Mediterranean Basin, which prepares benchmarks of possible desirable futures and sustainable development for the Mediterranean Basin by the year 2025. The Blue Plan has become a major partner of the Mediterranean Commission on Sustainable Development. It has structured its activities according to three main lines of action: the first covers analyses and evaluations of systems sustainability; the second concerns the structuring of data using information systems; and the third disseminates this work to actors of sustainable development. It also provides tools for monitoring ecological and social changes, using appropriate indicators. The Blue Plan's agenda developed in three phases: understand (1980–1984), explore (1985–1988), and suggest (since 1989). The work produced four contrasting types of development: (1) the continuation of present trends with increasing pressures on coastal areas (scenario T1); (2) a development with weak economic growth, harsh competition, and budgetary constraints, which would hamper development and the investment necessary for environmental protection (T2; the worst scenario); (3) a rapid growth but with insufficient environmental concern (T3); or (4) finally, a well-balanced development concerned with the environment, which should make economic and environmental conservation compatible (T4). This scenario corresponds to a logic of sustainable development and necessarily involves a subtle balance between demography, urbanization, tourism, agriculture, industry, energy, and transport, as well as the impact of these activities on soils, water, forests, conservation policies, and the sea. This scenario is the only one that can reconcile economic growth and environmental conservation in the long run.

Orientations of the Blue Plan include several crucial points, as follows:

1. Conservation of the coastal environment, which depends on the better control and planning of urbanization.
2. An important increase in food production. Efforts will have to focus on improving the efficiency of irrigation thanks to technical and institutional mechanisms for water saving and soil management and conservation, research in biotechnology, control of polluting agro-food industries, choice of appropriate crops, and conservation of biodiversity.
3. Controlling the development of industrial activities that are expected to increase tremendously in the south and east. In the sector of energy, the search for alternatives to firewood in the south and east and the development of solar energy will be favoured.
4. Tourism will be the major source of income in many countries. Much effort will have to be made for making tourists taking part in the effort for saving the natural and cultural heritage of the basin.

The alternative scenarios mentioned above recommend multilateral and bilateral intra-Mediterranean cooperation along two axes: the north–south axis under the impulse of the EU and the south–south axis at the initiative of Arab countries.

In order to inform and mobilize as many actors as possible, a series of booklets have been produced since 1990 on several crucial themes, namely fishing and fish farming, forestry, ecosystems, industry, islands, water, energy, tourism, transport, and natural hazards in the Mediterranean.

technology and planning schemes for sustainable and clean desalinization of sea water, among other approaches, that are integrated with energy management and conservation programs designed to address growing problems of local water scarcity (Harvey and Mercusot 2007; Flower and Thompson 2009). International cooperation schemes and agreements for the management of wildfire disasters in the Mediterranean region are also gaining momentum (Goldammer 2003). The Union for the Mediterranean deserves active support of the European Union and all Mediterranean nations as well (Martuscelli and Tolve 2002).

Finally, it is encouraging to note that The Economics of Ecosystems and Biodiversity (TEEB) initiative of the European Union and the UNDP (European Communities 2008) is now moving into phase two and intends to push forward international cooperation, policy change, and serious investment in biodiversity conservation and the restoration of natural capital.

13.6 Alternative futures

The most promising development in the struggle to preserve 'nature' and biological diversity in the Mediterranean area is the realization that natural capital—biodiversity and functioning ecosystem—is the fundamental basis not only of sustainable regional development but of all economies. Although sustainability has recently become a widely accepted concept, there is no generally accepted guidance on how to define and assess it, and far less on how to achieve it. Without such guidance, two problems arise. First, many uses will continue to contribute to depletion of species and degradation of ecosystems. Second, uses with social and conservation benefits have to struggle against hostile policies and private interests. Therefore the most urgent task is to develop and define guidelines on the sustainable use of wildlife and ecosystems. Such guidelines will have to be flexible enough to take into account the extraordinary diversity of cultures, traditions, and land use practices in the Mediterranean Basin. Any guideline on sustainability will be inefficient if local peoples with their practical experience and cultural values are not involved in its definition and do not agree with it.

The large number of political and cultural frontiers in the region acts as brakes to partnership, cooperation, and global management. As a result of this—and of the recent growth of several forms of fundamentalism, which is the exact opposite of what is needed for sustainable economic development, social justice, and environmental protection—there is much ingrained, irrational resistance and rejection to any action plan that would integrate the many regions of the basin into a single unit.

As a final word, promoting sustainability requires raising collective consciousness of the importance of preserving natural and cultural heritages that should be shared by all who live around this luminous sea. Only in that way can we possibly create a common space of cooperation and trade, a kind of 'Mediterranean Cultural Community' that Albert Jacquard (1991) has called for. Science can perhaps help us learn how to conserve 'nature' but the politics of nature conservation is a social and cultural affair to be taken up by 'nations'. Thus the ultimate challenge will be to take fully into account and preserve insofar as possible the common biological and cultural heritage shared by all the regions and nations within the basin, including the regional and ethnic specificities, another facet of biodiversity, which make up the essence of the Mediterranean world.

The Mediterranean has always been more than a simple geographical unit. For millennia, people in the Mediterranean have met and fought, but they are linked together by an unrivalled heritage and a common environment. The Mediterranean Basin has been a humanist and spiritual forum, the home of Plato, Aristotle, and Hippocrates, Moses and Maimonides, Jesus and Mohammed, Augustine and Averroès. This spiritual and cultural heritage still has profound influence on living standards, economy, and relationships between humans and nature in all parts of the basin. Long prior to the modern era, the basin had already undergone several periods of cultural and economic globalization. High levels of trade and cultural exchange pursued by the ancient Egyptians, Phoenicians, Greeks, Romans, Carthaginians, Genovese, and Venetians, and others superimposed new living standards and cultural diversity without destroying pre-existing ones. To achieve this effect in our day, a gigantic effort will be necessary because of the tremendous disparity in all that makes up the lives of the various societies around the Mediterranean. While it is true that the Mediterranean no longer plays the pivotal role in international relations it once did in the age of the great sea-faring ships, its long history as a biological and cultural crossroads, as well as its position as undisputed cradle of western civilization, surely makes of it as good a place as any to start this long and indispensable process.

Summary

What makes the Mediterranean Basin particularly sensitive is the combination of population growth and various components of global change, which potentially threaten biodiversity and ecosystems more severely than in many other parts of the world. The Mediterranean Basin is an area of both division and convergence, presenting the full range of major socio-economic, political, and environmental problems that are faced by the whole planet. The 24 states which encircle the basin constitute two sharply contrasting worlds. People in the northern countries enjoy a much higher yearly income than the average income of people inhabiting countries of the eastern and southern shores of the Sea. Although the status of biodiversity is still challenging and sometimes really at risk, many efforts have been made in recent decades to reverse these negative trends. Conservation programmes include active protection of species, reinforcement and reintroduction of populations, preservation of ancient varieties of plants and animals, and erection of preserves and various kinds of protected areas. Concurrently, active programmes are under way for restoring, managing, and reintegrating the main habitat types, as well as traditional cultural landscapes. Fisheries are not expected to increase the amount of fish extracted from the sea because of a strong trend of decline of the stocks of wild fish. But mariculture is rapidly developing and needs to be carefully organized to reduce its negative effects on the marine environment. Finally, the launching of international programmes such as the Blue Plan opens the way to make the Mediterranean region a pilot area for sustainable development and the restoration of natural capital. The ultimate goal should be to find a sustainable, just, desirable, and generous future for all Mediterranean peoples. Learning to better manage and maintain the remarkable biodiversity of Mediterranean lands and the sea is a prerequisite for achieving this goal.

Glossary

abyssopelagic Pelagic animals (plankton and nekton) that live at great depths over the abyssal plains.
allogamous Mating system in which plants must be cross-pollinated.
allopatric Taxa having non-overlapping distribution areas.
allopatric speciation Differentiation process whereby two or more species arise from a mother species as a result of isolation by a barrier to dispersal.
anecic Large earthworms living in galleries in the soil and feeding on organic material.
anemochoric Plant species with seeds that are dispersed by wind.
aphyllous Leafless plants in which green stems are photosynthetically active.
autogamous Mating system in which plants are self-pollinated.
autopolyploidy Polyploidy in which all the chromosomes come from the same species.
autotetraploid A form of polyploidy in which the nucleus includes four times the haploid number of chromosomes coming from the same species.
bath'a Type of low matorral common in the northeastern Mediterranean quadrant.
bathypelagic Pelagic animals (plankton and nekton) that live at the level of the continental slope (bathyal zone of the benthos).
benthic The ecological region at the lowest level of a body of water, such as an ocean or a lake.
benthos Organisms living on, in, or near the sea (or lake) bottom.
bioremediation Process whereby living organisms actively contribute to remove toxic chemicals from soils.

bioturbation The stirring or mixing of sediment or soil by organisms, especially by burrowing or boring.
boreal Major life zone covering the northern part of the continents of the northern hemisphere.
calciphobe Plant that does not support active limestone.
Cenozoic The most recent geologic era which includes the Tertiary and the Pleistocene periods, including the Holocene (65 mya to present).
ceras (plural: cerata) Bludgeon-shaped dorsal structure of some nudibranchs with a terminal cnidosac containing a toxic mixture to fend off attackers.
chasmophyte A plants that grows in cracks and crevices of cliffs or walls, independent of surface soil.
chlorophyll *a* The most common photosynthetic pigment of the plant kingdom. Present in all terrestrial and aquatic plants. Its concentration is used as an indicator of phytoplankton biomass.
cladogenetic Branching of lineages during phylogenetic processes of speciation.
commensal Species living in close association, for example in the same burrow, shell or house, without mutual influence (i.e. not symbiotic).
cool summergreen Deciduous broad-leaved plants living in temperate and northern parts of the northern hemisphere.
Coriolis force Phenomenon of deviation of masses in movement in contact with the rotation of the Earth (towards the right in the northern hemisphere and towards the left in the southern hemisphere). On a large scale, it acts on the movements of air and water masses.
Cyanophyceae (also called Cyanobacteria) They differ from bacteria by the presence of chlorophyll *a* and some other pigments, and they are

capable of photosynthesis. The eutrophization of water favours their pullulation in coloured colonies or filaments. Some emit poisonous toxins.

dehesa Category of land-use system consisting of a mixture of woodland, crops, and pastures.

diadromous Species (usually fish) which move between fresh and salt water in relation to stages of their life cycle, usually breeding or reproductive stages.

diaspore Dispersal organ of plants or animals, for example seeds, fruits, eggs, or spores.

dioecious Unisexual; that is, male and female reproductive organs borne on different individuals.

diploid Having chromosomes in pairs in a nucleus. Chromosomes in a pair are homologous so that twice the haploid number is present.

disharmony Refers to changes in the relative proportions of different taxa or trophic levels among communities on islands compared with those on the nearby mainland.

ecomorphology Form and shape of organisms in relation to their ecology.

ecosystem goods and services (also known as natural goods and services) Foods, fuels, or other products of economic or cultural value that are supplied by ecosystems, and various economically valuable services that ecosystems provide to people, such as flood-water retention and erosion control, all without costs of production or maintenance. Services include key processes, such as pollination, seed dispersal, etc.

ecosystem trajectory An alternative or complementary concept to the 1930s–1970s' concept of more or less linear successional ecosystem development towards a climax. It is the path described by an ecosystem (or projectile) whose dynamics are driven by various forces, internal or external.

ecotone Mixed habitat formed by the overlapping areas or transition zone between two habitats. Transitional strip separating two communities.

ecotypic variation Variation of the subunits of a species in relation to variation of environmental conditions. Ecotypic variation may be but not necessarily genetically determined.

edaphic Refers to the physical and chemical conditions of a soil and their influence on the growth of plants.

eddy Local movement of rotation of a body water. For large rotary movements in the oceans and seas, the word gyre is also used.

endemic Indigenous or native in a restricted locality, area, or region.

endogeic Invertebrates living in soils, for example earthworms, and actively contributing to bioturbation.

endoreic marsh Body of water, often temporary, that does not empty into any river or larger body of water.

entomogamous Plants whose mating systems require pollination by insects.

ephemerals Very short-lived organisms, especially plants. Usually found in desert or arid Mediterranean regions, germinating especially after rain events exceeding approximately 25 mm.

eremic (eremean) Desert-dwelling, of desert origins. The eremean region is the wide arid belt stretching from Mauritania to Arabia and north-western India into central Asia.

euphotic (or photic; Greek for 'well lit') The depth of the water in a lake or ocean that is exposed to sufficient sunlight for photosynthesis.

euryhaline Species that is resistant to great changes in salinity over a season or life cycle.

eurythermic A term used for organisms that tolerate large variations in ambient temperature.

eustatic Variation of sea level resulting from climatic fluctuations.

evergreen Refers to perennial plants bearing living leaves all year round.

evergreenness The habit of plants having photosynthetically active leaves all year round.

evolutionary convergence Evolution producing an increasing similarity of morphology or any other phenotypic features between groups of organisms that are phylogenetically unrelated.

feedback systems Refers to the modification or control of a process or system by its results and effects.

feralization The process whereby domesticated animals escaping human control constitute wild populations, eventually returning to ancestral forms.

garrigue Type of vegetation composed of perennial readily coppicing shrubs and trees. Sometimes restricted to calcareous soils and usually seen as low scrubland with patches of bare ground.

geophyte Herbs with perennial buds below soil surface (class of Raunkiaer's life forms).

germplasm Term for plant material of any kind useful for propagation or long-term storage.

granulometry The measurement of the size distribution in a collection of grains or particles, such as those which make up a soft sea bottom.

growth form Type of morphological and physiological features resulting from evolutionary responses to particular bioclimates.

halophyte A plant that tolerates salty soils and water, at least for certain periods of life cycle; a condition typical of plants growing near the seashore and in coastal salt flats and river estuaries.

haplotype Series of genes located on a chromosome and which segregate together during meiosis. The corresponding genetic material may correspond to nuclear DNA, mitochondrial DNA (mtDNA) in animals, or chloroplastic DNA in plants.

hard-leaved evergreen A broad-leaved plant with hard sclerophyllous evergreen leaves.

hemicryptophyte A herb with perennial buds at or near soil level, protected during dry season by soil itself or by dry, dead portions of the plant (class of Raunkiaer's life forms).

herbivory The process whereby animals eat living parts of plants.

heterozygosity Having two or more alleles at a same genetic locus.

holobenthic Marine animals which spend the totality of their life on or near the sea bottom, without any pelagic larval phase.

Holocene Geological epoch following the Pleistocene and consisting of recent times since end of the last ice age, about 11 000 years ago.

hysteranthous Plant flowering in autumn in a leafless state. Usually found in geophytes.

inertia Property of matter or a system by which it continues in its existing state unless changed by an external force.

karst Limestone systems which are highly fissured as a result of calcareous dissolution.

krill Term used to designate large shoals of pelagic shrimps from the Euphausiacae family. It is the main foodstuff of whales.

leptophyll Growth-form class of plants having very reduced leaf surface area; common in alpine and desert habitats.

Levantine Intermediate Water (LIW) 'Mediterranized' Atlantic water moving west from the Levantine Basin towards the Atlantic Ocean in an intermediate depth; between the surface and the deep-sea waters.

life-form spectrum Range of vegetative growth forms that plant species can present, including trees, shrubs, herbs, forbs, grass, bulbs, etc. Life forms are often found to be closely linked to life history traits which are adaptive to conditions of a given environment.

life zone An ecological term to describe areas with similar plant and animal communities that occur, for example, at a given elevation on a series of mountains at a given latitude.

lithology Refers to the science of rock substrates.

Living Planet Index Biodiversity index based on long-term trends of population size of many animal species. Used, for example by the World Wildlife Fund (WWF) for assessing the state of the biosphere.

local specialization Process whereby a population evolves life-history traits that are tightly adapted it to its local environment. Local specialization is genetically determined.

macrobenthos Benthic metazoans exceeding 2 mm in size.

Maghreb The three countries of north-western Africa: Morocco, Algeria, and Tunisia. Belongs to the Palaearctic biogeographical realm.

maquis (or *macchia*) Kind of shrubland consisting of sclerophyllous evergreen plants. Sometimes used for shrublands occurring on siliceous substrates.

matorral Any kind of shrubby predominantly evergreen Mediterranean vegetation.

meiobenthos Small benthic metazoans between 0.1 and 2 mm, barely visible to the naked eye.

meroplankton Subset of planktonic larvae of benthic invertebrates.

Mesogean Refers to features that occurred before the establishment of the modern Mediterranean Sea.

Mesozoic The geologic era that includes the Triassic, Jurassic, and Cretaceous periods, from 250 to 65 mya.

metaclimax The whole set of habitats that are required to ensure the survival of all the species produced by the evolutionary history of biotas at the scale of a landscape.

metapopulation A series of local populations interconnected through processes of extinction and re-colonization.

Milankovitch cycles Astronomical events that are responsible for the alternation of glacial and interglacial cycles since the beginning of the Pleistocene.

mimicry Protective similarity in appearance of one species of animal to another (generally insects). In Batesian mimicry, the imitated species is poisonous and often conspicuously marked. In Müllerian mimicry, both species are protected from predators, gaining mutually from having the same warning coloration.

mitotype Subgroup of a main haplotype; set of closely linked mitochondrial DNA (mtDNA) or chloroplast DNA markers.

molecular clock Process whereby molecules evolve at an approximately constant rate through time for a group of organisms. The difference between the form of a molecule in two species is therefore proportional to the time elapsed since the species diverged from a common ancestor.

monoecious Having both male and female reproductive organs on the same individual.

monotypic A genus with only one species.

Natufian Epi-Palaeolithic human culture which developed in the Near East between 10 800 and 8200 BC.

natural capital An economic metaphor for the limited stocks of physical and biological natural resources found on Earth; also used to refer to functioning ecosystems—both natural and managed—and biodiversity.

nekton Includes all marine animals that move independently of the currents; that is, fish, squids, whales, and other cetaceans.

Neogene The second part of the Tertiary period, from the Miocene epoch (25 mya) onwards.

neotenic An animal which keeps larval characters throughout its life cycle. Neoteny is common in amphibians.

Nudibranchiata Slug-like marine gastropods, without a shell and almost always carnivorous.

oligotrophic Waters (or soils) that are low in nutrients and consequently low in primary productivity.

oogon Female sex organ of certain algae and fungi containing one or more oospheres.

outgroup A taxon that diverged from a group of other taxa before they diverged from each other. Used to root phylogenetic trees.

Palaeogene The first part of the Tertiary period, from the early Palaeocene epoch (65 mya) to the Oligocene, which ended c.25 mya.

Palaeozoic Geological eras between the Cambrian (540 mya) and the Permian (250 mya).

palynology The study of ancient floras and vegetations from fossil pollens that accumulate in soil profiles.

parthenogenetic A species which reproduces without paternal contribution of genes.

pedology The science of soils.

pelagic Living in open sea waters.

pelecypods Bivalves.

phenotype The way in which the genotype of an individual is expressed in its morphology, physiology, behaviour, or any life-history trait.

phenotypic plasticity Phenotypic variation expressed by a single genotype in different environments.

photic *see* euphotic

photophilous Liking (or needing) sunlight.

phyllode A flattened leaf-like petiole.

phylogeny The genealogy of a group of taxa. Study of the branching relationships among species deriving from a same ancestor. Phylogeny aims at reconstructing the evolutionary history of a lineage.

phylogeography The reconstruction from molecular markers (for example mitochondrial DNA (mtDNA) in animals and chloroplast DNA in plants) of the history of the spatial distribution of species and populations.

phytal Portion of the marine domain where the plants can grow.

phytophagy *see* herbivory

plasticity *see* phenotypic plasticity

ploidy Number of haploid number of chromosomes in a nucleus.

polyploid Having several times (three or more) the haploid number of chromosomes in a nucleus.

polytypic Species composed of several subspecies.

propagule Any part of an organism or stage in the life cycle (seeds, individuals, at least one pregnant female) that can reproduce the species and thus establish a new population.

psu (practical salinity unit) A unit of measurement of salinity similar to parts per thousand (ppt), that expresses the quantity of salt in 1 l of sea water.

refugia In historical biogeography, regions where species have persisted during harsh climatic periods (for example Pleistocene glaciations) while becoming extinct elsewhere.

resilience The property of an ecological system to recover after disturbance.

resistance The magnitude of the response of an ecological system to a disturbance event.

retamoid From the genus *Retama*, refers to broom-like plants with photosynthesizing stems and highly reduced leaves.

riparian Plants and vegetation types growing along rivers or streams and taking water mostly from ground water.

ruderals Plant species commensal to humans that are usually frequent in and around human-made habitats such as roadsides, ditches, and other frequently disturbed habitats; usually annuals (*see also* segetals).

sarmatic Organisms belonging to the coastal fauna that in late Tertiary times inhabited the shallow, brackish, or salt Sarmatic inland sea, which formed an eastern continuation of the Mediterranean Sea.

sciaphilous Shade-loving, as opposed to heliophilous (light-loving).

sclerophylly Leaf form which is generally evergreen, coriaceous, and often spiny.

sebkha A depression which contains brackish water after rainfall, but is dry and covered by salt incrustations in the summer (common in North Africa and the Near East).

segetals Plant species, mostly annuals, common in cereal crops fields and tied to this habitat.

sessile Species not able to move, fixed in a specific area, frequent after a larval mobile stage in aquatic invertebrates.

sierras Spanish name for mountains.

sorting processes Processes whereby not all species of a flora or a fauna remain in biotas through time. Sorting processes generally occur when a geological era is replaced by another.

speciation Mechanisms by which one mother species splits into two or more daughter species.

spinescence The habit of leaves or stems to be spiny.

steady state Synonym for quasi-equilibrium.

stenotherm An organism capable of living or growing over only a narrow range of temperatures.

summergreen Refers to perennial plants having leaves only during the growing season; that is, from spring to autumn (deciduous).

supersaturated Communities with more co-existing species than expected from biogeographical rules. Refers to island communities which are not at equilibrium.

Sverdrup Unit of measure used in oceanography to measure the flow of currents. It corresponds to $10^6 \text{ m}^3 \text{ s}^{-1}$ or $0.001 \text{ km}^3 \text{ s}^{-1}$.

sympatric Refers to two or more species or populations occurring in the same geographical area.

thermocline The region of greatest rate of vertical temperature change in a body of water, with warm water above and cold water below.

thermohaline Thermohaline circulation is the permanent circulation of the sea water on a large scale, generated by differences of salinity and temperature among the water masses.

trajectory *see* ecosystem trajectory

turnover Changes in species composition across a range of habitats, for example along ecological transects or between habitats within a landscape.

vicariance The geographical separation of a group of organisms resulting in differentiation of the original group into new varieties or species.

wadi Temporary rivers in arid regions. The term is mostly used in North Africa and the Near East.

References

Abbo, S., I. Zezak, E. Schwartz, S. Lev-Yadun, Z. Kerem, and A. Gopher. (2008). Wild lentil and chickpea harvest in Israel: bearing on the origins of Near Eastern farming. *Journal of Archaeological Science* **35**:3172–3177.

Abdoun, F., A. Jull, F. Guibal, and M. Thinon. (2005). Radial growth of the Sahara's oldest trees: *Cupressus dupreziana* A. Camus. *Trees–Structure and Function* **19**:661–670.

Abdul Riga, A. M. M., and M. Bouché. (1995). The eradication of an earthworm genus by heavy metals in southern France. *Applied Soil Ecology* **2**:45–52.

ACCOBAMS (Agreement on the Conservation of Cetaceans of the Black Sea, Mediterranean Sea and Contiguous Atlantic Area). (2006). *IUCN Red List of Cetaceans in the Mediterranean and Black Seas*. Resolution 3.19.

Achérar, M., and S. Rambal. (1992). Comparative water relations of four Mediterranean oak species. *Vegetatio* **99–100**:177–184.

Achérar, M., J. Lepart, and M. Debussche. (1984). La colonisation des friches par le pin d'Alep (*Pinus halepensis* Mill.) en Languedoc méditerranéen. *Oecologia Plantarum* **19**:179–189.

Aerts, R. (1995). The advantage of being evergreen. *Trends in Ecology & Evolution* **10**:402–407.

Affre, L., J. D. Thompson, and M. Debussche. (1997). Genetic structure of continental and island populations of the Mediterranean endemic *Cyclamen balearicum* (Primulacae). *American Journal of Botany* **84**:437–451.

Aguilar, A., and J. A. Raga. (1993). The striped dolphin epizootic in the Mediterranean Sea. *Ambio* **22**:524–528.

Aidoud, A., and D. Nedjraoui. (1992). The steppes of alfa (*Stipa tenacissima*) and their utilisation by sheep. In C. A. Thanos (ed.), *Plant-Animal Interactions in Mediterranean-Type Ecosystems*, pp. 62–67. University of Athens.

Al Hallani, F., R. Andreoni, M. Bassil, C. Combe, P. Magaud, and L. Seytre. (1995). La cédraie de Barouk (Liban). *Forêt Méditerranéenne* **16**:171–183.

Alcover, J. A., S. Moya-Sola, and J. Pons-Moya. (1981). *Les Quimeres Del Passat. Els Vertebras Fossils Del Plio-Quaternari de les Baleares i Pitiuses*. Palma de Mallorca.

Aldasoro, J. J., C. Aedo, C. Navarro, and F. Muñoz Garmendia. (1998). The genus *Sorbus* L. (Rosaceae) in Europe and the N of Africa. *Systematic Botany* **23**:189–212.

Almaça, C. (1976) Zoogeografia e especiacao dos ciprinideos da Peninsula Iberica. *Natura, Lisboa* **4**:3–28.

Alonso, D., J. Arizaga, R. Miranda, and M. A. Hernández. (2006). Morphological diversification of common crossbill *Loxia curvirostra* populations within Iberia and the Balearics. *Ardea* **94**:99–107.

Alonzo, S. H., and R. R. Warner. (1999). A trade-off generated by sexual conflict: Mediterranean wrasse males refuse present mates to increase future success. *Behavioral Ecology* **10**:105–111.

Al-Tardeh, S., T. Sawidis, B. E. Dianneildis, and S. Delivopoulos. (2008). Water content and reserve allocation patterns within the bulb of the perennial geophyte red squill (Liliaceae) in relation to the Mediterranean climate. *Botany-Botanique* **86**:291–299.

Amigues, S. (1980). Quelques aspects de la forêt dans la littérature grecque antique. *Revue Forestière Française* **32**:211–223.

Amigues, S. (1991). Le témoignage de l'antiquité classique sur des espèces en régression. *Revue Forestière Française* **33**:47–57.

Amiot, J., Y. Salmon, C. Collin, and J. D. Thompson. (2005). Differential resistance to freezing and spatial distribution in a chemically polymorphic plant *Thymus vulgaris*. *Ecology Letters* **8**:370–377.

Angerbjörn, A. (1985). The evolution of body size in mammals on islands: some comments. *The American Naturalist* **125**:304–309.

Anstett, M. C., G. Michaloud, and F. Kjellberg. (1995). Critical population size for fig/wasps mutualism in a seasonal environment: effect and evolution of the duration of female receptivity. *Oecologia* **103**:453–461.

Anton, C., M. Musche, V. Hula, and J. Settele. (2008). Myrmica host-ants limit the density of the ant-predatory large blue *Maculinea nausithous*. *Journal of Insect Conservation* **12**:511–517.

Apostolidis, A., D. Loukovitis, and C. Tsigenopoulos. (2008). Genetic characterization of brown trout (*Salmo trutta*) populations from the Southern Balkans using mtDNA sequencing and RFLP analysis. *Hydrobiologia* **600**:169–176.

Apostolopoulos, Y., and S. Sönmez. (2000). New directions in Mediterranean tourism: restructuring and cooperative marketing in the era of globalization. *Thunderbird International Business Review* **42**:381–392.

Araújo, M. B., and A. Guisan. (2006). Five (or so) challenges for species distribution modelling. *Journal of Biogeography* **33**:1677–1688.

Araújo, M. B., W. Thuiller, P. H. Williams, and I. Reginster. (2005). Downscaling European species atlas distributions to a finer resolution: implications for conservation planning. *Global Ecology and Biogeography* **14**:17–30.

Arévalo, J. R., J. D. Delgado, R. Otto, A. Naranjo, M. Salas, and J. M. Fernández-Palacios. (2005). Distribution of alien vs. native plant species in roadside communities along an altitudinal gradient in Tenerife and Gran Canaria (Canary Islands). *Perspectives in Plant Ecology, Evolution and Systematics* **7**:185–202.

Arnold, C. (2008). *Mediterranean Islands*. Survival Books, London.

Aronne, G., and V. De Micco. (2001). Seasonal dimorphism in the Mediterranean *Cistus incanus* L. subsp *incanus*. *Annals of Botany* **87**:789–794.

Aronson, J., and A. Shmida. (1992). Diversity along a Mediterranean-desert gradient in response to interannual fluctuations in rainfall. *Journal of Arid Environments* **23**:235–247.

Aronson, J., J. Kigel, A. Shmida, and J. Klein. (1992). Adaptive phenology of desert and Mediterranean populations of annual plants grown with and without water stress. *Oecologia* **89**:17–28.

Aronson, J., J. Kigel, and A. Shmida. (1993). Reproductive allocation strategies of desert and Mediterranean populations of annual plants grown with and without water stress.*Oecologia* **93**:336–342.

Aronson, J., S. J. Milton, and J. N. Blignaut (eds). (2007). *Restoring Natural Capital: Science, Business and Practice*. Island Press, Washington DC.

Aronson, J., J. S. Pereira, and J. G. Pausas (eds). (2009). *Cork Oak Woodlands on the Edge: Ecology, Adaptive Management, and Restoration*. Island Press, Washington DC.

Arrignon, J. (2004). *L'Ecrevisse et Son Elevage*, 4th edn. TEC & DOC, Paris.

Attenborough, D. (1987). *The First Eden. The Mediterranean World and Man*. Fontana/Collins, London.

Auerbach, M., and A. Shmida. (1985). Harmony among endemic littoral plants and adjacent floras in Israel. *Journal of Biogeography* **12**:175–187.

Augier, H. (1973). Les particularités de la mer Méditerranée: son origine, son cadre, ses eaux, sa flore, sa faune, ses peuplements, sa fragilité écologique. In *La Mer Méditerranée*, pp. 27–53. CIHEAM, Paris.

Aussenac, G. (2002). Ecology and ecophysiology of circum-Mediterranean firs in the context of climate change. *Annals of Forest Science* **59**:823–832.

Avise, J. C., and D. Walker. (1998). Pleistocene phylogeographic effects on avian populations and the speciation process. *Proceedings of the Royal Society of London Series B Biological Sciences* **265**:457–463.

Avishai, M. (1985). The role of botanic gardens. In C. Gomez-Campo (ed.), *Plant Conservation in the Mediterranean Area*, pp. 221–236. Dr. W. Junk, Dordrecht.

Axelrod, D. I. (1975). Evolution and biogeography of Madrean-Tethyan sclerophyll vegetation. *Annals of the Missouri Botanical Garden* **62**:280–334.

Babour, M. G., and R. A. Minnich. (1990). The myth of chaparral convergence. *Israel Journal of Botany* **39**:453–463.

Backlund, M., and M. Thulin. (2007). Revision of the Mediterranean species of *Plocama* (Rubiaceae). *Taxon* **56**:516–520.

Backlund, M., B. Bremer, and M. Thulin. (2007). Paraphyly of Paederieae, recognition of Putorieae and expansion of *Plocama* (Rubiaceae-Rubioideae). *Taxon* **56**: 315–328.

Badr, A., K. Müller, R. Schäeffer-pregl, H. E. Rabey, S. Effgen, H. H. Ibrahim, C. Pozzi, W. Rohde, and F. Salamini. (2000). On the origin and domestication history of barley (*Hordeum vulgare*). *Molecular Biology and Evolution* **17**:499–510.

Bairlein, F. (1991). Nutritional adaptations to fat deposition in the long-distance migratory garden warbler *Sylvia borin*. In *Proceedings of the XXth International Congress of Ornithology*, pp. 2149–2158. Christchurch, New Zealand.

Bairlein, F. (1992). Recent prospects on trans-Saharan migration of songbirds. *Ibis* **134**, suppl. 1:41–46.

Baker, H. G., and G. L. Stebbins. (1965). *The Genetics of Colonizing Species*. Academic Press, New York.

Baletto, E., and A. Casale. (1991). Mediterranean insect conservation. In N. M. Collins, and J. A. Thomas (eds), *The Conservation of Insects and their Habitats*, pp. 121–142. Academic Press, London.

Ballesteros, E., J.Garrabou, B. Heren, M. Zabala, E. Cebrian, and E. Sala. (2009). Deep-water stands of *Cystoseira zosteroides* C. Agardh (Fucales, Ochrophyta) in the Northwestern Mediterranean: insights into assemblage structure and population dynamics. *Estuarine, Coastal and Shelf Science* **82**:477–484.

Balmford, A., R. E. Green, and M. Jenkins. (2003). Measuring the changing state of nature. *Trends in Ecology & Evolution* **18**:326–330.

Banaigs, B., and J.-M. Kornprobst. (2007). La biodiversité marine et le médicament: espoirs, réalités et contraintes. *L'Actualité Chimique* **March 2007**:7–13.

Banarescu, P. (1973). Some reconsiderations on the zoogeography of the euro-mediterranean freshwater fish fauna. *Revue Roumaine de Biologie, Section Zoology* **18**:257–264.

Barbéro, M., A. Benabid, C. Peyre, and P. Quézel. (1981). Sur la presence au Maroc de *Laurus azorica* (seub.) Franco. *Anales del Jardin Botanico de Madrid* **37**:467–472.

Barbéro, M., R. Loisel, and P. Quézel. (1991). Sclerophyllous *Quercus* forests in the eastern Mediterranean area: ethological significance. *Flora et Vegetatio Mundi* **9**:189–198.

Barbéro, M., R. Loisel, and P. Quézel. (1992). Biogeography, ecology and history of Mediterranean *Quercus ilex* L. ecosystems. *Vegetatio* **99–100**:19–34.

Barbéro, M., R. Loisel, P. Quézel, D. M. Richardson, and F. Romane. (1998). Pines of the Mediterranean Basin. In D. M. Richardson (ed.), *Ecology and Biogeography of Pinus*, pp. 153–170. Cambridge University Press, Cambridge.

Barbraud, C., M. Lepley, V. Lemoine, and H. Hafner. (2001). Recent changes in the diet and breeding parameters of the purple heron *Ardea purpurea* in southern France. *Bird Study* **48**:308–316.

Bauchot, R., and M. L. Bauchot. (1964). *Les Poissons.* Que-sais-je n° 642. Presses Universitaires de France, Paris.

Bauchot, M. L., and A. Pras. (1980). *Guide des Poisons Marins d'Europe.* Delachaux et Niestlé, Paris.

Baumann, H., and S. Kuenkele. (1982). *Die wildwachsenden Orchideen Europas.* Kosmos Verlag, Stuttgart.

Beadle, N. C. W. (1966). Soil phosphate and its role in molding segments of the Australian flora and vegetation, with special reference to xeromorphy and sclerophylly. *Ecology* **47**:992–1007.

Beja-Pereira, A., P. R. England, N. Ferrand, S. Jordan, A. O. Bakhiet, M. A. Abdalla, M. Mashkour, L. Jordana, P. Taberlet, and G. Luikart. (2004). African origins of the domestic donkey. *Science* **304**:1781.

Bellan-Santini, D. (1985). The Mediterranean benthos: reflections and problems raised by a classification of the benthic assemblages. In M. Moraitou-Apostolopoulou, and V. Kiortsis (eds), *Mediterranean Marine Ecosystems*, pp. 19–48. Plenum Press, New York.

Bellot, J., A. Bonet, J. Peña, and J. Sánchez. (2007). Human impacts on land cover and water balances in a coastal Mediterranean county. *Environmental Management* **39**:412–422.

Benchekroun, F., and G. Buttoud. (1989). L'arganeraie dans l'économie rurale du sud-ouest marocain. *Forêt méditerranéenne* **11**:127–136.

Bennett, K. D., P. C. Tzedakis, and K. J. Willis. (1991). Quaternary refugia of north-European trees. *Journal of Biogeography* **18**:103–115.

Benoît, G., and A. Comeau (dir). (2005). *Méditerranée. Les Perspectives du Plan Bleu sur l'Environnement et le Développement.* Editions de l'Aube et Plan Bleu.

Berberoglu, S., K. T. Yilmaz, and C. Özkan. (2004). Mapping and monitoring of coastal wetlands of Çukurova Delta in the Eastern Mediterranean region. *Biodiversity and Conservation* **13**:615–633.

Bernhardt, P., and P. Burns-Balogh. (1986). Floral mimesis in *Thelymitra* (Orchidaceae). *Plant Systematics and Evolution* **151**:187–202.

Berrahmouni, N., P. Regato, M. Ellatifi, H. Daly-Hassen, M. Bugalho, S. Bensaid, M. Diaz, and J. Aronson. (2009). Ecoregional planning for biodiversity conservation. In Aronson, J., J. S. Pereira, and J. G. Pausas (eds), *Cork Oak Woodlands on the Edge: Ecology, Adaptive Management, and Restoration*, pp. 203–216. Island Press, Washington DC.

Berthold, P. (1993). *Bird Migration. A General Survey.* Oxford University Press, Oxford.

Bertrand, C., N. Fabre, C. Mouilis, and J.-M. Bessière. (2003). Composition of the essential oils in *Ruta corsica* DC. *Journal of Essential Oil Research* **15**:88–89.

Bertrand S., A. Camasses, and H. Escriva. (2007). L'amphioxus, ou comment devient-on un vertébré? *Journal de la Société de Biologie* **201**:51–57.

Besnard, G., R. Rubio de Casas, and P. Vargas. (2007). Plastid and nuclear DNA polymorphism reveals historical processes of isolation and reticulation in the olive tree complex (*Olea europaea*). *Journal of Biogeography* **34**:736–752.

Béthoux, J. P., and B. Gentili. (1999). Functioning of the Mediterranean Sea: past and present changes related to freshwater input and climate changes. *Journal of Marine Systems* **20**:33–47.

Béthoux, J. P., B. Gentili, J. Raunet, and D. Tailliez. (1990). Warming trend in the western Mediterranean deep water. *Nature* **147**:660–662.

Béthoux, J. P., P. Morin, and D. P. Ruiz-Pino. (2002). Temporal trends in nutrient ratios: chemical evidence of Mediterranean ecosystem changes driven by human activity. *Deep-Sea Research II* **49**:2007–2016.

Bhagwat, S. A., and K. J. Willis. (2008). Species persistence in northerly glacial refugia in Europe: a matter of chance or biogeographical traits? *Journal of Biogeography* **35**:464–482.

Bianchi, C. N., and C. Morri. (2000). Marine biodiversity of the Mediterranean Sea: situation, problems and prospects for future research. *Marine Pollution Bulletin* **40**:367–376.

Bianco, P. G. (1986). The zoogeographic units of Italy and western Balkans based on cyprinid species ranges (Pisces). *Biologia Gallo-Hellenica* **12**:291–299.

Bianco, P. G. (1990). Potential role of the palaeohistory of the Mediterranean and parathetys basins on the early dispersal of Euro-Mediterranean freshwater fishes. *Ichthyological Exploration of Freshwaters* **1**:167–184.

Bibby, C. J. (1992). Conservation of migrants on their breeding grounds. *Ibis* **134**:29–34.

Biggs, R., C. Raudsepp-Hearne, C. Atkinson-Palombo, E. Bohensky, E. Boyd, G. Cundill, H. Fox, S. Ingram, K. Kok, S. Spehar *et al.* (2007). Linking futures across scales: a dialog on multiscale scenarios. *Ecology and Society* **12**:17.

Biju-Duval, B., J. Dercourt, and X. Le Pichon. (1976). La genèse de la Méditerranée. *La Recherche* **7**:811–822.

Bino, R. J., A. Dafni, and A. D. J. Meeuse. (1982). The pollination ecology of *Orchis galilea* (Bornm. et Schulze) Schltr. (Orchidaceae). *New Phytologist* **90**:315–319.

BirdLife International. (2004) *State of the World's Birds 2004: Indicators for our Changing World*. BirdLife International, Cambridge.

Blanch, J., J. Peñuelas, and J. Llusià. (2007). Sensitivity of terpene emissions to drought and fertilization in terpene-storing *Pinus halepensis* and non-storing *Quercus ilex*. *Physiologia Plantarum* **131**:211–225.

Blancheton, J. P., and B. Canaguier. (1995). Bacterial and particulate materials in recirculating seabass (*Dicentrarchus labrax*) production system. *Aquaculture* **133**:215–224.

Blangy, S., and H. Mehta. (2006). Ecotourism and ecological restoration. *Journal for Nature Conservation* **14**: 233–236.

Block, B. A., S. Teo, A. Walli, A. Boustany, M. Stokesbury, C. Farwell, K. Weng, H. Dewar, and T. Williams. (2005). Electronic tagging and population structure of Atlantic bluefin tuna. *Nature* **434**:1121–1127.

Blondel, J. (1969). *Synécologie des Passereaux Résidents et Migrateurs dans un Echantillon de la Région Méditerranéenne Française*. Centre Régional de Documentation Pédagogique, Marseille.

Blondel, J. (1985). Historical and ecological evidence on the development of Mediterranean avifaunas. In *Acta XVIII Congressus Internationalis Ornithlogici*, vol. 1, pp. 373–386. Moscow.

Blondel, J. (1986). *Biogéographie Evolutive*. Masson, Paris.

Blondel, J. (1987). From biogeography to life history theory: a multithematic approach. *Journal of Biogeography* **14**:405–422.

Blondel, J. (1991). Birds in biological isolates. In C. M. Perrins, J. D. Lebreton, and G. Hirons (eds), *Bird Population Studies, their Relevance to Conservation and Management*, pp. 45–72. Oxford University Press, Oxford.

Blondel, J. (1995). *Biogéographie. Approche Ecologique et Evolutive*. Masson, Paris.

Blondel, J. (2000). Evolution and ecology of birds on islands: trends and prospects. *Vie et Milieu* **50**: 205–220.

Blondel, J. (2006). Man as 'designer' of Mediterranean landscapes: a millennial story of humans and ecological systems during the historic period. *Human Ecology* **34**:713–729.

Blondel, J. (2007). Coping with habitat heterogeneity: the story of Mediterranean blue tits. *Journal of Ornithology* **148**:S3–S15.

Blondel, J., and H. Farré. (1988). The convergent trajectories of bird communities in European forests. *Oecologia* **75**:83–93.

Blondel, J., and J. D. Vigne. (1993). Space, time, and man as determinants of diversity of birds and mammals in the Mediterranean region. In R. E. Ricklefs, and D. Schluter (eds), *Species Diversity in Ecological Communities*, pp. 135–146. Chicago University Press, Chicago.

Blondel, J., and J. Aronson. (1995). Biodiversity and ecosystem function in the Mediterranean basin: human and non-human determinants. In G. W. Davis, and D. M. Richardson (eds), *Mediterranean-Type Ecosystems. The Function of Biodiversity*, pp. 43–119. Springer-Verlag, Berlin.

Blondel, J., and J. Aronson. (1999). *Biology and Wildlife in the Mediterranean Region*. Oxford University Press, Oxford.

Blondel, J., and M. Cheylan. (2008). Lost species and animal survivors. In C. Arnold (ed.), *Mediterranean Islands*, pp. 41–45. Survival Books, London.

Blondel, J., and F. Médail. (2009). Biodiversity and conservation. In J. C. Woodward (ed.), *The Physical Geography of the Mediterranean Basin*, pp. 604–638. Oxford University Press, Oxford.

Blondel, J., F. Vuilleumier, L. F. Marcus, and E. Terouanne. (1984). Is there ecomorphological convergence among Mediterranean bird communities of Chile, California and France? *Evolutionary Biology* **18**:141–213.

Blondel, J., D. Chessel, and B. Frochot. (1988). Bird species impoverishment, niche expansion and density inflation in Mediterranean island habitats. *Ecology* **69**:1899–1917.

Blondel, J., P. Dias, M. Maistre, and P. Perret. (1993). Habitat heterogeneity and life history variation of Mediterranean blue tits. *The Auk* **110**:511–520.

Blondel, J., F. Catzeflis, and P. Perret. (1996). Molecular phylogeny and the historical biogeography of the warblers of the genus *Sylvia* (Aves). *Journal of Evolutionay Biology* **9**:871–891.

Blondel, J., P. C. Dias, P. Perret, M. Maistre, and M. M. Lambrechts. (1999). Selection-based biodiversity at a small spatial scale in an insular bird. *Science* **285**: 1399–1402.

Blondel, J., D. W. Thomas, A. Charmantier, P. Perret, P. Bourgault, and M. M. Lambrechts. (2006). A thirty-year study of phenotypic and genetic variation of blue tits in Mediterranean habitat mosaics. *BioScience* **56**: 661–673.

Blue Plan. (1999). *Mediterranean Vision on Water, Population and the Environment for the XXIst Century*. By J. Margat and D. Vallée. Contribution to the World Water Vision of the World Water Council and the Global Water Partnership prepared by the Blue Plan in the Framework of the MEDTAC/GWP.

Blue Plan. (2004). *L'Eau des Méditerranéens: Situation et Perspectives*. By J. Margat, with S. Treyer. UNEP, MAP, Blue Plan.

Bocquet, G., B. Widler, and H. Kiefer. (1978). The Messinian model – a new outlook for the floristics and systematics of the Mediterranean area. *Candollea* **33**: 269–287.

Bodin, P. (1977). Les peuplements de copépodes Harpacticoïdes (Crustacea) des sédiments meubles de la zone intertidale des côtes Charentaises (Atlantique). *Mémoires du Muséum National d'Histoire Naturelle, Nouvelle Série, Série A, Zoologie* **104**:1–120.

Bodiou, J.-Y. (1975). Copépodes harpacticoïdes (Crustacea) des sables fins infralittoraux de Banyuls-sur-Mer. I – Description de la communauté. *Vie et Milieu* **25**: 313–330.

Bodiou, J.-Y., and G. Boucher. (2006). Biologie de l'extrême en milieu marin (la méiofaune benthique). In R. Peduzzi, M. Tonolla, and R. Boucher-Rodoni (eds), *Milieux Extremes: Conditions de Vie en Milieu Marin et Milieu Alpin*, pp. 71–82. Edizioni Centro Biologia Alpina, Piora.

Boeuf, G. (2000). Present status of the French aquaculture. Japan Aquaculture Society, "The new paradigm III for aquaculture". *Suisanzoshoku* **48**:243–248.

Boeuf, G. (2002). Acclimatization of aquatic organisms in culture. In *Knowledge for Sustainable Development – An insight into the Encyclopedia of Life Support Systems*, vol. 3, UNESCO Publishing-EOLSS Publishers, Paris.

Boeuf, G. (2003). L'aquaculture dans le monde: quel avenir? *Perspectives, Conférences et Débats de l'Université de Perpignan* **2**:91–104.

Boeuf, G., and P. Y. Le Bail. (1999). Does light have an influence on fish growth? *Aquaculture* **177**:129–152.

Boeuf, G., and P. Payan. (2001). How should salinity influence fish growth? *Comparative Biochemistry and Physiology, Toxicology & Pharmacology* **130C**:411–423.

Boeuf, G, and J.-Y. Bodiou. (2008). Quel futur pour la biodiversité en Méditerranée? In *Le Patrimoine Méditerranéen*, pp. 181–202. Actes, Association Monégasque pour la Connaissance des Arts, Monaco.

Böhme, W., and H. Wiedl. (1994). Status and zoogeography of the herpetofauna of Cyprus, with taxonomic and natural history notes on selected species (genera *Rana*, *Coluber*, *Natrix*, *Vipera*). *Amphibia and Reptilia* **10**: 31–52.

Boix, D., J. Sala, X. D. Quintana, and R. Moreno-Amich. (2004). Succession of the animal community in a Mediterranean temporary pond. *Journal of the North American Benthological Society* **23**: 29–49.

Bond, W. J., and J. E. Keeley. (2005). Fire as a global 'herbivore': the ecology and evolution of flammable ecosystems. *Trends in Ecology & Evolution* **20**:387–394.

Bond, W. J., G. F. Midgley, and F. I. Woodward. (2003). The importance of low atmospheric CO^2 and fire in promoting the spread of grasslands and savannas. *Global Change Biology* **9**:973–982.

Bonet, A., and J. G. Pausas. (2007). Old field dynamics on the dry side of the Mediterranean Basin: patterns and processes in semiarid southeast Spain. In V. A. Cramer, and R. J. Hobbs (eds), *Old Fields. Dynamics and Restoration of Abandoned Farmland*, pp. 247–264. Island Press, Washington DC.

Bonhomme, F., A. Orth, T. Cucchi, E. Hadjisterkotis, J.-D. Vigne, and J.-C. Auffray. (2004). Découverte d'une nouvelle espèce de souris sur l'île de Chypre. *Comptes Rendus Biologies* **327**:501–507.

Bonhôte, J., and J.-L. Vernet. (1988). La mémoire des charbonnières. Essai de reconsitution des milieux forestiers dans une vallée marquée par la métallurgie (Aston, Haute-Ariège). *Revue Forestière Française* **40**:197–212.

Bonis, A., J. Lepart, and P. Grillas. (1995). Seed bank dynamics and coexistance of annuyal macrophytes in a temporary and variable habitat. *Oikos* **74**:81–92.

Borg-Karlson, A.-K. (1990). Chemical and ethological studies of pollination in the genus *Ophrys* (Orchidaceae). *Phytochemistry* **29**:1359–1387.

Borg-Karlson, A.-K., and J. Tengoe. (1986). Odour mimetism? Key substances in the *Oprhys lutea-Andrena* pollination relationship (Orchidaceae-Andrenidae). *Journal of Chemical Ecology* **12**:1927–1941.

Borras, A., J. Cabrera, and J. C. Senar. (2008) Local divergence between Mediterranean crossbills occurring in two different species of pine. *Ardeola* **55**:169–177.

Bosch, J., I. Martinez-Solano, and M. Garcia-Paris. (2001). Evidence of a chytrid fungus infection in the decline of the midwife toad in protected areas of Central Spain (Anura: Discoglossidae). *Biological Conservation* **97**: 331–337.

Bouby, L., and V. Léa. (2006). Exploitation de la vesce commune (*Vicia sativa* L.) au Néolithique moyen dans le Sud de la France. Données carpologiques du site de Claparouse (Lagnes, Vaucluse). *Comptes Rendus Palevol* **5**:973–980.

Bouché, M. (1972). Lombriciens de France, écologie et systématique. *Annales de Zoologie et d'Ecologie Animale*, **72**:suppl. 2. INRA, Paris.

Boudouresque, C. F. (1969).Etude qualitative et quantitative d'un peuplement algal à *Cystoseira mediterranea* dans la région de Banyuls-sur-mer (P. O.). *Vie et Milieu* **20**:437–452.

Boudouresque, C. F. (2004). The erosion of Mediterranean biodiversity. In C. Rodríguez-Prieto, and G. Pardini (eds), *The Mediterranean Sea. An Overview of its Present State and Plans for Future Protection*, pp. 53–112. Universitat de Girona Publication.

Boudouresque, C. F., and M. Perret-Boudouresque. (1979). Dénombrement des algues benthiques et rapport R/P le long des côtes françaises de la Méditerranée. *Rapport de la Commission Internationale sur la Mer Méditerranée* **25/26**:149–152.

Boudouresque, C. F., and A. Meinesz. (1982). *Découverte de l'Herbier de Posidonie*. Parc National de Port-Cros and Parc Naturel Régional de la Corse.

Boudouresque, C. F. and M. Perret-Boudouresque. (1987). *A Checklist of the Benthic Marine Algae of Corsica*. GIS posidonie, Marseille.

Boudouresque, C. F., M. Perret-Boudouresque, and M. Knoeppfler-Péguy. (1984). Inventaire des algues marines benthiques dans les Pyrénées Orientales (Méditerranée, France). *Vie et Milieu* **34**:41–59.

Boudouresque, C. F., A. Meinesz, M. A. Ribera, and E. Ballesterose. (1995). Spread of the green alga *Caulerpa taxifolia* (Caulerpales, Chlorophyta) in the Mediterranean: possible consequences of a major ecological event. *Scientia marina* **59**:21–29.

Bourgeois, K., and E. Vidal. (2008). The endemic Mediterranean yelkouan shearwater *Puffinus yelkouan*: distribution, threats and a plea for more data. *Oryx* **42**: 187–194.

Boury-Esnault, N., D. Aurelle, N. Bensoussan, P. Chevaldonné, J. Garrabou, J. G. Harmelin, L. Laubier, J. B. Ledoux, C. Lejeusne, C. Marschal et al. (2006). *Evaluation des Modifications de la Biodiversité Marine sous l'Influence du Changement Global en Méditerranée Nord Occidentale*. Rapport final. Institut Français de la Biodiversité, Paris.

Bowen, M. E., C. A. McAlpine, A. P. N. House, and G. C. Smith. (2007). Regrowth forests on abandoned agricultural land: a review of their habitat values for recovering forest fauna. *Biological Conservation* **140**:273–296.

Boydak, M. (1987). A new occurrence of *Phoenix theophrasti* in Kumluca - Karaöz. *Principes* **31**:89–95.

Boydak, M. (2003). Regeneration of Lebanon cedar (*Cedrus libani* A. Rich.) on karstic lands in Turkey. *Forest Ecology and Management* **178**:231–243.

Brandt, C. J., and J. B. Thornes (eds). (1996). *Mediterranean Desertification and Land Use*. Wiley and Sons, Chichester.

Braudel, F. (1949). *La Méditerranée et le Monde Méditerranéen à l'Epoque de Philippe II*. 2 vols. Armand Colin, Paris.

Braudel, F. (1985). *La Méditerranée. L'Espace et l'Histoire*. Flammarion, Paris.

Bretagnolle, F. (1993). *Etude de Quelques Aspects des Mécanismes de la Polyploidisation et de ses Conséquences Evolutives dans le Complexe Polyploide du Dactyle (Dactylis glomerata L.)*. Thesis. University of Paris XI, Orsay.

Breton, C., G. Besnard, and A. Bervillé. (2003). Using multiple types of molecular markers to understand olive phylogeography. In M. A. Zeder, D. Decker-Walters, D. Bradley, and B. Smith (eds), *Documenting Domestication: New Genetic and Archaeological Paradigms*, pp. 143–152. Smithsonian Press, Washington DC.

Breton, C., B. Guillaume, and A. Bervillé. (2006). Olive domestication: molecular evidence for different origins of cultivars. In M. A. Zeder, D. Decker-Walters, D. Bradley, and B. Smith (eds), *Documenting Domestication: New Genetic and Archaeological Paradigms*, pp. 143–152. Smithsonian Press, Washington DC.

Briand, F. (2002). Foreword to the collection. In F. Briand (ed.), *CIESM Atlas of Exotic Species in the Mediterranean, vol. 1. Fishes*, pp 7–9. CIESM Publishers, Monaco.

Briggs, J. C. (1974). *Marine Zoogeography*. McGraw-Hill, New York.

Brisebarre, A. M. (1978). *Bergers des Cévennes*. Berger-Levrault, Paris.

Britton, R. H., and A. R. Johnson. (1987). An ecological account of a Mediterranean Salina: the Salin de Giraud, Camargue (France). *Biological Conservation* **42**: 185–230.

Britton, R. H., and A. J. Crivelli. (1993). Wetlands of southern Europe and North Africa: Mediterranean wetlands. In D. F. Whigam (ed.), *Wetlands of the World, I*, pp. 129–194. Kluwer Academic Publishers, Dordrecht.

Brook, B. W., and D. M. J. S. Bowman. (2004). The uncertain blitzkrieg of Pleistocene megafauna. *Journal of Biogeography* **31**:517–523.

Brosset, A. (1990). L'évolution récente de l'avifaune du nord-est Marocain: pertes et gains depuis 35 ans. *Revue d'Ecologie (Terre et Vie)* **45**:237–244.

Brown, J. H., and A. C. Gibson. (1983). *Biogeography*. C.V. Mosby Company, Saint Louis, MO.

Brown, T. A., M. K. Jones, W. Powell, and R. G. Allaby. (2008). The complex origins of domesticated crops in the Fertile Crescent. *Trends in Ecology & Evolution* **24**: 103–109.

Brucet, S., D. Boix, R. López-Flores, A. Badosa, and X. D. Quintana. (2006). Size and species diversity of zooplankton communities in fluctuating Mediterranean salt marshes. *Estuarine, Coastal and Shelf Science* **67**: 424–432.

Brusca, R. C., and G. J. Brusca. (2002). *Invertebrates*, 2nd edn. Sinauer Associates, Sunderland, MA.

Bugalho, M., T. Plieninger, J. Aronson, M. Ellatifi, and D. G. Crespo. (2009). A diversity of uses (and overuses). In J. Aronson, J. S. Pereira, and J. G. Pausas (eds), *Cork Oak Woodlands on the Edge: Ecology, Adaptive Management, and Restoration*, pp. 33–45. Island Press, Washington DC.

Burke, J. M., S. J.Knapp, and L. H. Rieseberg. (2005). Genetic consequences of selection during the evolution of cultivated sunflower. *Genetics* **171**:1933–1940.

Burney, D. A., and T. F. Flannery. (2005). Fifty millennia of catastrophic extinctions after human contact. *Trends in Ecology & Evolution* **20**:395–401.

Bush, M. B., M. R. Silman, and D. H. Urrego. (2004). 48,000 years of climate and forest change from a biodiversity hotspot. *Science* **303**:827–829.

Butzer, K. W. (1997). Sociopolitical discontinuity in the Near East 2200 B.C.E.: scenarios for Palestine and Egypt. In H. N. Dalfes, G. Kukla, and H. Weiss (eds), *Third Millennium BC, Climate change and Old World Collapse*, pp. 245–296. Springer, Berlin.

Butzer, K. W. (2005). Environmental history in the Mediterranean world: cross-disciplinary investigation of cause-and-effect for degradation and soil erosion. *Journal of Archaeological Science* **32**:1773–1800.

Cabiddu, S., M. C. Follesa, A. Gastoni, C. Porcu, and A. Cau. (2008). Gonad development of the deep-sea lobster *Polycheles typhlops* (Decapoda: Polichelidae) from the central western Mediterranean. *Journal of Crustacean Biology* **28**:494–501.

Cabioc'h, J., J. Y.Floc'h, A. Le Toquin, C. F. Boudouresque, A. Meinesz, and M. Verlaque. (2006). *Guide des Algues des Mers d'Europe*. Delachaux et Niestlé, Paris.

Cabrero-Sañudo, F. J., and J. M. Lobo. (2006). Determinant variables of Iberian Peninsula Aphodiinae diversity (Coleoptera, Scarabaeoidea, Aphodiidae). *Journal of Biogeography* **33**:1021–1043.

Caccone, A., M. C. Milinkpvitch, V. Sbordoni, and J. R. Powell. (1994). Molecular biogeography: using the Corsica-Sardinia microplate disjunction to calibrate mitochondrial rDNA evolutionary rates in mountain newts (*Euproctus*). *Journal of Evolutionary Biology* **7**: 227–245.

Cadi, A., and P. Joly. (2004). Impact of the introduction of the red-eared slider (*Trachemys scripta elegans*) on survival of the European pond turtle (*Emys orbicularis*). *Biodiversity and Conservation* **13**:2511–2518.

Campos, P., and A. Caparrós. (2006). Social and private total Hicksian incomes of multiple use forests in Spain. *Ecological Economics* **57**:545–557.

Campos, P., P. Ovando, A. Chebil, and H. Daly-Hassen. (2009). Cork oak woodland conservation and household subsistence economy challenges in northern Tunisia. In Aronson, J., J. S. Pereira, and J. G. Pausas (eds), *Cork Oak Woodlands on the Edge: Ecology, Adaptive Management, and Restoration*, pp. 177–188. Island Press, Washington DC.

Cañadas, A., R. Sagarminaga, P. Marcos, and E. Urquiola. (2004). Sector Sur (Andalucia y Ceuta). In J. A. Raga, and J. Pantoja (eds), *Proyecto Mediterraneo. Zonas de Especial Interés para la Conservacion de los Cetaceos en el Mediterraneo Espanol*, pp. 133–190. Ministerio de Medio Ambiente, Naturaleza y parques nacionales.

Candela, L., K. Wallis, and R. Mateos. (2008). Non-point pollution of groundwater from agricultural activities in Mediterranean Spain: the Balearic Islands case study. *Environmental Geology* **54**:587–595.

Canestrelli, D., and G. Nascetti. (2008). Phylogeography of the pool frog *Rana* (*Pelophylax*) *lessonae* in the Itallian peninsula and Sicily: multiple refugia, glacial expansions and nuclear-mitochondrial discordance. *Journal of Biogeography* **35**:1923–1936.

Carcaillet, C., Y. Bergeron, P. J. H. Richard, B. Frechette, S. Gauthier, and I. T. Prairie. (2001). Change of fire frequency in the eastern Canadian boreal forests during the Holocene: does vegetation composition or climate trigger the fire regime? *Journal of Ecology* **89**:930–946.

Carcaillet, C., H. Almquist, H. Asnong, R. H. W. Bradshaw, J. S. Carrión, M.-J. Gaillard, K. Gajewski, J. N. Haas, S. G. Haberle, P. Hadorn *et al*. (2002). Holocene biomass burning and global dynamics of the carbon cycle. *Chemosphere* **49**:845–863.

Carine, M. A., S. J. Russell, A. Santos-Guerra, and J. Francisco-Ortega. (2004). Relationships of the Macaronesian and Mediterranean floras: molecular evidence for multiple colonizations into Macaronesia and back-colonization of the continent in Convolvulus (Convolvulaceae). *American Journal of Botany* **91**:1070–1085.

Carlquist, S. (1974). *Island Biology*. Columbia University Press, New York.

Carlton, J. T., J. B. Geller, M. I. Reaka Kudla, and E. A. Norse. (1999). Historical extinctions in the sea. *Annual Review of Ecology and Systematics* **30**:525–538.

Carrer, S., and R. Leardi. (2006). Characterizing the pollution produced by an industrial area: Chemometric methods applied to the Lagoon of Venice. *Science of the Total Environment* **370**:99–116.

Carriglio, D., R. Sandulli, S. Deastis, M. Gallo d'Addabbo, and S. Grimaldi de Zio. (2003). *Caulerpa racemosa* spread effects on the meiofauna of the Gulf of Taranto. *Biologia Marina Mediterranea* **10**:509–511.

Carrión, J. S. (2002). Patterns and processes of Late Quaternary environmental change in a montane region of southwestern Europe. *Quaternary Science Reviews* **21**:2047–2066.

Carrión, J. S., I. Para, C. Navarro, and M. Munueras. (2000). Past distribution and ecology of the cork oak (*Quercus suber*) in the Iberian Peninsula: a pollen-analytical approach. *Diversity and Distributions* **6**:29–44.

Casazza, G., E. Zappa, M. Mariotti, F. Médail, and L. Minuto. (2008). Ecological and historical factors affecting distribution patterns and richness of endemic plant specoes: the case of the Maritime and Ligurian Alps hotspot. *Diversity and Distributions* **14**:47–58.

Casevitz-Weulersse, J. (1992). Analyse biogéographique de la myrmécofaune Corse et comparaison avec celle des régions voisines. *Compte Rendus des Séances de la Société de Biogéographie* **68**:105–129.

Catsadorakis, G. (1997). The importance of Prespa National Park for breeding and wintering birds. *Hydrobiologia* **351**:157–174.

Catsadorakis, G. (2003) *Greece's Heritage from Nature*. WWF, Athens.

Catsadorakis, G., and A. J. Crivelli. (2001). Nesting habitat characteristics and breeding performance of Dalmatian Pelicans in Lake Mikri Prespa, NW Greece. *Waterbirds* **24**:386–393.

Celona, A. (2000). First record for a tiger shark *Galeocerdo cuvieri* in the Italian waters. *Annals for Istrian and Mediterranean Studies, series Historia Naturalis* **10**:207–210.

Chaline, J. (1974). Palingenèse et phylogenèse chez les campagnols (Arvicolidae, Rodentia). *Comptes-rendus de l'Académie des Sciences de Paris* **D278**:437–440.

Chapman, M. A., and J. M. Burke. (2007). DNA sequence diversity and the origin of cultivated safflower (*Carthamus tinctorius* L.; Asteraceae). *BMC Plant Biology* **7**:60.

Charpentier, B. (1998). *A Blue Plan for the Mediterranean Peoples*. UNEP, Nice.

Charrier, A. (1995). France maintains strong tradition of support for biodiversity activities worldwide. *Diversity* **11**:89–90.

Chauvet, J. M., E. Brunel Deschamps, and C. Hillaire. (1995). *La Grotte Chauvet*. Le Seuil, Paris.

Chevaldonné, P., and C. Lejeusne. (2003). Regional warming-induced species shift in north-west Mediterranean marine caves. *Ecology Letters* **6**:371–379.

Cheylan, G. (1984). Les mammifères des îles de Provence et de Méditerranéenne occidentale: un exemple de peuplement insulaire déséquilibré? *Revue d'Ecologie (Terre et Vie)* **39**:37–54.

Cheylan, G. (1991). Patterns of Pleistocene turnover, current distribution and speciation among Mediterranean mammals. In R. H. Groves, and F. di Castri (eds), *Biogeography of Mediterranean Invasions*, pp. 227–262. Cambridge University Press, Cambridge.

Cheylan, M., and F. Poitevin. (1994). Conservazione di rettili e anfibi. In X. Monbailliu, and A. Torre (eds), *La Gestione degli Ambienti Costieri Einsulari del Mediterraneo*, pp. 275–336. Edizione del Sole, Alghero.

Chouquer, G., M. Clavel-Lévêque, and F. Favory. (1987). Le paysage révélé: l'empreinte du passé dans les paysages contemporains. *Mappemonde* **4**:16–21.

Chrétiennot-Dinet, M. J., C. Courties, A. Vaquer, J. Neveux, H. Claustre, J. Lautier, and M. C. Machado. (1995). A new marine picoeucaryote: *Ostreococcus tauri* gen. et sp. nov. (Chlorophyta, Prasinophyceae). *Phycologia* **34**:285–292.

Christodoulakis, N. S., H. Tsimbani, and C. Fasseas. (1990). Leaf structural peculiarities in *Sarcopoterium spinosum*, a seasonally dimorphic subshrub. *Annals of Botany* **65**:291–296.

Chuine, I., and W. Thuiller. (2005). Impact du changement climatique sur la biodiversité. *Le Courrier de la Nature* **223**:20–26.

Chytrý, M., P. Pysek, J. Wild, J. Pino, L. S. Maskell, and M. Vilà. (2009). European map of alien plant invasions based on the quantitative assessment across habitats. *Diversity and Distributions* **15**:98–107.

CIESM. (2008). *Climate Warming and Related Changes in Mediterranean Marine Biota*. No. 35 in CIESM Workshop Monographs (F. Briand, ed.). CIESM Publishers, Monaco.

Cinelli, F. (1985). On the biogeography of the benthic algae of the Mediterranean. In M. Moraitou-Apostolopoulou, and V. Kiortsis (eds), *Mediterranean Marine Ecosystems*, pp. 49–56. Plenum Press, New York.

Clavero, M., and E. García-Berthou. (2006). Homogenization dynamics and introduction routes of invasive freshwater fish in the Iberian Peninsula. *Ecological Applications* **16**:2313–2324.

Clement, V. (2008). Les feux de forêt en Méditerranée: un faux procès contre Nature. *Forêt méditerranéenne* **29**:267–280.

Clewell, A., and J. Aronson. (2007). *Ecological Restoration: Principles, Values, and Structure of an Emerging Profession*. Island Press, Washington DC.

Cody, M. L. (1975). Towards a theory of continental species diversity: bird distribution over Mediterranean habitat gradients. In M. L. Cody, and J. M. Diamond (eds), *Ecology and Evolution of Communities*, pp. 214–257. Harvard University Press, Cambridge, MA.

Cody, M. L., and H. A. Mooney. (1978). Convergence versus nonconvergence in Mediterranean-climate ecosystems. *Annual Review of Ecology and Systematics* **9**: 265–321.

Coelho, I. S., and P. Campos. (2009). Mixed cork oak-stone pine woodlands in the Alentejo region of Portugal. In Aronson, J., J. S. Pereira, and J. G. Pausas (eds), *Cork Oak Woodlands on the Edge: Ecology, Adaptive Management, and Restoration*, pp. 153–161. Island Press, Washington DC.

Cohen, D., and A. Shmida. (1993). The evolution of flower display and reward. *Evolutionary Biology* **68**:81–120.

Cohen, D. M., T. Inada, T. Iwamoto, and N. Scialabba. (1990). *Gadiform Fishes of the World (Order Gadiformes). An Annotated and Illustrated Catalogue of Cods, Hakes, Grenadiers and other Gadiform Fishes Known to Date*. FAO Species Catalogue, vol. 10. FAO Fish. Synop.

Colas, B., I. Olivieri, and M. Riba. (2001). Spatio-temporal variation of reproductive success and conservation of the narrow-endemic *Centaurea corymbosa* (Asteraceae). *Biological Conservation* **99**:375–386.

Collignon, J. (1991). *Ecologie et Biologie Marines. Introduction à l'Halieutique*. Ed. Masson, Paris.

Collins, N. M., and J. A. Thomas (eds). (1991). *The Conservation of Insects and their Habitats*. Academic Press, London.

Comes, H. P., and R. J. Abbott. (1998). The relative importance of historial events and gene flow on the population structure of a Mediterranean ragwort, *Senecio gallicus*. *Evolution* **52**:355–367.

Comstock, J. P., and B. E. Mahall. (1985). Drought and changes in leaf orientation for two California chaparral shrubs: *Ceanothus megacarpus* and *Ceanothus crassifolius*. *Oecologia* **65**:531–535.

Cooper, J. H. (2000). First fossil record of Azure-winged Magpie *Cyanopica cyanus* in Europe. *Ibis* **142**:150–151.

Coppack, T., and C. Both. (2002). Predicting life-cycle adaptation of migratory birds to global climate change. *Ardea* **90**:369–378.

Corbacho, C., and J. M. Sánchez. (2001). Patterns of species richness and introduced species in native freshwater fish faunas of a Mediterranean-type basin: the Guadiana River (southwest Iberian Peninsula). *Regulated Rivers: Research & Management* **17**:699–707.

Cortez, J., and M. Bouché. (2001). Decomposition of mediterranean leaf litters by *Nicodrilus meridionalis* (Lumbricidae) in laboratory and field experiments. *Soil Biology and Biochemistry* **33**:2023–2035.

Courties, C., A. Vaquer, M. Trousselier, J. Lautier, M. J. Chrétiennot-Dinet, J. Neveux, C. Machado, and H. Claustre. (1994). Smallest eukaryotic organism. *Nature* **370**:255.

Couvet, D., A. Atlan, E. Belhassen, C. Gliddon, P. H. Gouyon, and F. Kjellberg. (1990). Co-evolution between two symbionts: the case of cytoplasmic male-sterility in higher plants. *Oxford Surveys in Evolutionary Biology* **7**:225–248.

Covas, R., and J. Blondel. (1998). Biogeography and history of the Mediterranean bird fauna. *Ibis* **140**: 395–407.

Cowling, R. M., P. W. Rundel, B. B. Lamont, M. K. Arroyo, and M. Arianoutsou. (1996). Plant diversity in mediterranean-climate regions. *Trends in Ecology & Evolution* **11**:362–366.

Cowling, R. M., F. Ojeda, B. B. Lamont, P. R. Rundel, and R. Lechmere-Oertel. (2005). Rainfall reliability, a neglected factor in explaining convergence and divergence of plant traits in fire-prone mediterranean-climate ecosystems. *Global Ecology and Biogeography* **14**: 509–519.

Cowx, I. G. (1997). Introduction of fish species into European waters: economic success or ecological disaster? *Bulletin Français de la Pêche et la Pisciculture* **344/345**: 57–77

Cox, N., J. Chanson, and S. Stuart. (2006). *Statut de Conservation et Répartition Géographique des Reptiles et Amphibiens du Bassin Méditerranéen*. IUCN.

Cozzolino, S., and A. Widmer. (2005). Orchid diversity: an evolutionary consequence of deception? *Trends in Ecology & Evolution* **20**:487–494.

Cozzolino, S., A. M. Nardella, S. Impagliazzo, A. Widmer, and C. Lexer. (2006). Hybridization and conservation of Mediterranean orchids: should we protect the orchid hybrids or the orchid hybrid zones? *Biological Conservation* **129**:14–23.

Crick, H. Q. P. (2004). The impact of climate change on birds. *Ibis* **146**:48–56.

Crick, H. Q. P., and P. J. Jones. (1992). The ecology and conservation of Palaearctic-African migrants. *Ibis* **134**: 1–132.

Crivelli, A. J. (1996). *The Freshwater Fish Endemic to the Northern Mediterranean Region*. Tour du Valat, Arles.

Crivelli, A. J., and P. S. Maitland. (1995). Endemic freshwater fishes of the northern Mediterranean region. *Biological Conservation* **72**:121–337.

Crocq, C. (1990). *Le Casse-Noix Moucheté*. Lechevallier and Chabaud, Paris.

Cronquist, A. (1981). *An Integrated System of Classification of Flowering Plants*. Columbia University Press, New York.

Cury, P., and S. Morand. (2004). Biodiversité marine et changements globaux: une dynamique d'interactions où l'humain est partie prenante. In *Biodiversité et Changements Globaux*, pp. 50–79. Adpfe, Ministère des Affaires Etrangères.

Cuzin, F. (1996). Répartition actuelle et statut des grands mammifères sauvages du Maroc (Primates, Carnivores, Artiodactyles). *Mammalia* **60**:101–124.

Dafni, A., and P. Bernhardt. (1990). Pollination of terrestrial orchids of Southern Australia and the Mediterranean region. Systematic, ecological and evolutionary implications. *Evolutionary Biology* **24**:193–252.

Dafni, A., and C. O'Toole. (1994). Pollination syndromes in the Mediterranean: generalisations and peculiarities. In M. Arianoutsou, and R. H. Groves (eds), *Plant-Animal Interactions in Mediterranean-Type Ecosystems*, pp. 125–135. Kluwer Academic Publishers, Dordrecht.

Dafni, A., and A. Shmida. (1996). The possible implications of the invasion of *Bombus terrestris* (L.) (Apidae) at Mt. Carmel, Israel. In International Bee Research Association (ed.), *The Conservation of Bees*, pp. 183–200. Linnean Society of London.

Dafni, A., P. Berhnardt, A. Shmida, Y. Ivri, S. Greenbaum, C. O'Toole, and L. Losito. (1990). Red bowl-shaped flowers: convergence for beetle pollination in the Mediterranean region. *Israel Journal of Botany* **39**:81–92.

Daget, P. (1977). Le bioclimat méditerranéen: caractères généraux, modes de caractérisation. *Vegetatio* **34**:1–20.

Dallman, P. R. (1998). *Plant Life in the World's Mediterranean Climates*. Oxford University Press, Oxford.

Damesin, C., S. Rambal, and R. Joffre. (1998). Co-occurrence of trees with different leaf habit: a functional approach on Mediterranean oaks. *Acta Oecologica* **19**:195–204.

Dapporto, L., and R. L. D. Dennis. (2008). Conservation biogeography of large Mediterranean islands. Butterfly impoverishment, conservation priorities and inferences of an ecological paradigm. *Ecography* **32**:169–179.

Darlington, P. J. (1957). *Zoogeography. The Geographical Distribution of Animals.* Wiley, New York.

Darnaud, J., M. Lecumberry, and R. Blanc. (1978a). Coléoptères Carabidae *Chrysocarabus solieri*. In *Iconographie Entomologique*, Planche 4.

Darnaud, J., M. Lecumberry, and R. Blanc. (1978b). Coléoptères Carabidae *Chrysocarabus rutilans*. In *Iconographie Entomologique*, Planche 2.

Darnaud, J., M. Lecumberry, and R. Blanc. (1981). Coléoptères Carabidae genre *Macrothorax*. In *Iconographie Entomologique*, Planche 13.

Darnaud, J., M. Lecumberry, and R. Blanc. (1984a). Coléoptères Carabidae genre *Procerus*. In *Iconographie Entomologique*, Planche 16.

Darnaud, J., M. Lecumberry, and R. Blanc. (1984b). Coléoptères Carabidae genre *Megodontus* Solier, *caelatus/croatricus*. In *Iconographie Entomologique*, Planche 17.

Darwin, C. (1862). *On the Various Contrivances by which British and Foreign Orchids are Fertilised by Insects, and on the Good Effects of Intercrossing*. John Murray, London (2nd edn 1877).

Darwin, C. (1881). *The Formation of Vegetable Mould, through the Action of Worms, with Observations on their Habits*. John Murray, London.

Davies, C. P., and P. L. Fall. (2001). Modern pollen precipitation from an elevational transect in central Jordan and its relationship to vegetation. *Journal of Biogeography* **28**:1195–1210.

Davis, M. A. (2003b). Biotic globalization: does competition from introduced species threaten biodiversity? *BioScience* **53**:481–489.

Davis, M. B. (1976). Pleistocene biogeography of temperate deciduous forests. *Geoscience Canada* **13**:13–26.

Davis, P. H. (1965). *Flora of Turkey and the East Aegean Islands, vol. 1*. Edinburgh University Press, Edinburgh.

Davis, S. J. M. (2003a). The zooarchaeology of Khirokitia (Neolithic, Cyprus) including a view from the mainland. In J. Guilaine, and A. Le Brun (eds), *Le Néolithique de Chypre*, pp. 253–268. Département des Antiquités de Chypre, Nicosie.

Dawson, T., and R. Fry. (1998). Agriculture in nature's image. *Trends in Ecology & Evolution* **13**:50–51.

Dayan, T. (1996). Weasels from the iron age of Israel: a biogeographic note. *Israel Journal of Zoology* **42**:295–298.

Daszak, P., A. A. Cunningham, and A. D. Hyatt. (2003). Infectious disease and amphibian population decline. *Diversity and Distributions* **9**:141–150.

De Beaulieu, J. L., Y. Miras, V. Andrieu-Ponel, and F. Guiter. (2005) Vegetation dynamics in north-western Mediterranean regions: instability of the Mediterranean bioclimate. *Plant Biosystems* **139**:114–126.

De Bonneval, L. (1990). *D'un Taillis à l'Autre. La Déshérance d'un Patrimoine Forestier Communal (Valliguières, Gard). 1820-1990.* INRA, Unité d'Ecodéveloppement, Montfavet.

Debussche, M., and P. Isenmann. (1992). A Mediterranean bird disperser assemblage: composition and phenology in relation to fruit availability. *Revue d'Ecologie (Terre et Vie)* **47**:411–432.

Debussche, M., and J. D. Thompson. (2003). Habitat differentiation between two closely related Mediterranean plant species, the endemic *Cyclamen balearicum* and the widespread *C. repandum*. *Acta Oecologica* **24**:35–45.

Debussche, M., G. Debussche, and L. Affre. (1995). La distribution fragmentée de *Cyclamen balearicum* Willk: analyse historique et conséquence des activités humaines. *Acta Botanica Gallica* **142**:439–450.

Debussche, M., M. Grandjanny, G. Debussche, and L. Affre. (1996). Ecologie d'une espèce endémique et rare à distribution fragmentée: *Cyclamen balearicum* Willk. en France. *Acta Botanica Gallica* **143**:65–84.

Debussche, M., E. Garnier, and J. D. Thompson. (2004). Exploring the causes of variation in phenology and morphology in Mediterranean geophytes: a genus-wide study of *Cyclamen*. *Botanical Journal of the Linnean Society* **145**:469–484.

Defleur, A., J. F. Bez, E. Crégut-Bonnoure, E. Desclaux, G. Onoratini, C. Radulescu, M. Thinon, and P. Vilette. (1994). Le niveau moustérien de la grotte de l'Adaouste (Jouques, Bouches-du-Rhône). Approche culturelle et paléoenvironnements. *Bulletin du Muséum Anthropologique et de Préhistoire de Monaco* **37**:29–35.

de Juana, E. (1997). Audouin's gull. In W. J. M. Hagemeijer, and M. J. Blair (eds), *The EBCC Atlas of European Breeding Birds. Their Distribution and Abundance*. T. & A.D. Poyser, London.

Delamare-Deboutteville, C. (1960). *Biologie des Eaux Souterraines Littorales et Continentales*. Hermann, Paris.

Delanoë, O., B. de Montmollin, and L. Olivier. (1996). *Conservation of Mediterranean Island Plants. 1. Strategy for Action*. IUCN, Gland.

De Lattin, G. (1967). *Grundriss der Zoogeographie*. F. Fischer, Stuttgart.

Delaugerre, M. (1988). Statut des Tortues marines de la Corse et de la Méditerranée. *Vie et Milieu* **37**: 243–264.

Delaugerre, M., and M. Cheylan. (1992). *Atlas de Répartition des Batraciens et Reptiles de Corse*. Parc Naturel Régional de la Corse, E.P.H.E., Ajaccio.

Del Barrio, J., M. Ortega, A. De La Cueva, and R. Elenarosselló. (2006). The influence of linear elements on plant species diversity of Mediterranean rural landscapes: assessment of different indices and statistical approaches. *Environmental Monitoring and Assessment* **119**:137–159.

De Lillis, M. (1991). An ecomorphological study of the evergreen leaf. *Braun-Blanquetia* **7**:1–127.

De Lope, F., J. Guerrero, and C. de la Cruz. (1984). Une nouvelle espèce à classer parmi les oiseaux de la Péninsule Ibérique: *Estrilda (Amandava) amandava* (Ploceidae, Passeriformes). *Alauda* **52**:312.

Dennis, R. L. H., T. G. Shreeve, and W. R. Williams. (1995). Taxonomic differentiation in species richness gradients among European butterflies (Papilionoidea, Hesperioidea): contribution of macroevolutionary dynamics. *Ecography* **18**:27–40.

De Stefano, D. (2004). *Freshwater and Tourism in the Mediterranean*. WWF Mediterranean Programme, Rome.

Deuve, T. (2004). Phylogénie et classification du genre *Carabus* Linné 1758. Le point des connaissances actuelles. *Bulletin de la Société Entomologique de France* **109**:5–39.

Diadema, K., F. Médail, and F. Bretagnolle. (2007). Fire as a control agent of demographic structure and plant performance of a rare Mediterranean endemic geophyte. *Comptes Rendus Biologies* **330**:691–700.

Diamond, J. (1992). Twilight of the pygmy hippos. *Nature* **359**:15.

Diamond, J. (2000). Blitzkrieg against moas. *Science* **287**:2170–2121.

Diamond, J. (2002). Evolution, consequences and future of plant and animal domestication. *Nature* **418**:700–707.

Diamond, J., and P. Bellwood. (2003). Farmers and their languages: the first expansions. *Science* **300**:597–603.

Dias, P., and J. Blondel. (1996). Local specialization and maladaptation in Mediterranean blue tits, *Parus caeruleus*. *Oecologia* **107**:79–86.

Dias, P. C., G. R. Verheyen, and M. Raymond. (1996). Source-sink populations in Mediterranean blue tits: evidence using single-locus minisatellite probes. *Journal of Evolutionary Biology* **9**:965–978.

Díaz, M., P. Campos, and F. J. Pulido. (1997). The Spanish dehesas: a diversity in land-use and wildlife. In D. J. Pain, and M. W. Pienkowski (eds), *Farming and Birds in Europe. The Common Agricultural Policy and its Implications for Bird Conservation*, pp. 178–209. Academic Press, London.

Diaz-Delgado, R., F. Lloret, X. Pons, and J. Terradas. (2002). Satellite evidence of decreasing resilience in Mediterranean plant communities after recurrent wildfires. *Ecology* **83**:2293–2303.

di Castri, F. (1981). Mediterranean-type shrublands of the world. In F. Di Castri, D. W. Goodall, and R. L. Specht (eds), *Mediterranean-Type Shrublands, Collection Ecosystems of the World*, vol. 11, pp. 1–52. Elsevier, Amsterdam.

di Castri, F. (1990). On invading species and invaded ecosystems: the interplay of historical chance and biological necessity. In F. Di Castri, A. J. Hansen, and M. Debussche (eds), *Biological Invasions in Europe and the Mediterranean Basin*, pp. 3–16. Kluwer Academic, Dordrecht.

di Castri, F. (1998). Politics and environment in mediterranean-climate regions. In P. W. Rundel, G. Montenegro, and F. Jaksic (eds), *Landscape Degradation and Biodiversity in Mediterranean-Type Ecosystems*, pp. 407–432. Ecological Series 136, Springer-Verlag, Berlin.

di Castri, F., and H. A. Mooney. (1973). *Mediterranean-Type Ecosystems. Origin and Structure*. Springer-Verlag, Heidelberg.

di Castri, F., and V. di Castri. (1981). Soil fauna of mediterranean-climate regions. In F. Di Castri, D. W. Goodall, and R. L. Specht (eds), *Mediterranean-Type Shrublands, Collection Ecosystems of the World*, vol. 11, pp. 445–478. Elsevier, Amsterdam.

di Castri, F., W. Goodall, and R. L. Specht (eds). (1981). *Mediterranean-Type Shrublands. Collection Ecosystems of the World*, vol. 11. Elsevier, Amsterdam.

di Castri, F., A. J. Hansen, and M. Debussche (eds). (1990). *Biological Invasions in Europe and the Mediterranean Basin*. Kluwer Academic Publishers, Dordrecht.

Djamali, M., J.-L.de Beaulieu, M. Shah-Hosseini, V. Andrieu-Ponel, P. Ponel, A. Amini, H. Akhani, A. S. Leroy, L. Stevens, H. Alizadeh, and S. Brewer. (2008a). A late Pleistocene long pollen record from Lake Urmia, NW Iran. *Quaternary Research* **62**:413–420.

Djamali, M., H. Kürschner, H. Akhani, J.-L. de Beaulieu, A. Amini, V. Andrieu-Ponel, P. Ponel, and L. Stevens. (2008b). Palaeoecological significance of the spores of the liverwort *Riella* (Riellaceae) in a late Pleistocene long pollen record from the hypersaline Lake Urmia, NW Iran. *Review of Palaeobotany and Palynology* **152**:66–73.

Drake, J. A., H. A. Mooney, F. di Castri, R. H. Groves, F. J. Kruger, M. Rejmanek, and M. Williamson. (1989). *Biological Invasions. A Global Perspective*. John Wiley and Sons, Chichester.

Driscoll, C. A., M. Menotti-Raymond, A. L. Roca, K. Hupe, W. E. E. Johnson, E. Geffen, E. H. Harley, M. Delibes, D. Pontier, A. C. Kitchener *et al.* (2007). The Near Eastern origin of cat domestication. *Science* **317**:519–523.

Dubar, M., J. P. Ivaldi, and M. Thinon. (1995). Feux de forêt méditerranéens: une histoire de pins. *La Recherche* **273**:188–189.

Dufaure, J.-J. (1984). La mobilité des paysages méditerranéens. *Revue Géographique des Pyrénées et du Sud-Ouest* **Suppl. 1**:1–387.

Duggen, S., K. Hoernie, P. van den Bogard, L. Rüpke, and J. P. Morgan. (2003). Deep roots of the Messinian salinity crisis. *Nature* **422**:602–606.

Duhem, C., P. Roche, E. Vidal, and T. Tatoni. (2008). Effects of anthropogenic food resources on yellow-legged gull colony size on Mediterranean islands. *Population Ecology* **50**:91–100.

Dulvy, N. K., Y. Sadovy, and J. D. Reynolds. (2003). Extinction vulnerability in marine populations. *Fish and Fisheries* **4**:25–64.

du Merle, P. (1978). Le massif du Ventoux, Vaucluse. Eléments d'une synthèse écologique. *La Terre et la Vie* **Suppl. 1**:1–314.

du Merle, P., P. Jourdheuil, J. P. Marro, and R. Mazet. (1978). Evolution saisonnière de la myrmécofaune et de son activité prédatrice dans un milieu forestier: les interactions clairière-lisière-forêt. *Annales de la Société Entomologique de France* **14**:141–157.

Dupouey, J. L., E. Dambrine, J. D. Laffite, and C. Moares. (2002). Irreversible impact of past land use on forest soils and biodiversity. *Ecology* **83**:2978–2984.

Du Puy, B., and P. W. Jackson. (1995). Botanic gardens offer key component to biodiversity conservation in the Mediterranean. *Diversity* **11**:47–50.

Eckert, R., D. Randall, W. Burggren, and K. French. (1999). Ionic and osmotic equilibrium. In D. Randall, W. Burggren, and K. French (eds), *Animal Physiology, Mechanisms and Adaptation*, 4th edn, pp. 571–626. DeBoeck University, Bruxelles.

Economidis, P. S. (1991). *Check List of Freshwater Fishes of Greece*. Hellenic Society for the Protection of Nature, Athens.

Edwards, C. J., R. Bollongino, A. Scheu, A. Chamberlain, A. Tresset, J.-D. Vigne, J. F. Baird, G. Larson, S. Y. W. Ho, T. H. Heupink *et al.* (2007). Mitochondrial DNA analysis shows a Near Eastern Neolithic origin for domestic cattle and no indication of domestication of European aurochs. *Proceedings of the Royal Society of London Series B Biological Sciences* **274**:1377–1385.

EEA (European Environment Agency). (2003). *Europe's Water: An Indicator-Based Assessment*. Report, EEA.

EEA (European Environment Agency). (2007). Marine and coastal environment. In *The Fourth Assessment*, pp. 208–250. Report of the EEA. www.eea.europa.eu/publications/state_of_environment_report_2007_1/chapter5.pdf.

Ehlers, B., and J. D. Thompson. (2004). Do co-occurring plant species adapt to one another? The respoins of *Bromus erectus* to the presence of different *Thymus vulgaris* chemotypes. *Oecologia* **141**:511–518.

Eisikowitch, D., Z. Gat, O. Karni, F. Chechik, and D. Raz. (1992). Almond blooming under adverse conditions. A compromise between various forces. In C. A. Thanos (ed.), *Plant-Animal Interactions in Mediterranean-Type Ecosystems*, pp. 234–240. University of Athens.

Ekman, S. (1957). *Zoogeography of the Sea*. Sidgwick & Jackson, London.

Elvira, B. (1995). Native and exotic freshwater fishes in Spanish river basins. *Freshwater Biology* **33**:103–108.

Emberger, L. (1930a). La végétation de la région Méditerranéenne. Essai d'une classification des groupements végétaux. *Revue Générale de Botanique* **42**:641–662.

Emberger, L. (1930b). Sur une formule climatique applicable en géographie botanique. *Comptes Rendus de l'Académie des Sciences, Paris* **191**:389–390.

Emig, C. C., and P. Geitsdoerfer. (2004). *The Mediterranean Deep-Sea Fauna: Historical, Evolution, Bathymetric Variations and Geographical Changes*. Notebooks in Geology.

Entwistle, A., S. Atay, A. Byfield, and S. Oldfield. (2002). Alternatives for the bulb trade from Turkey: a case study of indigenous bulb propagation. *Oryx* **36**:333–341.

Escarré, J., C. Houssard, M. Debussche, and J. Lepart. (1983). Evolution de la végétation et du sol après abandon cultural en région méditerranéenne: étude de la succession dans les garrigues du Montpéllierais (France). *Acta Oecologica - Ecologia Plantarum* **4**:221–239.

Espadaler, X., and L. Lopez-Soria. (1991). Rareness of certain Mediterranean ant species: fact or artifact? *Insectes Sociaux* **38**:365–377.

European Communities. (2008). *The Economics of Ecosystems and Biodiversity: An Interim Report*. European Communities, Brussels.

Evenari, M., L. Shanan, and N. Tadmor. (1982). *The Negev. The Challenge of a Desert*. Harvard University Press, Cambridge, MA.

Fabbio, G., M. Merlo, and V. Tosi. (2003). Silvicultural management in maintaining biodiversity and resistance of forests in Europe–the Mediterranean region. *Journal of Environmental Management* **67**:67–76.

FAO. (2003). *Forestry Outlook Study for Africa, Regional Report–Opportunities and Challenges towards 2020*. FAO Forestry Paper No. 141. FAO, Rome.

FAO. (2007). Capture production 2005. *FAO Yearbook of Fishery Statistics*, vol. 100/1. FAO, Rome.

FAO. (2008). *Fishery and Aquaculture Statistics. 2006*. FAO Yearbook. FAO, Rome.

Fernández, H., S. Hughes, J.-D. Vigne, D. Helmer, G. Hodgins, C. Miquel, C. Hänni, G. Luikart, and P. Taberlet. (2006). Divergent mtDNA lineages of goats in an Early Neolithic site, far from the initial domestication areas. *Proceedings of the National Academy of Sciences USA* **103**:15375–15379.

Fernandez Torquemada, Y., and J. L. Sánchez Lizaso. (2005). Effects of salinity on leaf growth and survival of the Mediterranean seagrass *Posidonia oceanica*. *Journal of Experimental Marine Biology and Ecology* **320**:57–63.

Fineschi, S., S. Cozzolino, M. Migliaccio, and G. G. Vendramin. (2004). Genetic variation of relic tree species: the case of Mediterranean *Zelkova abelicea* (Lam.) Boisser and *Z. sicula* Di Pasquale, Garfi and Quézel (Ulmaceae). *Forest Ecology and Management* **197**:273–278.

Finlayson, C., and J. S. Carrión. (2007). Rapid ecological turnover and its impact on Neanderthals and other human populations. *Trends in Ecology & Evolution* **22**:213–222.

Finlayson, C., F. G. Pacheco, J. Rodriguez-Vidal, D. A. Fa, J. M. Gutierrez Lopez, A. S. Perez, G. Finlayon, E. Allue, J. B. Preysler, I. Caceres et al. (2006). Late survival of Neanderthals at the sourthernmost extreme of Europe. *Nature* **443**:850–853.

Flahaut, C. (1937). *La Distribution Géographique des Végétaux dans la Région Méditerranéenne Française*. Lechevalier, Paris.

Flower, R., and J. Thompson. (2009). An overview of integrated hydro-ecological studies in the MELMARINA Project: monitoring and modelling coastal lagoons— making management tools for aquatic resources in North Africa. *Hydrobiologia* **622**:3–14.

Focardi, S., and A. Tinelli. (2005). Herbivory in a Mediterranean forest: browsing impact and plant compensation. *Acta Oecologica* **28**:239–247.

Fons, R. (1975). Premières données sur l'écologie de la Pachyure étrusque *Suncus etruscus* (Savi, 1822) et comparaison avec deux autres crocidurinae: *Crocidura russula* (Hermann, 1780) et *Crocidura suaveolens* (Pallas, 1811) (Insectivora Soricidae). *Vie Milieu* **25**:315–360.

Forel, J., and J. Leplat. (1995). *Les Carabes de France*. Editions Sciences Nat.

Foster, D. R. (2002). Conservation issues and approaches for dynamic cultural landscapes. *Journal of Biogeography* **29**:1533–1535.

Foster, J. B. (1964). The evolution of mammals on islands. *Nature* **202**:234–235.

Fox, B. J., and M. D. Fox. (1986). Resilience of animal and plant communities to human disturbance. In B. Dell, A. J. M. Hopkins, and B. B. Lamont (eds), *Resilience in Mediterranean-Type Ecosystems*, pp. 39–64. Dr. W. Junk, Dordrecht.

Fragman, O., and A. Shmida. (1998). *Bulbous Plants of the Holy Land*. Sutlands, London.

Francisco-Ortega, J., A. Santos-Guerra, S.-C. Kum, and D. J. Crawford. (2000). Plant genetic diversity in Canary Islands: a conservation perspective. *American Journal of Botany* **87**:909–919.

Fredj, G., and L. Laubier. (1985). The deep Mediterranean benthos. In M. Moraitou-Apostolopoulou, and V. Kiortsis (eds), *Mediterranean Marine Ecosystems*, pp. 109–145. Plenum Press, New York.

Fredj, G., D. Bellan-Santini, and M. Meinardi. (1992). Etat des connaissances sur la faune marine méditerranéenne. *Bulletin de l'Institut Océanographique de Monaco* **9**:133–145.

Frérot, H., C. Lefèbvre, W. Gruber, C. Collin, A. Dos Santos, and J. Escarré. (2006). Specific interactions between local metallicolous plants improve the phytostabilization of mine soils. *Plant and Soil* **282**:53–65.

Fritsch, P. W. (2001). Phylogeny and biogeography of the flowering plant genus *Styrax* (Styracaceae) based on chloroplast DNA restriction sites and DNA sequences of the Internal Transcribed Spacer region. *Molecular Phylogenetics and Evolution* **19**:387–408.

Fromentin, J. M., and J. E. Powers. (2005). Atlantic bluefin tuna: population dynamics, ecology, fisheries and management. *Fish and Fisheries* **6**:281–306.

Frontier, S., and D. Pichod-Viale. (1991). *Ecosystèmes: Structure, Fonctionnement, Evolution*. Collection d'Ecologie 21, ed. Masson, Paris.

Fuller, D. Q. (2007). Contrasting patterns in crop domestication and domestication rates: recent archaeobotanical insights from the Old World. *Annals of Botany* **100**: 903–924.

Fyllas, N., and A. Y. Troumbis. (2009). Simulating vegetation shifts in north-eastern Mediterranean mountain forests under climatic change scenarios. *Global Ecology and Biogeography* **18**:64–77.

Galil, B. S. (2007). Loss or gain? Invasive aliens and biodiversity in the Mediterranean Sea. *Marine Pollution Bulletin* **55**:314–322.

Galil, B., C. Frogilia, and P. Noel. (2002). *CIESM Atlas of Exotic Species in the Mediterranean, vol. 2. Crustaceans: Decapods and Stomatopods* (F. Briand, ed). CIESM Publishers, Monaco.

Gamisans, J., and J.-F. Marzocchi. (1996). *La Flore Endémique de la Corse*. Edisud, Aix-en-Provence.

Garcia-del-Rey, E., W. Cresswell, C. M. Perrins, and A. Gosler. (2006). Variable effects of laying date on clutch size in the Canary Island blue tits (*Cyanistes teneriffae*). *Ibis* **148**:564–567.

Garfì, G. (1997). On the flowering of *Zelkova sicula* (Ulmaceae): additional description and comments. *Plant Biosystems* **131**:137–142.

Garnier, E., S. Lavorel, P. Ansquer, H. Castro, P. Cruz, J. Dolezal, O. Eriksson, C. Fortunel, H. Freitas, C. Golodets *et al.* (2007). Assessing the effects of land use change on plant traits, communities and ecosystem functioning in grasslands: a standardized methodology and lessons from an application to 11 European sites. *Annals of Botany* **99**:967–985.

Garrabou, J., T. Perez, P. Chevaldonné, N. Bensoussan, O. Torrents, C. Lejeusne, J. C. Romano, J. Vacelet, N. Boury-Esnault, M. Harmelin-Vivien *et al.* (2003). Is global change a real threat for conservation of the NW Mediterranean marine biodiversity? *Geophysical Research Abstracts* **5**:10522.

Gass, I. G. (1968). Is the Troodos massif of Cyprus a fragment of Mesozoic ocean floor? *Nature* **220**:39–42.

Gassó, N., D. Sol, J. Pinol, E. D. Dana, F. Lloret, M. Sanz-Elorza, E. Sobrino, and M. Vilà. (2009). Exploring species attributes and site characteristics to assess plant invasions in Spain. *Diversity and Distributions* **15**: 50–58.

Gaston, K. (1996). Biodiversity congruence. *Progress in Physical Geography* **20**:105–112.

Gaussen, H. (1954). Théorie et classification des climats et microclimats. *VIIème Congrès International de Botanique*:125–130.

Gentry, H. S., A. J. Verbiscar, and T. F. Barigan. (1987). Red squill (*Urginea maritime*, Liliaceae). *Economic Botany* **50**:517–521.

Georgoudis, A. (1995). Animal genetic diversity plays important role in Mediterranean agriculture. *Diversity* **11**:16–19.

GIEC (Groupe d'Experts Intergouvernemental sur l'Evolution du Climat). (2007). *Bilan 2007 des Changements Climatiques*. Contribution des Groupes de travail I, II et III au quatrième Rapport d'évaluation du GIEC (publié sous la direction de Pachauri, R. K., et A. Reisinger). GIEC, Geneva.

Gilaldi, I., M. Segoli, and E. D. Ungar. (2008). The effect of shrubs on the seed rain of annuals in a semiarid landscape. *Israel Journal of Plant Sciences* **55**:83–92.

Gilot, F., and E. Rousseau. (2008) La Fauvette à lunettes *Sylvia conspicillata* en France: répartition, effectifs et évolution. *Alauda* **76**:47–58.

Gilpin, M. E., and M. E. Soulé. (1986). Minimum viable populations: the process of species extinctions. In M. E. Soulé (ed.), *Conservation Biology: The Science of Scarcity and Diversity*, pp. 13–34. Sinauer Associates, Sunderland, MA.

Ginocchio, R., and G. Montenegro. (1992). Effects of insect herbivory on plant architecture. In C. A. Thanos (ed.), *Plant-Animal Interactions in Mediterranean-Type Ecosystems*, pp. 7–21. University of Athens.

Gintzburger, G., J. J. Rochon, and A. P. Conesa. (1990). The French mediterranean zones: sheep rearing systems and the present and potential role of pasture legumes. In A. E. Osman, M. H. Ibrahim, and M. A. Jones (eds), *The Role of Legumes in the Farming Systems of the Mediterranean Area*, pp. 179–94. ICARDA, Aleppo.

Giorgi, F. (2006). Climate change hot-spots. *Geophysical Research Letters* **33**:L08707.

Giuffra, E., J. M. H. Kijasa, V. Amargera, Ö. Carlborga, J.-T. Jeona, and L. Anderssona. (2000). The origin of the domestic pig: independent domestication and subsequent introgression. *Genetics* **154**:1785–1791.

Godoy, O., D. M. Richardson, F. Valladares, and P. Castro-Diez. (2009). Flowering phenology of invasive alien plant species compared with native species in three Mediterranean-type ecosystems. *Annals of Botany* **103**:485–494.

Golani, D., L. Orsi-Relini, E. Massuti, and J. P. Quignard. (2002). *CIESM Atlas of Exotic Species in the Mediterranean, vol.1. Fishes*. (F. Briand, ed). CIESM Publishers, Monaco.

Goldammer, J. G. (2003). Towards international cooperation in managing forest fire disasters in the Mediterranean region. In H. G. Brauch, P. H. Liotta, A. Marquina, P. F. Rogers, and M. El-Sayed Selim (eds), *Security and Environment in the Mediterranean. Conceptualising Security and Environmental Conflicts*, pp. 907–915. Springer Verlag, Heidelberg.

Gomez, J. M., R. Zamora, J. A. Hodar, and D. Garcia. (1996). Experimental study of pollination by ants in Mediterranean high mountains and arid habitats. *Oecologia* **105**:236–242.

Gomez-Campo, C. (1985). *Plant Conservation in the Mediterranean Area*. Dr. W. Junk, Dordrecht.

Gomez-Campo, C., and J. M. Herranz-Sanz. (1993). Conservation of Iberian endemic plants: the botanical reserve of La Encantada (Villarrobledo, Albacete, Spain). *Biological Conservation* **64**:155–160.

Gondard, H., F. Romane, J. Aronson, and Z. Shater. (2003). Impact of soil surface disturbances on functional group diversity after clear-cutting in Aleppo pine (*Pinus halepensis*) forests in southern France. *Forest Ecology and Management* **180**:165–174.

González, J. L., P. Garzon, and M. Merino. (1990). Censo de la poblacion espanola de cernicalo primilla. *Quercus* **49**:6–12.

González-Sampériz, P., B. L. Valero-Garcés, J. S. Carrión, J. L. Peña-Monné, J. M. García-Ruiz, and C. Martí-Bono. (2005). Glacial and Lateglacial vegetation in northeastern Spain: New data and a review. *Quaternary International* **140–141**:4–20.

Gordon, I. (1956). *Paromola cuvieri* (Risso), a crab new to the Orkneys, Shetlands and Norway. *Nature* **178**:1184–1185.

Goren, M., and R. Ortal. (1999). Biogeography, diversity and conservation of the inland water fish communities in Israel. *Biological Conservation* **89**:1–9.

Goren, M., and B. S. Galil. (2005). A review of changes in the fish assemblages of Levantine inland and marine ecosystems following the introduction of non-native fishes. *Journal of Applied Ichthyology* **21**:364–370.

Goren-Inbar, N., C. S. Feibel, K. L. Verosub, Y. Melamed, M. E. Kislev, E. Tchernov, and I. Saragusti. (2000). Pleistocene milestones on the out-of-Africa corridor at Gesher Benot Ya'aqov, Israel. *Science* **289**:944–947.

Goren-Inbar, N., N. Alperson, M. E. Kislev, O. Simchoni, Y. Melamed, A. Ben-Nun, and E. Werker. (2004). Evidence of hominin control of fire at Gesher Benot Ya'aqov, Israel. *Science* **304**:725–727.

Goricki, S., and P. Trontelj. (2006). Structure and evolution of the mitochondrial control region and flanking sequences in the European cave salamander *Proteus anguinus*. *Gene* **378**:31–41.

Gotelli, N. (2002). Biodiversity in the scales. *Nature* **419**:575–576.

Gould, S. J., and R. C. Lewontin. (1979). The sprandels of San Marco and the Panglossian paradigm: a critique of the adaptationist programme. *Proceedings of the Royal Society of London Series B Biological Sciences* **205**:581–598.

Gouyon, P. H., and D. Couvet. (1987). A conflict between two sexes, females and hermaphrodites. In S. C. Stearns (ed.), *The Evolution of Sex and its Consequences*, pp. 245–261. Birkhauser Verlag, Berlin.

Gouyon, P. H., P. Vernet, J.-L. Guillerm, and G. Valdeyron. (1986). Polymorphisms and environments: the adaptive value of the oil polymorphisms in *Thymus vulgaris* L. *Heredity* **57**:59–66.

Graham, L. E., and L. W. Wilcox. (2000). *Algae*. Prentice-Hall, Upper Saddle River, NJ.

Gramond, D. (2002). *Dynamique de l'Occupation du Sol et Variation des Usages de l'Eau en Anatolie Centrale (Turquie) au cours du 20ème Siècle. Recherches Méthodologiques basées sur l'Analyse Diachronique de Données Satellites et Statistiques*. Université Paris VI-Sorbonne, Paris.

Granjon, L., and G. Cheylan. (1989). Le sort de rats noirs (*Rattus rattus*) introduits sur une île, révélé par radio-tracking. *Comptes Rendus de l'Académie des Sciences, Paris, Série III* **309**:571–575.

Grant, P. R. (1998). *Evolution on Islands*. Oxford University Press, Oxford.

Granval, P. (1988). *Approche Ecologique de la Gestion de l'Espace Rural: des Besoins de la Bécasse à la Qualité des Milieux*. Unpublished thesis. University of Rennes.

Granval, P., and B. Muys. (1992). Management of forest soils and earthworms to improve woodcock (*Scolopax* sp.) habitats: a literature survey. *Gibier Faune Sauvage* **9**:243–255.

Grenon, M., and Batisse, M. (1989). *Futures for the Mediterranean Basin: The Blue Plan*. Oxford University Press, Oxford.

Greuter, W. (1991). Botanical diversity, endemism, rarity, and extinction in the Mediterranean area: an analysis based on the published volumes of Med-Checklist. *Botanica Chronica* **10**:63–79.

Greuter, W. (1994). Extinction in Mediterranean areas. *Philosophical Transactions of the Royal Society London Series B* **344**:41–46.

Griffiths, D. (2006). Pattern and porocess in the ecological biogeography of European freshwater fish. *Journal of Animal Ecology* **75**:734–751.

Grillas, P., and J. Roché. (1997). *Végétation des Marais Temporaires. Ecologie et Gestion*. MeedWet, Tour du Valat, Arles.

Grime, J. P. (1979). *Plant Strategies and Vegetation Processes*. John Wiley and Sons, Chicester.

Grison-Pigé, L., J.-M. Bessière, C. J. Turlings, F. Kjellberg, J. Roy, and M. Hossaert-McKey. (2001). Limited intersex mimicry of floral odour in *Ficus carica*. *Functional Ecology* **15**:551–558.

Gritti, E. S., B. Smith, and M. T. Sykes. (2006). Vulnerability of Mediterranean basin ecosystems to climate change and invasion by exotic plant species. *Journal of Biogeography* **33**:145–157.

Grove, A. T., and O. Rackham. (2001). *The Nature of Mediterranean Europe. An Ecological History*. Yale University Press, New Haven, CT.

Groves, R. H. (1986). Invasion of mediterranean ecosystems by weeds. In B. Dell, A. J. M. Hopkins, and B. B. Lamont (eds), *Resilience in Mediterranean-Type Ecosystems*, pp. 129–145. Dr. W. Junk, Dordrecht.

Groves, R. H., and M. J. Kilby. (1991). Introduced flora of Mediterranean-climate regions. Convergence or divergence? In C. A. Thanos (ed.), *Plant-Animal Interactions in Mediterranean-Type Ecosystems*, pp. 351–356. University of Athens.

Guarini, J. M., L. Chauvaud, and J. Coston-Guarini. (2008). Can the intertidal benthic microalgal primary production account for the "Missing Carbon Sink"? *Journal of Oceanography, Research and Data* **1**:13–19.

Guarini, M., P. Gros, G. F. Blanchard, P. Richard, and A. Fillon. (2004). Benthic contribution to pelagic microalgal communities in two semi-enclosed, European-type littoral ecosystems (Marennes-Oleron Bay and Aiguillon Bay, France). *Journal of Sea Research* **52**:241–258.

Habsurgo-Lorena, A. S. (1983). Socioeconomic aspects of the crayfish industry in Spain. *Freshwater Crayfish* **5**:552–554.

Haffer, J. (1977). Secondary contact zones of birds in Northern Iran. *Bonner Zoologische Monographien* **10**:1–64.

Hagemeijer, W. J. M., and M. J. Blair. (1997). *The EBCC Atlas of European Breeding Birds*. T. & A.D. Poyser, Academic Press, London.

Hall, M. A., and G. P. Donovan. (2002). Environmentalists, fisherman, cetaceans, and fish: is there a balance and can science help to find it? In P. G. H. Evans, and J. A. Raga (eds), *Marine Mammals: Biology and Conservation*, pp. 491–521. Kluwer Academic/Plenum Publishers, New York.

Hampe, A. (2003). Large-scale geographical trends in fruit traits of vertebrate-dispersed temperate plants. *Journal of Biogeography* **30**:487–496.

Hampe, A., and J. Arroyo. (2002). Recruitment and regeneration in populations of an endangered South Iberian Tertiary relict tree. *Biological Conservation* **107**:263–271.

Harlan, J. R., and D. Zohary. (1966). Distribution of wild wheats and barley. *Science* **153**:1074–1080.

Harper, J. L. (1982). After description. In E. I. Newman (ed.), *The Plant Community as a Working Mechanism*, pp. 11–25. Blackwell Scientific Publications, Oxford.

Harvey, B., and M. Mercusot. (2007). Cooperation between Mediterranean countries of Europe and the southern rim of the Mediterranean. *Desalination* **203**:20–26.

Haubois, A. G., F. Sylvestre, J. M. Guarini, P. Richard, and G. F. Blanchard. (2005). Spatio-temporal structure of the epipelic diatom assemblage from an intertidal mudflat in Marennes-Oléron Bay, France. *Estuarine, Coastal and Shelf Science* **64**:385–394.

Hawkes, J. G. (1995). Centers of origin for agricultural diversity in the Mediterranean: From Vavilov to the present day. *Diversity* **11**:109–111.

Hays, J. D., J. Imbrie, and N. J. Shackleton. (1976). Variations in the Earth's orbit: pacemaker of the ice ages. *Science* **194**:1121–1132.

Haywood, A. M., and Valdes, P. J. (2004) Modelling Pliocene warmth: contribution of atmosphere, oceans and cryosphere. *Earth and Planetary Science Letters* **218**:363–377.

Heaney, L. R. (2007). Is a new paradigm emerging for oceanic island biogeography? *Journal of Biogeography* **34**:753–757.

Henkin, Z., L. Hadar, and I. Noy-Meir. (2007). Human-scale structural heterogeneity induced by grazing in a Mediterranean woodland landscape. *Landscape Ecology* **22**:577–587.

Henry, P. M. (1977). The Mediterranean: a threatened microcosm. *Ambio* **6**:300–307.

Hepper, N. (1981). *Bible Plants at Kew*. Her Majesty's Stationery Office, London.

Herbert, R. J. H., and S. J. Hawkins. (2006). Effect of the rock type on the recruitment and early mortality of the barnacle *Chthamalus montagui*. *Journal of Experimental Marine Biology and Ecology* **334**:96–108.

Herms, D. A., and W. J. Mattson. (1992). The dilemma of plants: to grow or defend. *Quarterly Review of Biology* **67**:283–335.

Herrera, C. M. (1984). A study of avian frugivores, bird-dispersed plants, and their interactions in Mediterranean scrublands. *Ecological Monographs* **54**:1–23.

Herrera, C. M. (1992). Historical effects and sorting processes as explanations for contemporary ecological patterns: character syndromes in Mediterranean woody plants. *The American Naturalist* **140**:421–446.

Herrera, C. M. (1995). Plant-vertebrate seed dispersal systems in the Mediterranean: Ecological, evolutionary, and historical determinants. *Annual Review of Ecology and Systematics* **26**:705–727.

Herrera, C. M. (2002). Correlated evolution of fruit and leaf size in bird-dispersed plants: species-level variance in fruit traits explained a bit further? *Oikos* **97**:426–432.

Hewitt, G. M. (1999) Post-glacial recolonisation of European biota. *Biological Journal of the Linnean Society* **68**:87–112.

Heywood, V. H. (1995). The Mediterranean flora in the context of world biodiversity. *Ecologia Mediterranea* **20**:11–18.

Higgins, L. G., and N. D. Riley. (1988). *A Field Guide to the Butterflies of Britain and Europe*. Collins, London.

Hildebrand, E. E. (1987). Die struktur von waldböden—ein gefährdetes fliessgleichgewicht. *Allgemeine Forst Zeitschrift* **16–17**:424–426.

Hillel, D. (1994). *Rivers of Eden. The Struggle for Water and the Quest for Peace in the Middle East*. Oxford University Press, New York.

Hobbs, R., D. M. Richardson, and G. W. Davis. (1995). Mediterranean-type ecosystems: opportunities and constraints for studying the function of biodiversity. In G. W. Davis, and D. M. Richardson (eds), *Mediterranean-Type Ecosystems. The Function of Biodiversity*, pp. 1–42. Springer-Verlag, Berlin.

Hockin, D. C. (1980). The biogeography of the Butterflies of the Mediterranean Islands. *Nota Lepidopterologica* **3**:119–125.

Hódar, J., and R. Zamora. (2004). Herbivory and climatic warming: a Mediterranean outbreaking caterpillar attacks a relict, boreal pine species. *Biodiversity and Conservation* **13**:493–500.

Hohmann, S., J. W. Kadereit, and G. Kadereit. (2006). Understanding Mediterranean-Californian disjunctions: Molecular evidence from chenopodiaceae-betoideae. *Taxon* **55**:67–78.

Holdridge, L. R. (1947). Determination of world plant formations from simple climatic data. *Science* **105**: 367–368.

Houston, J. M. (1964). *The Western Mediterranean World. An Introduction to its Regional Landscapes*. Frederick A. Praeger, New York.

Hsü, K. J. (1971). Origin of the Alps and western Mediterranean. *Nature* **233**:44–48.

Hsü, K., L. Montadert, D. Bernouilli, M. B. Cita, A. Erickson, R. E. Garrison, R. B. Kidd, F. Melieres, C. Müller, and R. Wright. (1977). History of the Mediterranean salinity crisis. *Nature* **267**:399–403.

Hulme, P. E., G. Brundu, I. Camarda, P. Dalias, P. Lambdon, F. Lloret, F. Médail, E. Moragues, C. M. Suehs, A. Traveset et al. (2008). Assessing the risks to Mediterranean islands ecosystems from alien plant introductions. In B. Tokarska-Guzik, J. H. Brock, G. Brundu, L. Child, C. C. Daehler, and P. Pysek (eds), *Plant Invasions: Human Perception, Ecological Impacts and Management*, pp. 39–56. Backhuys Publishers, Leiden.

Huntington, E. (1911). *Palestine and its Transformation*. Boston, Dutton.

Huntley, B. (1988). European post-glacial vegetation history: a new perspective. In H. Ouellet (ed.), *Acta XIX Congressus Internationalis Ornithologici, vol. 1*, pp. 1061–1077. National Museum of Natural Sciences, Ottawa.

Huntley, B. (1993). Species-richness in north-temperate zone forests. *Journal of Biogeography* **20**:163–180.

Huntley, B., and H. J. B. Birks. (1983). *An Atlas of Past and Present Pollen Maps for Europe: 0-13000 Years Ago*. Cambridge University Press, Cambridge.

Huston, M. A. (1994). *Biological Diversity: the Coexistence of Species on Changing Landscapes*. Cambridge University Press, Cambridge.

Huston, M. A. (1999). Local processes and regional patterns: appropriate scales for understanding variation in the diversity of plants and animals. *Oikos* **86**:393–401.

IPCC (Intergovernmental Panel on Climate Change). (2007). *Climate Change 2007: the Physical Science Basis*. Contribution of Working Group I to the Fourth Assessment Report of the IPCC. Cambridge University Press, Cambridge.

IUCN. (1985). *United Nations List of National Parks and Protected Areas*. IUCN, Gland.

IUCN. (2003). Dolphins, whales and purpoises, 2002-2010 conservation action plan for the world's cetaceans. In R. R. Reeves, B. D. Smith, E. A. Crespo, and G. Notarbartolo di Sciara (compilers), *IUCN/SSC Cetacean Specialist Group*, pp. 75–80. IUCN, Gland.

IUCN. (2008). *The IUCN Red List of Threatened Species*. IUCN. www.iucnredlist.org/.

Izhaki, I., P. B. Walton, and U. N. Safriel. (1991). Seed shadow generated by frugivorous birds in an eastern Mediterranean scrub. *Journal of Ecology* **79**:575–590.

Jacquard, A. (1991). *Voici le Temps du Monde Fini*. Seuil, Paris.

Jacques, G. (2006). *Ecologie du Plancton*. Lavoisier, Paris.

Jacquet, K. (2006). *Biodiversité et Perturbations: Dynamique de l'Avifaune après Incendie et ses Relations avec la Dynamique Végétale*. PhD Thesis, University of Montpellier II.

Jansen, T., P. Forster, M. A. Levine, H. Oelke, M. Hurles, C. Renfrew, J. R. Weber, and K. Olek. (2002). Mitochondrial DNA and the origins of the domestic horse. *Proceedings of the National Academy of Sciences USA* **99**:10905–10910.

Janzen, D. H. (1980). When is it coevolution? *Evolution* **34**:611–612.

Jiménez, P., D. Agúndez, R. Alía, and L. Gil. (1999). Genetic variation in central and marginal populations of *Quercus suber* L. *Silvae. Genetica* **48**:278–284.

Joffre, R., and S. Rambal. (1993). How tree cover influences the water balance of Mediterranean rangelands. *Ecology* **74**:570–582.

Joffre, R., J. Vacher, C. de los Llanos, and G. Long. (1988). The dehesa: an agrosilvopastoral system of the Mediterranean region with special reference to the

Sierra Morena area of Spain. *Agroforestry Systems* **6**: 71–96.

Johnson, A., and F. Cézilly. (2007). *The Greater Flamingo*. T. & A.D. Poyser, London.

Jolivet, L., J. P. Brun, B. Meyer, G. Prouteau, J. M. Rouchy, and B. Scaillet. (2008). *Géodynamique Méditerranéenne*. Vuibert, Paris.

Jorgensen, T. H., and J. M. Olesen. (2001). Adaptive radiation of island plants: evidence from *Aeonium* (Crassulaceae) of the Canary Islands. *Perspectives in Plant Ecology, Evolution and Systematics* **4**:29–42.

Juan, C., B. C. Emerson, P. Oromi, and G. M. Hewitt. (2000). Colonization and diversification: towards a phylogeographic synthesis for the Canary Islands. *Trends in Ecology & Evolution* **15**:104–109.

Julliard, R., F. Jiguet, and D. Couvet. (2004). Common birds facing global changes: what makes a species at risk? *Global Change in Biology* **10**:148–154.

Jürgens, K. D. (2002). Etruscan shrew muscle: the consequences of being small. *Journal of Experimental Biology* **205**:2161–2166.

Jürgens, K. D., R. Fons, T. Peters, and S. Sender. (1996). Heart and respiratory rates and their significance for connective oxygen transport rates in the smallest mammal, the Etruscan shrew *Suncus etruscus*. *Journal of Experimental Biology* **199**:2579–2584.

Kaniewski, D., J. Renault-Miskovsky, C. Tozzi, and H. de Lumley. (2005). Upper Pleistocene and Late Holocene vegetation belts in western Liguria: an archaeopalynological approach. *Quaternary International* **135**:47–63.

Kanyamibwa, S., A. Schierer, R. Pradel, and J.-D. Lebreton. (1990). Changes in adult survival rates in a western European population of the white stork *Ciconia ciconia*. *Ibis* **132**:27–35.

Kaplan, D. Y., and M. Gutman. (1989). Food composition of the mountain gazelle and cattle in the southern Golan. *Journal of Zoology* **36**:154.

Kark, S., and D. Sol. (2005). Establishment success across convergent Mediterranean ecosystems: an analysis of bird introductions. *Conservation Biology* **19**: 1519–1527.

Kayser, Y., M. Gauthier-Clerc, A. Béchet, G. Poulin, G. Massez, Y. Chérain, J. Paoli, N. Sadoul, E. Vialet, G. Paulus et al. (2008). Compte rendu ornithologique camarguais pour les années 2001–2006. *Revue d'Ecologie (Terre et Vie)* **63**:299–349.

Kazakis, G., D. Ghosn, I. N. Vogiatzakis, and V. P. Papanastasis. (2007). Vascular plant diversity and climate change in the alpine zone of Lefka Ori, Crete. *Biodiversity and Conservation* **16**:1603–1615.

Keefover-Ring, K., J. D. Thompson, and. Y. B. Linhart. (2009). Beyond six scents: defining a seventh *Thymus vulgaris* chemotype new to southern France by ethanol extraction. *Flavour and Fragrance Journal* **24**: 117–122.

Keeley, J. E. (1991). Seed germination and life history syndromes in the California chaparral. *Botanical Review* **57**:81–116.

Kéfi, S., M. Rietkerj, C. L. Alados, Y. Pueyo, V. P. Papanastasis, A. ElAich, and P. C. de Ruiter. (2007). Spatial vegetation patterns and imminent desertification in Mediterranean arid ecosystems. *Nature* **449**:213–218.

Kennedy, T. A., S. Naeem, K. M. Howe, J. M. H. Knops, D. Tilman, and P. Reich. (2002). Biodiversity as barrier to ecological invasion. *Nature* **417**:636–638.

Khadari, B., C. Grout, S. Santoni, I. Hochu, J.-P. Roger, M. Ater, U. Aksoy, and F. Kjellberg. (2005a). Etude préliminaire des origines de *Ficus carica* L. et de sa domestication. *Les Actes du BRG* **5**:53–65.

Khadari, B., C. Grout, S. Santoni, and F. Kjellberg. (2005b). Contrasting genetic diversity and differentiation among Mediterranean populations of *Ficus carica* L.: a study using mtDNA RFLP. *Genetic Resources and Crop Evolution* **52**:97–109.

Kilian, B., H. Ozkan, A. Walther, J. Kohl, T. Dagan, F. Salamini, and W. Martin. (2007). Molecular diversity at 18 loci in 321 wild and 92 domesticate lines reveal no reduction of nucleotide diversity during *Triticum monococcum* (einkorn) domestication: implications for the origin of agriculture. *Molecular Biology and Evolution* **24**:2657–2668.

Kim, S.-C., M. R. McGowen, P. Lubinsky, J. C. Barber, M. E. Mort, and A. Santos-Guerra. (2008). Timing and tempo of early and successive adaptive radiations in Macaronesia. *PLoS ONE* **3**:e2139.

Kislev, M. E. (1985). Early Neolithic horsebean from Yiftah'el, Israel. *Science* **228**:319–320.

Kislev, M. E., D. Nadel, and I. Carmi. (1992). Epipalaeolithic (19,000 BP) cereal and fruit diet at Ohalu II, Sea of Galilee, Israel. *Review of Palaeobotany and Palynology* **73**:161–166.

Kislev, M. E., E. Weiss, and A. Hartmann. (2004). Impetus for sowing and the beginning of agriculture: ground collecting of wild cereals. *Proceedings of the National Academy of Sciences USA* **101**:2692–2695.

Kislev, M. E., A. Hartmann, and O. Bar-Yosef. (2006). Early domesticated fig in the Jordan Valley. *Science* **312**:1372–1374.

Kjellberg, F., P. H. Gouyon, M. Ibrahim, M. Raymond, and G. Valdeyron. (1987). The stability of the symbiosis between dioecious figs and their pollinators: a study of *Ficus carica* L. and *Blastophaga psenes*. *Evolution* **91**: 117–122.

Klein, B., and W. Roether. (2001). Ozeanographie und Wasseraushalt. In R. Hofrichter (ed.), *Das Mittelmeer:*

Fauna, Flora, Ökologie, Bd 1. *Allgemeiner Teil*, pp. 258–287. Spektrum Akademischer Verlag, Heidelberg.

Klicka, J., and R. M. Zink. (1997). The importance of recent ice ages in speciation: a failed paradigm. *Science* **277**:1666–1669.

Kolars, J. (1982). Earthquake-vulnerable populations in modern Turkey. *Geographical Review* **72**:20–35.

Kouki, H. (2007). L'organisation forestière en Tunisie. *Nouvelles des forêts méditerranéennes* **8**:3–5.

Kowalski, K., and B. Rzebik-Kowlaska. (1991). *Mammals of Algeria*. Polish Academy of Sciences, Institute of Systematics and Evolution of Animals, Wroclaw.

Kraiem, M. M. (1983). Les poissons d'eau douce de Tunisie: Inventaire commenté et répartition géographique. *Bulletin de l'Institut National des Sciences et Techniques Océanographiques* **10**:107–124.

Krijgsman, W., E. J. Hilgen, I. Raffi, F. J. Sierro, and D. S. Wilson. (1999). Chronology, causes and progression of the Messinian salinity crisis. *Nature* **400**:652–655.

Krkosek, M., J. Ford, A. Morton, S. Lele, R. A. Myers, and M. Lewis. (2007). Declining wild salmon populations in relation to parasites from farm salmon. *Science* **318**:1772–1775.

Küchler, A. W. (1964). *Potential Natural Vegetation of the Conterminous United States*. American Geographic Society, New York.

Küçük, M. (2008). L'organisation forestière en Turquie. *Nouvelles des forêts méditerranéennes* **9**:3–5.

Kuhnholz-Lordat, G. (1938). *La Terre Incendiée. Essai d'Agronomie Comparée*. Maison Carrée, Nîmes.

Kuhnholz-Lordat, G. (1958). L'écran vert. *Mémoires du Muséum National d'Histoire Naturelle* **9**:1–276.

Kullenberg, B. (1961). Studies in *Ophrys* pollination. *Zoologiska Bidrag Uppsala* **34**:1–340.

Kullenberg, B. (1973). New observations on the pollination of *Ophrys* L. (*Orchidaceae*). *Zoon* **Suppl. 1**:9–13.

Kullenberg, B., and G. Bergström. (1976). The pollination of *Ophrys* orchids. *Botaniska Notiser* **129**:11–9.

Kurtz, C., and P. Luquet. (1996). The traffic in Mediterranean birds of prey. In *Biologia y Conservacion de las Rapaces Mediterraneas, 1994*. Monografias, no. 4. SEO, Madrid.

Kusler, J. A., and M. E. Kentula. (1990). *Wetland Creation and Restoration: The Status of the Science*. Island Press, Washington DC.

Kvist, L., J. Broggi, J. C. Illera, and K. Koivula. (2005). Colonisation and diversification of the blue tits (*Parus caeruleus teneriffae* group) in the Canary Islands. *Molecular Phylogenetics and Evolution* **34**:501–511.

Lacombe, H., and P. Tchernia. (1960). Quelques traits généraux de l'hydrologie Méditerranéenne. *Cahiers Océanographiques* **12**:527–547.

Lambdon, P. W., and P. E. Hulme. (2006). How strongly do interactions with closely-related native species influence plant invasions? Darwin's naturalization hypothesis assessed on Mediterranean islands. *Journal of Biogeography* **33**:1116–1125.

Lambdon, P. W., F. Lloret, and P. E. Hulme. (2008). Do nonnative species invasions lead to biotic homogenization at small scales? The similarity and functional diversity of habitats compared for alien and native components of Mediterranean floras. *Diversity and Distributions* **14**:774–785.

Lambrechts, M. M., J. Blondel, S. Hurtrez-Boussès, M. Maistre, and P. Perret. (1997). Adaptive inter-population differences in blue tit life-history traits on Corsica. *Evolutionary Ecology* **11**:599–612.

Lamont, B. B. (1982). Mechanisms for enhancing nutrient uptake in plants, with particular reference to mediterranean South Africa and Australia. *Botanical Review* **48**:597–689.

Lantoine, F. (1995). *Caractérisation et Distribution des Différentes Populations du Picoplancton (Picoeucaryotes, Synechococcus spp., Prochlorococcus spp.) dans Diverses Situations Trophiques (Atlantique Tropical, Golfe du Lion)*. PhD Thesis. University Pierre and Marie Curie, Paris VI.

Lanza, B., and M. Poggesi. (1986). *Storia Naturale delle Isole Satelliti della Corsica*. Istituto Geografico Militare, Florence.

Larsen, T. B. (1986). Tropical butterflies of the Mediterranean. *Nota Lepidopterologica* **9**:63–77.

Larson, G., K. Dobney, U. Albarella, M. Fang, E. Matisoo-Smith, J. Robins, S. Lowden, H. Finlayson, T. Brand, E. Willerslev et al. (2005). Worldwide phylogeography of wild boar reveals multiple centers of pig domestication. *Science* **307**:1618–1621.

Latham, R. E., and R. E. Ricklefs. (1993). Continental comparisons of temperate-zone tree species diversity. In R. E. Ricklefs, and D. Schluter (eds), *Species Diversity in Ecological Communities: Historical and Geographical Perspectives*, pp. 294–314. Chicago University Press, Chicago.

Laubier, L. (1966). Le Coralligène des Albères, monographie biocénotique. *Annales de l'Institut Océanographique* **18**:137–316.

Lavelle, P. (1988). Earthworm activities and the soil system. *Biology and Fertility of Soils* **6**:237–251.

Lavergne, S., W. Thuillier, J. Molina, and M. Debussche. (2005). Environmental and human factors influencing rare plant local occurrence, extinction and persistence: 115-year study in the Mediterranean region. *Journal of Biogeography* **32**:799–811.

Laville, L., and F. Reiss. (1992). The Chironomid fauna of the Mediterranean region reviewed. *Netherlands Journal of Aquatic Ecology* **26**:239–245.

Magnin, F., and T. Tatoni. (1995). Secondary successions on abandoned cultivation terraces in calcareous Provence. II-The gastropod communities. *Acta Oecologica* **16**: 89–101.

Magnin, G. (1987). *An Account of the Illegal Catching and Shooting of Birds in Cyprus during 1986*. International Council for Bird Preservation, Cambridge.

Magnin, G. (1991). Hunting and persecution of migratory birds in the Mediterranean region. In T. Salathé (ed.), *Conserving Migratory Birds*, pp. 59–71. International Council for Bird Preservation, Cambridge.

Magyari, E. K., J. C. Chapman, B. Gaydarska, E. Marinova, T. Deli, J. P. Huntley, J. R. M. Allen, and B. Huntley. (2008). The 'oriental' component of the Balkan flora: evidence of presence on the Thracian Plain during the Weichselian late-glacial. *Journal of Biogeography* **35**: 865–883.

Malcolm, J. R., C. Liu, R. P. Neilson, L. Hansen, and L. Hannah. (2006). Global warming and extinctions of endemic species from biodiversity hotspots. *Conservation Biology* **20**:538–548.

Manicacci, D., D. Couvet, E. Belhassen, P. H. Gouyon, and A. Atlan. (1996). Founder effects and sex ratio in the gynodioecious *Thymus vulgaris*. *Molecular Ecology* **5**: 63–72.

Mansion, G., G. Rosenbaum, N. Schoenenberger, G. Bacchetta, J. A. Rossello, and E. Conti. (2008). Phylogenetic analysis informed by geological history supports multiple, sequential invasions of the Mediterranean Basin by the angiosperm family araceae. *Systematic Biology* **57**:269–285.

Manzaneda, A. J., U. Sperens, and M. B. García. (2005). Effects of microsite disturbances and herbivory on seedling performance in the perennial herb *Helleborus foetidus* (Ranunculaceae). *Plant Ecology* **179**:73–82.

Marchand, H. (1990). *Les Forêts Méditerranéennes. Enjeux et Perspectives*. Les Fascicules du Plan Bleu, 2. Economica, Paris.

Margalef, R. (1974). *Ecología*. Ed. Omega, Barcelona.

Margalef, R. (1984). Le plancton de la Méditerranée. *La Recherche* **158**:1082–1094.

Margalef, R. (ed.). (1985). *Western Mediterranean. Series Key Environments*. Pergamon Press, London.

Margaris, N. S. (1981). Adaptive strategies in plants dominating mediterranean-type ecosystems. In F. di Castri, D. W. Goodall, and R. L. Specht (eds), *Mediterranean-Type Shrublands, Collection Ecosystems of the World, vol. 11*, pp. 309–316. Elsevier, Amsterdam.

Margaris, N. S., and D. Vokou. (1982). Structural and physiological features of woody plants in phryganic ecosystems related to adaptive mechanisms. *Ecologia Mediterranea* **8**:449–459.

Marinkovic, S., and B. Karadzic. (1999). The role of nomadic farming in the distribution of the Griffon vulture (*Gyps fulvus*) on the Balkan peninsula. *Contributions to the Zoogeography and Ecology of the Eastern Mediterranean Region* **1**:141–152.

Marrero, A., R. S. Almeida, and M. González-Martín. (1998). A new species of the wild dragon tree, *Dracaena* (Dracaenaceae) from Gran Canaria and its taxonomic and biogeographic implications. *Botanical Journal of the Linnean Society* **128**:291–314.

Marrero-Gómez, M. V., J. Ramón Arévalo, Á. Bañares-Baudet, and E. Carqué Álamo. (2000). Study of the establishment of the endangered *Echium acanthocarpum* (Boraginaceae) in the Canary Islands. *Biological Conservation* **94**:183–190.

Marsh, G. F. (1874). *The Earth as Modified by Human Action*. Arno, New York.

Martin, J., and P. Gurrea. (1990). The peninsular effect in Iberian butterflies (Lepidoptera: Papilionoidea and Hesperioidea). *Journal of Biogeography* **17**:85–96.

Martin, P. S. (1984). Prehistoric overkill: the global model. In P. S. Martin, and R. G. Klein (eds), *Quaternary Extinctions*, pp. 354–403. University Arizona Press, Tucson, AZ.

Martinez, I. M., and J. P. Lumaret. (2006). Las practicas agropecuarias y sus consecuencias en la entomofauna y el entorno ambiental. *Folia Entomologica Mexica* **45**: 57–68.

Martínez-Sánchez, M. J., M. C. Navarro, C. Pérez-Sirvent, J. Marimón, J. Vidal, M. L. García-Lorenzo, and J. Bech. (2008). Assessment of the mobility of metals in a mining-impacted coastal area (Spain, Western Mediterranean). *Journal of Geochemical Exploration* **96**:171–182.

Martuscelli, E., and F. Tolve. (2002). The Euro-Mediterranean Experts' Meeting to strengthen scientific and technological cooperation for the conservation, restoration and valorisation of the Euro-Mediterranean cultural heritage (Aswan, 24-26 February 2002). *Journal of Cultural Heritage* **3**:163–168.

Marty, P., J. Aronson, and J. Lepart. (2007). Dynamics and restoration of abandoned farmland and other old fields in Southern France. In V. A. Cramer, and R. J. Hobbs (eds), *Old Fields. Dynamics and Restoration of Abandoned Farmland*, pp. 202–224. Island Press, Washington DC.

Mathez, J., P. Quézel, and C. Raynaud. (1985). The Maghreb countries. In C. Gomez-Campo (ed.), *Plant Conservation in the Mediterranean Area*, pp. 141–157. Dr. W. Junk, Dordrecht, Boston, and Lancaster.

Matvejević, P. (1999). *Mediterranean: A Cultural Landscape*. University of California Press, Berkeley, CA.

Mauchamp, A., P. Chauvelon, and P. Grillas. (2002). Restoration of floodplain wetlands: Opening polders along a coastal river in Mediterranean France, Vistre marshes. *Ecological Engineering* **18**:619–632.

Maul, G. E. (1986). Trachichthyidae. In P. J. P. Whitehead, M.-L. Bauchot, J.-C. Hureau, J. Nielsen, and E. Tortonese (eds), *Fishes of the North-Eastern Atlantic and the Mediterranean*, pp. 749–752. UNESCO, Paris.

Mazzella, L., M. C. Buia, M. C. Gambi, M. Lorenti, G. Russo, M. B. Scipione, and V. Zupo. (1995). A review of the trophic organization in the *Posidonia oceanica* ecosystem. In F. Cinelli, E. Fresi, L. Lorenzi, and A. Mucedola (eds), *La Posidonia Oceanica, Revista marittima publ., Ital.* **12**:31–47.

Mazzoleni, S., G. di Pascale, M. Mulligan, P. di Martino, and F. Rego. (2004). *Recent Dynamics of Mediterranean Vegetation and Landscape*. John Wiley and Sons, London.

McCulloch, M. N., G. M. Tucker, and S. R. Baillie. (1992). The hunting of migratory birds in Europe: a ringing recovery analysis. *Ibis* **34**:55–65.

McGuire, A. F., and K. A. Kron. (2005). Phylogenetic relationships of European and African *Ericas*. *International Journal of Plant Science* **166**:311–318.

McKinney, M. L., and J. L. Lockwood. (1999). Biotic homogenization: a few winners replacing many losers in the next mass extinction. *Trends in Ecology & Evolution* **14**:450–453.

McKinney, M. L., and F. A. La Sorte. (2007). Invasiveness and homogenization: synergism of wide dispersal and high local abundance. *Global Ecology and Biogeography* **16**:394–400.

McNab, B. H. (1994). Energy conservation and the evolution of flightlessness in birds. *The American Naturalist* **144**:628–642.

McNab, B. K. (2002). Minimizing energy expenditure facilitates vertebrate persistence on oceanic islands. *Ecology Letters* **5**:693–704.

McNeil, J. R. (1992). *The Mountains of the Mediterranean World, an Environmental History*. Cambridge University Press, Cambridge.

Médail, F. (2008a). Mediterranean. In S. E. Jørgensen (ed.), *Encyclopedia of Ecology*. Elsevier, Amsterdam.

Médail, F. (2008b). A natural history of the islands' unique flora. In C. Arnold (ed.), *Mediterranean Islands*, pp. 27–33. Survival Books, London.

Médail, F., and P. Quézel. (1997). Hot-spots analysis for conservation of plant biodiversty in the Mediterranean Basin. *Annals of the Missouri Botanical Garden* **84**: 112–127.

Médail, F., and V. Verlaque. (1997). Ecological characteristics and rarity of endemic plants from S.E. France and Corsica. Implications for biodiversity conservation. *Biological Conservation* **80**:269–281.

Médail, F., and P. Quézel. (1999). Biodiversity hotspots in the Mediterranean Basin: setting global conservation priorities. *Conservation Biology* **13**:1510–1513.

Médail, F., and P. Quézel. (2003). Conséquences écologiques possibles des changements climatiques sur la flore et la végétation du bassin méditerranéen. *Bocconea* **16**:397–422.

Médail, F., and N. Myers. (2004). Mediterranean Basin. In R. A. Mittermeier, P. Robles Gil, M. Hoffmann, J. Pilgrim, T. Brooks, C. G. Mittermeier, J. Lamoreux, and G. A. B. da Fonseca (eds), *Hotspots Revisited: Earth's Biologically Richest and most Endangered Terrestrial Ecoregions*, pp. 144–147. CEMEX (Monterrey), Conservation International (Washington), and Agrupación Sierra Madre (Mexico).

Médail, F., and K. Diadema. (2006). Biodiversité végétale méditerranéenne et anthropisation: approches macro et micro-régionales. *Annales de Géographies* **651**: 169–192.

Médail, F., and K. Diadema. (2009). Glacial refugia influence plant diversity patterns in the Mediterranean Basin. *Journal of Biogeography* **36**:1333–1345.

Médail, F., H. Michaud, J. Molina, G. Paradis, and R. Loisel. (1998). Conservation de la flore et de la végétation des mares temporaires dulçaquicoles et oligotrophes de France méditerranéenne. *Ecologia Mediterranea* **24**:119–134.

Medrano, M., and C. M. Herrera. (2008). Geographical structuring of genetic diversity across the whole distribution range of *Narcissus longispathus*, a habitat-specialist, Mediterranean narrow endemic. *Annals of Botany* **102**:183–194.

MedWet. (2008). *Towards an Observatory of Mediterranean Wetlands. Evolution of Biodiversity from 1970 until today*. Tour du Valat, Arles.

Meigs, R. (1982). *Trees and Timber in the Ancient Mediterranean World*. Clarendon Press, Oxford.

Meliadou, A., and A. Troumbis. (1997). Aspects of heterogeneity in the distribution of diversity of the European herpetofauna. *Acta Oecologica* **18**:393–412.

Mengoni, A., A. J. M. Baker, M. Bazzicalupo, R. D. Reeves, N. Adigüzel, E. Chianni, F. Galardi, R. Gabbrielli, and C. Gonnelli. (2003). Evolutionary dynamics of nickel hyperaccumulation in *Alyssum* revealed by its nrDNA analysis. *New Phytologist* **159**:691–699.

Menzel, R., and A. Shmida. (1993). The ecology of flower colours and the natural colour vision of insect pollinators: the Israeli flora as a case study. *Biological Review* **68**:81–120.

Merlo, M., and P. Paiero. (2005). The state of Mediterranean forests. In M. Merlo, and L. Croitoru (eds),

Valuing Mediterranean Forests: Towards Total Economic Value, pp. 5–15. CABI Publishing, Wallingford.

Mesléard, F., J. Lepart, and L. Tan Ham. (1995). Impact of grazing on vegetation dynamics in former ricefields. *Journal of Vegetation Science* 6:683–390.

M'Hirit, O. (1999). Mediterranean forests: ecological space and economic and community wealth. *Unasylva* 50: 3–15.

M'Hirit, O., and P. Blerot (dir.) (1999). *Le Grand Livre de la Forêt Marocaine*. Mardaga, Sprimont.

Michaud, H., L. Toumi, R. Lumaret, T. X. Li, F. Romane, and F. di Giusto. (1995). Effect of geographical discontinuity on genetic variation in *Quercus ilex* L. (Holm oak). Evidence from enzyme polymorphism. *Heredity* 74: 590–606.

Michel, C., P. Lejeune, and J. Voss. (1987). Biologie et comportement des Labridés Européens (Labres, Crénilabres, Rouquiers, Vieilles et Girelles). *Revue Française d'Aquariologie Herpétologie* 14:1–80.

Mienis, H. (2003). Native marine molluscs replaced by Lessepsian migrants. *Tentacle* 11:15–16.

Mies, B. A., and G. B. Feige. (2003). Lichenisierte Ascomyceten (Flechten). In R. Hofrichter (ed.), *Das Mittelmeer: Fauna, Flora, Ökologie, Bd 2 (1). Bestimmungsführer*, pp. 172–203. Spektrum Akademischer Verlag, Heidelberg.

Mineur, F., M. P. Jonson, C. Maggs, and H. Stegenga. (2007). Hull fouling on commercial ships as a vector of macroalgal introduction. *Marine Biology* 151:1299–1307.

Mittermeier, R. A., P. Robles Gil, M. Hoffmann, J. Pilgrim, T. Brooks, C. G. Mittermeier, J. Lamoreux, and G. A. B. da Fonseca (eds). (2004). *Hotspots Revisited: Earth's Biologically Richest and most Endangered Terrestrial Ecoregions*. CEMEX (Monterrey), Conservation International (Washington), and Agrupación Sierra Madre (Mexico).

Molero, J., and A. M. Rovira. (1998). A note on the taxonomy of the Macaronesian *Euphorbia obtusifolia* complex (Euphorbiaceae). *Taxon* 47:321–332.

Møller, A. P., P. Berthold, and W. Fiedler. (2004). *Advances in Ecological Research. Birds and Climate Change, vol. 35*. Elsevier, Amsterdam.

Monard, A. (1935). Les Harpacticoïdes marins de la région de Salammbo. *Bulletin de la Station océanographique de Salammbo* 34:1–94.

Monk, C. D. (1966). An ecological significance of evergreenness. *Ecology* 47:504–505.

Mönkkönen, M. (1994). Diversity patterns in Palaearctic and Nearctic forest bird assemblages. *Journal of Biogeography* 21:183–195.

Montmollin, B., and W. Strahm (eds). (2005). *The Top 50 Mediterranean Island Plants. Wild Plants At the Brink of Extinction, and What is Needed to Save Them*. IUCN, Gland.

Mooney, H. A., and E. L. Dunn. (1970). Convergent evolution of Mediterranean climate evergreen sclerophyllous shrubs. *Evolution* 24:292–303.

Morand, A. (2001). *Conservation des Zones Humides Méditerranéennes. Amphibiens et Reptiles*. MedWet, Tour du Valat, Arles.

Moreau, R. E. (1972). *The Palaearctic-African Bird Migration System*. Academic Press, London.

Moreira, F., F. Catry, I. Duarte, V. Acácio, and J. Silva. (2009). A conceptual model of sprouting responses in relation to fire damage: an example with cork oak (*Quercus suber* L.) trees in Southern Portugal. *Plant Ecology* 201:77–85.

Moreno, J. M., and W. C. Oechel. (1994). Fire intensity as a determinant factor of postfire plant recovery in southern California chaparral. In J. M. Moreno, and W. C. Oechel (eds), *The Role of Fire in Mediterranean-Type Ecosystems*, pp. 26–45. Springer-Verlag, New York.

Moretti, M., M. Conedera, R. Moresi, and A. Guisan. (2006). Modelling the influence of change in fire regime on the local distribution of a Mediterranean pyrophytic plant species (*Cistus salviifolius*) at its northern range limit. *Journal of Biogeography* 33:1492–1502.

Moritz, C. (1994). Applications of mitochondrial DNA analysis in conservation: a critical review. *Molecular Ecology* 3:401–411.

Morrell, P. L., and M. T. Clegg. (2007). Genetic evidence for a second domestication of barley (*Hordeum vulgare*) east of the Fertile Crescent. *Proceedings of the National Academy of Sciences USA* 104:3289–3294.

Morri, C., and C. N. Bianchi. (2001). Recent changes in biodiversity in the Ligurian Sea (NW Mediterranean): is there a climatic forcing? In F. M. Faranda, L. Guglielmo, and G. Spezie (eds), *Structure and Processes in the Mediterranean Ecosystems*, pp. 375–384. Springer, Milan.

Morrone, J. J. (2001). Homology, biogeography and areas of endemism. *Diversity and Distributions* 7:297–300.

Mouillot, F., S. Rambal, and R. Joffre. (2002). Simulating climate change impacts on fire-frequency and vegetation dynamics in Mediterranean-type ecosystem. *Global Change Biology* 8:423–437.

Mouillot, F., J.-P.Ratte, R. Joffre, J. M. Moreno, and S. Rambal. (2003). Some determinants of the spatio-temporal fie cycle in a mediterranean landscape (Corsica, France). *Landscape Ecology* 18:665–674.

Mouillot, F., J.-P. Ratte, R. Joffre, D. Mouillot, and S. Rambal. (2005). Long-term forest dynamic after land abandonment in a fire prone Mediterranean landscape (central Corsica, France). *Landscape Ecology* 20:101–112.

Moya, D., J. De las Heras, F. López-Serrano, S. Condes, and I. Alberdi. (2009). Structural patterns and biodiversity in burned and managed Aleppo pine stands. *Plant Ecology* **200**:217–228.

Muller, M. H., J. M. Prosperi, S. Santoni, and J. Ronfort. (2003). Inferences from mitochondrial DNA patterns on the domestication history of alfalfa (*Medicago sativa*). *Molecular Ecology* **12**:2187–2199.

Muntaner, J., and J. Mayol. (1996). *Biologia y Conservacion de las Rapaces Mediterraneas, 1994*. Monografias no. 4. SEO, Madrid.

Musselman, J. L. (2007). *Figs, Dates, Laurel, and Myrrh. Plants of the Bible and the Quran*. Timber Press, Portland Pregon.

Myers, N., R. A. Mittermeier, C. G. Mittermeier, G. A. B. da Fonseca, and J. Kents. (2000). Biodiversity hotspots for conservation priorities. *Nature* **403**:853–858.

Naderi, S., H. R. Rezaei, P. Taberlet, S. Zundel, S. A. Rafat, H. R. Naghash, M. A. A. Elbarody, O. Ertugrul, and F. Pompanon. (2007). Large-scale mitochondrial DNA analysis of the domestic goat reveals six haplogroups with high diversity. *PLoS ONE* **2**:e1012.

Natali, A., and D. Jeanmonod. (1996). Flore analytique des plantes introduites en Corse. In D. Jeanmonod, and H. M. Burdet (eds), *Compléments au Prodrome de la Flore Corse*, pp. 1–211. Conservatoire et Jardin botaniques de Genève, Geneva.

Naveh, Z. (1974). Effects of fire in the Mediterranean region. In T. T. Kozlowski, and C. E. Ahlgren (eds), *Fire and Ecosystems*, pp. 401–434. Academic Press, New York.

Naveh, Z. (1999).The role of fire as an evolutionary and ecological factor on the landscape and vegetation of Mt. Carmel. *Journal of Mediterranean Ecology* **1**: 11–26.

Naveh, Z., and J. Dan. (1973). The human degradation of Mediterranean landscapes in Israel. In F. di Castri, and H. A. Mooney (eds), *Mediterranean-Type Ecosystems: Origins and Structure*, pp. 373–390. Ecological Studies, vol. 7. Springer-Verlag, Berlin.

Naveh, Z., and R. H. Whittaker. (1979). Structural and floristic diversity of shrublands and woodlands in northern Israel and other Mediterranean areas. *Vegetatio* **41**:171–190.

Nègre, M. (1931). Les reboisements du massif de l'Aigoual. *Bulletin de la Société d'Etude des Sciences Naturelles de Nîmes* 1–135.

Nevo, E., D. Zohary, A. H. D. Brown, and M. Haber. (1979). Genetic diversity and environmental associations of wild barley, *Hordeum spontaneum*, in Israel. *Evolution* **33**:815–833.

Nilsson, L. A. (1978). Pollination ecology and adaptation in *Platanthera chlorantha* (Orchidaceae). *Botaniska Notiser* **131**:35–51.

Noodt, W. (1955). Marine Harpacticoiden (Crust. Cop.) aus dem Marmara Meer. *Revue de la Faculté des Sciences, Université d'Istanbul* **20**:49–96.

Noy-Meir, I. (1973). Desert Ecosystems: Environment and Producers. *Annual Review of Ecology and Systematics* **4**:25–51.

Noy-Meir, I. (1988). Dominant grasses replaced by ruderal forbs in a vole year in undergrazed Mediterranean grasslands in Israel. *Journal of Biogeography* **15**: 579–587.

Noy-Meir, I., M. Gutman, and Y. Kaplan. (1989). Response of Mediterranean grassland plants to grazing and protection. *Journal of Applied Ecology* **77**:290–310.

Oberdorff, T., S. Lek, and J.-F. Guégan. (1999) Patterns of endemism in riverine fish of the Northern Hemisphere. *Ecology Letters* **2**:75–81.

Olden, J. D. (2006). Biotic homogenization: a new research agenda for conservation biogeography. *Journal of Biogeography* **33**:2027–2039.

Olden, J. D., and N. L. Poff. (2004). Ecological processes driving biotic homogenization: testing a mechanistic model using fish faunas. *Ecology* **85**:1867–1875.

Olivieri, I., D. Couvet, and P. H. Gouyon. (1990). The genetics of transient populations: research at the metapopulation level. *Trends in Ecology & Evolution* **5**:207–210.

Oosterbroek, P., and J. W. Arntzen. (1992). Area-cladograms of Circum-Mediterranean taxa in relation to Mediterranean palaeogeography. *Journal of Biogeography* **19**:3–20.

Orshan, G. (1964). Seasonal dimorphism of desert and Mediterranean chamaephytes and its significance as a factor in their water economy. In A. J. Rutter, and F. H. Whitehead (eds), *The Water Relations of Plants*, pp. 206–222. Blackwell, Edinburgh.

Orshan, G. (1972). Morphological and physiological plasticity in relation to drought. In *Proceedings of the International Symposium on Wildland Shrub Biology and Utilization*, pp. 245–254. Utah State University, Logan, UT.

Orshan, G. (ed.). (1989). *Plant Pheno-Morphological Studies in Mediterranean-Type Ecosystems*. Kluwer Academic Publishers, Dordrecht.

Ortega, M., R. Elena-Rosello, and J. M. García del Barrio. (2004). Estimation of Plant Diversity at Landscape Level: A methodological approach applied to three Spanish rural areas. *Environmental Monitoring and Assessment* **95**:97–116.

Ortolani, F., and S. Pagliuca. (2006). *Geoarchaeological Evidences of Cyclical Climatic-Environmental Changes in the Mediterranean Area (2500 BP-Present Day)*. GeoSed, Modena.

O'Toole, C., and A. Raw. (1991). *Bees of the World*. Blandford Press, London.

Ovando, P., P. Campos, J. L. Oviedo, and M. Gregorio. (2009). Cost–benefit analysis of cork oak woodland afforestation and facilitated natural regeneration in Spain. In J. Aronson, J. S. Pereira, and J. G. Pausas (eds). *Cork Oak Woodlands on the Edge: Ecology, Adaptive Management, and Restoration*, pp. 177–188. Island Press, Washington DC.

Ozenda, P. (1975). Sur les étages de végétation dans les montagnes du Bassin Méditerranéen. *Documents de Cartographie Ecologique* **16**:1–32.

Palacio, S., P. Millard, and G. Montserrat-Martí. (2006). Aboveground biomass allocation patterns within Mediterranean sub-shrubs: A quantitative analysis of seasonal dimorphism. *Flora - Morphology, Distribution, Functional Ecology of Plants* **201**:612–622.

Palamarev, E. (1989). Paleobotanical evidence of the Tertiary history and origin of the Mediterranean sclerophyll dendroflora. *Plant Systematics and Evolution* **162**:93–107.

Palumbi, S. (2001). Humans as the world's greatest evolutionary force. *Science* **293**:1786–1790.

Panou, A., J. Jacobs, and D. Panos. (1993). The endangered Mediterranean Monk Seal *Monachus monachus* in the Ionian Sea, Greece. *Biological Conservation* **64**:129–140.

Papachristou, T. G., A. S. Nastis, R. Mathur, and M. R. Hutchings. (2003). Effect of physical and chemical plant defences on herbivory: implications for Mediterranean shrubland management. *Basic and Applied Ecology* **4**: 395–403.

Papanastasis, V. P. (2007). Land abandonment and old field dynamics in Greece. In V. A. Cramer, and R. J. Hobbs (eds), *Old Fields. Dynamics and Restoration of Abandoned Farmland*, pp. 225–246. Island Press, Washington DC.

Papanastasis, V. P., S. Kyriakakis, and G. Kakais. (2002). Plant diversity in relation to overgrazing and burning in mountain mediterranean ecosystems. *Journal of Mediterranean Ecology* **3**:53–63.

Papayannis, T. (ed.). (2008). *Actions for Culture in Mediterranean Wetlands*. Med-INA, Athens.

Papazachos, B. C., and C. Papazachou. (2003). *The Earthquakes of Greece*. Ziti Publ., Thessaloniki.

Parmesan, C. (2006). Ecological and evolutionary responses to recent climate change. *Annual Revue of Ecology and Systematics* **37**:637–669.

Parmesan, C., and G. Yohe. (2003). A globally coherent finger-print of climate change impacts across natural systems. *Nature* **421**:37–42.

Parmesan, C., N. Ryrholm, C. Stefanescu, J. K. Hill, C. D. Thomas, H. Descimon, B. Huntley, L. Kaila, J. Kullberg, T. Tammaru *et al.* (1999). Polewards shifts in geographical ranges of butterfly species associated with regional warming. *Nature* **399**:579–583.

Pascal, M., O. Lorvelec, and J.-D. Vigne. (2006). *Invasions Biologiques et Extinctions*. Belin, Paris.

Pasquet, E. (1998). Plylogeny of the nuthatches of the *Sitta canadensis* group and its evolutionary and biogeographic implications. *Ibis* **140**:150–156.

Pasquet, E., and J. C. Thibault. (1997). Genetic differences among mainland and insular forms of the Citril Finch (*Serinus citrinella*). *Ibis* **139**:679–684.

Pastor, T., J. C. Garza, A. Aguilar, E. Tounta, and E. Androukaki. (2007). Genetic diversity and differentiation between the two remaining populations of the critically endangered Mediterranean monk seal. *Animal Conservation* **10**:461–469.

Pausas, J. G. (2004). Changes in fire and climate in the eastern Iberian Peninsula (Mediterranean Basin). *Climatic Change* **63**:337–350.

Pausas, J. G. (2006). Simulating Mediterranean landscape pattern and vegetation dynamics under different fire regimes. *Plant Ecology* **187**:249–259.

Pausas, J. G., C. Bladé, A. Valdecantos, J. Seva, D. Fuentes, J. Alloza, A. Vilagrosa, S. Bautista, J. Cortina, and R. Vallejo. (2004). Pines and oaks in the restoration of Mediterranean landscapes of Spain: New perspectives for an old practice — a review. *Plant Ecology* **171**:209–220.

Pausas, J. G., T. Marañón, M. Caldeira, and J. Pons. (2009). Natural regeneration. In J. Aronson, J. S. Pereira, and J. G. Pausas (eds), *Cork Oak Woodlands on the Edge: Ecology, Adaptive Management, and Restoration*, pp. 115–124. Island Press, Washington DC.

Pavlícek, T., and C. Csuzdi. (2006). Species richness and zoogeographic affinities of earthworms in Cyprus. *European Journal of Soil Biology* **42**:S111–S116.

Pearce, F., and A. Crivelli. (1994). *Caractéristiques Générales des Zones Humides Méditerranéennes*. MedWet, Arles.

Pearson, R. G., and T. P. Dawson. (2003). Predicting the impacts of climate change on the distribution of species: are bioclimatic envelope models useful? *Global Ecology and Biogeography* **12**:361–371.

Pellegrino, G., A. M. Palermo, M. E. Noce, F. Bellusci, and A. Musacchio. (2007). Genetic population structure in the Mediterranean Serapias vomeracea, a nonrewarding orchid group. Interplay of pollination strategy and stochastic forces? *Plant Systematics and Evolution* **263**:145–157.

Peng, C. H., J. Guiot, and E. Van Campo. (1998). Estimating changes in terrestrial vegetation and carbon storage: using palaeoecological data and models. *Quaternary Science Reviews* **17**: 719–735.

Peng, J., Y. Ronin, T. Fahima, M. S. Röder, Y. Li, E. Nevo, and A. Korol. (2003). Domestication quantitative trait loci in *Triticum dicoccoides*, the progenitor of wheat. *Proceedings of the National Academy of Sciences USA* **100**:2489–2494.

Peñuelas, J. (2008). Commentary: An increasingly scented world. *New Phytologist* **180**:735.

Peñuelas, J., and M. Boada. (2003). A global change-induced biome shift in the Montseny mountains (NE Spain). *Global Change Biology* **9**:131–140.

Peñuelas, J., I. Filella, and P. Comas. (2002). Changed plant and animal life cycles from 1952 to 2000 in the Mediterranean region. *Global Change Biology* **8**:531–544.

Pérès, J. M. (1967). The Mediterranean benthos. *Oceanography and Marine Biology: An Annual Review* **5**:449–533.

Pérès, J. M. (1985). History of the Mediterranean biota and the colonization of the depths. In R. Margalef (ed.), *Western Mediterranean*, pp. 198–232. Pergamon Press, Oxford.

Pérès, J. M., and J. Picard. (1964). Nouveau manuel de bionomie benthique de la Mer Méditerranée. *Recueil de Travaux de la Station Marine d'Endoume* **31**:5–137.

Pérez, T. (2008). *Impact of Climate Change on Biodiversity in the Mediterranean Sea*. UNEP-MAP-RAC/SPA, RAC/SPA, Tunis. www.rac-spa.org/dl/Perez%202008_en.pdf.

Pérez, T., J. Garrabou, S. Sartoretto, J. G. Harmelin, P. Francour, and J. Vacelet. (2000). Mortalité massive d'invertébrés marins: un événement sans précédent en Méditerranée nord-occidentale. *Comptes Rendus de l'Académie des Sciences de Paris (Sciences et Vie)* **323**: 853–865.

Pérez-Corona, M. E., M. C. P. Hernandez, and F. B. de Castro. (2006). Decomposition of alder, ash, and poplar litter in a Mediterranean riverine area. *Communications in Soil Science and Plant Analysis* **37**:1111–1125.

Perret-Boudouresque, M., and H. Seridi. (1989). *Inventaire des Algues Marines de l'Algérie*. GIS Posidonie, Marseille.

Petanidou, T., W. N. Ellis, N. S. Margaris, and D. Vokou. (1995). Constraints on flowering phenology in a phryganic (East Mediterranean shrub) community. *American Journal of Botany* **82**:607–620.

Petanidou, T., S. Athanasios, S. Kallimanis, J. Tzanopoulos, S. P. Sgardelis, and J. D. Pantis. (2008). Long-term observation of a pollination network: fluctuation in species and interactions, relative invariance of network structure and implications for estimates of specialization. *Ecology Letters* **11**:564–575

Peters, J., A. von den Driesch, and D. Helmer. (2005). The upper Eurphrates-Tigris basin: cradle of agro-pastoralism? In J. D. Vigne, J. Peters, and D. Helmer (eds), *The First Steps of Animal Domestication. New Archaeological Approaches*, pp. 96–124. Oxbow Books, Oxford.

Petit, C., and J. D. Thompson. (1998). Phenotypic selection and population differentiation in relation to habitat heterogeneity in *Arrhenatherum elatius* (Poaceae). *Journal of Ecology* **86**:829–840.

Petit, C., and J. D. Thompson. (1999). Species richness and ecological range in relation to ploidy level in the flora of the Pyrenees. *Evolutionary Ecology* **13**:45–66.

Petit, C., J. D. Thompson, and F. Bretagnolle. (1996). Phenotypic plasticity in relation to ploidy levels and corm production in the perennial grass *Arrhenaterum elatius*. *Canadian Journal of Botany* **74**:1964–1973.

Petit, R. J., E. Pineau, B. Demesure, R. Bacilieri, A. Ducouso, and A. Kremer. (1997). Chloroplast DNA foorptints of postglacial recolonization by oaks. *Proceedings of the National Academy of Sciences USA* **94**:9996–10001.

Petit, R. J., A. El Mousadik, and O. Pons. (1998). Identifying populations for conservation on the basis of genetic markers. *Conservation Biology* **12**:844–855.

Pfeffer, F. (1973). *Les Animaux Domestiques et leurs Ancêtres*. Bordas, Paris.

Pianka, E. R. (1989). Latitudinal gradients in species diversity. *Trends in Ecology & Evolution* **4**:223.

Piazzi, L., and D. Balata. (2008). The spread of *Caulerpa racemosa* var. *cylindracea* in the Mediterranean Sea: an example of how biological invasions can influence beta diversity. *Marine Environmental Research* **65**:50–61.

Pickett, S. T. A., and P. S. White (ed.). (1985). *The Ecology of Natural Disturbance and Patch Dynamics*. Academic Press, New York.

Pignatti, S. (1978). Evolutionary trends in mediterranean flora and vegetation. *Vegetatio* **37**:175–185.

Pillon, Y., M. F. Fay, A. B. Shipunov, and M. W. Chase. (2006). Species diversity versus phylogenetic diversity: a practical study in the taxonomically difficult genus *Dactylorhiza* (Orchidaceae). *Biological Conservation* **129**: 4–13.

Pineda, F. D., and J. Montalvo. (1995). Dehesa systems in the western Mediterranean. In P. Halladay, and D. A. Gilmour (eds), *Conserving Biodiversity Outside of Protected Areas. The Role of Traditional Agro-Ecosystems*, pp. 107–122. IUCN and AMA, Gland.

Pison, G. (2007). Tous les pays du monde (2007). *Bulletin Mensuel d'Information de l'Institut National d'Etudes démographiques, Population et Sociétés* **328**:1–5.

Plitmann, U. (1981). Evolutionary history of the old world lupines. *Taxon* **30**:430–437.

Pons, A. (1981). The history of the Mediterranean shrublands. In F. di Castri, D. W. Goodall, and R. L. Specht (eds), *Mediterranean-Type Shrublands, Collection Ecosystems of the World*, vol. 11, pp. 131–138. Elsevier, Amsterdam.

Pons, A. (1984). Les changements de la végétation de la région méditerranéenne durant le Pliocène et le Quaternaire en relation avec l'histoire du climat et de l'action de l'homme. *Webbia* **38**:427–439.

Pons, A., and P. Quézel. (1985). The history of the flora and vegetation and past and present human disturbance in the Mediterranean region. In C. Gomez-Campo (ed.), *Plant Conservation in the Mediterranean Area*, pp. 25–43. Dr. W. Junk, Dordrecht.

Por, F. D. (1964). A study of the Levantine and Pontic Harpacticoida (Copepoda, Crustacea). *Zoologische Verhandelingen* **64**:1–128.

Posillico, M., A. Meriggi, E. Pagnin, S. Lovari, and L. Russo. (2004). A habitat model for brown bear conservation and land use planning in the central Apennines. *Biological Conservation* **118**:141–150.

Potts, S. G., T. Petanidou, S. Roberts, C. O'Toole, A. Hulbert, and P. Willmer. (2006). Plant-pollinator biodiversity and pollination services in a complex Mediterranean landscape. *Biological Conservation* **129**:519–529.

Poulin, B., G. Lefebvre, and A. J. Crivelli. (2007). The invasive red swamp crayfish as a predictor of Eurasian bittern density in the Camargue, France. *Journal of Zoology* **273**:98–105.

Pourkheirandish, M., and T. Komatsuda. (2007). The importance of barley genetics and domestication in a global perspective. *Annals of Botany* **100**:999–1008.

Povz, M. D., D. Jesensek, P. Berrebi, and A. Crivelli. (1996). *The Marble Trout, Salmo trutta marmoratus, Cuvier 1817, in the Soca River Basin, Slovenia*. Tour du Valat Publication, Arles.

Preatoni, D., A. Mustoni, A. Martinoli, E. Carlini, B. Chiarenzi, S. Chiozzini, S. Van Dongen, L. A. Wauters, and G. Tosi. (2005). Conservation of brown bear in the Alps: space use and settlement behavior of reintroduced bears. *Acta Oecologica* **28**:189–197.

Prendergast, J. R., R. M. Quinn, J. H. Lawton, B. C. Eversham, and D. W. Gibbons. (1993). Rare species, the coincidence of diversity hotspots and conservation strategies. *Nature* **365**:335–337.

Presa, P., B. G. Pardo, P. Martinez, and L. Bernatchez. (2002). Phylogeographic congruence between mtDNA and rDNA ITS markers in brown trout. *Molecular Biology and Evolution* **19**:2161–2175.

Primack, R. B. (1993). *Essentials of Conservation Biology*. Sinauer Associates, Sunderland, MA.

Prodon, R., and J.-D. Lebreton. (1981). Breeding avifauna of a Mediterranean succession: the holm oak and cork oak series in the Eastern Pyrénées. I. Analysis and modelling of the structure gradient. *Oikos* **37**:21–38.

Prodon, R., R. Fons, and F. Athias-Binche. (1987). The impact of fire on animal communities in Mediterranean area. In L. Trabaud (ed.), *The Role of Fire in Ecological Systems*, pp. 121–157. SPB Academic Publishing, The Hague.

Provan, J., and K. D. Bennett. (2008). Phylogeographic insights into cryptic glacial refugia. *Trends in Ecology & Evolution* **23**:564–571.

Pullin, A. S. (ed.). (1995). *Ecology and Conservation of Butterflies*. Chapman and Hall, London.

Quéro, J. C., and J. J. Vayne. (1998). *Les Fruits de la Mer et Plantes Marines des Pêches Françaises*. Encyclopédies du naturaliste, Delacahaux et Niestlé/IFREMER éditeurs, Paris.

Questiau, S. (1999). Phytogeographical evidence of gene flow among common Crossbill (*Loxia curvirostra*) populations at the continental level. *Heredity* **83**:196–205.

Quézel, P. (1976a). Le dynamisme de la végétation en région méditerranéenne. *Collana Verde* **39**:375–391.

Quézel, P. (1976b). Les forêts du pourtour méditerranéen. *Note Technique MAB* **2**:9–34.

Quézel, P. (1978). Analysis of the flora of Mediterranean and Saharan Africa. *Annals of the Missouri Botanical Garden* **65**:479–534.

Quézel, P. (1985). Definition of the Mediterranean region and origin of its flora. In C. Gomez-Campo (ed.), *Plant Conservation in the Mediterranean Area*, pp. 9–24. Dr. W. Junk, Dordrecht.

Quézel, P. (1995). La flore du bassin méditerranéen: origine, mise en place, endémisme. *Ecologia Mediterranea* **21**:19–39.

Quézel, P. (2004). Large-scale post-glacial distribution of vegetation structures in the Mediterranean region. In S. Mazzoneli, G. di Pasquale, M. Mulligan, P. di Martino, and F. Rego (eds), *Recent Dynamics of the Mediterranean Vegetation and Landscape*, pp. 3–12. John Willey & Sons, Chichester.

Quézel, P., and F. Médail. (1995). La région circumméditerranéenne, centre mondial majeur de biodiversité végétale. In *Actes des 6èmes Rencontres de l'Agence Régionale pour l'Environnement, Provence-Alpes-Côte d'Azur*, pp. 152–160. Gap.

Quézel, P., and F. Médail. (2003). *Ecologie et Biogéographie des Forêts du Bassin Méditerranéen*. Elsevier, Paris.

Quézel, P., M. Barbéro, G. Bonin, and R. Loisel. (1990). Recent plant invasion in the Circum-Mediterranean region. In F. di Castri, A. J. Hansen, and M. Debussche (eds), *Biological Invasions in Europe and the Mediterranean Basin*, pp. 51–60. Kluwer Academic Publishers, Dordrecht.

Raga, J. A., and J. Pantoja (eds). (2004). *Proyecto Mediterráneo. Zonas de Especial Interés para la Conservación de los Cetáceos en el Mediterráneo Español*. Ministerio de Medio Ambiente, Organismo Autónomo Parques Nacionales, Madrid.

Rajabi-Maham, H., A. Orth, and F. Bonhomme. (2007). Phylogeography and post-glacial expansion of Mus musculus domesticus inferred from mitochondrial DNA coalescent, from Iran to Europe. *Molecular Ecology* **17**:627–641.

Ramade, F. (1997). *Conservation des Ecosystèmes Méditerranéens*. Economica, Paris.

Rambal, S. (1987). Evolution de l'occupation des terres et ressource en eau en région méditerranéenne karstique. *Journal of Hydrology* **93**:339–357.

Randi, E. (1996). A mitochondrial cytochrome B phylogeny of the *Alectoris* partridges. *Molecular Phylogenetics and Evolution* **6**:214–227.

Rasmont, P., A. Coppee, D. Michez, and T. De Meulemeester. (2008). An overview of the *Bombus terrestris* (L. 1758) subspecies (Hymenoptera: Apidae). *Annales De La Societe Entomologique De France* **44**:243–250.

Raunkiaer, C. (1934). *The Life Form of Plants and Statistical Plant Geography*. Oxford University Press, Oxford.

Raven, P. H. (1964). Catastrophic selection and edaphic endemism. *The American Naturalist* **98**:336–338.

Raven, P. H. (1973). The evolution of Mediterranean floras. In F. di Castri, and H. A. Mooney (eds), *Mediterranean-Type Ecosystems, Origin and Structure*, pp. 213–224. Springer-Verlag, Heidleberg.

Raven, P. H., and D. I. Axelrod. (1974). Angiosperm biogeography and past continental Movements. *Annals of the Missouri Botanical Garden* **61**:539–673.

Raven, P. H., and D. I. Axelrod. (1978). Origin and relationships of the California flora. *University of California Publications in Botany* **72**:1–115.

Razouls, C., F. de Bovée, J. Kouwenberg, and N. Desreumaux. (2005–9). *Diversité et répartition géographique chez les Copépodes planctoniques marins*. http://copepodes.obs-banyuls.fr.

Read, J., and G. D. Sanson. (2003). Characterizing sclerophylly: the mechanical properties of a diverse range of leaf types. *New Phytologist* **160**:81–99.

Real, R., A. L. Marquez, A. Estrada, R. Munoz, and J. M. Vargas. (2008). Modelling chorotypes of invasive vertebrates in mainland Spain. *Diversity and Distributions* **14**:364–373.

Reese, D. (1990). Tale of the pygmy hippo. *Cyprus View* **6**:50–53

Reifenberg, A. (1955). *The Struggle Between the Desert and the Sown*. The Publishing Department of the Jewish Agency, Jerusalem.

Reijnders, P. J. H. (1997). A mass mortality hits the already critically endangered Mediterranean Monk Seal. *Species* **29**:49–50.

Reille, M. (1992). New pollen-analytical researches in Corsica: the problem of *Quercus ilex* L. and *Erica arborea* L., the origin of *Pinus halepensis* Miller forests. *New Phytologist* **122**:359–278.

Reille, M., H. Triat, and J.-L. Vernet. (1980). Les témoignages des structures actuelles de végétation méditerranéenne durant le passé contemporain de l'action de l'homme. *Naturalia Monspeliensia* **23**:79–87.

Rey Benayas, J. M., J. M. Bullock, and A. C. Newton. (2008). Creating woodland islets to reconcile ecological restoration, conservation, and agricultural land use. *Frontiers in Ecology and the Environment* **6**:329–336.

Reyjol, Y., B. Hugueny, D. Pont, P. G. Bianco, U. Beier, N. Caiola, F. Casals, I. Cowx, A. Economou, T. Ferreira *et al.* (2006). Patterns in species richness and endemism of European freswater fish. *Global Ecology and Biogeography* **16**:65–75.

Rezaei, H. R. (2007). *Phylogénie Moléculaire du Genre Ovis (Mouton et Mouflon), Implications pour la Conservation du Genre et pour l'Origine de l'Espèce Domestique*. PhD Thesis. University of Grenoble.

Rhazi, M., P. Grillas, A. Charpentier, and F. Médail. (2004). Experimental management of Mediterranean temporary pools for conservation of the rare quillwort *Isoetes setacea*. *Biological Conservation* **118**:675–684.

Richardson, D. M., P. Pysek, M. Rejmánek, M. G. Barbour, F. D. Panetta, and C. J. West. (2000). Naturalization and invasion of alien plants: concepts and definitions. *Diversity and Distributions* **6**:93–107.

Richardson, D. M., P. W. Rundel, S. T. Jackson, R. O. Teskey, J. Aronson, A. Bytnerowicz, M. J. Wingfield, and Å. Procheåÿ. (2007). Human impacts in pine forests: past, present, and future. *Annual Review of Ecology, Evolution, and Systematics* **38**:275–297.

Roché, B. (1988). Bilan des premiers inventaires ichtyologiques du réseau hydrographique de la Corse. Remarques sur les espèces d'eaux courantes. *Bulletin d'Ecologie* **19**:235–245.

Roché, B., and J. Mattei. (1997). Les espèces animales introduites dans les eaux douces de Corse. *Bulletin Français de la Pêche et la Pisciculture* **344/345**:233–239.

Rodriguez, C. F., E. Bécares, M. Fernandez-Alaez, and C. Fernandez-Alaez. (2005). Loss of diversity and degradationof wetlands as a result of introducing exotic crayfish. *Biological Invasions* 7:75–85.

Rodríguez-Sánchez, F., and J. Arroyo. (2008). Reconstructing the demise of Tethyan plants: climate-driven range dynamics of *Laurus* since the Pliocene. *Global Ecology and Biogeography* 17:685–695.

Roll, U., T. Dayan, D. Simberloff, and M. Goren. (2007). Characteristics of the introduced fish fauna of Israel. *Biological Invasions* 9:813–824.

Romanyà, J., P. Casals, J. Cortina, P. Bottner, M.-M. Coûteaux, and V. R. Vallejo. (2000). CO2 efflux from a Mediterranean semi-arid forest soil. II. Effects of soil fauna and surface stoniness. *Biogeochemistry* 48:283–306.

Ros, J. D., J. Romero, E. Ballesteros, and J. M. Gili. (1985). Diving in blue water. The benthos. In R. Margalef (ed.), *Western Mediterranean*, pp. 233–295. Pergamon Press, Oxford.

Rosenbaum, G., G. S. Lister, and C. Duboz. (2002a). Reconstruction of the tectonic evolution of the western Mediterranean since the Oligocene. *Journal of the Virtual Explorer* 8:107–130.

Rosenbaum, G., G. S. Lister, and C. Duboz. (2002b). Relative motions of Africa, Iberia and Europe during Alpine orogeny. *Tectonophysics* 359:117–129.

Rosenberg, A. A. (2008). Aquaculture: The price of lice. *Nature* 451:23–24.

Ross, J. D., and C. Sombrero. (1991). Environmental control of essential oil production in Mediterranean plants. In J. B. Harborne, and F. A. Tomas-Barberan (eds), *Ecological Chemistry and Biochemistry of Plant Terpenoids*, pp. 83–94. Clarendon Press, Oxford.

Rouchy, J. M., and A. Caruso. (2006). The Messinian salinity crisis in the Mediterranean basin: A reassessment of the data and an integrated scenario. *Sedimentary Geology* 188–189:35–67.

Rouzaud, P. (1982). Selected aspects of coastal aquaculture development. In A. G. Coche (ed.), *Coastal Aquaculture: Development Perspectives in Africa and Case Studies from Other Regions*. CIFA Technical Paper - CIFA/CPCA/T9.

Roy, J. (1990). In search of the characteristics of plant invaders. In F. di Castri, A. J. Hansen, and M. Debussche (eds), *Biological Invasions in Europe and the Mediterraean Basin*, pp. 333–352. Kluwer Academic Publishers, Dordrecht.

Roy, J., and L. Sonié. (1992). Germination and population dynamics of *Cistus* species in relation to fire. *Journal of Applied Ecology* 29:647–655.

Rubio, J. L., and A. Calvo (eds). (1996). *Soil Degradation and Desertification in Mediterranean Environments*. Geoforma Ediciones, Logroño.

Rubolini, D., R. Ambrosini, M. Caffi, P. Brichetti, S. Armiraglio, and N. Saino. (2007). Long-term trends in first arrival and first egg laying dates of some migrant and resident bird species in northern Italy. *International Journal of Biometeorology* 51:553–563.

Rundel, P. W., G. Montenegro, and F. M. Jaksic. (1998). *Landscape Disturbance and Biodiversity in Mediterranean-Type Ecosystems*. Springer-Verlag, Berlin.

Saarma, U., S. Y. W. Ho, O. G. Pybus, M. Kaljuste, I. L. Tumanov, I. Kojola, A. A. Vorobiev, N. I. Markov, A. P. Saveljev, H. Valdmann et al. (2007). Mitogenetic structure of brown bears (*Ursus arctos* L.) in northeastern Europe and a new time frame for the formation of European brown bear lineages. *Molecular Ecology* 16:401–413.

Sablin, M. V., and G. A. Khlopachev. (2002). The earliest ice age dogs: evidence from Eliseevichi 1. *Current Anthropology* 43:795–799.

Sadoul, N. (1996). *Dynamique Spatiale et Temporelle des Colonies de Charadriiformes dans les Salins de Camargue: Implications pour la Conservation*. PhD Thesis. University of Montpellier II.

Saied, A., J. Gebauer, K. Hammer, and A. Buerkert. (2008). *Ziziphus spina-christi* (L.) Willd.: a multipurpose fruit tree. *Genetic Resources and Crop Evolution* 55:929–937.

Sakshaug, E., A. Bricaud, Y. Dandonneau, P. G. Falkowski, D. A. Kiefer, L. Legendre, A. Morel, J. Parslow, and M. Takahashi. (1997). Parameters of photosynthesis: definitions, theory and interpretation of results. *Journal of Plankton Research* 19:1637–1670.

Sala, O. E., F. S. Chapin, III, J. J. Armesto, E. Berlow, J. Bloomfield, R. Dirzo, E. Huber-Sanwald, L. F. Huenneke, R. B. Jackson, A. Kinzig et al. (2000). Global biodiversity scenarios for the year 2100. *Science* 287:1770–1774.

Salamini, F., H. Özkan, A. Brandolini, R. Schäfer-Pregl, and M. Martin. (2002). Genetics and geography of wild cereal domestication in the Near East. *Nature Reviews Genetics* 3:429–441.

Salat, J., and M. Pascual. (2002). The oceanographic and meteorological station at L'Estartit (NW Mediterranean). In F. Briand (ed.), *Tracking Long-Term Hydrological Change in the Mediterranean Sea*, pp. 31–34. CIESM Workshop Series No. 16. CIESM, Monaco.

Salleo, S., A. Nardini, and M. A. Gullo. (1997). Is sclerophylly of Mediterranean evergreens an adaptation to drought? *New Phytologist* 135:603–612.

Sami, A. (2004). Competitive asymmetry, foraging area size and coexistence of annuals. *Oikos* 104:51–58.

Sánchez Goñi, M. F., M. F. Loutre, M. Crucifix, O. Peyron, L. Santos, J. Duprat, B. Malaizé, J. L. Turon, and J. P. Peypouquet. (2005). Increasing vegetation and climate gradient in Western Europe over the Last Glacial Inception (122-110 ka): data-model comparison. *Earth and Planetary Science Letters* 231:111–130.

Santos-Guerra, A. (2001). Flora vascular nativa. In J. M. Fernandez-Palacios, and J. Martın-Esquivel (eds), *Naturaleza de las Islas Canarias*, pp. 185–192. Turquesa, Santa Cruz de Tenerife.

Sanz, J. J., J. Potti, J. Moreno, S. Merino, and O. Frias. (2003). Climate change and fitness components of a migratory bird breeding in the Mediterranean region. *Global Change Biology* **9**:461–472.

Sarà, M., E. Bellia, and A. Milazzo. (2006). Fire disturbance disrupts co-occurrence patterns of terrestrial vertebrates in Mediterranean woodlands. *Journal of Biogeography* **33**:843–852.

Sari, D. (1977). *L'Homme et l'Erosion dans l'Ouarsenis*. University of Alger.

Sarrazin, F., and R. Barbault. (1996). Reintroduction: challenge and lessons for basic ecology. *Trends in Ecology & Evolution* **11**:474–478.

Sattout, E. J., S. N. Talhouk, and P. D. S. Caligari. (2007). Economic value of cedar relics in Lebanon: An application of contingent valuation method for conservation. *Ecological Economics* **61**:315–322.

Sax, D. F., and S. D. Gaines. (2008). Species invasions and extinction: The future of native biodiversity on islands. *Proceedings of the National Academy of Sciences USA* **105**:11490–11497.

Scarascia-Mugnozza, G., G. Oswald, H. Piussi, and K. Radoglou. (2000). Forest of the Mediterranean region: gaps in knowledge and research needs. *Forest Ecology and Management* **132**:97–109.

Schaefer, H.-C., W. Jetz, and K. Böhning-Gaese. (2008). Impact of climate change on migratory birds: community reassembly versus adaptation. *Global Ecology and Biogeography* **17**:38–49.

Schemske, D. W. (1981). Floral convergence and pollinator sharing in two bee pollinated tropical herbs. *Ecology* **62**:946–954.

Schiller, G., S. Cohen, E. D. Ungar, Y. Moshe, and N. Herr. (2007). Estimating water use of sclerophyllous species under East-Mediterranean climate: III. Tabor oak forest sap flow distribution and transpiration. *Forest Ecology and Management* **238**:147–155.

Schluter, D., and R. E. Ricklefs. (1993). Convergence and the regional component of species diversity. In R. E. Ricklefs, and D. Schluter (eds), *Species Diversity in Ecological Communities*, pp. 230–240. University of Chicago Press, Chicago.

Schmekel, L., and A. Portmann. (1982). *Opisthobranchia des Mittelmeeres: Nudibranchia und Saccoglossa*. Springer-Verlag, Berlin.

Schneider, S. H., and T. L. Root. (1996). Ecological implications of climate change will include surprises. *Biodiversity and Conservation* **5**:1109–1119.

Schulze, E. D. (1982). Plant life forms and their carbon, water and nutrient relations. In O. L. Lange, P. S. Nobel, C. B. Osmond, and H. Ziegler (eds), *Physiological Plant Ecology II. Water Relations and Carbon Assimilation, Encyclopedia of plant physiology, New Series, vol. 12 B*, pp. 615–677. Springer-Verlag, Heidelberg.

Scott, J. M., B. Csuti, and F. Davis. (1991). Gap analysis: an application of Geographic Information Systems for wildlife species. In D. J. Decker, M. E. Krasny, R. Goff, C. R. Smith, and D. W. Gross (eds), *Challenges in the Conservation of Biological Resources: A Practitioner's Guide*, pp. 167–179. Westview Press, Boulder, CO.

Sealy, J. R. (1949). *Arbutus unedo*. *Journal of Ecology* **37**:365–388.

Seligman, N. G., and A. Perevolotsky. (1994). Has intensive grazing by domestic livestock degraded Mediterranean Basin rangelands? In M. Arianoutsou, and R. H. Groves (eds), *Plant-Animal Interactions in Mediterranean-Type Ecosystems*, pp. 93–104. Kluwer Academic Publishers, Dordrecht.

Seoane, J., and L. M. Carrascal. (2008). Interspecific differences in population trends of Spanish birds are related to habitat and climatic preferences. *Global Ecology and Biogeography* **17**:111–121.

SER (Society for Ecological Restoration). (2002). *The SER Primer on Ecological Restoration*. SER, Science & Policy Working Group, Tucson, AZ. www.ser.org.

Sezik, E. (1989). Turkish orchids and saleps. In *Orchidées Botaniques du Monde Entier*, pp. 181–189. Société Française d'Orchidophilie, Paris.

Shay, C. T., J. M. Shay, and J. Zwiazek. (1992). Paleobotanical investigations at Kommos, Crete. In C. A. Thanos (ed.), *Plant-Animal Interactions in Mediterranean-Type Ecosystems*, pp. 382–389. University of Athens.

Sher, A., D. E. Goldberg, and A. Novoplansky. (2004). The effect of mean and variance in resource supply on survival of annuals from Mediterranean and desert environments. *Oecologia* **141**:353–362.

Shirihai, H., G. Gargallo, and A. J. Helbig. (2001). *Sylvia Warblers*. Helm, London.

Shitzer, D., I. Noy-Meir, and D. Milchunas. (2008). The role of geologic grazing refuges in structuring Mediterranean grassland plant communities. *Plant Ecology* **198**:135–147.

Shmida, A. (1981). Mediterranean vegetation in California and Israel: similarities and differences. *Israel Journal of Botany* **30**:105–123.

Shmida, A., and S. Ellner. (1983). Seed dispersal on pastoral grazers in open Mediterranean chaparral, Israel. *Israel Journal of Botany* **32**:147–159.

Shmida, A., and R. Whittaker. (1984). Convergence and non-convergence of mediterranean type communities in the old and the new world. *Tasks for Vegetation Science* **13**:5–11.

Shmida, A., and J. Aronson. (1986). Sudanian elements in the flora of Israel. *Annals of the Missouri Botanical Garden* **73**:1–28.

Shmida, A., and A. Dafni. (1989). Blooming strategies, flower size and advertising in the 'lily-group' geophytes in Israel. *Herbertia* **45**:111–123.

Shmida, A., and M. J. A. Werger. (1992). Growth form diversity on the Canary Islands. *Vegetatio* **102**:183–199.

Siles, G., P. J. Rey, J. M. Alcántara, and J. M. Ramírez. (2008). Assessing the long-term contribution of nurse plants to restoration of Mediterranean forests through Markovian models. *Journal of Applied Ecology* **45**:1790–1798.

Siljak-Yakovlev, S., V. Stevanovic, M. Tomasevic, S. C. Brown, and B. Stevanovic. (2008). Genome size variation and polyploidy in the resurrection plant genus *Ramonda*: cytogeography of living fossils. *Environmental and Experimental Botany* **62**:101–112.

Simberloff, D. (1981). Community effects of introduced species. In M. H. Nitecki (ed.), *Biotic Crises in Ecological and Evolutionary Time*, pp. 51–81. Academic Press, New York.

Simberloff, D. (2003). Confronting introduced species: a form of xenophobia? *Biological Invasions* **5**:179–192.

Simmons, A. H. (1988). Extinct pygmy hippopotamus and early man in Cyprus. *Nature* **333**:554–557.

Simmons, A. H. (1991). Humans, island colonization and Pleistocene extinctions in the Mediterranean: the view from Akrotiri Aetokremnos, Cyprus. *Antiquity* **65**:857–868.

Simon, A., D. Thomas, J. Blondel, P. Perret, and M. M. Lambrechts. (2004). The effects of ectoparasites (*Protocalliphora* spp.) and food abundance on metabolic capacity of nestlings. *Physiological and Biochemical Zoology* **77**:492–591.

Simpson, B. B., and J. L. Neff. (1981). Floral rewards: alternatives to pollen and nectar. *Annals of the Missouri Botanical Garden* **68**:301–322.

Sindaco, R., and V. K. Jeremcenko. (2008). *The Reptiles of the Western Palearctic*. Societas Herpetologica Italica – I, Ed. Belvedere, Latina.

Sirami, C., L. Brotons, I. Burfield, J. Fonderflick, and J. L. Martin. (2008). Is land abandonment having an impact on biodiversity? A meta-analytical approach to bird distribution changes in the north-western Mediterranean. *Biological Conservation* **141**:450–459.

Sket, B. (1997). Distribution of *Proteus* (Amphibia: Urodela: Proteidae) and its possible explanation. *Journal of Biogeography* **24**:263–280.

Smith, K. G., and W. R. T. Darwall. (2006). *The Status and Distribution of Freshwater Fish Endemic to the Mediterranean Basin*. IUCN, Gland.

Snogerup, S. (1971). Evolutionary and plant geographical aspects of chasmophytic communities. In P. H. Davis, P. C. Harper, and I. C. Hedge (eds), *Plant Life of Southwest Asia*, pp. 157–170. The Botanical Society of Edinburgh, Edinburgh.

Socias i Company, R. (1990). Breeding self-compatible almonds. *Plant Breeding Review* **8**:313–338.

Sondaar, P. Y. (1977). Insularity and its effects on mammal evolution. In P. C. Goody, and B. M. Heckt (eds), *Major Patterns of Vertebrate Evolution*, pp. 671–707. Plenum Publishing Corporation, New York.

Sonnante, G., D. Pignone, and K. Hammer. (2007). The domestication of artichoke and cardoon: From Roman times to the genomic age. *Annals of Botany* **100**:1095–1100.

Soulier, A. (1993). *Le Languedoc pour Héritage*. Presses du Languedoc, Montpellier.

Specht, R. L., and P. W. Rundel. (1990). Sclerophylly and foliar nutrient status of mediterranean-climate plant communities in southern Australia. *Australian Journal of Botany* **38**:459–474.

Stamps, J. A., and M. Buechner. (1985). The territorial defense hypothesis and the ecology of insular vertebrates. *Quarterly Review of Biology* **60**:155–181.

Stebbins, G. L. (1971). *Chromosomal Evolution in Higher Plants*. E. Arnold, London.

Stebbins, G. L., and D. Zohary. (1959). Cytogenetic and evolutionary studies in the genus *Dactylis*. I. Morphology, distribution and interrelationships of the diploid subspecies. *University of California at Berkeley Publication in Botany* **31**:1.

Stefanescu, C., J. Peñuelas, and I. Filella. (2003). Effects of climatic change on the phenology of butterflies in the northwest Mediterranean Basin. *Global Change Biology* **9**:1494–1506.

Stevens, J., and P. R. Last. (1998). Sharks, rays and chimaeras. In J. R. Paxton, and W. N. Eschmeyer (eds), *Encyclopedia of Fishes*, pp. 60–69. Academic Press, San Diego, CA.

Stiner, M., N. D. Munro, T. A. Surovell, E. Tchernov, and O. Bar-Yosef. (1999). Paleolithic population growth pulses evidenced by small animal exploitation. *Science* **283**:190–194.

Suc, J. P. (1980). *Contribution à la Connaissance du Pliocène et du Pléistocène Inférieur des Régions Méditerranéennes d'Europe Occidentale par l'Analyse Palynologique des Dépôts du Languedoc-Roussillon (Sud de la France) et de la Catalogne (Nord-Est de l'Espagne)*. PhD Thesis. University of Montpellier II.

Suc, J. P. (1984). Origin and evolution of the Mediterranean vegetation and climate in Europe. *Nature* **307**: 429–432.

Sunding, P. (1979). Origins of the Macaronesian flora. In D. Bramwell (ed.), *Plants and Islands*, pp. 13–40. Academic Press, London.

Svenning, J. C., and F. Skov. (2007). Ice age legacies in the geographical distribution of tree species richness in Europe. *Global Ecology and Biogeography* **16**:234–245.

Taberlet, P., and J. Bouvet. (1994). Mitochondrial DNA polymorphism, pylogeography and conservation genetics of the Brown Bear *Ursus arctos* in Europe. *Proceedings of the Royal Society of London Series B Biological Sciences* **225**:195–200.

Taberlet, P., and R. Cheddadi. (2002). Quaternary refugia and persistence of biodiversity. *Science* **297**:2009–2010.

Taberlet, P., L. Fumagali, A. G. Wust-Saucy, and J. F. Cosson. (1998). Comparative phylogeography and postglacial colonization routes in Europe. *Molecular Ecology* **7**:453–464.

Taberlet, P., A. Valentini, H. R. Rezaei, S. Naderi, F. Ponpanon, R. Negrini, and P. Ajmone-Marsan. (2008). Are cattle, sheep, and goats endangered species? *Molecular Ecology* **17**:275–284.

Tamisier, A. (1971). Le régime alimentaire des sarcelles d'hiver *Anas crecca* L. en Camargue. *Alauda* **19**:1–31.

Tamisier, A., and P. Grillas. (1994). A review of habitat changes in the Camargue: an assessment of the effects of the loss of biological diversity on the wintering waterfowl community. *Biological Conservation* **70**: 39–47.

Tanno, K., and G. Willcox. (2006). How fast was wild wheat domesticated? *Science* **311**:1886.

Tarayre, M., J. D. Thompson, J. Escarré, and Y. B. Linhart. (1995). Intra-specific variation in the inhibitory effects of *Thymus vulgaris* (Labiatae) monoterpenes on seed germination. *Oecologia* **101**:110–118.

Tarayre, M., P. Saumitou-Laprade, J. Cuguen, D. Couvet, and J. D. Thompson. (1997). A comparison of the spatial genetic structure of cytoplasmic (cpDNA) and nuclear (allozymes) markers within and among populations of the gynodioecious *Thymus vulgaris* L. (Labiatae) in southern France. *American Journal of Botany* **84**:1675–1684.

Tchernov, E. (1984). Faunal turnover and extinction rate in the Levant. In P. S. Martin, and R. G. Klein (eds), *Quaternary Extinctions*, pp. 528–552. University Arizona Press, Tucson, AZ.

Tchernov, E. (1992). Eurasian-African biotic exchanges through the Levantine corridor during the Neogene and Quaternary. *Courier Forschungsinstitut Senckenberg* **153**:103–123.

Terral, J.-F., N. Alonso, B. Capdevilla, N. Chatti, L. Fabre, G. Fiorentino, P. Marinval, G. Perez Jorda, B. Pradat, N. Rovira, and P. Alibert. (2004). Historical biogeography of olive domestication (*Olea europaea* L.) as revealed by geometrical morphometry applied to biological and archaeological material. *Journal of Biogeography* **31**: 63–77.

Terrasse, M. (1996). Réintroduction de rapaces dans l'aire Méditerranenne de répartition. In J. Muntaner, and J. Mayol (eds), *Biologia y Conservacion de las Rapaces Mediterraneas, 1994*, pp. 251–259. Monografias, no. 4. SEO, Madrid.

Thanos, C. A. (1992). Theophrastus on plant-animal interactions. In C. A. Thanos (ed.), *Plant-Animal Interactions in Mediterranean-Type Ecosystems*, pp. 1–5. University of Athens.

Thébaud, C., A. C. Finzi, L. Affre, M. Debussche, and J. Escarré. (1996). Assessing why two introduced *Coniza* differ in their ability to invade Mediterranean old fields. *Ecology* **77**:791–804.

Thellung, A. (1908–10). La flore adventice de Montpellier. *Bulletin de la Société des Sciences Naturelles de Cherbourg* **38**:57–72.

Thevenot, M., R. Vermon, and P. Bergier. (2003). *The Birds of Morocco*. B.O.U. Check-List no. 20, British Ornithologists' Union & British Ornithologists' Club.

Thirgood, J. V. (1981). *Man and the Mediterranean Forest*. Academic Press, New York.

Thomas, C. D., A. Cameron, R. E. Green, M. Bakkenes, L. J. Beaumont, Y. C. Collingham, B. F. N. Erasmus, F. Ferreira de Siqueira, A. Grainger, L. Hannah *et al.* (2004). Extinction risk from climate change. *Nature* **427**: 145–148.

Thomas, J. A., G. W. Elmes, J. C. Wardlaw, and M. Woyciechowski. (1989). Host specificity among *Maculinea* butterflies in *Myrmica* ant nests. *Oecologia* **79**:452–457.

Thompson, J. D. (1999). Population differentiation in Mediterranean plants: insights into colonization history and the evolution and conservation of endemic species. *Heredity* **82**:229–236.

Thompson, J. D. (2001). How do visitation patterns vary among pollinators in relation to floral display and floral design in a generalist pollination system? *Oecologia* **126**:386–394.

Thompson, J. D. (2002). Population structure and the spatial dynamics of genetic polymorphism in thyme. In E. Stahl-Biskup, and F. Saez (eds), *Thyme: The Genus Thymus*, pp. 44–74. Taylor and Francis, London.

Thompson, J. D. (2005). *Plant Evolution in the Mediterranean*. Oxford University Press, Oxford.

Thompson, J. D., and M. Tarayre. (2000). Exploring the genetic basis and proximate causes of variation in female fertility advantage in gynodioecious *Thymus vulgaris*. *Evolution* **54**:1510–1520.

Thompson, J. D., D. Manicacci, and M. Tarayre. (1998). Thirty five years of thyme: a tale of two polymosphisms. *BioScience* **48**:805–815.

Thompson, J. D., P. Gauthier, J. Amiot, B. Ehlers, C. Collin, J. Fossat, V. Barrios, F. Arnaud-Miramont, K. Keefover-Ring, and Y. Linhart. (2007). Ongoing adaptation to Mediterranean Climate extremes in a chemically polymorphic plant. *Ecological Monographs* **77**:421–439.

Thuiller, W. (2007). Biodiversity: climate change and the ecologist. *Nature* **448**:550–552.

Thuiller, W., L. Brotons, M. B. Araújo, and S. Lavorel. (2004). Effects of restricting environmental range of data to project current and future species distributions. *Ecography* **27**:165–172.

Thuillier, W., D. M. Richardson, P. Pysek, G. F. Midgley, G. O. Hugues, and M. Rouget. (2005). Niche-based modelling as a tool for predicting the risk of alien plant invasions at a global scale. *Global Change Biology* **11**:2234–2250.

Tielbörger, K., and R. Kadmon. (2000). Indirect effects in a desert plant community: is competition among annuals more intense under shrub canopies? *Plant Ecology* **150**:53–63.

Tomaselli, R. (1981). Main physiognomic types and geographic distribution of shrub systems related to mediterranean climates. In F. di Castri, D. W. Goodall, and R. L. Specht (eds), *Mediterranean-Type Shrublands, Collection Ecosystems of the World, vol. 11*, pp. 95–106. Elsevier, Amsterdam.

Torre, I., and M. Díaz. (2004). Small mammal abundance in Mediterranean post-fire habitats: a role for predators? *Acta Oecologica* **25**:137–142.

Tortonese, E. (1985). Distribution and ecology of endemic elements in the Mediterranean faunas (fishes and echinoderms). In M. Moraitou-Apostolopoulo, and V. Kiortsis (eds), *Mediterranean Marine Ecosystems*, pp. 57–83. Plenum Press, New York.

Toumi, L., and R. Lumaret. (1998). Allozyme variation in cork oak (*Quercus suber* L.): the role of phylogeography, genetic introgression by other Mediterranean oak species and human activities. *Theoretical Applied Genetics* **97**:647–656.

Toumi, L., and R. Lumaret. (2001). Allozyme characterisation of four Mediterranean evergreen oak species. *Biochemical Systematics and Ecology* **29**:799–817.

Toutain, F., A. Diagne, and F. Le Tacon. (1988). Possiblités de modification du type d'humus et d'amélioration de la fertilité des sols à moyen terme en hêtraie par apport d'éléments minéraux. *Revue Forestière Française* **40**:99–107.

Traba, J., S. Arrieta, J. Herranz, and M. C. Clamagirand. (2006). Red fox (*Vulpes vulpes* L.) favour seed dispersal, germination and seedling survival of Mediterranean hackberry (*Celtis australis* L.). *Acta Oecologica* **30**: 39–45.

Trabaud, L. (1981). Man and fire: impacts on mediterranean vegetation. In F. di Castri, D. W. Goodall, and R. L. Specht (eds), *Mediterranean-type Shrublands*, pp. 523–537. Collection Ecosystems of the World, vol. 11. Elsevier, Amsterdam.

Trabaud, L., and J. Oustric. (1989). Heat requirements for seed germination of three *Cistus* species in the garrigue of southern France. *Flora* **183**:321–325.

Trabaud, L., and R. Prodon. (2002.) *Fire in Biological Processes*. Backhuys Publishers, Leiden.

Trabaud, L. V., N. L. Christensen, and A. M. Gill. (1993). Historical biogeography of fire in temperate and mediterranean ecosystems. In P. J. Crutzen, and J. G. Goldammer (eds), *Fire in the Environment: Its Ecological, and Atmospheric Importance*, pp. 277–295. John Wiley and Sons, New York.

Travé, J. (2000). La Réserve naturelle de la Massane. Un exemple de forêt ancienne protégée. *Forêt méditerranéenne* **21**:278–282.

Traveset, A. (1992). Production of galls in *Phillyrea angustifolia* induced by cecidomyiid flies. In C. A. Thanos (ed.), *Plant-Animal Interactions in Mediterranean-Type Ecosystems*, pp. 198–204. University of Athens.

Traveset, A., G. Brundu, L. Carta, I. Mprezetou, P. Lambdon, M. Manca, F. Médail, E. Moragues, J. Rodríguez-Pérez, A.-S. Siamantziouras *et al.* (2008). Consistent performance of invasive plant species within and among islands of the Mediterranean basin. *Biological Invasions* **10**: 847–858.

Tremblay, I., D. W. Thomas, M. M. Lambrechts, J. Blondel, and P. Perret. (2003). Variation in blue tit breeding performance across gradients in habitat richness. *Ecology* **84**:3033–3043.

Triat-Laval, H. (1979). Histoire de la forêt provençale depuis 15000 ans d'après l'analyse pollinique. *Forêt méditerranéenne* **1**:19–24.

Tsigenopoulos, C. S., and P. Berrebi. (2000). Molecular phylogeny of North Mediterranean freshwater barbs (Genus *Barbus*: Cyprinidae) inferred from cytochrome b sequences: biogeographic and systematic implications. *Molecular Phylogenetics and Evolution* **14**:165–179.

Tsounis, G. (2009). Jewel of the deep. *Natural History* **April**:130–135.

Tsounis, G., S. Rossi, J. M. Gili, and W. Arntz. (2006). Population structure of an exploited benthic cnidarian: the case study of red coral (*Corallium rubrum* L.). *Marine Biology* **149**:1059–1070.

Tsounis, G., S. Rossi, J. M. Gili, and W. Arntz. (2007). Red coral fishery at the Costa Brava (NW Mediterranean): case study of an overharvested precious coral. *Ecosystems* **10**:975–986.

Tucker, G. M., and M. F. Heath. (1994). *Birds in Europe: Their Conservation Status*. BirdLife Conservation Series no. 3. BirdLife International, Cambridge.

Turley, C. (1999). The changing Mediterranean Sea – a sensitive ecosystem? *Progress in Oceanography* **44**: 387–400.

Tyler, P. A., and J. D. Gage. (1984). The reproductive biology of Echinothuriid and Cidarid sea urchins from the deep sea. *Marine Biology* **80**:63–74.

UNESCO. (1963). Bioclimatical map of the Mediterranean zone. *Arid Zone Research* **21**:1–60.

Urbieta, I. R., M. A. Zavala, and T. Marañón. (2008). Human and non-human determinants of forest composition in southern Spain: evidence of shifts towards cork oak dominance as a result of management over the past century. *Journal of Biogeography* **35**:1688–1700.

Valdiosera, C. E., N. Garcia, N. Anderung, L. Dalen, E. Crégure-Bonnoure, R.-D. Kalhke, M. Stiller, M. Brandström, M. G. Thomas, J. L. Arsuagua *et al.* (2007). Staying out in the cold: glacial refugia and mitochondrial DNA phylogeography in ancient European brown bears. *Molecular Ecology* **11**:5140–5148.

Valeria Ruggiero, M., and G. Procaccini. (2004). The rDNA ITS region in the Lessepsian marine angiosperm *Halophila stipulacea* (Forssk.) Aschers. (Hydrocharitaceae): intragenomic variability and putative pseudogenic sequences. *Journal of Molecular Evolution* **58**: 115–121.

Valéry, L., H. Fritz, J.-C. Lefeuvre, and D. Simberloff. (2008). In search of a real definition of the biological invasion phenomenon itself. *Biological Invasions* **10**:1345–1351.

Valière, N., L. Fumagalli, L. Gielly, C. Miquel, B. Lequette, M.-L. Poulle, J.-M. Weber, R. Arlettaz, and P. Taberlet. (2003). Long-distance wolf recolonization of France and Switzerland inferred from non-invasive genetic sampling over a period of 10 years. *Animal Conservation* **6**:83–92.

Vallauri, D. (1997). *Dynamique de la Restauration Forestière des Substrats Marneux avec Pinus nigra J.F. Arnold ssp. nigra dans le Secteur Haut Provençal*. PhD Thesis. University of Marseille 3–CEMAGREF.

Vallauri, D. (1998). Towards a long term ecological framework for forest restoration programs. An illustration from restoration for erosion control on badlands in south-western Alps (France). *Ecologie* **29**: 189–192.

Vallauri, D., J. Aronson, and M. Barbero. (2002). An Analysis of Forest Restoration 120 Years after Reforestation on Badlands in the Southwestern Alps. *Restoration Ecology* **10**:16–26.

Vallejo, R., and J. A. Alloza. (1998). The restoration of burned lands: the case of eastern Spain. In J. M. Moreno (ed.), *Large Forest Fires*, pp. 91–108. Backhuys Publishers, Leiden.

Vallejo, R., J. Aronson, J. G. Pausas, J. S. Pereira, and C. Fontaine. (2009). The way forward. In J. Aronson, J. S. Pereira, and J. G. Pausas (eds), *Cork Oak Woodlands on the Edge: Ecology, Adaptive Management, and Restoration*, pp. 235–246. Island Press, Washington DC.

Vanderpoorten, A., N. Devos, B. Goffinet, O. J. Hardy, and A. J. Shaw. (2008). The barriers to oceanic island radiation in bryophytes: insights from the phylogeography of the moss Grimmia montana. *Journal of Biogeography* **35**:654–663.

Vannière, B., D. Colombaroli, E. Chapron, A. Leroux, W. Tinner, and M. Magny. (2008). Climate versus human-driven fire regimes in Mediterranean landscapes: the Holocene record of Lago dell'Accesa (Tuscany, Italy). *Quaternary Science Review* **27**:1181–1196.

van Valen, L. (1973). A new evolutionary law. *Evolutionary Theory* **1**:1–30.

Vaudour, J. (1979). *La Région de Madrid. Altérations, Sols et Paléosols*. PhD Thesis. University of Aix-Marseille.

Vavilov, N. I. (1935). Botanical-geographical basis of breeding. In N. I. Vavilov (ed.), *Origin and Geography of Cultivated Plants*, pp. 288–333. Nauka, Leningrad.

Vavilov, N. I. (1951). The origin, variety, immunity and breeding of cultivated plants. *Chronica Botanica* **13**: 1–366.

Vazquez, A., J. M. Garcia Del Barrio, M. Ortega Quero, and O. Sanchez Palomares. (2006). Recent fire regime in peninsular Spain in relation to forest potential productivity and population density. *International Journal of Wildland Fires* **15**:397–405.

Véla, E., and S. Benhouhou. (2007). Évaluation d'un nouveau point chaud de biodiversité végétale dans le Bassin méditerranéen (Afrique du Nord). *Comptes Rendus Biologies* **330**:589–605.

Verdú, M., P. Dávila, P. García-Fayos, N. Flores-Hernández, and A. Valiente-Banuet. (2003). 'Convergent' traits of mediterranean woody plants belong to pre-mediterranean lineages. *Biological Journal of the Linnean Society* **78**:415–427.

Verdú, J. R., and E. Galante. (2002). Climatic stress, food availability and human activity as determinants of endemism patterns in the Mediterranean region: the case of dung beetles (Coleoptera, Scarabaeoidea) in the Iberian Peninsula. *Diversity and Distributions* **8**: 259–274.

Vergé-Franceschi, M. (2002). *Dictionnaire d'Histoires Maritimes*. Laffont.

Vergés, A., T. Alcoverro, and E. Ballesteros. (2009). Role of fish herbivory in structuring the vertical distribution of canopy algae *Cystoseira* spp in the Mediterranean Sea. *Marine Ecology Progress Series* **375**:1–11.

Verlaque, R., A. Aboucaya, M. A. Cardona, and J. Contandriopoulos. (1991). Quelques exemples de spéciation insulaire en Méditerranée occidentale. *Botanica Chronica* **10**:137–154.

Verlaque, R., F. Médail, and A. Aboucaya. (2001). Valeur prédictive des types biologiques pour la conservation de la flore méditerranéenne. *Comptes Rendus de l'Académie des Sciences de Paris (Sciences et Vie)* **324**:1157–1165.

Vernet, J.-L. (1973). Etude sur l'histoire de la végétation du sud-est de la France au Quaternaire d'après les charbons de bois principalement. *Paléobiologie Continentale* **4**: 1–73.

Vernet, P., J. L. Guillerm, and P. H. Gouyon. (1977). Le polymorphisme chimique de *Thymus vulgaris* L. (Labiée). I. Répartition des formes chimiques en relation avec certains facteurs écologiques. *Œcologia Plantarum* **12**:181–194.

Vernet, P., P. H. Gouyon, and G. Valdeyron. (1986). Genetic control of the oil content in *Thymus vulgaris* L.: a case of polymorphism in a biosynthetic chain. *Genetica* **69**: 227–231.

Vigne, J. D. (1983). Le remplacement des faunes de petits mammifères en Corse, lors de l'arrivée de l'homme. *Comptes Rendus de la Société de Biogéographie* **59**: 41–51.

Vigne, J. D. (1988). Apports de la biogéographie insulaire à la connaissance de la place des mammifères sauvages dans les sociétés néolithiques méditerranéennes. *Anthropozoologica* **8**:31–52.

Vigne, J. D. (1990). Biogeographical history of the mammals on Corsica (and Sardinia) since the final Pleistocene. In A. Azzaroli (ed.), *Biological Aspects of Insularity*, pp. 370–392. Academia Nazionale dei Lincei, Rome.

Vilà, C., P. Savolainen, J. E. Maldonado, I. R. Amorim, J. E. Rice, K. A. Crandall, J. Lundeberg, and R. K. Wayne. (1997). Multiple and ancient origins of the domestic dog. *Science* **276**:1687–1689.

Vilà, M., M. Tessier, C. M. Suehs, G. Brundu, L. Carta, A. Galanidis, P. Lambdon, M. Manca, F. Médail, E. Moragues *et al.* (2006). Local and regional assessments of the impacts of plant invaders on vegetation structure and soil properties of Mediterranean islands. *Journal of Biogeography* **33**:853–861.

Vilà, M., A-S. D. Siamantziouras, G. Brundu, I. Camarda, P. Lambdon, F. Médail, E. Moragues, C. Suehs, A. Traveset, A. Y. Troumbis, and P. E. Hulme. (2008). Widespread resistance of Mediterranean island ecosystems to the establishment of three alien species. *Diversity and Distribution* **14**:839–851.

Vitousek, P. M. (1994). Beyond global warming: ecology and global change. *Ecology* **75**:1861–1876.

Vitousek, P. M., H. A. Mooney, J. Lubchenco, and J. M. Melillo. (1997). Human domination on earth's ecosystems. *Science* **277**:494–499.

Vogiatzakis, I. N., A. M. Mannion, and G. H. Griffiths. (2006). Mediterranean ecosystems: problems and tools for conservation. *Progress in Physical Geography* **30**:175–200.

Vuilleumier, F. (1977). Suggestions pour des recherches sur la spéciation des oiseaux en Iran. *Terre et Vie* **31**: 459–488.

Wall, R., and L. Strong. (1987). Environmental consequences of treating cattle with the anti-parasitic dung ivermectin. *Nature* **327**:418–421.

Wallace, A. R. (1880). *Island Life*. Macmillan, London.

Walter, H. S. (1979) *Eleonora's Falcon: Adaptations to Prey and Habitat in a Social Raptor*. University of Chicago Press, Chicago.

Walter, H., and H. Lieth. (1960). *Klimadiagramm Weltatlas*. Fisher Verlag, Jena.

Walther, G. R., E. Post, P. Convey, A. Menzel, C. Parmesan, T. J. C. Beebee, J. M. Fromentin, O. Hoegh-Guldberg, and F. Bairlein. (2002). Ecological responses to recent climate change. *Nature* **416**:389–395.

Warwick, R. M., N. R. Collins, J. M. Gee, and C. L. George. (1986). Species size distribution of benthic and peagic metazoan: evidence for interactions? *Marine Ecology Progress Series* **34**:63–68.

Watson, L., and T. Ritchie. (1985). *Whales of the World, A Complete Guide to the World's Living Whales, Dolphins, and Porpoises*. Hutchinson, London.

Webb, T., and P. J. Bartlein. (1992). Global changes during the last 3 million years: climatic controls and biotic responses. *Annual Review of Ecology and Systematics* **23**:141–173.

Weeden, N. F. (2007). Genetic changes accompanying the domestication of *Pisum sativum*: Is there a common genetic basis to the 'Domestication Syndrome' for legumes? *Annals of Botany* **100**:1017–1025.

Weiss, E., W. Wetterstrom, D. Nadel, and O. Bar-Yosef. (2004). The broad spectrum revisited: evidence from plant remains. *Proceedings of the National Academy of Sciences USA* **101**:9551–9555.

Weiss, E., M. E. Kislv, and A. Hartmann. (2006). Autonomous cultivation before domestication. *Science* **312**:1608–1610.

Westman, W. E. (1988). Species richness. In R. L. Specht (ed.), *Mediterranean-Type Ecosystems, A Data Source Book*, pp. 81–92. Kluwer Academic Publishers, Dordrecht.

Whale and Dolphin Conservation Society. (2008). *Mediterranean Whales and Dolphins in Crisis*. Report. www.wdcs.org/wesailforthewhale/en/Mediterranean_whales_and_dolphins_in_crisis.pdf.

White, K. D. (1970). *Roman Farming*. Thames and Hudson, London.

Whitmore, T. M., B. L. Turner, D. L. Johnson, R. W. Kates, and T. R. Gottschang. (1990). Long-term population change. In B. L. Turner (ed.), *The Earth Transformed by Human Action*, pp. 25–40. Cambridge University Press, Cambridge.

Whittaker, R. H. (1972). Evolution and measurement of species diversity. *Taxon* **21**:213–251.

Whittaker, R. J., and J. M. Fernandez-Palacios. (2007). *Island Biogeography. Ecology, Evolution and Conservation.* Oxford University Press, Oxford.

Wick, L., G. Lemcke, and M. Sturm. (2003). Evidence of Lateglacial and Holocene climatic change and human impact in eastern Anatolia: high-resolution pollen, charcoal, isotopic and geochemical records from the laminated sediments of Lake Van, Turkey. *The Holocene* **13**:665–675.

Wiens, J. A. (1989). Spatial scaling in ecology. *Functional Ecology* **3**:385–397.

Wilcove, D. S., D. Rothstein, J. Dubow, A. Phillips, and E. Losos. (2000). Leading threats to biodiversity. In B. A. Stein, L. S. Kutner, and J. S. Adams (eds), *Precious Heritage. The Status of Biodiversity in the United States*, pp. 239–254. Oxford University Press, Oxford.

Williams, P. H., and K. J. Gaston. (1994). Measuring more of biodiversity: can higher taxon richness predict whole sale species richness? *Biological Conservation* **67**:211–217.

Wilson, E. O. (1992). *The Diversity of Life*. Allen Lane, The Penguin Press, London.

Wilson, R., D. Gutierrez, J. Gutierrez, and V. Monserrat. (2007). An elevational shift in butterfly species richness andcomposition accompanying recent climate change. *Global Change Biology* **13**:1873–1887.

Woldhek, S. (1979). *Bird Killing in the Mediterranean*. European Committee for the Prevention of Mass Destruction of Migratory Birds. Zeist, Netherlands.

Woodward, F. I. (1993). How many species are required for a functional ecosystem? In E. D. Schulze, and H. A. Mooney (eds), *Biodiversity and Ecosystem Function*, pp. 271–291. Ecological Series 99, Springer-Verlag, Berlin.

Worm, B., E. B. Barbier, N. Baumont, J. E. Duffy, C. Folke, B. S. Halpern, J. B. Jackson, H. K. Lotze, F. Micheli, S. R. Palumbi *et al.* (2006). Impacts of biodiversity loss on ocean ecosystem services. *Science* **314**:787–790.

Würsig, B., and P. G. H. Evans. (2001). Cetaceans and humans: influence of noise. In P. G. H. Evans, and J. A. Raga (eds), *Marine Mammals: Biology and Conservation*, pp. 565–587. Kluwer Academic/Plenum Press, New York.

Würsig, B., R. R. Reeves, and J. G. Ortega-Ortiz. (2001). Global climate change and marine mammals. In P. G. H. Evans, and J. A.Raga (eds), *Marine Mammals: Biology and Conservation*, pp. 589–608. Kluwer Academic/Plenum Press, New York.

WWF. (2001). *The Mediterranean Forests, A New Conservation Strategy*. WWF Mediterranean Programme, Rome.

Yasuda, Y., H. Kitagawa, and T. Nakagawa. (2000). The earliest record of major anthropogenic deforestation in the Ghab Valley, northwest Syria: a palynological study. *Quaternary International* **73–74**:127–136.

Yesson, C., N. H. Toomey, and A. Culham. (2009). *Cyclamen*: time, sea and speciation biogeography using a temporally calibrated phylogeny. *Journal of Biogeography* **36**:1234–1252.

Yom-Tov, Y., and H. Mendelssohn. (1988). Changes in the disribution and abundance of vertebrates in Israel during the 20th century. In Y Yom-Tov, and E. Tchernov (eds), *The Zoogeography of Israel*, pp. 515–547. Dr. W. Junk, Dordrecht.

Zaballos, M., A. López-López, L. Ovreas, S. Galán Bartual, G. D'Auria, J. C. Alba, B. Legault, R. Pushker, F. L. Daae, and F. Rodirguez-Valera. (2006). Comparison of prokaryotic diversity at offshore oceanic locations reveals a different microbiota in the Mediterranean Sea. *FEMS Microbiol Ecology* **56**:389–405.

Zavala, M. A. (2004). Integration of drought tolerance mechanisms in Mediterranean sclerophylls: a functional interpretation of leaf gas exchange simulators. *Ecological Modelling* **176**:211–226.

Zeder, M. A. (2008). Domestication and early agriculture in the Mediterranean Basin: Origins, diffusion, and impact. *Proceedings of the National Academy of Sciences USA* **105**:11597–11604.

Zeder, M. A., and B. Hesse. (2000). The initial domestication of goats (*Capra hircus*) in the Zagros Mountains, 10,000 years ago. *Science* **287**:2254–2257.

Zenetos, A., S. Gofas, G. Russo, and J. Templado. (2003). Volume 3: Molluscs. In F. Briand (ed.), *CIESM Atlas of*

Exotic Species in the Mediterranean. CIESM Publishers, Monaco.

Zohary, D. (1969). The progenitors of wheat and barley in relation to domestication and agricultural dispersal in the Old World. In P. J. Ucko, and G. W. Dimpley (eds), *The Domestication of Plants and Animals*, pp. 47–66. Aldine, Chicago.

Zohary, D. (1983). Wild genetic resources of crops in Israel. *Israel Journal of Botany* **32**:97–100.

Zohary, D., and P. Spiegel-Roy. (1975). Beginnings of fruit-growing in the Old World. *Science* **187**:319–327.

Zohary, D., and U. Plitmann. (1979). Chromosome polymorphism, hybridization and colonization in the *Vicia sativa* group (Fabaceae). *Plant Systematics and Evolution* **131**:143–156.

Zohary, D., and M. Hopf. (2000). *Domestication of Plants in the Old World*, 3rd edn. Oxford University Press, New York.

Zohary, M. (1961). On the oak species of the Middle East. *Bulletin of the Research Council of Israel* **9D**:161–186.

Zohary, M. (1962). *Plant Life in Palestine*. Ronald Press, New York.

Zohary, M. (1971). The phytogeographical foundations of the Middle East. In P. H. Davis, P. C. Harper, and I. C. Hedgel (eds), *Plant Life of South-West Asia*, pp. 43–52. The Botanical Society of Edinburgh, Edinburgh.

Zohary, M. (1973). *Geobotanical Foundations of the Middle East*. Gustav Fisher Verlag, Stuttgart.

Zohary, M., and N. Feinbrun. (1966–86). *Flora Palaestina*. Parts 1-4. The Israel Academy of Sciences and Humanities, Jerusalem.

Index

Abies 30, 122
  A. alba 102
  A. alba cephalonica 102
  A. borisii-regis 57, 102
  A. bornemulleriana 57
  A. cephalonica 57, 101, 120–1
  A. cilicica 57, 111, 121
  A. equi-trojani 57
  A. maroccana 57, 120
  A. nebrodensis 57, 121
  A. numidica 57, 121
  A. pinsapo 57, 120
Abra alba 135
acacia 100, 232
Acacia 21, 35, 112, 254, 269
  A. dealbata 269
  A. gummifera 100, 120
  A. raddiana 113
Acanthodactylus 49
  A. pardalis 109
  A. schreiberi 109
  A. scutellatus 109
accentor 127
Acer 20, 35
  A. campestre 101, 154
  A. hyrcanum 154
  A. monspeliensis 101
  A. monspessulanum 119, 120
  A. negundo 269
  A. obtusifolium 101
  A. obtusifolium subsp. *syriaca* 154
  A. opalus 101
  A. platanus 154
  A. sempervirens 101, 120, 154
  A. tataricum 154
Acetabularia 85
  A. acetabulum 83
  A. mediterranea 83
Achillea 166
Acinonyx 43
  A. jubatus 71
Acipenseridae 62, 272
Acipenser sturio 94
Acis 259
Acmaea virginea 198

Acomys 71
  A. cilicicus 72
  A. minous 72
Aconitum 174
Addax nasomaculatus 71
Adonis cyllenea 242
Aegilops tauschii 209
Aegolius funereus 104
Aegypius monachus 145
Aelopus 125
Aeluropus 129
Aeonium 55
African ass 250
African collared dove 275
African eagle owl 249, 250
African elephant 238
African horse 250
Agama sinaita 112
agame 112
Agaonideae 177
Agapornis fischeri 274
Agapornis personatus 274
Agavaceae 55
agave 269
Agave americana 269
Aglia tau 38
Ailanthus altissima 269
Alca torda 95
Alcephalus 43
  A. busephalus 71, 250
Alcyonium
  A. alcum 198
  A. palmatum 199
alder 127
Alectoris 40, 68
  A. barbara 274
  A. chukar 274
Aleppo pine 100–1, 109, 138, 154, 156, 160, 173, 218–20, 254, 259–60, 288, 301
alfa 125, 248, 291
alfalfa 123, 209–10, 251
algae 79, 81–5, 93, 128, 130, 136, 161, 188, 190–3, 195, 197, 199, 201, 262, 265, 278–9

Algerian oak 58
Algyroides 64
  A. fitzingeri 67
Allium 208
  A. commutatum 293
  A. permixtum 241
  A. porrum 210
Allysum 58
almond tree 36, 111, 171, 176, 208, 212, 221, 232, 251–2, 297
Alnus 35–6, 38, 127
Alopex lagopus 44
Alopias vulpinus 91
Alosa fallax 272
alose 272
Alpheus audouini 280
  A. dentipes 280
  A. glaber 280
  A. inopinatus 280
  A. rapacida 280
alpine accentor 127
alpine chough 250
alpine swift 127
Alyssum 54, 58
  A. purpureum 175
Alytes 246
  A. muletensis 52, 66–7, 274
  A. obstetricans 49, 132, 246
  A. talayoticus 274
Amandava amandava 274
Amaranthus
  A. albus 269
  A. retroflexus 269
amaryllis 169
Amaryllis 108
Amelanchier 37
American bear 43
American cotontail rabbit 277
American mink 276
Ammophila 128
  A. arenaria 108
Ammotragus lervia 250
Amorpha fruticosa 269
Amphibia 67
amphipod 89, 130, 136, 191, 195, 199

Amphitrite variabilis 198
Amygdalus 111
    A. dulcis 171
Anabasis articulata 113
Anacamptis 176
Anacardiaceae 113
Anaecypris 48
Anagasta kuehniella 148
Anas
    A. angustirostris 68, 130
    A. clypeata 161
    A. crecca 130, 161
    A. penelope 161
    A. platyrhynchos 161
    A. strepera 161
Anatololacerta troodica 67
anchovy 91, 96, 189, 201, 246, 265
Andropogon distachyos 33
Androsace 102–3
Anethum graveolens 173, 208
Anguidae 64, 112
Anguilla anguilla 94, 129, 272
Anguillicola crassus 272
Anguillidae 272
anise 208
annelid 79, 85, 89, 191–2, 194
Anser anser 161
ant 38, 59, 114, 143, 148, 175, 243
Antedon mediterranea 198
antelope 43–4, 71–2, 226, 250, 276
Anthemis abrotanifolia 241
Anthias anthias 93
Anthocarinae 38
Anthriscus cereifolium 208
Anthropoides virgo 247
Anthus
    A. berthelotii 69, 141
    A. trivialis 107
Antirrhinum majus 126
Anura 67
aoudad 250, 277
Aphaenogaster gibbosa 148
Aphanius 62
    A. dispar 279
    A. fasciatus 279
aphid 179
Aphodiidae 183
Aphodius 182–3
Aphrodite aculeate 199
aphyllanthe 168
Aphyllanthes 169
    A. monspeliensis 168
Apiaceae 37
Apium graveolens 208
Apodemus 275
Apogon imberbis 93
Apollonius barujana 55

Apomatres similis 198
apple 36, 101, 111, 212, 221
apricot 212
apterygote insect 191
Apus
    A. melba 127
    A. pallidus 127
    A. unicolor 69, 141
Aquifoliaceae 33
Aquila
    A. adalberti 69, 241
    A. chrysaetos 69
    A. clanga 161
    A. heliaca 69
Aquilegia 35
Arabian bustard 247
Aracaceae 34
Arachnida 52
arar 57
Arbutus 38, 57–8, 123
    A. andrachne 57–8
    A. canariensis 57–8, 122
    A. parvarii 57–8
    A. unedo 17, 57–8
Arceuthos drupacea 121
Archaeolacerta 66, 121
    A. bedriagae 67
Archicarabus alysidotus 60
Arctostaphyllos 107
Ardea
    A. alba 273
    A. cinerea 273
    A. purpurea 248
Ardeidae 273
Ardeola ralloides 273
Arecaceae 33
Arenaria 102
    A. tetraquetra 175
Argania spinosa 34, 120
argan tree 34, 100, 232–3
Argyrosomus regius 279
Aristida sieberiana 109
Aristolochia 294
Aristolochiaceae 33
aroid 34
Arrhenatherum
    A. elatius 150
    A. elatius subsp. *elatius* 150
    A. elatius subsp. *sardoum* 150
Artemia 130
    A. salina 129–30
Artemisia 21, 35, 38, 109
    A. dracunculus 210
    A. herba-alba 26, 112
Arthrocnemum 129, 133
arthropod 79, 90, 244
artichoke 208, 210

Arum 268
Arvicanthis 44
Arvicola sapidus 42
ascidian 89, 199, 200
Asclepiadaceae 35
ash 11, 20, 36, 101, 106, 111, 113, 119, 212, 223, 225, 269
Asian elephant 238
Asio capensis 68
Asparagus 33, 168
    A. aphyllus 180
    A. officinalis 208
asphodel 168, 173, 179, 260
asparagus 208
Asphodelus 168, 292
    A. aestivus 179
Asplenium petrarchae 125
ass 70, 72, 214, 250
Asteraceae 30, 37, 55, 58, 113, 125–6, 173, 242
Asterina
    A. burtoni 280
    A. gibbosa 280
Astragalus 58, 103, 111–2, 122, 260
Astropecten 194
Athanas nitescens 198
Athene
    A. angelis 239
    A. cretensis 239
    A. noctua 265
    A. cf. noctua 239
Atherina boyeri 129
Atherinidae 272
Atlantic bonito 246
Atlantic salmon 94
Atlas cedar 242, 266, 301
Atriplex 21, 30, 125, 254, 259, 269
aubergine 210
Aubretia 103
Audouin's gull 69, 95, 250
auk 29
Aulopyge 48
auroch 29, 44–5, 203, 214, 226, 237–8
Australian wattle 269
Austrian black pine 301–2
autumn crocus 111
Avena 208
    A. byzantina 209
    A. fatua 210
    A. nuda 209
    A. sativa 209
    A. sterilis 209
avocado 55
avocet 130–1, 255
Axinella polypoides 198
Aythia ferina 161

Balaena glacialis 95
Balaenoptera
  B. acutorostrata 96
  B. borealis 96
  B. musculus 96
  B. physalus 96
Balanites 35, 37
Balanus 192
bald ibis 247–50, 297
Balearic cyclamen 144, 150
baleen 95–7
Balistes carolinensis 279
Balitoridae 62
Balkan pine 101
Barbary deer 241
Barbary falcon 140
Barbary fig 297
Barbary macaque 44, 70
Barbary partridge 69
Barbary thuja 57, 100, 122, 236, 251, 259
barbel 244
Barbus 63
  B. barbus 63
  B. meridionali 63
  B. meridionalis petenyi 63
  B. meridionalis subsp. *peloponnesius* 63
barley 208–10, 226, 231, 251, 297
barnacle 190–1
barn owl 239, 249
barn swallow 143, 161–2
basil 173
basking shark 91
bass 92, 129, 136, 272–3, 305–6
bat 70–1, 127, 141, 143, 175
Batrachochytrium dendrobatidis 246, 274
bay tree 17, 55, 121–3
bean 210
bear 29, 42–3, 45, 70, 207, 237, 250, 294
bee 59, 123, 160, 175–7, 179, 184, 221, 242, 253, 271
beech 20, 30, 35, 38, 101, 103, 106–7, 117–8, 223, 282, 301
bee-eater 161
bee-fly 175
beet 208, 210
beetle 38, 59–61, 131–2, 174–5, 182–4, 219, 227, 242, 254, 257, 264, 271, 273, 293–4, 297, 303
Bellevalia 111
bellflower 178
bermuda buttercup 269
Berthelot's pipit 69

Beta 208
  B. vulgaris 210
betoum tree 125, 221, 248–9
Betula 20, 35
Bifora 37
birch 20
bison 29, 43, 226, 237
Bison 29
  B. bonasus 226
  B. priscus 44
  B. schoetensacki 43
Bispira volutacornis 198
bittern 273
bivalve 136, 186, 191–4, 199, 278
blackbird 181
blackcap 40, 105, 147, 181
black cumin 208, 210
black-eared wheatear 69, 104, 140
black francolin 274
black-headed bunting 69
black-headed gull 131, 255
black hellebore 174
black iris 111
black kite 249–50
black locust 269
black-mouthed dogfish 91
black pine 101, 111, 154, 160, 219, 301–2
black rat 45, 144
black redstart 127
black scorpion 105
black-spotted smooth-hound 91
black stork 241
black-throated diver 95
black vulture 145, 257, 294
black wheatear 249
black-winged kite 257
black woodpecker 107, 252
bladder senna 113
Blastophaga psenes 177
blennie 62, 93, 192
Blenniidae 62–3, 92–3, 272
blind cave salamander 66, Plate 5a
blite 208
Blitum 208
blue algae 190
blue butterfly 148
blue chaffinch 69, 140, Plate 6c
bluefin tuna 92, 98, 245–7, 306
bluefish 246
blue magpie 275
blue mussel 246
blue rock thrush 68, 127, 161, Plate 6b
blue shark 91
blue tit 15, 69, 70, 137, 144, 156–9, 219
blue whale 96–7

blue whiting 246
bluntnose six-gill shark 91
boar 43, 44, 70, 173, 207, 213, 250, 257, 275
bogue 246
Boissier's oak 110, 112
Bombus 271
  B. terrestris 271
Bombyx mori 231
Bonellia viridis 198
Bonelli's eagle 126, 250
bonito 246, 247
booted eagle 250
bony fish 79, 91–4
borage 210
Borago officinalis 210
Bos
  B. indicus 214
  B. primigenius 29, 44
  B. taurus 214
Bosnian pine 101, 154
Botaurus stellaris 273
bottlenose dolphin 97, 308
bovid 43, 295
Bovidae 44
boxwood 20, 166–7, 223
brachiopod 197, 199
Brachydontes pharaonis 279
Brachypodium pinnatum 152
brackish water-crowfoot 129
branchiopod 129
Branchiostoma lanceolatum 90
Branchipus 131
Brassica 208
  B. cretica 126
  B. oleracea 210
Brassicaceae 37, 54, 126
bream 92, 136, 194, 279
brine shrimp 129–30
brittle star 194
broad bean 210
broadbill 92
broadtail shortfin squid 200
Bromus 102, 172
  B. erectus 152
broom 123, 168, 173, 175
Broussonetia papyrifera 269
brown algae 79
brown bear 42–3, 45, 70, 250, 294
brown crow 248–9
brown trout 63, 94, 108
Bryophyta 52
bryozoan 79, 89, 192, 195, 197
bubal antelope 71
Bubalis bubalis 214
Bubas 182–3
  B. bubalus 183

Bubo
    B. ascalaphus 249
    B. bubo 29
    B. insularis 239
Bubulcus ibis 273
buckthorn 101, 167
Buddleja davidii 269
buffalo 214, 295
Bufo
    B. bufo 49
    B. calamita 132
    B. viridis 64
bull 358
bulbul 180
bullfrog 273–4, 285
bulrush 128
Bulweria bulwerii 95
bumblebee 271
Bunias 37
bunting 69, 147, 249, 252
Buprestidae 38
bustard 104, 247, 249
Butcher's broom 168
Buteo 145
    B. lagopus 145
    B. rufinus 249
Buthus occitanus 105
buttercup 132, 269
butterfly 25, 38–9, 59, 60, 105, 140, 142–3, 145, 148, 175–6, 241–2, 283, 294
butterfly bush 269
Buxus sempervirens 20
buzzard 145, 249–50, 252

cabbage 208, 210
cactus 269
cade juniper 167, 223
Calabrian pine 100–1, 109, 138, 154, 156, 219, 259
Calandrella brachydactyla 104
Calepina 37
Calidris canutus 26
California quail, 274
Callianassa truncata 199
Callichiton achatinus 198
Callicnemis latreillei 242
Calligonum comosum 113
Callionymus
    C. filamentosus 279
    C. pusillus 279
    C. risso 279
Callitris 57
Calluna 123, 167
Calonectris diomedea 95
Calosoma 60

Calycotome
    C. spinosa 124
    C. villosa 101, 124
Campanula 178
canary 69
Canary grass 208
Canary Island pine 154, 232, 259
Canary Islands strawberry tree 122
canid 71
Canidae 44
Canis 45
    C. aureus 45
    C. lupus 43–4
    C. lupus ssp. *dingo* 213
caouan 80
capelin 96
Cape owl 68
caper 126
Capitella capitata 136
Capitellidae 192
Capparis 33
    C. spinosa 126
Capra 44, 214
    C. aegagrus 214
    C. aegagrus cretica 276
    C. agrimi 214
    C. hircus 44, 214–5
    C. ibex 29, 44, 215
Capreolus 70
    C. capreolus 43–4
carab beetle 293–4
carabid 38, 59, 61, 131, 242, 294
Carabidae 38, 60
Carabus 60
    C. alysidotus 293
    C. clathratus subsp. *arelatensis* 61, 242, 293
    C. olympiae 61, 293–4
    C. rutilans 60
    C. solieri 60, 293–4
Caralluma 113, 126
Carcharodon carcharias 91
Carcinus mediterraneus 136
cardinal fish 93
Cardium 131
    C. lamarcki 136
cardoon 208, 210
Cardopatium 37
Carduelis
    C. cannabina 147
    C. carduelis 143
    C. citrinella 141
Carduus rugulosus 241
Caretta caretta 80
carob 100, 109–10, 122–3, 166–7, 208, 212, 221, 254
carp 129, 266, 272, 305

carpenter bee 176
carpet shell 307
Carpinion
    C. adriaticum 102
    C. selgecum 102
Carpinus 27, 119
    C. betulus 121
    C. orientalis 101, 120
Carpobrotus 266
    C. acinaciformis 269
    C. edulis 269
carrot 210
Carthamnus 211
    C. oxyacanthus 211
    C. palaestinus 211
    C. persicus 211
    C. tinctorius 208, 211
cartilaginous fish 79, 91, 244
Carya illinoinensis 254
Castanea 27
    C. sativa 119–20
Castor fiber 44
cat 45, 71, 216, 249, 275
cattle egret 273
Caulerpa 278
    C. racemosa 85
    C. racemosa var. *cylindracea* 278
    C. taxifolia 85, 278
cave bear 43
cave salamander 66, Plate 5a
Ceanothus 168
cecidomyiid fly 174
cedar 101, 109, 111, 122, 156, 221, 236, 242, 254, 266
Cedrus 122
    C. atlantica 121, 301
    C. brevifolia 120
    C. libani 111, 121
celery 208, 210
Cellana 191
    C. rota 280
Celtis 36
    C. australis 120, 222, 260
    C. tournefortii 120
Centaurea 30, 37, 58
    C. corymbosa 242
central Asian wildcat 216
Centrostephanus longispinus 279
Cephalantera 169, 177
    C. longifolia 178
    C. rubra 178
cephalocordate 90
cephalopod 96–7, 189, 194, 199, 200
Cerabratulus 198
cerambicid 242
Cerambycidae 254, 271
Cerambyx 294

Ceratonia
  C. siliqua 100
Ceratotherium simum 44
Cercis siliquastrum 36
Cerianthus membranaceus 198
cerithiid 280
Cerithium
  C. lividulum 280
  C. scabridum 280
  C. vulgatum 280
Ceroglossus 60
Cervidae 44
Cervus 70
  C. elaphus 43–5
  C. elaphus barbarus 241
cetacean 96–8, 188, 307–8
Cetorhinus maximus 91
Cetti's warbler 69, 161
Chaetomorpha 83
  C. linum 129
chaffinch 30, 69, 140, 147, 161, Plate 6c
Chalcides 64, 66
  C. sexlineatus 67
  C. simonyi 67
  C. viridanus 67
chalcid wasp 177
Chama
  C. gryphoides 198, 280
  C. pacifica 280
Chamaeleo
  C. africanus 64, Plate 4b
  C. chamaeleon 64
Chamaerops 33
  C. humilis 100
chameleon 64, Plate 4b
chamois, 257
Characeae 130, 161
Charadrius
  C. alexandrinus 130
  C. dubius 284
Charaxes jasius 38
cheetah 43
Chelonia mydas 80
Chelon labrosus 272
Chenopodiaceae 37–8, 125, 158
chervil 28
chestnut 119, 231
chickpea 208, 210
chicory 208
chiffchaff 30, 161
Chilean flamingo 274
chimaera 199
Chinese desert cat 216
Chironomidae 38, 59, 130
chironomid 38, 39
Chlamydotis undulata 249

Chlamys septemradiata 50
Chlorobionta 278
Chlorophyceae 129
Chondrichthyes 244
Chondrus giganteus 279
chordate 90
Choriotis arabs 247
chough 68, 250
Christ's thorn 221
Chromis chromis 92, 192
Chromobionta 278
Chrysocarabus 60
  C. auronitens 60
  C. hispanus 60
  C. lateralis 60
  C. lineatus 60
  C. punctatoauratus 60
  C. rutilans 60–1
  C. solieri 60–1
  C. splendens 60
Chrysomelidae 38, 254
Chthamalus 190–1
  C. montagui 191
  C. stellatus 191
Chysomelidae 271
chytrid fungus 246
cicada 105–6, 184, 242
Cicada orni 242
Cicer 208
  C. arietinum 210
cichlid 273
Cichlidae 62
Cichorium 208
Ciconia
  C. ciconia 248
  C. nigra 241
Cidaris cidaris 197
Cilician fir 111
Cinclus cinclus 104
Circaetus gallicus 249
cirriped 192
Cirripeda 192
cistude 65
Cistus 37–8, 123, 259, 292
  C. corsicus Plate 2a
  C. incanus subsp. *incanus* 169
  C. ladaniferus 123
  C. salviifolius 178, 259
citril finch 108, 141
Citrullus lanatus 210
Cladocera 133
clam 246, 305
Clangula hyemalis 95
Clariidae 62
Clavagella melitensis 198
clematis 127
Clematis 127

Cletocamptus retrogresses 130
Cliona viridis 198
clover 208
Clupeidae 62, 91, 272
cluster pine 100, 154, 172, 224, 301
Cneorum 166
Cnidaria 50, 89, 196
cnidarian 79, 197, 199
coal tit 144
Cobitidae 61–2
Cock's head 208
cocklebur 269
Codium, 83
Coelodonta antiquitatis 43–4
Colchicum 111
Coleoptera 52, 247
Colias 38
collared dove 275
collembolan 191
Colpomenia 83
Colubridae 67
Columba
  C. bollii 46, 69, 141
  C. junionae 46, 69, 141
  C. oenas 126
Colutea istria 113
colza 208
common cranes 241
common crossbill 108, 159–60
common dentex 307
common dolphin 97, 308
common eel 272
common juniper 108
common newt 64, 132
common pandora 307
common reed 128
common rue 208
common sole 93, 129, 246
common spadefoot 64
common spiny lobster 200
common tern 131, 255
common toad 49
common trout 244
common vetch 210
common waxbill 274
conger eel 196
Conyza 268–9
  C. canadensis 267
  C. sumatrensis 267
coot 161
copepod 81, 87–8, 130, 132, 188, 194
Coprinae 183
Copris 183
Coracias garrulus 104
coral 89, 199, 200, Plate 8a
Corallina 83
Corallinaceae 83

Corallium rubrum 89, 198, Plate 8a
coriander 173, 210
Coriandrum sativum 173
Coriaria 166
Coris julis 192
Corixidae 133
cork oak 17, 58, 100, 116, 119, 139, 147, 156, 164, 231–2, 236, 241–2, 254, 258–9, 282, 300, Plate 9b
cormorant 94–5, 128
corn bunting 147
corn-flag 174
corn spurrey 208
Cornus 212
    C. mas 171
    C. sanguinea 180
Corsican blue tit 144
Corsican boar 275
Corsican mouflon 275–6
Corsican nuthatch 40, 69
Corsican swallowtail butterfly 140, 294
corvid 227
Corvus
    C. corax 126
    C. ruficollis 249
Corylus 20, 27, 119
    C. avellana 35
Coryphaena hippurus 97
Coryphaenoides rupestris 199
Cotinus 123
Cottidae 62
cottontail rabbit 45, 277
cottony cochineal insect 219
Cotyledon 126
cow 109, 182, 264
cow bream 279
coypu 45, 276
crab 50, 80, 82, 97, 136, 189, 194, 200
crag martin 249
Crambe maritima 208
crane 161, 203, 241, 247
Crangonidae 194
Crassostrea gigas 307
Crassulaceae 55, 126
Crataegus 36, 111
    C. laciniata 221
crayfish 246–7, 257, 266, 273, Plate 5b
creeper 269
Crenilabrus 92
Crepis pygmaea 108
cress 208
crested lark 161
crested newt 39, 40, 64
crested tit 144
Cretan date plam 241

Cretan goat 276
cricket 105, 131, 184
Crithmum maritimum 108
Crocidura 52, 70–1, 275
    C. cypria 239
    C. leucodon 42
    C. russula 72
    C. sicula 72, 239
    C. suaveolens 72
    C. zimmermanni 72, 239
crocodile 64
Crocodylus niloticus 64, 246
crocus 111, 123, 169
Crocus 170
Crocuta 29
    C. spelaea 44
crossbill 108, 159–60
crow 249
Crustacea 50, 284
crustacean 85, 89, 90, 96, 129–32, 187–8, 190–2, 194–7, 199, 200, 256, 265, 278
Cucumaria saxicolia 198
Cucumis melo 210
cumacean 199
cumin 208, 210
Cuminum cyminum 208
Cuon alpinus 44
Cupressus 37
    C. dupreziana 58
    C. sempervirens 109, 120–1
cuttelfish 200, 246
Cuvier's beaked whale 97
Cyanistes
    C. caeruleus 15
    C. teneriffae 69, 141, 156
Cyanophyceae 136, 190
Cyanopica cyanopica 275
Cychrus 60
cyclamen 126, 144, 150–1, 169–70
Cyclamen 126, 170
    C. balearicum 144
    C. repandum 144
Cydonia oblonga 212
Cygnus
    C. columbianus 161
    C. olor 274
Cymodocea 86
    C. nodosa 85–6, 278
Cymodoceaceae 85
Cynara 208
    C. cardunculus 210
    C. cardunculus ssp. *scolymus* 210
    C. cardunculus var. *altilis* 210
    C. cardunculus var. *sativa* 210
    C. cardunculus var. *scolymus* 210
Cynotherium 239

cypress 33, 58, 100, 104, 109, 119, 122, 156, 218, 259
cyprinid 48
Cyprinidae 61–3, 272
Cyprinodontidae 61–2, 272
ciprinoid 48
Cyprinus carpio 129, 272
Cyprus warbler 69
Cyprus wheatear 69
Cystoseira 83, 192
    C. compressa 193
    C. crinitae 192
    C. meditarranea 192–3
    C. zosteroides 192
Cytisus 123

Dactylis
    D. glomerata 149
    D. glomerata subsp. *glomerata* 149
    D. glomerata subsp. *hispanica* 149–50
    D. glomerata subsp. *marina* 149
Dactylorhiza 169, 178, 250
daffodil 108
dalmatian pelican 128
Dama dama 44
damascene rose 208
Damasonium stellatum 132
damselfish 92, 192
Danaus plexippus 38
Danube salmon 63
daphne 103
Daphne 123, 166
    D. gnidium 180
    D. oleoides 103
    D. pontica 103
date palm 35, 112, 211, 241, 297
Daucus carota 210
decapod 89, 246, 279
decapoda 201
deer 29, 43–5, 70, 74, 203, 207, 226, 237–9, 241, 250, 257, 275, 277, 295
Delphinus delphis 97
demoiselle crane 247
Dendrocopos
    D. major 143
    D. syriacus 46
dentex 307
Dentex dentex 307
Dermochelys coriacea 80
Derocheilocaris remanei 90
desert hedgehog 45
Desidiopsis racovitzai 191
Dianthus multinervis 241
diatom 81–2, 88
Dicentrarchus labrax 92

Dicerorhinus hemitoechus 44
Dicrotendipes collarti 39
Dictyota 83
dill 173, 208, 210
dingo 213
dinoflagellate 251, 265
Diospyros kaki 297
Diplotaxis siettiana 241
dipper 104, 108
Diptera 38, 52
dipteran 59, 130, 176
Discoglossidae 66–7, 246
Discoglossus 65
    D. montalentii 52, 67
    D. nigriventer 246
    D. sardus 67
ditch grass 129
Ditiscidae 247, 273
diver 95
dog 43, 45, 71, 213, 239, 275, 277
dogfish 91
dogwood 171, 212
Dolichopodidae 130
dolphin 80, 95, 97–8, 298, 308
dolphin fish 97
domestic cat 275
domestic goat 214, 276
domestic sheep 215, 275
Donax 194
donkey 213–4, 275
Doronicum 35
Dorycnium jordani 129
Douglas fir 266
dove 126, 143, 161, 275
downy oak 46, 101, 104–5, 119, 148, 156–7, 219, 223, 301
Dracaena 55
    D. draco 55–6
dragonet 279
dragonfly 38, 131–2, 242, 247, 257, 273
dragon tree 55–6
Dreissena polymorpha 268
Dromia 200
    D. dromia 50
    D. vulgaris 50
Dryocopus martius 107–8
duck 68, 95, 130–1, 160–1, 203, 239, 257, 268, 275, 303
Dunaliellaceae 129
Dunaliella salina 129
dung beetle 182–3, 227, 264
dunnock 107, 161–2
Dupont's lark 69
dusky grouper 192, 196, 279
dwarf deer 238
dwarf elephant 238–9

dwarf hippo 145, 238–9
dwarf palm 123
dynastid 242

eagle 69, 126, 145, 161, 249–50, 257
eagle owl 29, 126–7, 160, 249–50
earthworm 162, 182, 242–3, 301
eastern imperial eagle 69
eastern savin 111–2
Echinaster seposirus 198
echinoderm 78–9, 89, 199, 200, 278
Echinodermata 50
echinothuriid 197
Echium 55
eel 94, 98, 129, 194, 196, 272
eggplant 210
egret 273, 284
Egretta garzetta 273
Egyptian vulture 126, 145, 250
einkorn wheat 209
Elaeagnus angustifolia 212
Elanus caeruleus 257
elasmobranch 91
Electra posidoniae 195
Eledone cirrhosa 200
elegant cuttlefish 200
Eleonora's falcon 94, 127, 145, 257
elephant 43–4, 71, 145, 216, 238–9, 250
elephant shrew 71
Elephas maximus 238
Eliomys 275
Ellobius 44
elm 20, 127, 242
Emberiza cia 249
Emex 172
emmer wheat 209
Emys orbicularis 65
endive 208
English stonecrop 175
Engraulidae 91
Enteromorpha 83
entognatha 191
Eobison 43
Ephedra 30, 35, 38
Ephydra 130
Ephydridae 130
Epimedium 35
Epinephelus marginatus 192, 295
Epipactis 176
    E. consimilis 179
Equidae 44
Equus 29, 43, 44
    E. africanus 213, 250
    E. asinus 250
    E. caballus 238

    E. ferus 213
    E. germanicus 44
    E. onager 250
Erebia 60
Eretmochelys imbricata 80
Erica 34, 123, 167
    E. arborea 33
Ericaceae 57
Erinacea 103
Eriobotrya japonica 297
Erithacus rubecula 147
Erodium 172
erse 208
Eruca sativa 208
Ervum 208
Eryngium 124
    E. maritimum 108, Plate 1b
Erysimum 126
Erythrean spiny oyster 280
Esocidae 272
Esox lucius 129
Estrilda astrild 274
Etruscan bear 43
Etruscan shrew 72
eucalyptus 119, 232, 254, 271
Eucalyptus 232
    E. camaldulensis 254
    E. gomphocephala 254
Euleptes 66, 141
Eunice torquata 198
Eunicella stricta 198
Euonymus 174
Euphorbia 58, 124, 168
Euproctes 272
Euproctus 39, 55, 272
    E. asper 39
    E. montanus 39, 66–7
    E. platycephalus 39, 66–7
    E. waltl 39
Euraphia 190
    E. depressa 190
Eurasian bittern 273
Eurasian red squirrel 277
Eurasian wild horse 213, 226
European anchovy 246
European ash 101, 111, 269
European bison 226
European carp 272
European hackberry 260
European hake 245, 279
European lynx 297
European mink 276
European pilchard 246
European pond terrapin 65, 246
European robin 147
European roller 104
European sprat 246

European swallowtail 141
European tree-frog 64
European wildcat 216
European wild horse 238
Eurynebria 242
Euscorpius flavicaudis 105
evening primrose 269, 273
Exoacantha 37

faba 208
Fabaceae 33, 259
Fagus 20, 27, 35
    F. moesiaca 101–2
    F. sylvatica 102, 120
Falco
    F. biarmicus 145
    F. eleonorae 94
    F. naumanni 104
    F. pelegrinoides 140
    F. peregrinus 94
    F. subbuteo 249
    F. tinnunculus 126
falcon 94, 126–7, 140, 145, 249–50, 257
fallow deer 44, 203, 277, 295
false indigo 269
false killer whale 97
false olive 30, 100, 174, 207, 259
fan-tailed warbler 161
felid 43, 71
Felidae 44
Felis 45
    F. libyca 275
    F. serval 250
    F. sylvestris 44, 275
    F. sylvestris bieti 216
    F. sylvestris libyca var. *reyi* 275
    F. sylvestris lybica 275
    F. sylvestris ornata 216
    F. sylvestris sylvestris 216
fennel 173, 208, 210
fern 53, 125–6, 132, 143
fescue 103, 210
Festuca 102–3
Ficedula hypoleuca 163
Ficus carica 126, 297
fig tree 126, 176–8, 208, 211–2, 221, 297
fin whale 96
finch 108, 127, 141, 249
fire salamander, 49
fir 30, 35, 57, 101, 103, 109, 111, 117–8, 122, 156, 236
Fischer's lovebird 274
Flabellina 89
flamingo 68, 104, 128–9, 131, 133, 256, 274

flatfish 92, 136, 194
flat oyster 307
flax 208, 211, 229–30
flea 191
fleabane 267–8
flour moth 148
fly 25, 58, 174–6, 179, 182, 190–1
flycatcher 161, 163, 180
fodder pea 208
Foeniculum vulgare 173
foetid hellebore 172
foraminifera 195
forget-me-not 103
Formica gagates 148
fowl 68
fox 45, 179, 275
francolin 274
Francolinus francolinus 274
Frangula 27
Fratercula arctica 95
Fraxinus 20, 36, 38
    F. ornus 101
    F. syriaca 113
frigate 249
fringed water-lily 128
fringe-toed sand lizard 109
Fringilla
    F. coelebs 30
    F. teydea 69, 141, Plate 6c
frog 25, 42, 49, 52, 64, 132, 246, 273–74
Fucelia maritima 190
Fulica 161
Fumana 123
fungus 246, 266, 274

Gadidae 92
gadwall 161
Galeocerdo cuvieri 91
Galeorhinus galeus 91
Galeus melastomus 91
Galliotia
    G. atlantica 67
    G. bravoana 67
    G. caesaris 67
    G. galloti 67
    G. intermedia 67
    G. simonyi 67
    G. stehlini 67
gall oak 111, 119, 174
gallinule 257
Gallotia 66, 145
    G. auaritae 246
Gallotiinae 66
gamba 189
Gambusia affinis 272
Gammarus locusta 130

gannet 29, 95
garden chervil 208
garden rocket 208
garden warbler 40, 161, 180
garlic 208
Garrulus glandarius 30
Gasterosteidae 62, 272
gastropod 89, 93, 136, 186, 189, 191–2, 194–6, 244, 278, 280
gastrotrich 194
Gavia
    G. arctica 95
    G. stellata 95
Gazella 43
    G. cuvieri 238, 250
    G. dama 71
    G. dorcas 238, 250, 295
    G. gazella 257, 295
    G. rufina 250
gazelle 43, 70–1, 203, 238, 250, 257, 295
gecko 246
Gekkonidae 67
genet 45, 71, 216
Genetta genetta 45
Genista 37, 103, 122–3, 166, 168
    G. acanthoclada 101, 124
    G. linifolia 259
Gentiana 35
Geocaryum bornmuelleri 241
Geotrupes 182–3
gerbil 70–2
Gerbillus 44
Geronticus eremita 247
ghost bat 71
giant deer 29
giant lizard 64
gilt-head 129
Ginkoaceae 168
gladiolus 111
Gladiolus 111
    G. segetum 174
Glareola pratincola 257
glass eel 94
glirid 71
Glis 275
globe artichoke 210
Globicephala melas 96
Globularia 166
Glossodoris luteorosea 198
Glycera 194
Glycytthiza glabra 129
goat 44, 70–1, 109, 111–2, 115, 146, 174, 182, 214–6, 222, 226, 230–2, 264, 266, 275–6, 295–6
goat's thorn 103
goatfish 279

gobie 92–3, 192, 246
Gobiesocidae 92
Gobiidae 61–2, 92–3
goldband goatfish 279
golden eagle 69, 145, 250, 257
goldfinch 143, 161
Goneplax rhomboides 199
goose 133, 161
gorgon 89
gorgonian 89
gorse 124, 208
goshawk 68
Graellsia isabellae 294
Grampus griseus 97
grape 127, 211–2, 230, 251, 264, 296–7
grasshopper 105–6, 131
grayling 108
great auk 29
greater flamingo 256
great owl 239
great right whale 95–6
great spotted woodpecker 143
great tit 144
great whale 95
great white egret 273
great white shark 91
Greek juniper 111, 138
Greek oregano 179
green algae 79, 129–30, 136, 192, 278
green crab 136
green sea turtle 80
green toad 64
green woodpecker 161
grenadier 199
grey heron 273
grey squirrel 277
grey trigger fish 279
grey wagtail 108
greylag goose 161
greyshrimp 194
griffon vulture 126, 145, 160, 226–7, 257, 294
grooved carpet shell 307
ground beetle 60, 303
grouper 192, 196, 279, 295
Grus grus 241
Guenther's vole 45
guild poppy 108, Plate 1a
guillemot 95
guinea fowl 68, 247
gull 29, 68–9, 94–5, 129–31, 163, 255–6
gull-billed tern 131, 255
Gulo spelaeus 44
gundi 71
Gypaetus barbatus 68

Gyps fulvus 126
Gypsophila struthium 175

hackberry 222, 260
Hadrocarabus genei 60
Haematopus ostralegus 130
hagfish 94
hairy poppy 108
hairy rockrose Plate 2a
hake 245–6
Haliaeetus albicilla 161
Halimeda 83
Halocynthia papillosa 198
Halophila 278
 H. stipulacea 85, 86, 278
Haloxylon 35, 125
Hamamelidaceae 35, 119
hammerhead 91
hamster 70–1
harbour porpoise 97, 308
hard wheat 209
hare 44, 71–2, 203
hawksbill sea turtle 80
hawthorn 36, 111, 212, 221
hazelnut 20, 26–7, 35, 101, 212
heath 57, 123, 258
heather 57, 123, 167, 207, 282
Hedera helix 166
hedgehog 45, 71, 103, 111
Hedysarum coronarium 208
Helianthemum 37, 123, 166
Helichrysum 34, 166
Helix aspera 152
hellebore 172, 174
Helleborus
 H. cyclophyllus 174
 H. foetida 172
helmet guinea fowl 247
Hemiechus auritus 45
Hemimysis
 H. margalefi 284
 H. speluncola 284
Hemitragus 44
 H. cedrensis 44
henna 211
Heptranchias perlo 91
Hermann's tortoise 29, 65
Herminium 176
hermit crab 194
heron 128, 160, 248, 257, 273, 275, 284
Herpestes 71
 H. auropunctatus 276
 H. edwardsi 276
 H. ichneumon 45, 276
herring 96
Hesperiidae 60
Hexanchus griseus 91

Hieraaetus fasciatus 126
Hierophis cypriensis 66–7
Himantoglossum 176
Himantopus himantopus 130
Himantura uarnak 91
Hipparchia neomiris 59
hippo 145, 238–9
Hippocampus 93
 H. hippocampus 93, 195
 H. ramulosus 93, 195
Hippodiplosoa fascialis 198
Hippolais 68
Hirundo
 H. rupestris 249
 H. rustica 143
hobby 249
holly 20, 107, 260
holm oak, 17, 38, 46, 100, 104–5, 107, 109, 115–6, 122–3, 126, 138–40, 147, 156, 158, 164, 166–7, 218, 223–4, 231, 236, 253–4, 258, 282, 292, 301
Holocephali 199
holothurian 199
Homo 202
 H. erectus 202
Homotherium 43
honey bee 176, 271
honey buzzard 252
honeycomb stingray 91
honeysuckle 101, 269
hop 127, 208
hop hornbeam 35
Hoplosthetus atlanticus 199
Hordeum 172, 209
 H. sativum 208
 H. spontaneum 209
 H. vulgare 209
hornbeam 35, 101, 111, 119
horn beetle 294
horned octopus 200
horse 29, 43, 174, 203, 213, 226, 237–8, 250, 275, 303–4
horse mackerel 246
horse parsley 208
horseradish tree 112
horseweed 267–8
houbara bustard 249
house mouse 45, 52
house sparrow 143
hoverfly 175, 179
huchen 63
Hucho hucho 63
humpback whale 96
Humulus lupulus 127, 208
hunting dog 71
huntress cricket 106

hyacinth 111
Hyaenidae 44
Hydrobates pelagicus 95, 249
Hydrobia 130
Hydrocharitaceae 85
hydroid 89, 192, 195, 200
Hydrophilidae 247, 273
hyena 29, 45, 237
Hyla
    H. arborea 64
    H. meridionalis 49
Hymenoptera 52
hymenopteran 175
Hyparrhenia hirta 33
Hyperoodon ampullatus 97
Hypophthalmichthys molitrix 272
hyrax 71
hyssop 208
Hyssopus officinalis 208
Hystricidae 44
Hystrix cristata 70

Iberian imperial eagle 241, 257
Iberian lynx 241
Iberian spadefoot 64, 132
Iberic pig 231
Iberis candolleana 108
Iberolacerta 246
ibex 29, 44, 214–5, 237, 257, 295
ibis 128, 203, 247–50, 275, 297
ice plant 266, 269–70
Ictaluridae 272
icterine warbler 161
Idothea hectica 195
Ilex aquifolium 20
Illex coindetii 200
Impatiens 269
Imperian eagle 241, 249, 257
Indian silverbill 274
Iridaceae 111
iris 111, 128, 169, 208
Iris 259
    I. chrysographes 111
    I. pallida 208
    I. pseudacorus 128
ironwood 35–6, 119, 122
Isoetes setacea 132
isopod 136, 190–1, 195
Isopoda 52
Isurus oxyrhinchus 91
ivy 166
Ixobrychus minutus 273

jackal 45
jackdaw 250
Janetiella thymicola 174
Japanese carpet shell 307

Japanese honeysuckle 269
Japanese oyster 307
Japanese persimmon 297
jasmine 211
Jasminum 33, 211
    J. odoratissimum 55
jay 30, 70, 174
jellyfish 80
jerboa 45, 70–2
Jerusalem sage 168
jewel box oyster 280
Judas tree 36, 260
Juglans 27
    J. regia 35
jujube 113, 208
Juncaceae 85
Juncus 132
juniper 29, 30, 101–2, 108–9, 111–3,
    122, 126, 128, 138, 154, 156,
    167, 229, 236, 288
Juniperus 29, 37, 102, 166
    J. communis 108, 121
    J. drupacea 111
    J. excelsa 111, 121
    J. foetidissima 121
    J. oxycedrus 111, 120, 167
    J. phoenicea 111, 113, 126
    J. thurifera 121, 138
    J. turbinata 121
Jynx torquilla 143

Kabyle nuthatch 40, 46
Katsuwonus pelamis 247
Kemp turtle 80
Kentish plovers 130
kermes oak 101, 104, 112, 115, 123–4,
    138–9, 166–7, 207, 218–9,
    258–9, 292
Kermococcus vermilio 219
kestrel 104, 126, 143, 145, 161, 248–50
killer seaweed 278
killer whale 97
killifish 279
kinglet 162, 248
kite 249, 257
Kleinia 55
knot 26
Kruper's nuthatch 40, 46

Labiatae 173
Labridae 92–3
Labrus 92
    L. merula 92
    L. viridis 92, 195
Lacerta 64
    L. lepida Plate 4c
Lacertid 145

Lacertidae 49, 64, 67, 109
Lactuca sativa 210
Lamiaceae 126, 173, 179
Laminaria 83
    L. japonica 85
    L. ochroleuca 82
lammergeier 68, 145, 249, 257, 294
Lamna nasus 91
Lampetra fluviatilis 94
lamprey 61, 94
lancelet 90, 194
land tortoise 52
land turtle 111
Languedocien scorpion 105
Lanius 265
lanner falcon 145, 249–50
lantern fish 97, 199
lappet-faced vulture 247
larch 160, 301
large blue butterfly 148
large-scaled scorpionfish Plate 8b
Larix 301
lark 69, 104, 161
Larus 29
    L. audouinii 95
    L. cachinnans 95
    L. genei 68
    L. melanocephalus 68
    L. ridibundus 131
Lasioglossum marginatum 179
Lathyrus 208, 210
Lauraceae 32–3, 55, 119–20
laurel 33, 55, 100, 119, 123, 166, 173,
    179, 210, 260
laurel pigeon 69
lauristinus 166, 181
Laurus 32, 123, 282
    L. azorica 55
    L. nobilis 17, 120
Lavandula 123, 166
    L. angustifolia 151
    L. latifolia 151
    L. stoechas 151
    L. vera 208
Lavatera 37
lavender 123, 151, 168, 173, 208, 211,
    300
Lawsonia inermis 211
Lebanon cedar 111, 242
Lebanon oak 110
leek 208, 210
legume 33, 103, 123, 129, 168, 172,
    210, 259
leiothrix 274
Leiothrix lutea 274
Lens 208
    L. culinaris 210

lentil 208, 210
lentisk 17, 30, 38, 100–1, 109, 123, 166, 181, 211, 258
leopard 71
Lepidasthenia elegans 198
Lepidium
    L. sativum 208
    L. subulatum 175
Lepidochelys kempii 80
lepidoptera 38, 52
Lepomis gibbosus 244
Leptobos 43
Leptopsammia pruvoti 198
Leptothorax unifasciatus 148
Lepus 45
    L. granatensis 72
lesser butter and eggs 108
lesser flamingo 274
lesser kestrel 104, 126, 145, 161, 248, 250
lettuce 210
Leucojum fabrei 184
Libythea celtis 38
lichen 81–2, 190
licorice 129
Ligia italica 190
lignum vitae 37
Lilford's wall lizard 145
Liliaceae 111, 173, 179
lily 111, 128, 168–70
Lima hians 198
lime 11, 20, 130
Limonium 55
    L. catanense 241
limpet 82, 191, 193, 279–80
limule 132
Linaria 37, 102
    L. supina 108
linden 20
linnet 147
Linum 208
    L. usitatissimum 211
lion 29, 43, 45, 207, 237, 250
Lissa chiragra 198
Lithodomus lithophaga 198
Lithophyllum 83
    L. byssoides 83–4
    L. incrustans 84
    L. lichenoides 83
    L. yessoense 279
Lithothamnion
    L. calcareum 85
    L. coralloides 85
little bittern 273
little bustard 104
little egret 273, 284

little owl 239, 242, 265
little ringed plover 284
little tern 255
Littorina neritoides 190
Liza
    L. aurata 272
    L. ramada 272
lizard 49, 64, 66, 76, 109, 111, 131, 145, 246, 274, Plate 4c
lizardfish 279
lobster 106, 189, 200, 245
locust 248, 269
Locustella luscinioides 257
Loligo forbesi 200
Lolium 210
Lonchura malabarica 274
long-finned pilot whale 96
long-legged buzzard 249–50
long-tailed duck 95
Lonicera
    L. etrusca 180
    L. implexa 101
    L. japonica 269
loosestrife 132
Lophortyx californicus 274
loquat 297
longicorn Plate 4a
Loranthus europaeus 174
Lotus 210
Louisiana crayfish 246, 257, 266, 273, Plate 5b
lovebird 274
Loxia
    L. curvirostra 108
    L. leucoptera 160
Loxodonta africana 238
Lucanus 294
lucerne 209
Ludwigia 273
Lullula arborea 147
lupine 208, 210
Lupinus 208, 210
lute turtle 80
Lycaenidae 60
lycaon 45
Lycaon 45
    L. pictus 71
Lygos 38, 168
lynx 45, 207, 241, 275, 297
Lynx 45
    L. pardinus 241
    L. spelaea 44
Lysimachia minoricensis 241
Lysmata
    L. seticaudata 50
    L. ternatensis 50
Lythrum 132

Macaca sylvanus 44
macaque 44, 70–1
Macedonian pine 101, 224
mackerel 92, 188, 246, 279
Macrothorax 60
    M. aumonti 60–1
    M. celtibericus 60
    M. morbillosus 60–1
    M. planatus 60–1
    M. rugosus 60–1, 242
Macrouridae 199
Macrovipera schweizeri 67
Maculinea 148
    M. arion 148
madder 208, 230
madrepore 196
Maenidae 92
magicienne dentelée 106
magpie 275
mahi-mahi 97
Majorcan midwife toad 274
mako 91
mallard 161
Malus 241
    M. sylvestris 212
mammoth 29, 43, 237
Mammuthus 29, 43
    M. meridionalis 43
    M. primigenius 44
    M. trogontherii 43
Maniola nurag 59
mantise 105
mantis shrimps 199
maple 20, 101, 106, 119, 121, 123, 153–6, 229, 301
maple ash 269
marble trout 63
marbled duck 68
marbled newt 39, 40, 64, 132
marbled polecat 45
marbled spinefoot 279
marbled teal 69, 130, 257
marginated tortoise 65
marine worm 130
maritime pine 100, 154, 224, 254, 282
Marmora's warbler 69
Marmotta marmotta 44
marram grass 128
marsh frog 64
marsh tit 252
Marsilea strigosa 132
Marsupenaeus japonicus 280
martin 249
masked lovebird 274
Mastomys 70–1
Matsucoccus feytaudi 172

Mauremys
  M. caspica 65
  M. leprosa 65
meadow pipit 161
meager 279, 307
Medicago 171–2, 254
  M. sativa 209
Mediterranean dwarf palm 100
Mediterranean flour moth 148
Mediterranean gull 68, 255–6
Mediterranean hake 246
Mediterranean mussel 307
medlar 212
Megaceros giganteus 44
Megaloceros 239
  M. algericus 45
  M. giganteus 29
Meganthereon 43
Megaptera novaeangliae 96
Megathyris detruncata 198
Megodontus
  M. caelatus 60–1
  M. croaticus 60–1
Melanitta
  M. fusca 95
  M. nigra 95
Melanocorypha calandra 104
Melaraphe neritoides 190
Meles meles 44
Melicertus
  M. japonicus 280
  M. kerathurus 280
Melierax metabates 68
melilot 210
Melilotus 210
Mellivora 71
melodious warbler 161
melon 210
Menispermaceae 35
Mentha 151
  M. piperita 208
Meriones 44
Merluccius merluccius var.
    mediterraneus 245
Mesophyllum lichenoides 83–4
Mespilus 212
Metapenaeus monoceros 280
Micromeria 126
Micropterus salmoides 272
microturbellaria 130
Microtus
  M. guentheri 45
  M. socialis 172
midwife toad 49, 274
Miliaria calandra 147
Milvus migrans 249
mimosa 269, 277

mink 276
minke whale 96
Minorcan midwife toad 274
mint 123, 151, 168, 173
mistle thrush 174
mistletoe 174, 301
mite 271
mole 71, 174
mole cricket 131
mole-rat 71
mollusc 50, 60, 65, 79, 80, 82, 130,
    136, 197, 200, 239, 268, 279,
    305, 307
Mollusca 52
Monachus monachus 250
monarch 38
mongoose 45, 71, 276
monk parakeet 274
monk seal 250–1, 295, 297
montane pine 108, 154, 160, 301
Monterey pine 254
Monticola solitarius 68, Plate 6b
Montifringilla nivalis 127
Montpellier maple 119, 121, 154
Moraceae 33
Moringa aptera 112
Moronidae 272
Moroteuthis robusta 96
Morus 212
mosquito 272
mosquito-fish 272
Motacilla cinerea 108
moth 38, 148, 172, 176, 242, 294
mouflon 226, 275–7, 295
mountain ash 36, 101, 106, 119, 212,
    223
mountain gazelle 257
mouse 45, 52, 71–2, 113, 216
Moussier's redstart 69, 249
moustached warbler 161
Mugil cephalus 272
Mugilidae 92, 129, 272
mugo pine 101
mulberry 212, 231, 269
mullet 129, 136, 192, 246, 272, 279
Mullidae 92
Mullus barbatus 279
munia 276
murid 45
Mus 275
  M. cypriacus 239
  M. musculus 45
musk ox 43
muskrat 45, 65, 276
mussel 187, 192–3, 237, 246, 268,
    279–80, 307
mustard 126, 208, 210

Mustela
  M. lutreola 276
  M. nivalis 275
  M. vison 276
mustelid 71
Mustelidae 44
Mustelus 44
  M. asterias 91
  M. mustelus 91
  M. punctulatus 91
mute swan 274
Myocastor coypus 45
Myopsitta monachus 274
Myosotis stolonifera 103
Myotragus balearicus 276
myriapod 72, 87, 191
Myriapore truncata 198
Myrica faya 55, 122
Myricaceae 55
Myriophyllum 129
  M. verticillatum 128
Myrmica specioides 148
Myrtaceae 32–3
myrtle 166, 173, 179, 181, 211,
    Plate 2b
Myrtus 38, 123
  M. communis 166, 180, Plate 2b
mysid 284
Mysidacea 284
Mysidae 244
mystacocarid 90, 192
Mysticetes 95–7
Mysticeti 95, 96
Mytilaster minimus 280
Mytilus galloprovincialis 307
Myxine glutinosa 94

Narcissus 37
narrow-barred mackerel 279
Naticidae 194
Natrix
  N. maura 274
  N. natrix 49
Neanderthals 203
Near Eastern wildcat 216
nematode 130, 150, 194, 266
nemertian 194
Nemoderma 83
  N. tingitanum 191
nenuphar 128
Neogoniolithon mamillosum 83
Neophron percnopterus 126
Nephrops norvegicus 200
Nephthys 194
Neptune grass 82, 85–7, 92–3, 98,
    191–7, 200–1, Plate 7b
Nereidae 136, 192

Nereis diversicolor 130
Nerium 33
    N. oleander 113
Netta rufina 130, 161
nettle-tree butterfly 38
Neumayer's rock nuthatch 140
newt 25, 39, 40, 42, 64–6, 132, 246, 272–3
Nigella 269
    N. sativa 208
night heron 273
nightingale 143, 147, 161, 163
nightjar 69
nightshade 174
Noea mucronata 113
Noemacheilus 62
northern bottlenose whale 97
Norway lobster 200, 245
Norway rat 45
Norway spruce 101
Nostoc 136
Nucifraga caryocatactes 180
nudibranch 89, 196
Numida meleagris 68, 247
nutcracker 180
nuthatch 40, 46, 52, 69, 140
Nyctea scandiaca 29
Nyctereutes 43
    N. procyonoides 277
Nymphaea alba 128
Nymphalidae 60
Nymphoides peltata 128

oak 17, 26, 29, 30, 33, 35–8, 46, 58, 69, 100–1, 104–8, 110–2, 114–6, 118–9, 122–4, 126–7, 138–40, 144, 147–8, 150, 153–8, 163, 166–7, 172, 174, 207, 218–21, 223–4, 229, 231–2, 236, 241–2, 254, 258–60, 282, 292, 301, 304, Plate 9b
oat 208–10, 231
Oceanodroma castro 95
ocellated lizard Plate 4c
Ochradenus baccatus 113
Ocimum basilicum 173
Ocotea foeten 55
Octocorallia 89
octopus 189, 200
Octopus 246, Plate 8c
    O. macropus 50
    O. vulgaris 198
    O. variabilis 50
Odobenus rosmarus 97
Odontocete 95–7
Odontoceti 96–8

Oenanthe
    O. cypriaca 69, 141
    O. hispanica 104
    O. leucura 249
    O. oenanthe 108
    O. pleschanka 140
Oenothera biennis 269
Olea 33–4, 38, 102, 120, 241
    O. europaea 16, 212
    O. europaea subsp. *oleaster* 100
    O. europaea var. *europaea* 211
    O. europaea var. *sylvestris* 211
    O. oleaster 211
Oleaceae 33, 55
oleander 113, 167
oleaster 212
olive tree 16–7, 33–4, 100–1, 106, 109, 119, 122, 166–7, 179, 181, 208, 211–2, 221, 230, 232, 251, 297
onager 250
Oncorhynchus mykiss 244, 266, 272
Ondatra zibethicus 45
onion 208, 293
Onobrychis 102
Ontophagus 183
Onychocella marioni 198
Opeatogenys gracilis 92
Ophidia 67
Ophiopsila aranea 198
Ophisaurus apodus 112
Ophiothrix fragilis 198
Ophrys 169, 176, 178–9, 250–1
    O. apifera 176, Plate 3a
    O. araneifera 176, Plate 3b
    O. bertolonii Plate 3c
    O. bicolor Plate 3e
    O. fusca 176, 250
    O. lutea 250, Plate 3d
    O. scolopax Plate 3f
    O. speculum 179
    O. sphegodes Plate 3g
    O. tenthredinifera 176, Plate 3h
Opuntia 269, 197
orange roughy 199
orchard grass 149–50, 210
orchid 123–4, 169, 176–9, 240, 250, Plate 3
Orchidaceae 123
Orchis 169, 178, 250, 259
    O. caspia 179
    O. galilea 179
Orcinus orca 97
oregano 151, 173, 179, 210
Oriental plane tree 35, 127
oriental silkworm 231
Origanum 151
ornate wrasse 279

Ornithogallum 168
Ornithopus 210
    O. sativus 208
orphean warbler 105
orthopteran 106
Oryctolagus cuniculus 45, 277
Oryx damah 71
Oscarella lobularis 198
osprey 94
ostracod 130, 132, 194
Ostrea edulis 307
Ostreococcus tauri 81
Ostrya 27, 35, 102, 119
    O. carpinifolia 101, 120
Osyris 169
    O. alba 168, 180
Otanthus maritima 108
Otis tetrax 104
Otus scops 242
Ovibos pallantis 43
Ovis
    O. ammon 215
    O. ammon musimon 275
    O. aries 215, 276
    O. gmelini 277
    O. orientalis 215, 226
    O. vignei 215
Oxalis pes-caprae 269
Oxyura
    O. jamaicensis 275
    O. leucocephala 68
oyster 186, 279–80, 305, 307
oyster plant 208
oystercatcher 130
owl 29, 68, 104, 108, 127, 160–1, 238–9, 242, 249–50

Pagellus erythrinus 307
Palaemon xiphias 195
Palaeoloxodon
    P. antiquus 43–4, 238
    P. falconeri 228
Palestine viper 111
Palinurus
    P. elephas 200
    P. mauritanicus 200
    P. vulgaris 198
pallid swift 69, 127
palm 33, 100, 112, 123, 241
Palmae 32
Palmate newt 64–5
Pamborus 60
Pancratium maritimum 108
panda 43
Pandion haliaetus 94
pandora 307
panther 29, 45, 237, 250

Panthera 29
　P. leo 29
　P. pardus 44
　P. spelaea 43–4
Papaver
　P. alpinum 108
　P. aurantiacum 108, Plate 1a
　P. rhaeticum 108
paper mulberry 269
Papilio
　P. alexanor 294
　P. hospiton 59, 141
　P. machaon 141
Papilionidae 60
Paracentrotus lividus 193, 195
Parailurus 43
Paralcyonium elegans 198
Paramuricea clavata 198
Parapholis marginata 293
Paratendipes striatus 39
Parietaria judaica 126
Parnassus 38
Paromola cuvieri 200
Paronychia bornmuelleri 241
Parrotia persica 35
parsley 173, 208
parsley frog 132
parsnip 208
partridge 40, 68–9, 203, 274–5
Parus 107
　P. lugubris 46
　P. palustris 252
Paspalum 129
Passer domesticus 143
passerine 241
Pastinaca sativa 208
Patella caerulea 280
pea 208, 210
peacock-of-the-night 242
pear 36, 111, 212, 221, 251
pearl oyster 279–80
pecacarid crustacean 194–5, 199
pecan 254
Peganum 37
Pelagodroma marina 95
Pelargonium 268
Pelecanus
　P. crispus 128
　P. onocrotalus 128
pelican 128, 203
pellet 247
Pelobates 65
　P. cultripes 64
　P. fuscus 64
pelocypod 89
Pelodytes punctatus 132
Peloponnesian pheasant's-eye 242

pen shell 192–3
Penaeid 280
Penaeus semisulcatus 280
Pennatula 199
peppermint 208
Perca fluviatilis 272
perch 272
Percidae 62, 272
peregrine falcon 94, 126–7, 140, 145, 249–50, 257
Pernis apivorus 252
Perophoropsis herdmanni 198
Persea indica 55
Petodoris atromaculata 198
Petrarch's fern 125
petrel 94, 249
Petromyzon marinus 94
Petromyzontidae 61–2
Petroselinum crispum 173, 208
Petrosia filiformis 198
Peyssoneliaceae 84
Phagnalon rupestre 126
Phalacrocorax
　P. aristotelis 95
　P. carbo 95
　P. pygmaeus 95
Phalaris canariensis 208
Phanourios minutus 154
Phascolosoma granulatum 198
pheasant 274
pheasant's eye 242
Pheidole pallidula 148
Phellia elongata 198
Phillyrea 33, 38, 100, 123, 288
　P. angustifolia 30, 180
Phleum 102
Phlomis 168
　P. fruticosa 168
Phocoena phocoena 97
Phoenician juniper 126, 128
Phoeniconaias minor 274
Phoenicopterus
　P. roseus 68, 256
　P. chilensis 274
Phoenicurus
　P. ochruros 127
　P. moussieri 249
Phoenix 241
　P. dactylifera 35
　P. theophrasti 241
Phragmites australis 128
Phyllodactylus 66, 141
phyllopod 130, 132
Phylloscopus
　P. collybita 30
　P. trochilus 161
Physeter macrocephalus 96

Picconia excelsa 55
Picea 35
　P. abies 101
Picidae 143
pied avocet 255
pied flycatcher 161, 163, 180
pied wheatear 140
Pieridae 60
Pieris 38
pig 71, 109, 139, 145, 213, 231, 275
pigeon 46, 69
pigweed 269
pike 129
pike-perch 129, 272
piked whale 96
pilchard 246
pilot whale 96
Pilularia 132
Pilumnus hirtellis 198
Pimpinella anisum 208
Pinaceae 57
Pinctada
　P. margaritifera 280
　P. radiata 279–80
pine 29, 35–6, 69, 100–1, 106–9, 111, 115, 117, 119, 122, 128, 138, 150, 153–6, 160, 172–3, 180, 212, 218–21, 224, 232, 254, 259–60, 282, 288, 292, 301–2, 304, Plate 9a
Pinguinus impennis 29
pink spiny lobster 200
Pinna
　P. nobilis 192
　P. rudis 193
pin-tailed sandgrouse 104
Pinus 29, 36–7, 120, 122
　P. brutia 101, 120, 138, 155
　P. canariensis 120, 154, 155
　P. cembro 180
　P. halepensis 100, 120, 138, 155
　P. hamiltonii 155
　P. heldreichii 101–2, 121, 155
　P. leucodermis 154
　P. maghrebiana 155
　P. mugo 101–2, 121
　P. nigra 101
　　P. nigra subsp. *clusiana* 120, 154
　　P. nigra subsp. *dalmatica* 155
　　P. nigra subsp. *laricio* 120, 155
　　P. nigra subsp. *mauretanica* 121, 154–5
　　P. nigra subsp. *nigra* 120, 155, 301
　　P. nigra subsp. *pallasiana* 121, 155

Pinus (cont.)
   P. nigra subsp. *salzmannii* 120, 154–5
   P. peuce 101–2
   P. pinaster 100, 155
      P. pinaster subsp. *hamiltonii* 156
   P. pinea 100, 120, 155, Plate 9a
      P. pinea subsp. *mesogeensis* 120
   P. radiata 254
   P. sylvestris 101, 121, 155
   P. uncinata 101, 108, 121, 155
pipe-fish 93, 195
pipit 69, 107, 161, 252
pistachio tree 111–2, 153–5, 174, 179, 181, 207, 211, 221, 254, 297
Pistacia 35, 38, 111, 123, 153, 288
   P. atlantica 113, 120, 125, 221
   P. lentiscus 17, 120, 180
   P. palaestina 110, 113, 140, 221
   P. saportae 153
   P. terebinthus 110, 180
Pisum 28
   P. sativum 210
Pitymys 72
plane tree 35
Platanthera 176
   P. chlorantha 176
Platanus orientalis 35
plebeian cicada 106
Plethodontidae 67
Plocama 56, 242
   P. brevifolia 56
   P. calabrica 56
   P. pendula 56
plover 130, 284
Poa 102, 112
pochard 161
Podarcis 64, 246
   P. filfolensis 67
   P. gaigeae 67
   P. lilfordi 67, 145, 274
   P. melissellensis 274
   P. milensis 67
   P. pityusensis 67
   P. raffonei 67
   P. sicula 274
   P. tiliguerta 67, 274
   P. wagleriana 67
Poeciliidae 272
polychaete 89, 136, 192, 194–6, 199, 200, 278
Polycheles typhlops 200
Polychelidae 200
Polygala myrtifolia 269
Polyommatinae 38
Polyphylla fullo 242
Pomacentridae 92

pomegranate 208, 212, 221–2, 251, 297
pondweed 129, 161
Pontogenia chrysocoma 198
pool frog 42
poplar 127
poppy 108, 123, Plate 1a
Populus alba 127
porbeagle 91
porcupine 44, 70–2, 276
porgie 92, 189, 192–3, 305
Porphyra 83
   P. yezoensis 279
Porphyrio porphyrio 257
porpoise 95, 97–8, 308
Portulaca oleracea 208
Posidonia 92, 278
   P. australis 50
   P. oceanica 50, 82, 278, Plate 7b
Posidoniaceae 85
Potamogeton pectinatus 129
Potamogetonaceae 85, 161
pratincole 257
Prespa trout 63
prickly pear cactus 269
primate 71
primrose 269
Primula 102–3
Primulacae 144, 169
Prionace glauca 91
Proboscidae 44
Procambarus clarkii 246, Plate 5b
Procavia 71
Procerus 60
   P. duponcheli 60–1
   P. gigas 60–1
   P. scabrosus 60–1
   P. syriacus 60–1
processionary moth 172
Procyon lotor 277
Prolagus 145, 239
Prosopis farcta 129
Proteidae 65
Proteus anguinus 65–6, Plate 5a
Prunella
   P. collaris 127
   P. modularis 107
Prunus 36, 241
   P. armeniaca 212
   P. mahaleb 180
Psammodromus 64, 66
Psammomys
   P. meriones 172
   P. obesus 172
Psammophis schokari 49
Pseudolithophyllum expansum 83
Pseudorca crassidens 97

Pseudothyone raphanus 199
Psittacula krameri 274
Psychodidae 38
Pterocarya 35
Pterocles alchata 104
Pterodroma
   P. feae 95
   P. madeira 94–5
puffin 95
Puffinus
   P. assimilis 95
   P. mauretanicus 95
   P. yelkouan 250
Punica
   P. granatum 212
   P. protopunica 222
purple gallinule 257
purple heron 248, 273
purslane 208
Pycnonotidae 180
pygmy cormorant 95, 128
Pyrola 107
Pyrrhocorax pyrrhocorax 68
Pyrus 36, 241
   P. elaeagrifolia 221
   P. syriaca 111

quail 274
Quercus 26–7, 35, 38
   Q. aegilops 120
   Q. afares 101, 120, 153
   Q. alnifolia 120, 139, 153
   Q. anatolica 153
   Q. aucheri 139, 153
   Q. baloot 153
   Q. boissieri 101, 110, 153
   Q. brantii 153
      Q. brantii subsp. *look* 111
   Q. calliprinos 17, 100, 109–10, 113, 120, 122, 138–40, 153, 156, 224
   Q. canariensis 58, 120, 153
   Q. cedororum 111
   Q. cerris 101, 120, 153
   Q. coccifera 17, 101, 120, 139, 153, 219, 260
   Q. faginea 101, 120, 153, 231
   Q. frainetto 101, 120
   Q. hartwissiana 153
   Q. humilis 46, 120, 153
   Q. ilex 17, 102, 120, 138, 153
      Q. ilex subsp. *ilex* 138
      Q. ilex subsp. *rotundifolia* 138
   Q. infectoria 101, 111, 120, 153
   Q. ithaburensis 46, 110, 153
   Q. libani 110, 153
      Q. libani subsp. *look* 111
   Q. macedonia 101, 120

Quercus (*cont.*)
   Q. macrolepis 221
   Q. pyrenaica 153, 231
   Q. robur 153
   Q. rotundifolia 153
   Q. suber 17, 120, 153, Plate 9b
   Q. trojana 101, 120
quillwort 132
quince 212, 221

rabbit 45, 69, 71, 203, 239, 277
rabbitfish 279
raccoon 277
raccoon dog 43, 277
radish 210
Rafetus euphraticus 246
ragwort 31
rainbow trout 266, 271
rainbow wrasse 192
Ralfsia 83
   R. verrucosa 191
Rana 246
   R. catesbeiana 273
   R. cerigensis 67
   R. cretensis 67
   R. lessonae 42
   R. perezi 132
   R. ridibunda 64
Rangifer tarandus 29, 44
Ranidae 67
Ranunculus 128
   R. laterifolius 132
   R. baudotii 129
rape seed 208
Raphanus sativus 210
rat 45, 144, 172
Rattus 275
   R. norvegicus 45
   R. rattus 45
raven 126–7, 250
ray 91, 93, 134
razorbill 95
Recurvirostra
   R. avosetta 130, 255
red algae 79, 191
red coral 89, 90, Plate 8a
red deer 43–5, 250, 257, 295
red juniper 113, 236
red mullet 246, 279
red munia 274
red scorpionfish 196
red seaweed 83
red squill 173
red squirrel 277
red-billed leiothrix 274
red-crested pochard 161
red-eared slider 273–4, 285

red-necked nightjar 69
red-throated diver 95
redshank 130–1
redstart 69, 127, 161, 163, 249
reed 128
Reeve's pheasant 274
Regulus 162
reindeer 29, 43, 237
Relictocarabus meurguesae 59
Retama 38, 168
   R. sphaerocarpa 175
   R. raetam 113
Reynoutria japonica 269
Rhagamys 145
Rhamnaceae 168
Rhamnus 35, 123, 167, 288
   R. alaternus 101, 180
Rhanterium 125
rhinoceros 29, 43–4, 71, 203, 237
Rhinocerotidae 44
Rhinoclavis kochi 208
Rhodobionta 278
Rhododendron 35
Rhodopechys githaginea 249
rhubarb 123
Rhus 123
   R. tripartita 113
   R. coriaria 208
Ribes uva-crispa 230
rice 104
Ridolfia 37
ring-necked parakeet 274
ringed plover 284
ringed snake 49
Riparia rupestris 127
Risso's dolphin 97
Rissoella verruculosa 82, 191
robin 147, 161, 181
Robinia pseudoacacia 269
rock bunting 249
rock nuthatch 69, 140
rock pine 180
rock swallow 127
rock thrush 68, 127, 161, 163, Plate 6b
rocket 208
rockrose 37, 111–2, 123–4, 168–9, 178,
   207, 211, 222, 229, 259, Plate 2a
rodent 39, 44–5, 70-72, 145, 172–3,
   238–9, 276
roe deer 43–4, 295
roller 104, 160–1, 163, 242
Rosa damascena 208
Rosaceae 36, 113
Rosalia alpina 38, Plate 4a
Rosalia longicorn Plate 4a
rose 168, 208
rose shrimp 248

rosemary 101, 123, 151, 173, 208, 210
Rosmarinus 166, 292
   R. officinalis 101
Rosularia lineata 126
rough pen shell 192–3
rough-toothed dolphin 97
roughy 199
round-seeded broom 175
roundnose grenadier 199
Rousettus aegyptiacus 71
Rubia tinctorum 208
Rubiaceae 33, 56, 242
ruddy duck 275
ruddy shelduck 130, 133
Ruditapes
   R. philippinarum 307
   R. semidecussatus 307
rue 151, 173, 208, 210
Rumex acetosa 208
Rupicapra 44
   R. rupicapra 44, 257
Ruppia 129
Ruscus aculeatus 168
rush 132, 168
Russian olive 212
Ruta 151, 173
   R. corsica 173
   R. graveolus 208
rye 209
ryegrass 210

Saccorhiza polychides 82
sacred ibis 275
sacred scarab beetle 184, 242
safflower 208, 211
saffron 111
Saga pedo 105–6
sage 168, 173, 208
sago pondweed 129
saiga antelope 226
Saiga tatarica 226
Saint John's bread tree 100
salamander 49, 65–6, 246, Plate 5a
Salamandra
   S. corsica 67
   S. salamandra 49, 246
Salamandridae 67
Salaria fluviatilis 62
salema 193, 195
Salicornia 133, 136
Salix 29
Salmo
   S. salar 94
   S. trutta 63, 272
      S. trutta fario 63
      S. trutta macrostigma 63, 272

Salmo (cont.)
    S. trutta marmoratus 63
    S. trutta peristericus 63
salmon 94, 247, 271, 306
salmonid 47, 272, 306
Salmonidae 61–2, 272
salp 80
salsify 208
Salsola 35, 125
saltgrass 125, 129, 133
saltbush 30, 37, 112, 125
Salvadora persica 35
Salvadoreaceae 33
Salvelinus 272
    S. fontinalis 272
Salvia 34, 37, 168, 173
    S. fruticosa 179
    S. officinalis 208
samphire 108
sand eel 194
sand flea 191
sand rat 172
sand-smelt 129
sandalwood 168
Sander lucioperca 129
sandpiper 160
sandwich tern 94, 131, 255
sandwort 175
sang-foin 210
Sapotaceae 33, 55, 100
Sarcopoterium 260
    S. spinosum 101, 124, 168
Sarcopterium 260
sardine 91, 96, 189, 201, 265
Sardinella 246
Sardinian warbler 69, 104
Sargassum muticum 85, 278
Sarpa salpa 193, 279
Satureja 124, 168
Saturnia pyri 242
Satyridae 60
Sauria 67
Saurida undosquamis 279
Savi's warbler 257
savin 111–2
sawfish 91
sawfly 176
Saxicola dacotiae 69, 141
Saxifraga 102, 103
    S. exarata 108
    S. oppositifolia 108
saxifrage 108
scalop 305
scarab beetle 184, 242
Scarabaeidae 38
Scarabaeinae 183
scarabeid 242

Scarabeus 182–3
    S. sacer 184
    S. semipunctatus 184
Scarites 242
Schismopora avicularis 198
Schizoma phillyreae 174
Schokar sand snake 49
Scincidae 67
Scirpus lacustris 128
Sciuridae 44
Sciurus
    S. carolinensis 277
    S. vulgaris 277
Scleractinia 196
Sclerotheca 243
Scolymus hispanicus 208
Scomberomorus commerson 279
Scombridae 91–2
scops owl 161, 242, 265
Scorpaena
    S. elongata 196
    S. scrofa 196, Plate 8b
scorpion 25, 105
scorpionfish 196, Plate 8b
Scorzonera 208
scoter 95
Scots pine 101, 106, 156, 160
Scyliorhinus
    S. canicula 91
    S. stellaris 91
sea anemone 192
sea bass 92, 129, 136, 305–6
sea bream 92, 136, 194, 305–6
sea daffodil 108
sea horse 93, 129, 189, 195
sea lion 97
sea mouse 199
sea onion 173
sea squill 169
sea samphire 108
sea turtle 80
sea urchin 193–7, 279
seabird 80, 94–5, 97
sea-kale 208
sea-snail 187
seagrass 79, 81, 85–7, 93, 98, 133,
    192–3, 196
seal 97, 250–1, 295, 297
seaperch 93
seaweed 81–3, 85, 92–3, 98, 190,
    192–3, 195, 278
Secale 209
    S. cereale 209
    S. montanum 209
Sedum 126, 168
    S. anglicum 175
sei whale 96

Senecio
    S. angulatus 269
    S. gallicus 31
    S. inaequidens 269
senna 113
Sepia elegans 200
Sepiolidae 50
Serapias 177, 250, 259
    S. cordigera Plate 3i
serin 161
Serinus
    S. canaria 69, 141
    S. citrinella 108
serradela 208
Serranidae 92
Sertella 198
serval 250
serviceberry 37, 212
Sesleria 102
seven-gill shark 91
shag 95
shark 91, 96–7, 188–9
she-ass 214
shearwater 94–5, 249–50
sheatfish 244, 272
sheep 71, 109, 112, 115, 146, 174, 182,
    207, 214–6, 226, 230–1, 258,
    264, 266, 268, 275–6, 291, 295
shelduck 129, 130, 133
shi drum 307
short-snouted sea horse 93
short-toed eagle 249
shoveler 161
showy hairy pink 293
shrew 42, 52, 70–2, 239, 276
shrike 242, 265
shrimp 50, 129–30, 189, 194–5, 199,
    200, 245, 248, 280, 305–6
Sicilian zelkova 242
Sideroxylon 55
Siganidae 279
Siganus
    S. rivulatus 279
    S. luridus 279
Silene 37
    S. velutina 293
silkworm 231
Siluridae 62, 272
Silurus glanis 244
silver carp 272
silverbill 274
Sinai agame 112
Sinapis 210
    S. alba 208
Sisyphus 183
Sitta 40
    S. kruperi 40

Sitta (cont.)
S. ledanti 40, 52
S. neumayer 140
S. tephronota 140
S. whiteheadi 40, 141
skate 91, 96–7
skipjack 249
slender-billed gull 68–9, 129–31, 255–6
slider 273–4, 285
small toad 66
smaller little tern 131
smilax 166, 181
Smilax 33
S. aspera 166, 180
smoke bush 123
smooth-hound 91
Smyrniopsis 37
Smyrnium olusatrum 208
snail 46, 60, 72, 114, 152, 162, 187
snake 49, 64, 66, 109, 111–2, 246, 274
snapdragon 126
snow finch 127
snowy owl 29
social vole 172
Solanum melongena 210
Soldanella 103
sole 93, 129, 246
Solea vulgaris 93
solitary bee 176–7, 179, 253, 271
sombre tit 46, 69
Sonchus 55
song thrush 107, 161, 181, 252
Sorbus 36
sorrel 208
southern swallowtail 294
sowbug 191
spadefoot 64
Spanish festoon 294
Spanish hare 72
Spanish imperial eagle 69
Spanish juniper 138
Spanish oak 101, 119, 254
Spanish oyster plant 208
Spanish sandwort 175
Spanish terrapin 65
sparid 192, 305, 307
Sparidae 92
sparrow 143
Spartina maritima 133
Spartium 123, 168
Sparus aurata 92, 129
spectacled warbler 105, 252
Speleomantes
S. flavus 67
S. genei 67

S. imperialis 67
S. supramontis 67
Spergula arvensis 208
sperm whale 96–7, 308
Sphaer ectinus granularis 198
Sphyrna 91
spider 59, 87, 191
spider crab 82, 200
spiked magician 105
spindle bush 174
spinefoot 279
spiny cocklebur 269
spiny lobster 200
spiny mouse 71–2
spiny oyster 280
Spirographis spallanzanii 198
Spondylus
S. gaederopus 280
S. spinosus 280
sponge 79, 89, 197
sponge crab 200
spoonbill 128
spotless starling 69
spotted eagle 161
sprat 246
spruce 35, 101, 160, 301
spur-thighed tortoise 65
spurge 58, 124
spurry 208
squacco heron 273, 284
squid 96–8, 188–9, 200–1, 247, 306
Squilla mantis 199
squirrel 71, 277
Stachys 37, 126
stag beetle 294
Staphylinidae 38
star-of-Bethlehem 168
starfish 97, 192, 194, 280
starling 69
Stenella coeruleoalba 97
Steno bredanensis 97
Sterna
S. albifrons 131
S. hirundo 131
S. nilotica 131
S. sandvicensis 131
Sternbergia 111
Stichopus regalis 199
sticky-weed 126
stilt 130, 133
stingray 91
Stipa 112, 172
S. tenacissima 125
Stipagrostis lanatus 108
stock dove 126
stone pine 100, 109, 119, 128, 221, 224, Plate 9a

stonechat 69
stonecrop 175
stonefly 59
storax 36, 110–1, 123, 211, 221
stork 163, 241, 248
storm petrel 249
Stramonita haemastoma 280
strawberry tree 17, 37, 57–8, 105, 107, 122–3, 258
Streptopelia
S. roseogrisea 275
S. turtur 143
stripe-necked terrapin 65
striped dolphin 97
stripeless tree frog 49, 64
sturgeon 62, 94, 98, 271
Styracaceae 110, 222
Styrax 123, 222
S. officinalis 36, 110, 120
Suaeda 35, 125
subalpine warbler 161, Plate 6a
subtropical melon 210
succulent creeper 269
sunflower 173
Suidae 44
Sula bassana 29, 95
sumac 208
summer asphodel 179
Suncus 275
S. etruscus 72
surmullet 246
Sus
S. scrofa 43–4
S. strozzi 43
swallow 127, 143, 161–2
swallowtail butterfly 93, 140–1, 294
swallowtail seaperch 93
swan 161, 274
swede 208
swift 69, 127, 143
swordfish 92, 188–9, 246, 265
Sylvia 40–1, 68
S. atricapilla 40–1, 107
S. borin 40–1
S. cantillans 41, 107, Plate 6a
S. communis 41, 180
S. conspicillata 41, 105, 107
S. curruca 41
S. deserticola 41
S. hortensis 41, 105, 107
S. leucomelaena 41
S. melanocephala 41, 104, 107
S. melanothorax 41, 69, 141
S. minula 41
S. mystacea 41
S. nana 41

Sylvia (cont.)
  S. nisoria 41
  S. rueppelli 41
  S. sarda 41, 69, 141
  S. undata 41, 107
  S. whiteheadi 141
Sylvilagus floridanus 45
Symphodus 92
  S. ocellatus 193
  S. rostratus 92, 195
Syngnathidae 92–3, 195
Syngnathus 93
  S. abaster 129
Syrian ash 113
Syrian woodpecker 46, 69
Syrmaticus reevesii 274
syrphid 176
Syrphidae 130, 179

Tabor oak 46, 110
Tadorna 129
  T. ferruginea 130
  T. tardorna 130
tamarisk 113, 129
Tamarix 113, 259
  T. africana 129
  T. canariensis 129
  T. gallica 129
  T. tetranda 129
Taphozous nudiventris 71
tapir 43
Tapirus 43
Tarentola 66
  T. angustimentalis 67
  T. boettgeri 67
  T. delalandi 67
  T. gomerensis 67
tarpan 213, 226
tarragon 210
tawny pipit 161
Taxus baccata 20
teal 69, 130, 161, 257
Teline 123, 168
Tellina 194
tenebrionid 131
Tengmalm's owl 104, 108
terebinth 110, 140, 166, 174, 221
tern 94, 130–1, 255
Ternstroemiaceae 55
terrapin 65, 246, 274
terrestrial turtle 246
Testudinidae 64
Testudo
  T. graeca 65, 111, 274
  T. hermanni 29
  T. kleinmanni 246
  T. marginata 65, 274

T. weissingeri 52
T. werneri 246
Tetraclinis articulata 57, 120
Tettigetta pygmaea 242
Tettigoniidae 106
Teucrium 37, 126
  T. aristatum 132
Thalasseus sandvicensis 94
Thalassoma pavo 279
Thaumetopoea
  T. pinivora 172
  T. pityocampa 172
Theaceae 33
thistle 112
thresher 91
Threskiornis aethiopica 275
thuja 57, 100, 122, 156, 236, 251, 259
thrumwort 132
thrush 68, 107, 127, 161–3, 174,
  180–1, 252, Plate 6b
Thunnus
  T. albacares Plate 8d
  T. thynnus 92
Thymallus thymallus 108
thyme 123–4, 137, 148, 151–2, 168,
  173–4, 208, 210, 300
Thymelaea 166
  T. hirsuta 30, 113
Thymus 37, 124, 166
  T. vulgaris 151
Thysanura 87
Tibicen plebejus 105–6, 242
Tichodroma muraria 127
tiger shark 91
tilapia 3054
Tilia 20, 27, 38
tit 15, 46, 69, 70, 107, 137, 144, 156–9,
  219, 252
toad 49, 64–6, 132, 246, 274
toothed whale 95
tope 91
Torgos tracheliotus 247
tortoise 29, 52, 64–5, 203, 238, 246
Trachemys scripta 273
Tracheophyta 52
Trachitchthyidae 199
Tragopogon porrifolius 208
Trebizond date 212
tree heath 33
tree pipit 107, 161
tree-of-heaven 269
trefoil 210
Tribulus 37
Trifolium 172, 208, 210
trigger fish 279
Trigonella 210
Tringa

T. glareola 160
T. totanus 130
Triops 131
  T. cancriformis 132
triplefin 93, 192
Tripterygiidae 92–3
Tripterygion
  T. delaisi 93
  T. tripteronotus 93
Triticum 209
  T. aestivum 209
  T. boeticum 209
  T. dicoccoides 209
  T. dicoccon 209
  T. monococcum 209
  T. timopheevii 209
  T. turgidum 209
Triturus 42, 246
  T. cristatus 39
  T. helveticus 64
  T. marmoratus 39
  T. vulgaris 64
Trogonophidae 64
Tropnyx triunguis 64
trout 63, 94, 108, 244, 266, 271–2
true bison 43
true horse 43
true lavender 151
trumpeter finch 249
tulip 111, 123, 169
Tulipa 111, 259
tuna 92, 96, 98, 134, 188–9,
  245–7, 265, 306–8,
  Plate 8d
tunicates 79
turbellarian 194
Turdus
  T. merula 181
  T. philomelos 107
  T. viscivorus 174
Turkey oak 101, 110
turnip 208
Turritella communis 136
Tursiops truncatus 97
turtle 80, 111, 189, 201, 203, 246,
  265, 274, 307
turtle dove 143, 161
two-tailed pasha 38, 105, 107
Typha latifolia 128
Tyto cf. *alba* 239

Ulex 260
  U. europaeus 123, 208
Ulmaceae 242
Ulmus 20, 27, 35, 38
Ulva 83, 129
Umbrina cirrosa 307

Undaria pinnatifida 85, 279
Upeneus moluccensis 279
urchin 80, 192–7, 279
Urginea 170
Uria 29
　U. aalge 95
Urodela 67
ursid 71
Ursidae 44
Ursus
　U. americanus 43
　U. arctos 42, 44
　U. etruscus 43
　U. spelaeus 43–4
　U thibetanus 44

Vaccinium 29
Valenciidae 62
Vallonea oak 221
Valonia 83
Varanus griseus 64
Varroa 271
Varthemia iphionoides 126
veined squid 200
Veretillium 199
Verrucaria amphibian 82
vetch 58, 208, 210
Viburnum 123
　V. tinus 166, 180
Vicia 102, 208, 210
　V. altissima 259
　V. faba 210
　V. melanops 259
　V. sativa 210
Viguieriotes ewardsii 198
Vinca minor 230
Viola 102
viper 111
Vipera 64
　V. palestina 111
Viperidae 64, 67
Viscum
　V. album 174
　　V. album subsp. *austriacum* 301
Visnea mocanera 55, 122
Vitaceae 33
Vitis silvestris 127
vole 42, 45, 52, 71–2, 172
Vormela peregusna 45
Vulpes 45
　V. vulpes 44
vulture 126, 145, 160, 226–7, 247, 250, 257, 294

Wader 130, 133, 160, 249, 257
wagtail 108
wall lizard 145
wallcreeper 127
wallflower 126
walnut 35, 212, 221, 254
walrus 97
warbler 40–1, 47, 68–9, 104–5, 107, 147, 161, 163, 180, 252, 257, 284, Plate 6a
wasp 175–9, 184
water buffalo 214
water crowfoot 129
water fern 132
water milfoil 128–9
water pipit 161
water vole 42
watermelon 210
wattle 269
waxbill 274
weasel 44, 275
whale 91, 95–8, 188–9, 201, 307
whalebone 95
wheat 16, 104, 204, 208–10, 213, 231, 251, 297
wheatear 69, 104, 108, 140, 161, 249
whelk 280
white coral 199
white egret 273
white fir 101
white mustard 208
white pelican 128
white rhino 44
white stork 248
white-headed duck 68, 275
white-tailed eagle 161
white-toothed shrew 42
whitethroat 180
whiting 246
whorl-leaf water milfoil 128
whortleberries 29
wigeon 161
wild almond 111
wild apple 212
wild barley 209
wild boar 43–4, 70, 213, 250, 257
wild date palm 112
wild dog 213
wild donkey 213
wild emmer 209
wild goat 214, 226
wild grape 127
wild horse 213, 226, 238
wild licorice 129
wild olive tree 100, 109, 122, 221
wild onion 293
wild pear 111

wild sheep 215
wild snapdragon 126
wild thyme 151
wildcat 216, 275
willow 29, 127
willow warbler 161, 180
winkle 187, 190
witchhazel 35, 119
wolf 43, 213, 258, 275
wolly rhino 43
wood sandpiper 160
wood warbler 161
woodlark, 147
woodpecker 48, 69, 107–8, 143, 161, 252
worm 60, 72, 130
wormwood 21, 26, 30, 112, 125
wrasse 92–3, 192–3, 195, 279
wren 161–2
wrinkle 193
wryneck 143

Xanthium spinosum 269
Xerus 70–1
Xiphias gladius 92
Xiphiidae 92

yelkouan shearwater 250
yellow coral 199
yellowfin tuna Plate 8d
yellow iris 128
yellow-legged gull 95, 131, 163, 255
yellow-tailed black scorpion 105
yew 20, 107, 251, 260

Zannichellia 129
zebra mussel 268
zeen oak 101
Zelkova 35, 119
　Z. abelicea 120, 242
　Z. sicula 120, 242
Zerinthia rumina 294
Zino's petrel 94
Ziphius cavirostris, 97
Ziziphus 21, 35, 112–3
　Z. spina-christi 221
Zoantharia 89
Zostera
　Z. marina 85–6
　Z. nana 85, 133
　Z. noltii 86, 133
Zosteraceae 85
Zygophyllaceae 37
Zygophyllum 21, 37
　Z. dumosum 113